MINERAL PROCESSING
SECOND EDITION

S.K. JAIN

Reader in Metallurgical Engineering,
Malaviya Regional Engineering College, Jaipur

CBSPD

CBS Publishers & Distributors Pvt Ltd

New Delhi • Bengaluru • Chennai • Kochi • Kolkata • Lucknow • Mumbai
Hyderabad • Jharkhand • Nagpur • Patna • Pune • Uttarakhand

Mineral Processing

ISBN: 978-81-239-0753-6

Copyright © Author and Publisher

First Edition: 1985

Second Edition: 2001

Reprint: 2003, 2007, 2010, 2011, 2012, 2014, 2016, 2020, 2024

Published by Satish Kumar Jain and produced by Varun Jain for

CBS Publishers & Distributors Pvt Ltd

4819/XI Prahlad Street, 24 Ansari Road, Daryaganj, New Delhi 110 002, India.
Ph: 23289259, 23266861   Website: www.cbspd.com
                                 e-mail: delhi@cbspd.com
Corporate Office: 204 FIE, Industrial Area, Patparganj, Delhi 110 092
Ph: 4934 4934   Fax: 4934 4935   e-mail: publishing@cbspd.com; publicity@cbspd.com

Branches

- **Bengaluru:** Seema House 2975, 17th Cross, K.R. Road, Banasankari 2nd Stage, Bengaluru 560 070, Karnataka
  Ph: +91-80-26771678/79   Fax: +91-80-26771680   e-mail: bangalore@cbspd.com
- **Chennai:** 7, Subbaraya Street, Shenoy Nagar, Chennai 600 030, Tamil Nadu
  Ph: +91-44-26680620, 26681266   Fax: +91-44-42032115   e-mail: chennai@cbspd.com
- **Kochi:** 42/1325, 1326, Power House Road, Opp KSEB, Ernakulam 682 018, Kochi, Kerala, India
  Ph: +91-484-4059061-65   Fax: +91-484-4059065   e-mail: kochi@cbspd.com
- **Kolkata:** 147, Hind Ceramics Compound, 1st Floor, Nilgunj Road, Belghoria, Kolkata 700 056, West Bengal, India
  Ph: +91-33-25633055/56   e-mail: kolkata@cbspd.com
- **Lucknow:** Basement, Khushnuma Complex, 7-Meerabai Marg (Behind Jawahar Bhawan), Lucknow 226 001, UP, India
  Ph: +0552-4000032   e-mail:tiwari.lucknowl@cbspd.com
- **Mumbai:** PWD Shed. Gala no. 25/26, Ramchandra Bhatt Marg, Next to JJ Hospital Gate no. 2, Opp. Union Bank of India, Noorbaug, Mumbai 400 009, Maharashtra, India
  Ph: 022-66661880/89   e-mail: mumbai@cbspd.com

Representatives

- **Hyderabad** 0-9885175004 • **Jharkhand** 0-9811541605 • **Nagpur** 0-8692091830
- **Patna** 0-9334159340 • **Pune** 0-9664372571 • **Uttarakhand** 0-9716462459

Printed at Neekunj Print Process, Kundli, Haryana, Delhi

Preface to the Second Edition

This book is intended to serve as a bridge between the study of the basic principles of ore processing and their application in industry. Modern trends in ore processing have been highlighted particularly in the context of declining grade of ores and their reserves. The work is the outcome of long periods spent in research and teaching of mineral processing.

The book has been planned to serve as a textbook for undergraduate students of Metallurgical, Chemical and Mining Engineering. It will also serve as a useful reference book for the postgraduate students of Ore Processing/ Mineral Dressing and Applied Geology. The subject matter is updated and thoroughly revised. Various corrections are also incorporated.

The subject matter has been arranged in 20 chapters. The first four chapters deal with minerals and background of mineral processing. Next two chapters deal with comminution (crushing and grinding) being an essential step in all ore processing flow sheets. Chapter 7 deals with movement of solids in fluids and classification, the understanding of which is essential in the design of solid-liquid separation and classification units. Gravity concentration methods based on three different principles, i.e. heavy media separation, concentration in vertical currents and concentration in streaming currents have been discussed in three chapters.

Due consideration is given to the importance of flotation in chapters 11, 12 and 13 to cover the history and theory, flotation reagents and their action, and flotation practice and machines.

Chapter 14 deals with magnetic and electrical separation being important in the concentration of magnetic and conductive minerals. An entire chapter has been devoted to solid-fluid separation as it is of great importance since the products obtained in ore processing are usually in the form of slurry and suspensions. Chemical treatment of ores has been discussed in chapter 16.

Chapter 17 details the application of microscopy in ore processing. The industrial treatment of ores is given in chapter 18. Chapter 19 highlights the increasing use of computers and automation. Finally chapter 20 covers location, layout and selection of equipment.

The author is indebted to the University Grants Commission, New Delhi, for sponsoring this work and providing the financial assistance for the preparation of the manuscript. The author is grateful to Dr. Dharmendra Kumar, Professor of Metallurgical Engineering, Malaviya Regional Engineering College, Jaipur and other friends who have given encouragement and constructive suggestions. The author is particularly grateful to Mr. U.S. Khandelwal, Hindustan Zinc Ltd., Udaipur for his help in completion of this work. Above all, is a profound debt of gratitude to Saroj Jain for her invaluable help and moral support.

S.K. JAIN

Contents

CHAPTER 1

Properties and Types of Minerals

A mineral may be regarded as a naturally occurring chemical compound having a definite chemical composition and crystal structure. An ore is natural aggregation of minerals from which a metal or metallic compound can be recovered with profit on large scale. When the percentage of metal/valuable is too low for profitable extraction/recovery, the rock ceases to be an ore. The physical properties of minerals play the most important role in the economic processing of various ores.

1.1 Physical Properties of Minerals

The physical properties of minerals can be determined without the use of chemical tests. They depend upon the kinds and arrangement of atoms in their crystal structures. The various physical properties of minerals include transparency, lustre, colour, specific gravity, hardness, cleavage, fracture, magnetic properties, electrical properties, radioactive, and optical properties. Most of these properties can be used for quick identification of minerals, although this should be confirmed with additional chemical and crystallographic information.

Ore processing makes use of many physical properties of minerals and rocks and sometimes chemical properties too. The difference in behaviour between the valuable and waste minerals affords the methods for the separation of the former from the latter. Physical properties of interest in or processing are discussed in this section.

1.1.1 TRANSPARENCY

This is the term used to describe the ease with which we can see through a mineral. Three terms for transparency are in common use, i.e. opaque, transparent and translucent. The opaque minerals are those through which no light can be seen. Transparent minerals are those which can be seen through clearly. Translucent minerals are those through which a little light can be seen, i.e. intermediate between the terms opaque and transparent. For most purposes the term transparency is descriptive and is rarely measured. Since minerals are rarely pure, different samples of the same mineral may show different degrees of transparency. The examples of transparent minerals are quartz, calcite, beryl, and diamond, though it is essential to have pure, clean crystals to show them transparent. Most of these examples of transparent minerals may be translucent also, especially if they are impure, or are not perfect crystals. Lightly coloured minerals as well as colourless minerals, may be transparent. Some examples of opaque minerals are silver, copper, galena, pyrite, hematite, magnetite, etc.

1.1.2 LUSTRE

This may be defined as the amount and quality of the reflection of light from the mineral surface. The lustre of mineral refers to its surface appearance. Like transparency, it is not usually measured. There are two major categories, i.e. (a) minerals which look like metals and are said to have metallic lustre, and (b) all other minerals having non-metallic lustre. There are no other descriptive words for the metallic minerals. The non-metallic minerals, however, are referred to by various terms which describe their surface appearance. The examples are pearly (gypsum), vitreous or glassy (quartz), greasy (nepheline), silky (asbestos), waxy (serpentine), resinous (sphalerite), etc. Other terms are adamantine, referring to a brilliant lustre like that of diamond; earthy, used for minerals like kaolin and the clay minerals, which appear dull. Like transparency, lustre depends upon the ways in which the atoms and their arrangement affect the incoming light waves. For example, both diamond and graphite are composed of carbon atoms, but graphite is metallic and diamond is adamantine. The difference in lustre is due to the arrangement of atoms in these two mineral structures.

Lustre is important in hand picking. For example, the brass-yellow of chalcopyrite, the pale-yellow of pyrite, the white of arsenopyrite, vitreous of quartz, the resinous of sphalerite, the adamantine of diamond and cerussite, the dull of chalk, and the pearly of talc furnish valuable aids in hand picking and also in judging the mill products.

1.1.3 COLOUR

In most cases the colour of mineral is due to absorption of certain wavelengths of light energy by the atoms making up the crystal. The remaining wavelengths of the light that are not absorbed give the sensation of colour to the eye. In general, few minerals have characteristic colours by which they can be recognised. Many minerals show a wide variety of colours and shades of colours. Combinations of wavelengths give rise to other intermediate colours.

Although usually descriptive, colour can be measured by determining the wavelengths of light transmitted by the specimen, with the help of *spectrometer*.

Metallic minerals, generally are of constant colour unless some surface alteration or tarnish has occurred. Thus samples of molybdenite from all over the world are bluish-steely-grey in colour. Bornite, however, is rarely untarnished and it is necessary to chip off a small fragment to reveal the true colour. The colour of the tarnish, however, may be characteristic, and some metallic minerals may be recognised by the colour of their tarnish. Non-metallic minerals usually vary in colour. Fluorite, for instance, can be blue, purple, green, pink, brown, orange, yellow, colourless,or almost black. Quartz, calcite,and many other minerals also have wide range of colours. A few non-metallic minerals do show nearly constant colour, e.g. sulphur (yellow) and graphite (black).

Certain minerals show colour due to the presence of small amounts of impurities or inclusions. The purple colour of amethyst (a variety of quartz) may be due to the presence of minute amounts of titanium and manganese atoms in the frame structure of silicon and oxygen atoms. Colourless calcite containing tiny needles of green hornblende, will be green. If the foreign crystals are large enough, close examination will show that the colour is due to them. Some minerals contain extremely minute inclusions so evenly distributed that it is difficult to see that they are the cause of colour.

Less variable than the colour of the actual specimen is the colour of the powdered mineral. The specimen is rubbed across a white unglazed porcelain plate, and the *smear* is a sample of the powdered mineral. The colour of this powder can be described and is known as streak of the mineral. The streaks of some minerals, however, may vary. Sphalerite (ZnS) when pure, gives a pale-yellow streak, whereas with increasing amounts of iron (replacing zinc atoms in the structure), the streak as well as the colour darkens to brown. The variation in colour of streak may also be due to the presence of tiny crystals of another mineral in the host mineral.

1.1.4 LUMINESCENCE

This refers to the emission of light by a mineral which is not the direct result of incandescence. Luminescence in most minerals is faint and can be seen only in the dark. Minerals which luminesce during exposure to ultraviolet light and X-rays, are called *fluorescent*. When the luminescence continues after the exciting rays are cut-off, the minerals are said to be *phosphorescent*. There is no sharp distinction between fluorescence and phosphorescence.

Fluorescence is produced when the energy of the short wave radiation is absorbed by the minerals/ions and released as longer wave radiation (visible light). Fluorescence is an unpredictable property as some specimens of a mineral show it, whereas other apparently similar specimens even from the same locality do not show. The fluorite mineral from which the property derives its name, will fluoresce. Its usual blue fluorescence may be due to the presence of organic material or rare earth ions. Other minerals that frequently but not invariably fluoresce include scheelite, willemite, calcite, diamond, autunite, eucryptite, scapolite, and hyalite.

The property luminescence is used in identification and sorting of minerals.

1.1.5 SPECIFIC GRAVITY

Specific gravity of a particular mineral is practically constant, it may vary a little with the presence of some impurities. The difference in specific gravities affords one of the surest means of separating minerals from each other, and has been put to practical use. Simple washing in water affects an efficient separation of gold grains from quartz sand, whereas use of heavy liquids affects the separation of lighter coal from heavier shale (ash). In Table 1.1 the specific gravities of some important minerals are given.

Table 1.1. Composition, hardness, and specific gravity of important minerals

Name of mineral	Composition	Hardness	Specific gravity
1	2	3	4
Anglesite	$PbSO_4$	2.8-3.0	6.1-6.4
Argentite	Ag_2S	2.0-2.5	7.2-7.4
Arsenopyrite	$FeAsS$	5.5-6.0	5.9-6.3
Azurite	$2CuCO_3 \cdot Cu(OH)_2$	3.5-4.0	3.8-3.9
Barite	$BaSO_4$	2.5-3.5	4.3-4.6
Bauxite	$Al_2O_3 \cdot 3H_2O$	1.0-3.0	2.6
Beryl	$3BeO \cdot Al_2O_3 \cdot 6SiO_2$	7.5-8.0	2.6-2.8
Calamine	$H_2Zn_2SiO_5$	4.5-5.0	3.4-3.5
Calcite	$CaCO_3$	3.0	2.7
Cassiterite	SnO_2	6.0-7.0	6.8-7.1
Celestite	$SrSO_4$	3.0-3.5	3.9-4.0
Cerrusite	$PbCO_3$	3.0-3.5	6.5-6.6
Chalcocite	Cu_2S	2.5-3.0	5.5-5.8
Chalcopyrite	$CuFeS_2$	3.5-4.0	4.1-4.3
Chromite	$FeO \cdot Cr_2O_3$	5.5	4.3-4.6
Chrysocolla	$CuSiO_3 \cdot 2H_2O$	2.0-4.0	2.0-2.2
Cinnabar	HgS	2.0-2.5	8.0-8.2
Cobaltite	$CoAsS$	5.5	6.0-6.3
Columbite	(Fe, Mn) $(Cb, Ta)_2O_6$	6.0	6.3
Corundum	Al_2O_3	9.0	3.9-4.1
Covellite	CuS	1.5-2.0	4.6
Cryolite	Na_3AlF_6	2.5	3.0
Cuprite	Cu_2O	3.5-4.0	5.9-6.2
Dolomite	$CaCO_3 \cdot MgCO_3$	3.5-4.0	2.8-2.9
Galena	PbS	3.0	7.4-7.6
Gypsum	$CaSO_4 \cdot 2H_2O$	1.5-2.0	2.3
Hematite	Fe_2O_3	5.5-6.5	4.9-5.3
Ilmenite	$FeO \cdot TiO_2$	5.0-6.0	4.5-5.0
Magnesite	$MgCO_3$	4.0-4.5	3.1
Magnetite	Fe_3O_4	5.5-6.5	5.2
Malachite	$CuCO_3 \cdot Cu(OH)_2$	3.5-4.0	4.0
Millerite	NiS	3.0-3.5	5.3-5.7
Molybdenite	MoS_2	1.0-1.5	4.7-4.8
Monazite	$(RE, Th)_3 (PO_4)_4$	5.0-5.5	4.9-5.3
Pentlandite	$(Fe, Ni)S$	3.5-4.0	4.6-5.0
Phosphate rock	$Ca_3 (PO_4)_2$	5.0	3.2

1	2	3	4
Pyrite	FeS_2	6.0-6.5	5.0
Pyrolusite	MnO_2	1.0-2.5	4.8
Quartz	SiO_2	7.0	2.65-2.66
Rhodochrosite	$MnCO_3$	3.5-4.5	3.5-3.6
Rutile	TiO_2	6.0-6.5	4.2
Scheelite	$CaWO_4$	4.5-5.0	5.9-6.1
Siderite	$FeCO_3$	3.5-4.0	3.9
Sphalerite	ZnS	3.5-4.0	3.9-4.1
Spodumene	$LiAl(SiO_3)_2$	6.5-7.0	3.1-3.2
Stibnite	Sb_2S_3	2.0	4.5-4.6
Talc	$H_2Mg_3(SiO_3)_4$	1.0-1.5	2.7-2.8
Tantalite	$FeTa_2O_6$	6.3	5.3-7.3
Uraninite	$UO_3 \cdot UO_2$	5.5	9.0-9.7
Wolframite	$(Fe, Mn)WO_4$	5.0-5.5	7.2-7.5
Zircon	$ZrSiO_4$	7.5	4.2-4.7

Based on Ref. 30.

Difference in specific gravities forms the basis of a class of ore-dressing processes known as 'gravity concentration methods'. Specific gravity differences between minerals cause differences in their behaviour when settling in water, air or other fluids, when agitated in stratified beds, when subjected to washing and shaking forces on smooth inclined surfaces, or when subjected to more complex conditions combining some of these. The quantitative behaviour of particles under such conditions are governed by various physical laws of motion, among which laws of settling are most important. For example, of two particles of the same size and shape, the heavier will settle faster whereas of the two particles of same specific gravity and shape, the larger will settle faster. Particles differing only in shape also behave differently, e.g. a rounded grain will settle faster than a tabular grain. In separation by heavy solution, the particles of lower specific gravity will float and those of higher specific gravity sink irrespective of particle size and shape.

1.1.6 HARDNESS

This may be defined as the ability of a mineral to resist scratching. This is different from the ease with which it can be broken. Diamond is one of the hardest materials known, but it can be shattered easily. The hardness of minerals is a characteristic property and can be measured. It is a valuable guide in identification of minerals. Like the other physical properties, the hardness is dependent on the kinds and arrangements of atoms in mineral structures. Traditionally, the hardness is recorded as the scratch hardness, measuring the capacity of a sharp corner of one mineral to scratch a smooth surface of another. The basis of the test is a set of minerals selected by the Austrian mineralogist, F. Mohs in 1974 numbered 1 to 10 in order of increasing hardness, each of which will scratch the one below it in the scale and will not scratch the one above. The set of hardness points may be prepared by mounting the mineral fragments in convenient shapes. The list of hardnesses of set are as following:

Mineral	Relative hardness
Talc	1
Gypsum (finger nail)	2
Calcite (copper coin)	3
Fluorite	4
Apatite (steel knife blade)	5
Orthoclase (steel file)	6
Quartz	7
Topaz	8
Corrundum	9
Diamond	10

Mohs hardness scale is relative, which means that calcite is not exactly three times as hard as talc. The true hardness of these minerals is quite surprising, and is shown in Fig. 1.1. It may be noted that diamond is about four times as hard as corundum. The true and absolute hardness of minerals can be measured with delicate instruments.

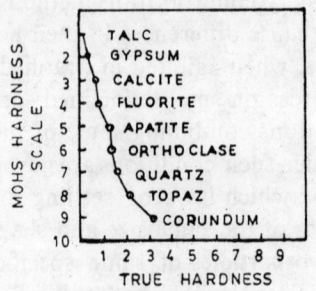

Fig. 1.1. True hardness of minerals.

The hardness of minerals depends upon the kinds of atoms in them, and the ways in which they are arranged. Both diamond and graphite are composed of the same kind of atoms, but they differ greatly in hardness. The explanation rests in the arrangement of the carbon atoms in these two minerals, whereas the atoms in diamond are in a close packed structure, the atoms in graphite are weakly bonded. Hardness of important minerals are given in Table 1.1.

In most crystals, the arrangement of the atoms is slightly different in different directions. Because of this, a mineral may be harder in one direction than in another.

The hardness of minerals has practical uses. Diamond is used as an abrasive and cutting material. Garnet sand is used to make sand paper. Soft minerals like talc are used in finely ground form, as a base for face powders and cosmetics, because of their non-abrasive qualities. Hardness affects the wear of crushing machines. i.e. harder the mineral the greater the wear. It does not necessarily affect the tendency of the minerals to produce fines or slimes in crushing.

1.1.7 TENACITY, BRITTLENESS AND FRIABILITY

This refers to the breaking strength of minerals, not to the hardness. Gold is a soft mineral, but it requires great force to break it. This includes the reactions of minerals to shock, crushing, cutting, and bending actions.

Various terms are used in mineralogy to describe the tenacity of minerals. Malleable minerals are those which can be beaten into thin sheets without breaking, e.g. native metals such as copper, silver, and gold. In contrast, brittle minerals such as diamond and quartz shatter/crumble to pieces under light blows or pressures. Some soft minerals, such as argentite (a silver mineral), can be cut with knife into thin shavings. Elastic minerals like muscovite can be bent but will spring back to the original shape when the bending force is removed. Flexible minerals such as thin plates of gypsum and flakes of molybdenite can be bent, but remain distorted after the force is removed.

Tenacity can have various practical applications. Small pieces of gold can be hammered into extremely thin sheets to make gold leaf, which is used for decoration and sign-writing. Some minerals such as native copper, mica, gypsum, etc., are soft but very tough, and this makes them difficult to break.

Hardness together with toughness, are the main determining factors in crushing and grinding. Hardness affects the brittleness and friability of a mineral, whereas toughness determines its resilience and elasticity. The interplay of these two qualities in a homogeneous piece of rock determines its response to the compressive forces (crushing) and abrasive forces (grinding) in comminution. Assessment of its response to comminuting methods is therefore empirical. Liberation of values from gangue is as an essential preliminary for concentration process. Comminution for this purpose is normally the most expensive stage in treatment. The hardness and toughness of an ore body are of critical importance, and are assessed by grindability test methods applied to representative samples.

Toughness is influenced by the manner in which the crystals of an ore are interlocked and by their grain size. A very fine (crystalline) structure is usually more resistant than a coarsely crystallised rock. Many ores consist mainly of valuable metallic sulphides and unwanted silica minerals, the former often being friable and the latter tough. There is a tendency for the small particles to be richer than the bigger lumps, and this can lead to losses in treatment, if it is ignored. It can also lead to sampling errors. Broadly, a micro-crystalline structure with the values disseminated evenly through the gangue minerals is far more resistant to crushing than one which is coarsely crystalline.

1.1.8 CLEAVAGE

This refers to the property of some minerals to break evenly along one or more special directions, leaving a smooth, flat surface. Minerals of the mica group are excellent examples showing a perfect cleavage. Mica can be split easily into very thin sheets. Cleavage directions are usually described in terms of their positions relative to the crystal faces of mineral. In some minerals possessing more than one cleavage direction, angles between these directions and their relation to the crystal faces can be measured.

Cleavage results due to a plane of weakness existing between the planes of atoms in a crystal structure. Since all crystal structures are regulated by the rules of repetition of atoms in three dimensions, the cleavage directions must also be regular and be closely related to those arrangements. However, mineral specimens with many parallel cleavages possess only one cleavage direction. It may be noted that the number of different directions of cleavage is important rather than the number of cleavages. The minerals possessing one good cleavage direction are graphite, mica group minerals, talc, topaz, etc. The minerals possessing two directions of cleavage are pyroxene group minerals (augite), and amphibole group minerals (hornblende), which have similar chemistry of appearance, but can be distinguished from each other by their cleavage directions. The minerals possessing three directions of cleavage are galena, halite (common salt), calcite, etc. There are minerals even with four cleavage directions (fluorite) and six cleavage directions (sphalerite).

A practical use of cleavage in minerals is well illustrated in diamond, which possesses perfect cleavage parallel to faces of an octahedron. Thus the diamond cutter can select a flawless, perfect part from natural crystal by cleaving. A bladed instrument (known as cleaver) is placed along the precise direction of the cleavage and tapped sharply. If the position of cleaver is correct, the diamond will break easily heaving a smooth, flat surface. If there is a slight error in placing the cleaver, the diamond may shatter into pieces, since it is very brittle.

1.1.9 FRACTURE

Fracture describes the appearance of the surfaces of minerals which have been broken across the planes of atoms. When a mineral is broken in a direction other than a cleavage plane, the nature of the fractured surface may be distinctive. In contrast to cleavage, fracture is not related to special directions within a crystal structure.

Fracture describes the appearance of the broken surface and cannot be measured. However, various descriptive terms may be used, e.g. concoidal fracture, hackly fracture, splintery fracture, etc. A *conchoidal* fracture (shell like) can be recognised by concentric sets of cracks, which are dish-shaped and either concave or convex. This type of fracture is shown by broken pieces of thick glass, quartz, olivine, flint, etc. The metallic elements such as gold, silver and copper, when broken reveal a *hackly* fracture. Other terms of fractures are self-explanatory.

Minerals not possessing cleavage (e.g. quartz) will show an appearance of curved fractures due to absence of preferential direction of breakage. The hackly fracture of some of the native metals arises from the fact that the bonds between the atoms are able to adjust to the new positions when under strain. Other less distinctive fractures are described as even and uneven.

The shape into which a mineral will break, is often determined by the presence or absence of crystalline structure. For example, galena, feldspar, mica, magnetite, and chalcopyrite tend to break into cubes, elongated fragments, flat scales, rounded grains, and conchoidal fragments, respectively. All these shapes directly affect the ability of the particles to settle in water or to move on a plane surface.

1.1.10 MAGNETIC PROPERTIES

Magnetic properties refer to the ways in which minerals react when placed in a magnetic field. Some minerals may be strongly attracted/repelled to the magnet, whereas others may be attracted/repelled less. The behaviour of all crystalline materials in a magnetic field may be classified under the various groups, i.e., *diamagnetic, paramagnetic, ferromagnetic, antiferromagnetic* or *ferrimagnetic*.

a) *Diamagnetic*

The minerals which are slightly repelled are called *diamagnetic* minerals. The examples are fluorite, calcite, and quartz.

b) *Paramagnetic*

The minerals which are weakly attracted, are *paramagnetic*. The examples are hematite, biotite, beryl, and hornblende.

c) *Ferromagnetic*

The minerals which are strongly attracted by a magnet, are termed as *ferromagnetic*. The examples are magnetite and pyrrhotite.

d) *Antiferromagnetic*

The interaction of adjacent atoms may be such as to align the spins in parallel but opposite directions, called *antiparallel* spins. This gives rise to *antiferromagnetism*. The two sets of magnetic moments cancel each other, and there is no permanent magnetic moment.

e) *Ferrimagnetic*

These are the cases of antiparallel alignment in which the components in opposed directions are not equal. It results into a permanent magnetic moment. *Ferrimagnetism* is shown by the important mineral magnetite ($Fe^{+2}Fe_2^{+3}O_4$) which is a member of spinel group.

In general, it is not possible to determine these differences without the use of a powerful electromagnet. However, use of a small hand magnet can indicate, whether the mineral is attracted to it or not. The minerals which are attracted are ferromagnetic, whereas the minerals not attracted, are either paramagnetic or diamagnetic. The latter two cannot be distinguished, and are usually termed as nonmagnetic.

The magnetic properties of minerals have some important practical applications in ore beneficiation. From a mixture of strongly magnetic, weakly magnetic and non-magnetic material, separation can be carried out by employing an electromagnetic separator with a provision for varying the field intensity. By careful adjustments of the instrument settings, most minerals can be separated even from complex mixtures of many minerals.

1.1.11 ELECTRICAL PROPERTIES

Electrical properties refer to the behaviour of minerals when affected by electricity, or to the ways in which minerals (under certain conditions) may produce electric currents. Most minerals are either conductors or non-conductors of electricity. Some non-conductors have the power to produce electricity when heated or put under pressure. Some native minerals (gold, platinum, and copper) and some of the sulphide and oxide minerals are good conductors. The minerals such as quartz, tourmaline, calcite, etc., are non-conductors.

Tourmaline crystals, when subjected to a temperature change, positive and negative charges are developed at opposite ends of the crystal. This property is called *pyroelectricity*. On the other hand, quartz crystals produce an electric current when the crystal is pressed and the property is known as *piezoelectricity*. The opposite is also true, i.e. if electricity is applied to quartz crystals, they will bend slightly.

Out of the various electrical properties, electrical conductivity of minerals is the most important in mineral dressing. Due to the fact that some minerals are relatively good conductors compared to others, it could be possible to have a commercial separation of two or more minerals by applying this principle.

1.1.12 RADIOACTIVE PROPERTIES

The minerals that emit high-energy radiation are called radioactive. The property depends on the presence of unstable atoms which are giving off radiation (α-rays, β-rays, and γ-rays) continuously in order to achieve a more stable state. The presence of general radiation can be detected and measured with various devices such as the *scintillation counter*.

Radioactive minerals possess this property due to the presence of radioactive elements such as uranium in uraninite, and thorium in thorite. Uranium and thorium decay spontaneously by emitting first an α-particle (identical with nucleus of He atom) from the nucleus. Afterwards they loose a β-particle (an electron), and by successive emission of further α- and β-particles, they pass through a series of unstable daughter products to yield finally, a stable isotope of lead. Simultaneously, the disintegration yields energy in the form of γ-rays. Radioactive materials are widely used as a source of energy.

1.1.13 OPTICAL PROPERTIES
The most important optical property is the refractive index of minerals. The other properties may be of polarisation, absorption behaviour of reflected light, etc. The optical properties of minerals can be determined by special instruments. Like the other properties, optical properties are also dependent on the kinds and arrangements of atoms in the structures of the minerals. Various optical properties help the identification of different minerals when examined under microscope.

1.1.14 FRICTION
The ability of particles to move or slide on a surface will be affected by their shapes and by the coefficient of friction as determined by the nature of the surfaces. The automatic slate pickers used in cleaning coal take advantage of this principle.

1.1.15 MINERAL AGGREGATION
This is the physical form in which the valuable mineral exists in the ore body. It is easier to recover the valuable mineral in an ore when this occurs in pure and relatively large masses rather than, when it occurs in very fine grains distributed through the ore body or when it occurs in a laminated or tabular forms.

1.2 Physicochemical Properties
Chemical properties utilised in ore processing are necessarily limited due to the mechanical nature of ore processing operations. The properties utilised are those which affect the physical behaviour of minerals in an ore processing operation without giving a chemical change to the bulk of the minerals or are directly applicable in chemical benefication. Some important cases are described below:

1.2.1 CHANGE OF POROSITY BY HEAT
Certain minerals when heated, loose a part of their volatile constituents and become porous or spongy. The pores are filled by the air and the mineral shows a lower apparent specific gravity, which many times aids in their separation.

1.2.2 DECREPITATION
Some minerals, when put over a hot plate, decrepitate or fly to pieces. This happens due to the unequal expansion which overcomes the cohesion of the molecules. The examples are calcite, fluorite, and barite. A mineral which decrepitates, may be separated from one which does not, by employing screening, e.g. concentration of barite.

1.2.3 CHANGE OF MAGNETISM BY HEAT

Certain minerals, particularly those of iron, when heated, loose oxygen, carbonic acid, or sulphur and are transferred from non-magnetic or slightly magnetic to strongly magnetic. The magnet may then be employed for separating them from non-magnetic minerals.

1.2.4 SURFACE PROPERTIES

The surface properties of a mineral determine its behaviour at an interface or surface between two phases. The various surface properties include greasiness, adhesion, wettability, contact angle, polarity, and surface tension. These surface properties may be changed by the addition of various reagents due to adsorption or chemical reactions.

Separation of minerals may be accomplished, if a mineral or a group of minerals adhers to or consolidates with a surface of some kind, presented equally to all the minerals. Air bubbles attach themselves to some minerals in flotation while not to others. Gold adheres to an amalgamated plate surface, whereas, quartz and other common minerals do not. Similarly diamond adheres to a greasy surface, while quartz does not. Therefore, this selective adhesion can be effectively employed for commercial separation of minerals.

In flotation, surface properties of minerals are varied by the addition of chemical reagents, and by this means, the valuable minerals are given the ability to adhere to the air–water surfaces of air bubbles. The selective adhesion of oil for certain minerals had also been used in certain older flotation processes.

1.2.5 SELECTIVE DISSOLUTION

Valuables from various ores of minerals can be recovered by selective dissolution of either desired or undesired constituents. For example, in cyanidation process, gold is selectively dissolved in cyanides. In treatment of uranium ores, uranium is selectively brought into solution by a suitable leaching agent. On the other hand, in beneficiation of high phosphorus manganese ores, phosphorus is selectively leached out with hydrochloric acid. Similarly, the ilmenite can be beneficiated by selective leaching of iron values with acids.

1.3 Minerals and Their Classification

Most minerals are crystalline and possess definite chemical compositions. Although the chemical composition of a mineral may vary slightly, it is only within certain limits. Every crystalline mineral has a regular three-dimensional arrangement of atoms, which is called the crystal structure. The minerals may be grouped into the various chemical classes on the basis of the anionic part of the formula. The important groups include, (a) native elements (copper, gold, silver, bismuth, diamond, etc.), (b) sulphides and sulphosalts (galena, PbS; sphalerite, ZnS; chalcopyrite, $CuFeS_2$; chalcocite,

Cu_2S, etc.), (c) oxides and hydroxides (cuprite, Cu_2O; corundum, Al_2O_3; hematite, Fe_2O_3; rutile, TiO_2; brucite, $Mg(OH)_2$; gibbsite, $Al(OH)_3$ etc), (d) carbonates (calcite, $CaCO_3$; magnesite, $MgCO_3$; cerussite, $PbCO_3$; etc.), (e) halides (halite, $NaCl$; fluorite, CaF_2; cryolite, $Na_3Al\ F_6$; etc.), (f) nitrates (soda nitre, $NaNO_3$; and nitre, KNO_3), (g) borates (borax $Na_2B_4O_7.10\ H_2O$), (h) sulphates (barytes, $BaSO_4$; anglesite, $PbSO_4$; gypsum, $CaSO_4 \cdot 2H_2O$, etc.), (i) chromates (crocoite, $PbCrO_4$), (j) phosphate (apatite, $Ca_5(O_4)_3$ (F, Cl, OH); monazite, $CePO_4$, etc.), (k) arsanates (mimetite, $Pb_5\ (AsO_4)_3\ Cl$; erythrite, $Co_3\ (AsO_4)_2\ 8H_2O$), (l) vanadates (vanadinite, $Pb_5(VO_4)_3\ Cl$; carnotite $K_2(UO_2)_2\ (VO_4)_2 \cdot 3H_2O$, (m) molybdate (wulfenite, $PbMoO_4$), (n) tungstate (scheelite, $CuWO_4$, wolframite, $(Fe, M_n)WO_4$, (o) silicates (quartz, SiO_2, zircon, $ZrSiO_4$); sillimanite, $Al_2O \cdot SiO_4$; nephaline, $Na\ (AlSiO_4)$; etc.

CHAPTER 2

Economics of Mineral Processing

2.1 Introduction

An ore is considered to be an aggregation of minerals from which a metal or metallic compound can be recovered economically on a commercial scale. When the percentage of metal or valuable in the ore is too low for profitable recovery, the rock ceases to be an ore.

The processing of ores or mineral dressing is commonly regarded as the direct treatment of raw minerals to yield marketable products and waste. The term *ore dressing* is usually confined to the mechanical separation of valuable minerals from the valueless material of an ore, whereby the valuable minerals are collected into smaller bulk without affecting the physical and chemical identity of the minerals. However, with the chemical processing of lean and complex ores, it is not possible to restrict the treatment of ores only by mechanical methods, as various chemical methods for the treatment of some ores are invariably employed for the economic recovery of valuables. Thus, the subject may be covered under a broad title known as ore processing, which includes any treatment physical or chemical, given to raw ores directly, in order to effect the economic recovery of valuables.

The minerals have been utilised by man for a long time and through the use of these, man has entered from the stone age barbarism to the modern complex synthetic civilisation. The increased requirements of various metals and other products forced the engineers and geologists to search new deposits of ores and to create mechanical facilities for the increased production.

The curve shown in Fig. 2.1 represents the world production of minerals with time. This may be considered as the representation of population, civilisation, standard of living, and technological advance, as all these are sustained by minerals.

Supply of required minerals is one of the most important fundamental factors on which the progress of a country or man is dependent. However, the minerals should be consumed in a very economic way, since the minerals are natural resource and are not renewable.

2.2 History of Mineral Processing

The oldest technique of concentrating an ore was undoubtedly *hand*

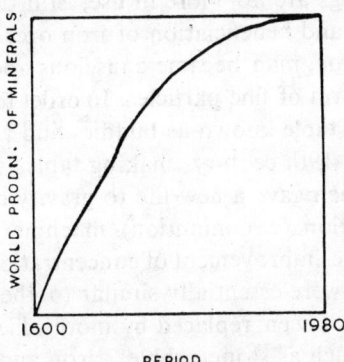

Fig. 2.1. A graph showing the world production of minerals with time.

sorting of valuable lumps based on appearance and weight (density) of the lump. This method played an important role in supplying the highest grade of ore required in smelter. Hand sorting was carried out underground, at the surface, or both places. In all probability, *the washing* was the next process evolved in concentration of minerals. This results in separation of fine sandy grains and slimy material which must have been noticed by primitive man. However, in the beginning the use of washing could have been used without the knowledge of chemical dissimilarity between coarse and fine particles, whereas in the latter periods this technique was employed with the knowledge of chemical dissimilarity. Up to 1940, washing was quite prominent as a concentration method and it is still used for the treatment of iron ores.

It was discovered quite early that the valuable minerals generally occur in relatively intimate aggregation with other gangue minerals and separation of dissimilar grains could be affected more efficiently after crushing the ore. The earliest crushing methods were of the sledge-hammer type, operated by man which were superseded by the crushing appliances, in which the animals and horses were employed to perform the work. The power driven machinery has been employed only recently which made the picture of the mineral dressing entirely different.

The washing operations combined with the force of gravity resulted in the use of slant surfaces which are evidenced by the stone tables found in ruins of some silver mines. In this case, water was allowed to flow down the slope of the table carrying away lighter and finer particles and leaving the valuable sediment for scrapping. This was the beginning of tabling, a process of gravity concentration.

In the Middle Ages (1000–1400 A.D.), it was discovered that concentration of granular material could be carried out by jigging it on a submerged sieve. Most of the developmental work in this technique occurred in the Harz Mountains of north-central Germany. Later on, these jigs became power-operated which played a great role in processing of minerals. Though for

various purposes, the jigs are no more in use, still they are of great importance in coal cleaning and beneficiation of iron ores.

From Agricola's time, man became conscious about the losses of valuable minerals in the form of fine particles. In order to minimise this loss, the
modifications of stone table known as buddles and vanners were developed.
Toward the end of the 19th century, shaking tables were developed by A.R.
Wilfley of Denver, which gave a new life to gravity concentration.

Modern size reduction (comminution) machinery has been developed
simultaneously with the improvement of concentration methods for the fine
ores. Early appliances were essentially similar to the equipments used for
milling flour which have been replaced by more efficient devices consisting
of grinding medium, such as stone pebbles, iron and steel balls, steel rods,
etc., contained within a rotating vessel.

The development of crushing and grinding operations required the development of techniques for the sizing of the crushed or ground ore for increasing the efficiency of crushing and grinding. Stationary screens were the
first used for sizing, which were replaced by the rotating cylindrical screens,
known as trommels. Later on trommels were replaced by vibrating screens.
For sizing fine particles, classifiers (based on settling velocity of solids in
fluid) are preferred.

In recent years (1930–1950), magnetism has been employed for the concentration of some iron ores and slightly magnetic minerals. However, the
application of this method was limited due to its high initial cost, particularly for less magnetic minerals.

The most important method of mineral concentration, known as flotation (based on surface properties of mineral) has been developed very recently in historical sense. In this process the particles sticking to air are
collected in froth whereas the particles sticking to water remain in the pulp.
In 1860, first time an Englishman, William Haynes successfully floated the
sulphide mineral from gangue by agitating the mixture of ore and water with
10–20 per cent fats or oils. Later on several people observed this phenomenon and by the end of the 19th century, flotation was well established and
offered remarkable advantages over the older methods of concentration. At
present, flotation is universally employed for the concentration of sulphide
ores and there is no alternate method to it. The application of flotation to
the treatment of non-sulphide ores has also achieved a great success.

In the current century, a number of processes based on specific characteristics (such as electrical conductivity, colour, etc.) have been developed
and used with success. Finally, the chemical methods (e.g. cyanidation, ion
exchange, selective leaching, segregation roasting, etc.) for processing of
specific ores have been developed for the recovery of values from the ores,
which are not amenable for concentration by any physical method. The recent
developments in mineral processing are towards the large capacity and
higher efficiency of the processes.

2.3 Scope and Objectives of Mineral Processing

The mill receives its raw materials from the ore areas by mining and quarrying operations. In ancient time, mining and quarrying operations were carried out to take out the deposits containing highest possible content of desired mineral. The mineral processing was either not carried out or was carried out in a very crude manner, which required selective mining.

With improvements in mineral processing techniques, it has become possible to carry out the mining of low-grade and complex ores employing bulk mining (mining of desired mineral along with unwanted gangue at much lower cost). In general, the scope of mineral processing is three fold, i.e., (a) elimination of undesired chemical species, (b) elimination of undesired size or structure, or (c) selective separation of desired material. The first two are commonly considered to be more important in mineral processing.

2.3.1 OBJECTIVES OF MINERAL PROCESSING

The basic objectives of mineral processing are of two kinds, i.e., (a) technical and (b) economic. In order to make the product or concentrate marketable, it has to be brought into suitable technical condition as required by the customer, i.e. undesired constituents of the original ore, must be removed or reduced below some specified limits. The product should conform to the various requirements, such as, particle size, assay grade, moisture content, etc. If the ore contains more than one valuable mineral, separation of those valuable minerals will be essential in order to market them separately, in a better price or because of the reason that a purchaser can handle them economically. The smelter or purchaser usually imposes penalties on all concentrates, failing the agreed grade in order to protect himself from financial losses.

A complete separation of ore's constituent minerals is not usually carried out in mineral processing. The process is stopped when an optimum economical or technical stage of concentration/physical condition has been reached. If an ore body consists of one valuable mineral along with gangue minerals, it may be decided that 90 per cent recovery of concentrate assaying 42 per cent of the desired constituent or metal will yield a higher profit than would a concentrate assaying 41 per cent but with 92 per cent recovery. However, for other similar deposits, the figures may be quite different, since, the factors determining the economics vary with local conditions.

The concentration process is rarely pushed to its limit, since the costs involved in regrinding, retreatment, and loss of value in tailings, cannot be taken beyond the economic limits. The processing of minerals is therefore restricted to confirm the schedule of cost, percentage recovery, and assay grade. In some cases, the physical condition may be important to the purchaser. For example, a very finely powdered dry concentrate can be treated in reverberatory furnace; whereas it cannot be treated in blast or shaft furnaces. In many instances, even the colour, size and crystal form may affect the selling, e.g. in pharmaceutical and paper industries the barite required should be white, whereas the colour is not important while employed as drilling muds. Similarly, the asbestos is priced,depending on the length and strength of fibre, along with freedom from retained dust. Similarly, the price of mica,is dependent on its transparency, size,and cleavage. Therefore, the main objectives of mineral processing,are to obtain the product or concentrate, as per the specifications laid down by the customer, in the most economical way.

2.3.2 ECONOMICS OF MINERAL PROCESSING

The various economic factors influencing the choice of a process and efficiency of operations are: (a) required reduction in bulk of the ore for transportation, (b) standardisation of product required for selling, and (c) a proper balance of processing cost with the market values of products/ concentrates.

The ore mined as such,can rarely be used for smelting or other applications. Since smelting operation is more expensive than ore processing, direct smelting can be used only for sufficiently rich ores,which can bear the high cost of smelting. Mineral processing may be regarded as the first stage of metal extraction operations, applied to run-of-mine ore as it provides the cheapest way to eliminate the unwanted material. The point at which the product comes to smelting,is determined by: (a) relative treatment cost, (b) transport cost involved in sending the material to smelter, and (c) technical requirements of the material to be used in smelter. Moisture and impurities along with the product,should be sent only under the circumstances, when their removal is not required for the smelter and the cost involved in removal is more than the freight to be paid. The losses of metals in smelter varies with the amount of slag, which in turn depends upon the extent of impurities to be removed. Therefore, pyrometallurgical benefits (technically as well as economically) can be achieved by receiving the concentrates of optimum grade with respect to metal content and level of impurities.

The cost of smelting depends on two factors, i.e. (a) grade of concentrate, and (b) tonnage of the concentrate treated. Deductions for slag and volatilisation losses, bonuses for additional desirable constituents, and penalties for unwanted impurities should be taken into consideration, while calculating the cost of smelting. The valueless tonnage sent to smelter,increases the freight charges. Since, higher grade of concentrate, results into higher tailing loss, and greater treatment cost, considerations should be balanced out at the time of deciding the optimum plant performance.

2.3.3 ADVANTAGES OF MINERAL PROCESSING

Compared with hydrometallurgical processing, pyrometallurgical processing, or refining processes, mineral dressing is inexpensive. The advantages of mineral processing for separating/concentrating the valuable constituents can be listed as following:

a) Cheaper physical/chemical method of rejecting the waste material is substituted for the more expensive chemical/metallurgical methods, such as smelting, refining, etc.

b) The rejected waste material is not transported,which saves freight, as no freight has to be paid on the waste discarded by the ore processing operations.

c) In the case of non-metalliferous ores, such as graphite, emery, and precious stones, the mechanical methods can only work in separation/concentration.

d) Separation of two valuable minerals associated in the ore,fetches more money than selling together in one ore. The mineral of less prominence goes with no return or sometimes even it is treated as detrimental, whereas the mineral of more prominence gains more value in selling due to the better grade achieved in separation and concentration.

e) Mineral processing results,in reduced losses at the smelter,due to the reduction in amount of metal-bearing slag produced in smelting.

f) The total smelting cost is reduced due to reduction in tonnage of ore to be smelted.

2.4 Choice of a Mineral Processing Method

A method of ore processing should be selected which is suitable to a particular ore. It is based on certain general principles. In general, three types of ore complex, i.e. massive, intergrown, and disseminate, are available. Coal and some bedded deposits,such as iron seams, are the examples of massive type, and the ore processing operations are either simple or unnecessary in these cases. On the other hand, most lead–zinc ores, are intergrown and require controlled grinding. The grinding operation is carried out to a point,where galena is sufficiently freed from sphalerite, but very fine particles are not produced,which are difficult to be treated by the concentration process. Today, disseminated values distributed sparsely through a valueless rock matrix,are characteristic of maximum tonnage to be milled. The whole of the ore has to be reduced by grinding to a very fine size,before carrying out the desired concentration. Thus, in such cases the choice of method is dependent on the grinding cost and liberation efficiency.

The essential factor,in all ore processing methods,is the correct liberation, which is the key to success. The valuable mineral should be just freed and the overgrinding of ore should be avoided, since it consumes power without giving further liberation. Too fine particles also make the efficient recovery of valuables more difficult.

2.5 Various Mineral Processing Methods Based on Properties of Minerals

The liberated particles caused by comminution should possess sufficient difference in their physical, electrical, magnetic or chemical properties to respond to an appropriate differentiating force. Physical differentiation makes use of the properties such as shape, size, surface area, specific gravity, porosity, colour, and gliding friction, in separation of the mixed species.

Electrical treatment is based on the magnetic susceptibility (natural or induced), conductivity, and radioactivity, whereas chemical attack depends on the reactivity of minerals with specific reagents. In processing of certain ore, one or more of these properties, may be employed. Table 2.1 represents the important exploitable characteristics along with possible applications.

Table 2.2. Important exploitable characteristics used in separation of minerals

Characteristic	Type of separating force	Techniques employed
Colour, lustre	Visual, manual, automated	Hand sorting of graded ore Fluorescent light or impulses triggered by reflected light may be used
Specific gravity	Differential movement in fluids	Jig, sluice, shaking table, spiral Heavy media separation
Ferromagnetism	Magnetic	Magnetic separators
Conductivity	Electrostatic charge	Separation by high tension separators based on differences in conductivity
Shape	Frictional	Sliding action to remove slate from coal
Texture	Crushing, screening, classifying	Techniques based on characteristic shapes and surfaces, which are developed during comminution
Radioactivity	α or β rays	Separating or picking devices used on the basis of activation by signals from emissions
Chemical reactivity	Reaction with suitable chemicals	Leaching of ores, separation of dissolved compounds by solvent extraction and ion exchange, precipitation, etc.
Surface reactivity	Differential surface tension in water	Separation of relatively aerophilic mineral as froth from aerated pulp by froth flotation

Based on Ref. 8, 20, 24, and 40.

Chemical treatment of ores is gaining more importance day-by-day, due to the demand for utilising those ores, which cannot be economically treated by physical methods. The cyanidation of gold and silver ores is one of the oldest chemical processes in the treatment of ores. The first step in uranium recovery from its ore, is the chemical attack by acids or alkalis. Other elements treated by chemical attack (leaching), include copper, aluminium, manganese, nickel, tungsten, and zinc. Normally, mineral processing, is confined to the operation of separation of ore constituents, without affecting their physical state. However, in the modern technology it is difficult to restrict the field to a specific point and the overlapping of two fields i.e. mineral processing and metallurgy or chemical technology is unavoidable.

In hydrometallurgy, the processes are conveniently employed to ores directly without affecting concentration. Sometimes pyrometallurgy is combined with mineral dressing. Segregation roasting of refractory copper ores followed by flotation is an example. In another case, a mixture of smelted metal sulphides is separated by froth flotation. Other such examples are the electro-chemical separation of antimony and zinc from their ores, extraction of magnesium from sea water, removal of iron values from ilmenite by selective leaching, etc.

2.6 Principal Steps of Mineral Processing

The principal steps involved in processing of ore to yield the suitable product from a chemical stand point are: (a) liberation or freeing of dissimilar particles from each other, i.e., valuable minerals from gangue, and (b) actual separation of liberated chemically or physically dissimilar particles into the various final concentrates and waste products. Therefore, in every instance of ore processing, two main steps are involved, i.e. (a) a size reducing or liberating operation or group of operations as a first step, and (b) separating operation or group of operations as a second step.

After the concentration or separation operations are performed, the products are sent to the smelter or other consumers. Figure 2.2 represents a simplified flowsheet for the various stages involved in processing of ores.

In general, there are four principal types of operations to which a number of auxiliary operations such as storage, conveying, feeding, pumping, etc., may be added. These auxiliary operations do not directly affect the liberation or separation. The four kinds of principal operations are comminution, sizing, concentration, and dewatering.

2.6.1 COMMINUTION

Comminution is the term used for reduction of ore to a smaller size. Comminution of an ore can be performed in dry or wet condition, depending upon the required size of material. The operations involved in comminution are regarded as crushing and grinding. Crushing is almost always conducted on dry ore, whereas grinding may be wet or dry. The extent of comminution depends on the mineralogical nature of the associated gangue minerals. With many ores mined today, grinding to complete liberation is not feasible either economically or technically. Therefore, a compromise is made by grinding the ore just to the point at which the return by the further grinding or liberation would be less than the extra cost involved. The extent of comminution is regulated by the particle size required for the subsequent concentration operation.

The primary purpose of comminution is the unlocking of values in the ore, but crushing may be combined with sorting, and performed in stages.

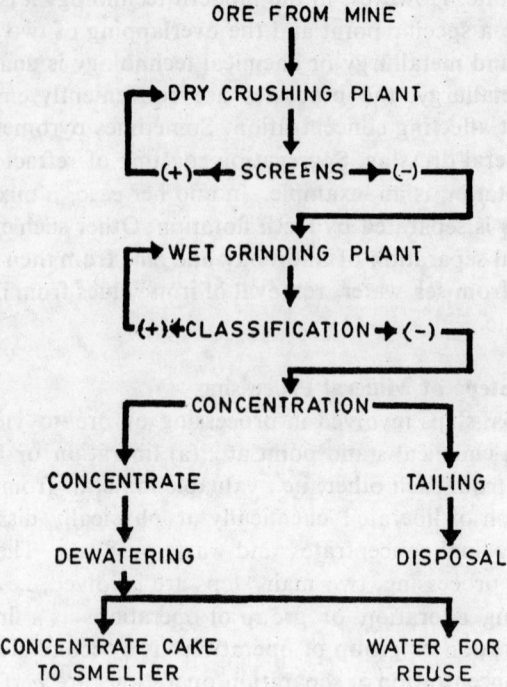

ORE FROM MINE

DRY CRUSHING PLANT

(+)◄— SCREENS —►(-)

WET GRINDING PLANT

(+)◄CLASSIFICATION ➔(-)

CONCENTRATION

CONCENTRATE   TAILING

DEWATERING   DISPOSAL

CONCENTRATE CAKE   WATER FOR
TO SMELTER   REUSE

Fig. 2.2. Simplified flowsheet for the various stages involved in ore processing.

2.6.2 SIZING

Sizing is the separation of material or product into various fractions characterised by difference in size. Sizing can be performed by screening or classification. The latter depends on the settling velocity of particles in a fluid (air or water).

Screening is also used to prevent oversized lumps of ore from entering the machine for which it is not built. Further, sizing is performed to prevent the overgrinding, in order to make the operation efficient and economic. The ground material may be sent directly to the concentration plant, in which case the system is called "open circuit grinding". But in most cases, the ground ore (pulp in case of wet grinding) is passed through a sizing device, which returns oversized ore for further grinding and sends on the ore which has been sufficiently ground. This system is called "closed circuit grinding" and is widely practised.

Grinding can be performed in one or more stages of size reduction. The larger the tonnage of ore treated, greater is the possibility of breaking down of every operation into several stages (each is applied within restricted limits of size range).

2.6.3 CONCENTRATION

Concentration may be regarded as collection of valuable minerals in a small bulk. Concentration or separation may be accomplished by a variety of appliances depending upon the type of mineral and its specific characteristics. Different equipments employed for concentration are washers, sluice boxes, shaking tables, flotation cells, magnetic separators, electrostatic separators, etc. The actual concentration operation gives two or more final products. During concentration, various unfinished products of intermediate grade are produced and must be retreated to make final disposition into concentrates (saleable products) and tailings (waste products).

2.6.4 DEWATERING

The products from the concentration plant leave either as moist gravel or sand, or as a fluid pulp. In case of tailing, the moist sand or pulp can be directly run to waste, whereas in case of concentrate, it has to be dewatered before its final use in the smelter or shipment. Dewatering is usually carried out only to the extent of producing a damp cake. This can be accomplished in two steps, i.e. first (a) in thickeners to remove most of the water, and then (b) in filters which receive the thickened pulp and yield the damp cake.

2.7 Quantitative Representation of Mineral Processing Operations

In order to have a proper control of various operations, a measure of the effectiveness of various ore processing operations is essential. There is no measure which has been accepted universally, but several measures are widely used to express the results of various operations. The important ones are discussed below:

2.7.1 DIRECT MEASUREMENTS

The simplest method to express the results is the measurement of the weight of various products obtained from a given weight of feed, along with their assay values as well as percentage of mineral and gangue whose separation was desired. For example, the results of concentrating a lead ore by flotation may be recorded as the following:

Product	Wt.%	Pb content in %	Gangue content in %
Feed	100	3.5	82.5
Concentrate	5.5	60.5	12.6
Tailing	94.5	0.2	94.6

These results though simple in presentation, do not give convenient interpretation. In order to know the advantages of the operation, eight results/data are to be kept in mind. In case the gangue content is of secondary importance, five results/data are to be kept in mind. In order to reduce this number of results, the operations may be quantified by expressing the results in other terms, such as "ratio of concentration", and "recovery of the valuable constituents".

2.7.2 RATIO OF CONCENTRATION

The ratio of the weight of feed (F) and the weight of concentrate (C) is the measure of the concentration in weight that has been affected. This ratio is termed as the 'ratio of concentration' (K) which may be analytically expressed as the following:

$$K = \frac{F}{C} \qquad (2.1)$$

If it is difficult to measure the weights of concentrate and tailing, the ratio of concentration K may be expressed in terms of assay value of feed (f), assay value of concentrate (c), and assay value of tailing (t) as the following:

$$K = \frac{c-t}{f-t} \qquad (2.2)$$

Equation (2.2) is obtained from the following two basic equations, i.e.

$$F = C + T \qquad (i)$$

which states that the weight of feed is equal to the sum of the weights of the products, and

$$Ff = Cc + Tt \qquad (ii)$$

which states that the weight of one constituent in the feed is equal to the sum of the weights of that constituent in the products.

If Eq. (i) is multiplied by t, then the equation becomes

$$Ft = Ct + Tt \qquad (iii)$$

The terms Tt in this equation and Eq. (ii) are the same.

On substracting Eq. (iii) from Eq. (ii), the relationship of Eq. (2.2) is obtained as follows:

$$F(f-t) = C(c-t)$$

or

$$\frac{F}{C} = \frac{c-t}{f-t}$$

or

$$K = \frac{c-t}{f-t}$$

The ratio of concentration, K alone is not much useful, since it does not disclose anything about the quality of the concentrate and tailing. It further loses its significance when more than one concentrate is obtained. However, the ratio of concentration is very useful in conjunction with recovery.

2.7.3 RECOVERY

The fraction of a certain metal contained in the feed represented in percentage is called recovery (*R*). This can be figured in terms of assay values of feed, concentrate and tailing. The recovery may be represented as the following (using the same notations as above):

$$R = 100 \cdot \frac{Cc}{Ff} \tag{2.3}$$

From Eqs. (2.2) and (2.3), *R* can be represented in terms of assay values of feed (*f*), concentrate (*c*) and tailing (*t*), as follows:

$$R = 100 \cdot \frac{c}{f} \cdot \frac{C}{F} = 100 \frac{c}{f} \cdot \frac{1}{K}$$

$$= 100 \cdot \frac{c}{f} \frac{(f-t)}{(c-t)} \tag{2.4}$$

The measurement of recovery is equally useful when more than one concentrate or middlings are obtained. Thus, if a feed is treated to obtain several concentrates of weights C_1, C_2, C_3, ..., and grades c_1, c_2, c_3, ..., the fundamental equation stating the recovery of any of the products will be

$$R_n = 100 \cdot \frac{C_n}{F} \frac{c_n}{f}$$

and the ratio of concentration may be stated as

$$K_n = \frac{F}{C_n}$$

In some cases, the values of R_n and K_n may be expressed in terms of assay values only, excluding weights, but the equations are somewhat complicated.

Representation of recovery and ratio of concentration are the usual methods of expressing the metallurgical results, particularly in case of metal ores. In case of mining of non-ferrous metal ores, recovery of metal is considered to be the most important criterion with grade of concentrate next. In case of coal washing, ratio of concentration is more important than recovery.

2.7.4 EFFICIENCY OF ORE-PROCESSING OPERATION

When various operations or tests are compared, it is quite difficult to make a selection of the best operation or test, particularly while considering recovery, ratio of concentration, and grade of concentrate at the same time. In order to assess the efficiency of an operation or test, various indices have been proposed. One of the simplest and most appropriate is the metallurgical efficiency proposed by R.W. Diamond. This may be defined as the arithmetical mean of the recoveries of the main constituents of each product (including tailing). For example, in separation of galena, sphalerite, and gangue, if the lead concentrate contains 92 per cent of the lead, the zinc concentrate contains 89 per cent of zinc, and the tailing contains 96 per cent of the gangue, the metallurgical efficiency will be equal to $(92 + 89 + 96)/3 = 92.3$ per cent. In general, the metallurgical efficiency, E, may be represented as follows:

$$E = \frac{\sum R_n N}{n} \tag{2.5}$$

where $R_n N$ represents the recovery of constituent n in the N product. However, the metallurgical efficiency is not zero when there is no separation.

A more perfect way of representing the efficiency, is through *selectivity index*. The selectivity index is the geometrical mean of the relative recoveries and relative rejections of two minerals, metals, or group of metals or minerals. If R_a is the recovery of a in product A, R_b is the recovery of b in A, J_a is the rejection of a in product B, and J_b is the rejection of b in B, the relative recovery of a to b is R_a/R_b and the relative rejection of b to a is J_b/J_a. Then according to the definition, the selectivity index (S.I.) may be represented as the following:

$$\text{S.I.} = \sqrt{\frac{R_a}{R_a} \cdot \frac{J_b}{J_a}} \tag{2.6}$$

But $J_a = 100 - Ra$, and $R_b = 100 - J_b$
Thus

$$\text{S.I.} = \sqrt{\frac{R_a J_b}{(100 - R_b)(100 - J_b)}} \tag{2.7}$$

For example, if lead recovery in a lead concentrate is 94 per cent and gangue rejection in tailing is 97 per cent then

$$\text{S.I.} = \sqrt{\frac{94 \times 97}{6 \times 3}} = 22.5$$

In some cases, it is easier to calculate selectivity indices from grades than from recoveries. Therefore, if assay values for substances a and b are X and Y in the concentrate, and x and y in the tailing, the selectivity index may be represented as follows:

$$\text{S.I.} = \sqrt{\frac{X}{x} \cdot \frac{y}{Y}} \tag{2.8}$$

In case of lead concentrate (example of Sec. 2.7.1), the selectivity index is

$$\text{S.I.} = \sqrt{\frac{60.5}{0.2} \cdot \frac{94.6}{12.6}} = 47.6$$

If the grade of concentrate and tailing is the same (no separation), the selectivity index will be unity. On the other hand, if the concentrate is completely free from waste constituents, and the tailing is completely free from valuable mineral, the selectivity index will be infinite. However, both the extremes are not reached in practice. The usual range of selectivity indices is 4 to 40, and only exceptionally poor or good results fall outside this range.

Selectivity index represents an accurate measure of efficiency of a concentration/separation. However, it fails to measure the value of successive improvements in a proper economic way. Thus, if the index is doubled, the returns from the sale of the concentrate will never be just double. It may be more than double in some case, whereas in the other cases, the increase may be slight, and the effect which makes the index double, is quite unpredictable.

2.7.5 ECONOMIC RECOVERY OR EFFICIENCY

A comprehensive measurement of the economic effectiveness of an ore processing operation cannot be made by either recoveries, or ratio of concentration, or selectivity indices. To make the measurement of economic effectiveness, a new quantity called *economic recovery or efficiency,* may be used. This is defined as the percentage ratio of the actual value of the concentrate obtained per tonne of ore to the value of that weight of concentrate, theoretically obtainable in mineralogically pure form from one tonne of ore. The following example will illustrate the economic recovery:

A sample of 0.13 tonne of lead concentrate worth Rs. 6000 per tonne can be obtained in practice from an ore that theoretically should yield 0.12 tonne of pure galena concentrate worth Rs. 8000 per tonne. In this case, though the actual lead recovery may be as high as 96 per cent, the economic recovery is only

$$\frac{0.13}{0.12} \times \frac{16000}{18000} \lambda \times 100 = 65 \text{ per cent}$$

2.7.6 SCREEN AND CLASSIFIER EFFICIENCY

The screen and classifier efficiencies are obtained from the following equations:

For screens, the efficiency can be written as

$$E = \frac{10,000 \ U}{uF} \tag{2.9}$$

where E is the screen efficiency, U is the tonnage passing through the screen for each F tonne of feed, and u is the percentage of undersize in feed as determined by test screening. And for classifier, the efficiency can be given as

$$E = 100 \ \frac{c(f-c)}{f(c-t)} \tag{2.10}$$

where c, f and t are the content of $-x$ mesh material in the overflow, feed, and underflow (x is any size for which the value of any of the quantities c, f and t is not zero).

The details are discussed in the respective chapters. These formulae are identical to the recovery formula [Eq. (2.4)]. In words, the efficiency of a classifier or screen may be defined as the ratio of the attained to the attainable in a size separation operation.

CHAPTER 3

Pre-treatment and Sampling of Ores

The run-of-mine (r.o.m.) ore contains a variety of valueless and even harmful impurities along with the desired mineral. Earlier, laborious hand sorting was used to produce a small, but fairly clean tonnage of concentrate. With the present mechanised mining of low-grade deposits, it is not usually possible to have careful discrimination between valuable lode-material and waste country rock. The mined ore may also carry some iron or steel pieces lost from the machines, which may damage the crushers, if fed in with the ore. Further, the various other materials such as wood, unexploded explosives, lubricating oil, grease, etc., also find their way into the run-of-mine ore. Some deposits may contain clays and primary slimes having no value and causing problems in the subsequent treatment. All these materials should be removed before feeding the ore to mill.

Preliminary rejections of waste rock and very low-grade ores may be carried out without decreasing an appreciable amount of the valuable minerals. When the ore comes to the mill, its quality and weight should be known. This involves the weighing of incoming ore and taking control samples with adequate accuracy. Sampling is required to check the assay grade of the rejected part as well as the ore accepted for treatment, to ensure the efficient working of the ore processing unit.

3.1 Weighing the Input Ore

The widely used weighing devices are those which register and record the weight of ore passing over a section of a belt conveyor. These devices are compact and quite accurate (within 0.5 per cent), provided the belt is kept clean, and regular checking and servicing of weighing mechanism is carried out.

In order to keep a close check on ore processing operations, it is necessary to determine the weight of ore delivered to or passing through various sections of the plant. In case the ore is being purchased from a seller, the weight of the ore is to be determined as accurately as possible, whereas, in case of processing the ore of the same company, the weight is to be determined only approximately, required for rough checking. Various important schemes for weighing are described below:

3.1.1 TRACK SCALES

These systems are of beam-scale type and are employed for car-load lots. An operator is required to balance the load on the scale beam and he records the weight. This method gives reasonably correct results (within the error of 0.5 per cent) but has the disadvantage of being dependent on the care and accuracy of the operator. In case of recording scales, the operator only balances the load and then with the help of a screw, automatic record of the weight is obtained by the punching action on the card. Some track scales weigh a train of ore automatically, as it passes slowly over the platform. In such a scale, the personal error is eliminated, but requires occasional standardisation.

3.1.2 AUTOMATIC DUMP SCALES

These scales consist of a hopper which receives ore until full and at that time the weight is recorded and the material discharges automatically. This type of scale is quite accurate in weighing. However, the scale requires considerable space and the operation is intermittent. For weighing a continuous stream of material, two units will be used so that one will be discharging while the other is getting filled.

3.1.3 CONVEYING WEIGHERS

These weighing equipments are attached to conveyors which automatically record the weight of material being carried by the conveyor. These require a small space and need not be attended. Occasional standardisation and adjustments may be needed to maintain the accuracy (within 0.5 per cent) of the weighers. These are employed to supply the feed at a constant rate.

3.1.4 PLATFORM SCALE

These are used similar to track scales when wagons or trucks deliver the material. The weight should be determined accurately. The load should be centred as nearly as possible to both loaded and unloaded weighings, or both weighings should be carried out in the same position.

3.1.5 ELECTRIC WEIGHING MACHINES

These operate on the principle that the amount of current flowing through an electric circuit is proportional to the product of voltage and the conductivity of the circuit. Voltage proportional to the speed of conveyor is produced by running a constant-field dynamo by gears or chains and sprocket attached to the driving shaft of the conveyor. Conductivity is varied with the change of load by a rheostat operated by a plunger in a mercury dashpot. An ampere-hour meter in circuit is graduated to give the units of weight.

3.1.6 Approximate Methods of Weighing

These methods are employed when great accuracy is not essential or the investment for a scale is not desired. Car loads or train loads are counted and the weight is calculated from the volume of the cars/wagons and the weight of a unit volume of the bulk ore. In some cases, car loads or train loads are occasionally weighed and this weight is used as a basis for determining the total weight from the number of wagons/cars or train loads counted. Measurements of bin volume are generally used as a rough check on other methods or for inventory at the end of statement periods. Tonnage of the material may also be determined by counting the number of strokes or revolutions of feeding devices, where the weight of material fed in one stroke or revolution is determined experimentally.

3.1.7 Weighing of Wet Pulp

This is accomplished usually by volume measurements and moisture determinations. The whole stream of pulp (flowing in pipes) is diverted into a suitable container for a particular time and the volume is determined. The proportion of solids is determined from a small dip sample taken out while the container is being filled. In case of small streams, the whole amount diverted may be used for the determination of solids in the pulp. The percentage of solids may be determined either by (i) weighing, dewatering, drying and weighing the dry solids, or (ii) by calculation from the specific gravity of wet pulp and specific gravity of the dry material. The first method is more accurate, but takes more time.

The tonnage of material in an agitating tank (where the volume is known) may be calculated from a dipper sample. However, the accuracy will depend on the degree of pulp uniformity caused due to agitation. The pulp flowing through pipes is sometimes measured by solution-meters and tonnage calculated from small specific gravity sample.

3.2 Washing and Scrubbing of Ore

Run-of-mine ore is washed to remove the obscuring dust and the dirt from the surface of the ore pieces. The washing action facilitates the recognition of minerals in performing hand picking. The primary slimes may be troublesome in the plant as it clings in dry-crushing machines and interferes with the performance of various equipments. Limited beneficiation of some ores such as clay ores containing limonites or manganese nodules, and tin bearing gravel may be obtained due to the gentle disintegration during washing or scrubbing action. If the ore is slimy, the system becomes slippery. Sometimes it is desirable to wash away soluble salts resulting due to weathering effect to prevent their entry into chemical reactions in the flotation or cyanide process.

The systems employed in washing or scrubbing are usually simple and work with gentle force of water and very mild agitation. The usual types of washers or scrubbers are discussed below:

3.2.1 Jet Washers

A jet washer or disintegrator is composed of a horizontal cylinder closed at one end. The feed hopper is placed on the top at a point about 60 cm from the closed end. A charge of ore is fed in through the hopper and a jet is delivered from a stream under a pressure of about 4 kg/cm^2. The jet plays in at the open end and the final disintegration to the clay is obtained. The size and velocity of the jet to be employed depends upon size of the ore and the method of holding the ore. Jet scrubbing is most effective when the ore is held in place. In a simple washing, a jet of water from a nozzle may play on ore at any suitable point while it is in transit.

3.2.2 Trough Washers

These are wooden troughs with or without the steel plate lining having a slope of 0-10° (upward or downward) and closed at one end. The upward inclination is more desirable since it allows larger charges to be worked and prevents the loss of certain small sizes of rich ores which would be carried off in case of downward slope. Water is supplied at the head end in sufficient amounts from a height of about 30 cm which exerts a considerable washing action. The run-of-mine ore or the product of some crusher or concentrator (less than 10 cm in size), is fed in at the tail end, shoveled over, and worked toward the head end until the fine material is removed. Finally it is shoveled to a gravel screen, where sometimes even the waste rock may be removed from the mineral in addition to the fine material. These trough washers can be stirred with hand tools or mechanical means. The latter type find some application in coal washing. The trough washer usually results into three products, i.e. (a) coarse sand left in washer, (b) fine sand in a small tailing tank, and (c) clayey waste.

3.2.3 Log Washers

These are mechanical trough washers and are employed for heavy work, where excessive tumbling action is required. This works as a shearing disintegrator, a tumbling device, washer, and a classifier. It consists of an inclined tank in which two logs in the form of interrupted spirals are attached with stirring blades. The blades rise inward and force the ore up-slope against a down current of water. The pulp consisting of slime escapes over a weir at the lower end, whereas the ore is washed upward against the water stream.

3.2.4 Wash Trommels or Drum Washers

These are hollow revolving cylinders or cones which are set with their axes horizontal. The interior of the trommel is a cylindrical heavy screen through which the ore tumbles and gets washed by jets of water. The disintegrating blades rotating about a central axis may also be provided to remove the clay. The ore is conveyed forward in the cylindrical form by oblique blades, acting on the principle of a propeller, or by continuous screw threads, whereas in conical form, the ore moves forward by gravity. The wash trommels basically are of two types, i.e. (a) having partially closed discharge end, in which lumps are immersed in a pool of water for washing. and (b) having the discharge end completely open, in which the ore is washed either by a stream or sprays of water, or by both.

3.2.5 TELSMITH SUPER SCRUBBER

This is development of drum washer. This rotates faster and gives a vigorous tumbling and rubbing action. A scrubber of 2.5 m diameter can accept the ore up to 20 cm in size and can handle up to 200 tonnes/hr with a water consumption of 5000 litres.

3.2.6 WASHING PANS

These are the large circular ring-shaped pans in which the ore is disintegrated by revolving blades, or by rollers and scrappers. The ore is fed with water at the periphery, the clay and fine sand overflow at the centre while the heavy material collects at the bottom of the pan. Such pans have been employed in the diamond field of south Africa to free the weathered diamond bearing 'Blue ground' from the fines, sand and mud.

3.2.7 OTHER WASHING METHODS

Gentle agitation together with washing is sometimes carried out on washing screens which are fitted with spraying arrangements to perform their work in water. Washing of ore may also be carried out on a rising conveyor belt to remove some slime. The belt must be troughed and should have the provision for discharging the resulting pulp clear of the pulley mechanism. Washing on conveyor belt does not result into complete removal of slime and dirt, but the cleaning of the ore surfaces helps in recognition and hand picking along the belt.

3.3 Hand Sorting or Hand Picking of Ores

Hand sorting is a common practice which is adopted during the course of mining minerals like asbestos, barytes, manganese ore, diamond, mica, etc. In case of chrysotile asbestos, the run of mine containing long staple fibre is sorted out manually from the mill grade fibre. Hand sorting is an universal practice. In India, in the case of barytes mining, three grades differentiation, namely, snow-white, white, and off-grade colour is done entirely by hand sorting. In the case of manganese ore, battery grade ore is sorted out from other grades from the r.o.m. by manual operations. Similarly in the case of diamond, the gem and industrial varieties are sorted out by hand-picking only. In the case of mica, hand sorting is the only means known so far for grading mica.

These days much advancement has been made in putting into operation the mechanical device for sorting minerals which take into advantage of opacity and the colour reflctance in sorting out minerals.

3.4 Mechanical Sorting of Ores

Low-grade ore deposits can only be economically exploited on a large scale, and in dealing such a large amount of ore, the sorting by visual recognition is impracticable. Thus visual recognition is needed to be replaced by mechanical sensing devices which typically assess a specific quality in a passing particle of ore and send a signal to a mechanical or electrical device which removes that particle from the passing material. These devices may be solenoid-operated or worked by an air-blast, deflecting device or trap. In mechanical sorting, photoelectric cells would take the place of eyes, vacuum tubes and magnetic relays of the brain, and compressed air or other mechanical device of hand. In order to have the efficient separation by mechanical sorting, following requirements should be met:

a) Some specific property in the sorting mineral differentiating it sharply from the main ore minerals.

b) Sufficient cleansing of all passing ore particles to prevent dimness of the signal due to superficial contamination.

c) Sufficiently close size of feed.

d) Presentation of particles one by one to the sorting device.

e) Provision for special excitation in certain cases where the signal is dull.

f) Simplicity and cheapness of operation.

In addition to the process cost, the economic factors considered in mechanical sorting, as in hand sorting, should also consider saving on transport, reduced mill head accepted for treatment, and disposal of rejects. The various characteristic properties of minerals which may activate a suitable sensing device, are (a) conductivity, (b) specific gravity, (c) reflection and refraction of a light beam, (d) light sorption, (e) radioactivity, natural or induced, and (f) ferromagnetism. Presently photoelectric sorting is very much in use, particularly for radioactive minerals.

La Pointe Picker can be used for sorting of moderately coarse radioactive concentrates, prepared by jigging (adherent radioactive dust is absent). In this, a single line of ore particles moves on a narrow belt, and passes below a Geiger-Muller tube. If radioactive emission exceeds a set level, a sorting device is operated which removes the signalling particle from the moving line.

K and H Equipment makes use of photoelectric cell. This is based on the triggering action by the particle's own radioactivity. The mineral lumps (5–20 cm) are delivered one by one, through three successive scanning devices, i.e. (a) neon light working on a.c. at high frequeney (its beam falls on a photoelectric cell after getting interrupted by the falling particle), (b) a scientillometer which determines whether the particle is above or below the present level of radioactivity, and (c) a series of elecronically controlled air jets to blow off the low-grade pieces.

3.5 Removal of Tramp Iron

The iron and steel pieces known as *tramp iron* inadvertently find their way into the stream which may damage the crushing machine and production may be held up during repairs. Though these iron and steel pieces can be removed by hand during the operations as described above, these can be removed more effectively by hanging an electromagnet over the belt conveyor or other suitable position. This arrangement may not be effective to catch small pieces such as nails and bolts which are held under the ore, and these can be removed by furnishing the belt conveyor with an electromagnetic head pulley. Iron thus held next to the belt is not discharged at the delivery end of the conveyor but is carried round and dropped sufficiently distant from the ore stream after leaving the magnetic zone.

Another device is the use of search coil which is very sensitive. This can detect a slight concentration of magnetic flux arising due to the passage of even feebly magnetic material through the coil. Some alloy-steels used in the mine are non-magnetic, and there is always a danger of pinning down a metal piece. In general, three search coils are used on the feed conveyors. When a piece of iron or other metal passes the first of these coils, a current is generated which can be amplified and caused to trip the motor of the feeder supplying the ore to the belt conveyor. The second search coil detects the same piece of metal and stops the conveyor. The distance between the two coils is so adjusted that the crusher becomes empty before the conveyor stops. Thus the inconvenience of ore build-up on a stationary conveyor is avoided. The third search coil is an extra, which would operate in the event of failure of either of the first two coils, or if the worker fails to remove all the metal pieces. Normally, it locates the tramp iron, picks it off, and then restarts the belt.

In another type of induced current detector, as soon as an alarm sounds, the driving motor trips out. If the appliance continues a short distance before stopping, the trouble can be quickly located. The use of guard magnets and hand picking is sometimes combined.

3.6 Preliminary Breaking of Ores
The size of the r.o.m. sometimes may not be receptible to crushers. In that event, the help of the secondary blasting or hammer mill is taken. These days a great advancement has been made in manufacturing big-size jaw crusher, gyratory crusher and cone crusher, which can accommodate appreciably large-size ore resulting out of the mine.

3.7 Sampling of Ores and Products
Sampling is an art of securing a representative fraction in a small weight or sample from a relatively large lot. In order to make the sample representative, it must be selected in a fair way. Ore sampling falls into two categories, i.e. exploratory sampling and control sampling. The former type is used in case of prospecting, proving, and developing a mine. In routine exploitation of the ore and control of the mill operation, samples are taken to check the efficiency of the work. The second category of sampling

is of main concern in mineral processing industries. In a given sampling system, the quality of the material is checked at the desired point with sufficient accuracy. This may include check of lump size, moisture content, assay grade, purity, chemical state and other desired informations. Sampling and assaying in ore processing industries are required to determine the quantity of valuables in the ore, losses of valuables in tailing, quality of middlings, and the value of concentrate. Consequently, the method adopted for sampling is of utmost importance. This work must be carried out in such a way that no slime or dust belonging to the ore product should be lost, as fine particles are often rich and a larger error would be caused by their loss than the loss of weight alone.

Despite the great importance of good sampling as a means of technical and economical control, it is rarely carried out with full satisfaction. The physical difficulties involved in selecting a truly representative sample and reducing it to the weight of few kilograms needed for assay are formidable. A sample drawn from the finely disseminated ore of fairly even grade may satisfactorily represent the whole lot, whereas it is quite difficult to obtain a truly representative sample from the ores containing precious metal values in the native form (existing in small scattered particles) or the material of very high grade existing in an irregularly distributed form. When the native metal or the high-grade mineral exists in large masses or crystals, it is much more difficult to obtain a true representative sample compared to the ore in which values are present in finely disseminated form. Therefore, sampling consists in obtaining from a lot of ore, a small portion representing as perfectly as possible the exact portion of the constituents in the original batch of the ore. This involves two operations which proceed by alternate stages, i.e. (a) cutting down or reducing the weight of sample, and (b) reducing the size of particles.

3.7.1 CUTTING DOWN OF SAMPLE

The control sample cut from the original feed known as *head sample*, may be taken by hand or mechanical means depending on the way in which the ore comes to the crushing plant. An elementary method is to make a *grab* sample in which a small portion is taken from the contents of each passing car-load of ore. Since the valuable minerals are usually more friable than the gangue, there is risk of washing down the rich ores and the sample taken from the top of car-load will not be representative. In case of ores having evenly distributed values, this system of taking the sample will not present any difficulty, but there may be other objections to hand sampling. If the reliable data for technical control is required, the system of sampling should be fool proof. Taking out an adequate sample by hand and its accurate reduction involve consciousness, hard labour and monotonous repetition.

3.7.2 HAND SAMPLING PROCEDURES

Various hand sampling procedures are used in selection of samples.

Commonly used procedures are discussed below:

a) *Fraction Selection by Shovel*

When ore is shovelled either to load or unload, every 5th, 10th or 12th shovelful (depending on the richness of the ore and distribution of minerals) may be thrown to one side for a sample. This sample is crushed to reduce the size of the lumps and again reduced in quantity by setting aside alternate shovelfuls as a sample. This operation is repeated several times until the sample needs crushing again. When a shovelful becomes too large a proportion of the sample, a small shovel should be employed.

b) *Coning and Quartering*

The crushed ore is brought to the sampling floor and collected evenly in a large ring. Then the ore is shovelled into a conical heap in the centre, while walking slowly around the ring. Too much of ore should not be shovelled from one position. Every shovelful should drop systematically upon the apex of the cone so that all sides of the cone receive contributions from each shovelful of ore. The ore is then raked out into a new ring by a shovel or a hoe and re-coned, or it may be directly shovelled into a new cone on another part of the floor. This process of reconing is repeated until the whole lot is thoroughly mixed. The cone is then systematically flattened by raking the ore out from the apex, while walking around the ring. The flattened cone is marked off into quarters with a stick or board on edge, along two diameters at right angles to each other. Two opposite quarters are taken away for the sample, taking all the cares. If the lumps are sufficiently small, the sample may be mixed as earlier and again cut down. If the ore contains particles of varying sizes, the fines tend to separate from the coarse and thus prevent thorough mixing. This tendency of separation is sufficiently decreased in case of damp ores, provided the moisture is not sufficient to make the ore ball-up into large masses. The floor used for the work should be smooth, clean and free from cracks. It is advisable to cover the floor with steel plates.

c) *Fractional Selection by Split Shovel, Riffle and Jones Sampler*

The split shovel is a fork in which prongs are separate scoops. Each scoop has the same width as the space between the scoops. It is laid on the floor and a shovelful of ore is spread over its surface. The shovel is moved back and forth across the scoops while the ore is sliding off. The split shovel is lifted leaving on the floor the material passing through the spaces between the scoops. The material going into the scoops is emptied on a heap. The riffle is also identical to split shovel except that it is larger and has a small handle on each side instead of a shovel handle on one side. The riffle is also used in the same way as split shovel. The riffle or the split shovel is very useful in the assay laboratory when the ore is in small quantities and of fine size.

The Jones sampler is a riffle consisting of two sets of scoopes sloping in opposite directions, instead of alternate scoops and spaces. It is shown in Fig. 3.1. The scoops discharge the ore as soon as it is poured into them. In one side the scoops are in even number whereas on other side in odd number. For either of these devices, the width of the scoop should be at least four times that of largest ore particle.

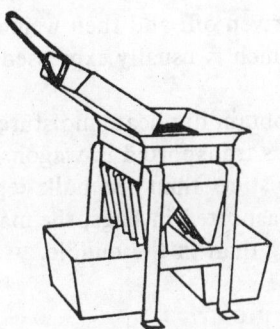

Fig. 3.1. Jones sampler.

Other methods include pipe samples, grab samples, etc. Pipe sampling consists in driving a scoop sampler or a pipe into the ore in the bins and a portion of the ore is removed for a sample. Grab samples are taken by dividing the surface of the ore pile into squares and approximately equal quantities of ore are removed from all corners of the squares and collected for sample. Running samples of the products coming from individual machines or tailing are sometimes taken by a dipper, a bucket, or some other similar arrangement. For this, the full stream is taken away for a definite period, say 30 sec at definite intervals (15 or 30 min).

3.7.3 MOISTURE SAMPLES

Since the assays of ore are always conducted on dry samples (to get the constant weight and value), and since all ores are usually damp when sampled, it is essential to take out moisture samples to determine the weight loss in drying. The moisture samples should be taken either just before or just after the ore is weighed to avoid the errors due to evaporation or subsequent wetting. The sample should be taken out by a rapid method which does not require crushing, further cutting down, or other handling before testing. Therefore, the moisture sample should be independent of the regular sample, unless the latter is small enough to be dried as a whole with convenience. Some form of grab sampling is frequently used so that the sample can be quickly collected and placed in tight containers. The assumption is that error due to crudeness of method is less than the error introduced by longer exposure of material during more elaborate sampling. Grab samples for moisture are frequently taken from the end of a conveying weigher. Each portion taken for the sample should be immediately kept into tight containers to avoid any evaporation before the test. As soon as possible, the sample is rapidly mixed, and a portion is weighed out to be dried; preferably the whole sample should be dried. The difference between the wet and dry weights divided by the wet weight gives the percentage of moisture in the wet ore. Calculation may be simplified by taking samples in 100 g or 1,000 g or some multiples of these weights.

Samples are weighed when wet, dried at a suitable temperature until all hygroscopic moisture is driven off, and then weighed again. The difference represents the moisture which is usually expressed as percentage of the wet weight.

It is quite difficult to obtain duplicate moisture samples that check within close limits. When ore is transported in wagons, the outer or top layers contain more or less moisture than the bulk depending upon the climate conditions. If the ore is trasported in bags, the material will usually be drier near the outside of the bag than in the middle.

3.7.4 Mechanical Sampling

A mechanical sampler should possess the following essential features:

a) The sampler should take the whole stream of ore (wet or dry) at a time and not part of the stream all the time, as the values are never distributed evenly across the stream.

b) For cutting the sample, scoop must move fully across and out of the stream in one direction at each cut.

c) The scoop must move at a uniform rate in order to take equal proportions from all parts of the stream.

d) The scoop should be adjustable to take larger or smaller proportions of the ore.

e) The interval of time between cuts should be constant.

f) The scoop should be deep and broad enough so that the ore entering the scoop, does not fall out.

g) The machine used in sampling should be simple and accessible for cleaning to avoid the contamination in subsequent sampling.

In mechanical sampling, various appliances known as sampler, are employed. Some of the important ones are described below:

Snyder Sampler

Figure 3.2 represents a snyder sampler. This consists of a circular pan with flaring sides, set edgewise on a revolving horizontal shaft. A spout s_1 projecting through the flaring side passes under the feed spout s_2 at each revolution of the sampler and delivers a sample into the sample spout s_3. During the rest of the revolution, the ore is diverted into the spout s_4.

Vezin Sampler

This is represented in Fig. 3.3. It consists of two hollow truncated cones bolted together at their large bases. They are attached to a cast iron spider

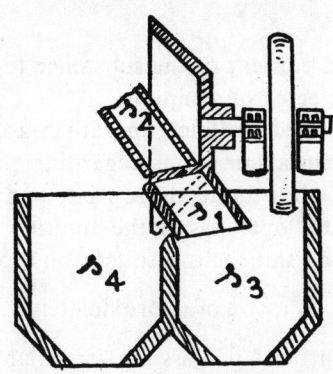
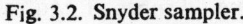
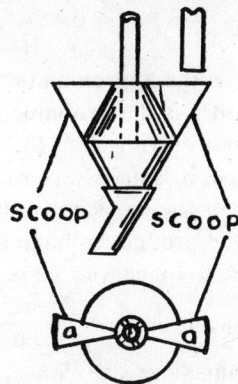

Fig. 3.2. Snyder sampler.   Fig. 3.3. Vezin sampler.

which is keyed to a vertical shaft. This shaft is supported by a collar at the upper end held in place by two guide boxes. The shaft is driven at a constant speed in one direction by beveled gears at the upper end. The upper cone carries one or more scoops the openings of which have the form of sectors of a circle. The ore is fed from the spout and the portion that enters the scoops, passes into the interior of the cones and is carried to the sample bin, or to the crusher, if further crushing is required. The main portion of the ore falls into a hopper and is spouted to storage bins or to wagons.

Electric Samplers

Self-contained and electrically operated sampling units are widely used for automatic sampling of both wet and dry ores. The cutter is moved horizontally across the ore stream by the threaded shaft. A small reversible motor is coupled to the screw shaft by a chain and sprocket drive. An adjustable electric time switch turns the motor on and off and regulates the frequency of the cuts. The type of cutter used depends on the nature of material to be sampled and on the nature of the sampling installations. The electric samplers are quite accurate and reliable. These are capable of easy adjustment to meet the operating requirements suiting to size and frequency of sampling. The electric samplers are compact and readily installed in existing flow sheets of the plant.

3.8 Size of Sample to be Taken

The minimum number of particles in a sample depends upon (a) the approximate content of the substance to be analysed, (b) required accuracy of sampling, and (c) whether the ore consists of free or locked particles, and in case of locked particles the nature and character of locking will govern the number of particles.

Taking theory of errors into consideration, the number of particles n (assumed that all the particles are of same grade, and free) required can be given according to the following equation:

$$n = 0.45 \frac{x}{y^2} \qquad (3.1)$$

where, x is the approximate volumetric content of the substance to be assayed, and y is the probable volumetric error of sampling.

When applying Eq. (3.1) to a crushed product finer than a limiting screen size, it is necessary to make certain assumptions regarding the average size of the particles in the product. It is quite appropriate and safe to assume the product to have the size of the openings in the limiting screen. However, it appears to be reasonable and satisfactory to take an average of volume mean or $D = \dfrac{\sum nd^4}{\sum nd^3}$, which gives a size of approximately 0.7 times the limiting sieve size. Since, crushed particles possess the irregular shape, averaging in volume to one-half the volume of cubes of equivalent size is quite reasonable. The effective number of particles, n' per gram according to the following equation is an acceptable approximation:

$$n' = \frac{6}{\Delta a^3} \qquad (3.2)$$

where, Δ is the specific gravity of the particles, and a is the opening of the limiting screen, expressed in cm. Following examples illustrate the use of Eqs. (3.1) and (3.2).

Example 3.1: The assay of gold ore reduced to pass a 104 micron screen is about 6 g/tonne. Find out the weight of sample required to assure correctness of sampling to 0.02 g/tonne, if the gold is free. The specific gravity of the rock is 3.2.

The weight proportion of gold to gangue is about 1 in 1,65,000, the volume proportion is then approximately 1 in 11,00,000. Hence, in Eq. (3.1), $x = 0.000,000,9$, and $y = 0.000,000,03$. Thus, substituting the values of x and y,

$$n = 0.45 \times \frac{0.000,000,9}{(0.000,000,03)^2} = 0.45 \times 10^9$$

and from Eq. (3.2), the number of particles per gram is

$$n' = \frac{6}{3.2 \times (0.0104)^3} = 1.7 \times 10^6$$

Thus, the weight required for sampling is

$$w = \frac{0.45 \times 10^9}{1.7 \times 10^6} = 265 \text{ g.}$$

Example 3.2: A flotation tailing (sp.gr. 3.0), reduced to pass a 104 micron screen contains about 0.1 per cent copper (as chalcophyrite). Find out the weight of sample required to assure correctness of sampling to 0.005 per cent copper. Assume that chalopyrite is free and the specific gravity of chalcopyrite is 4.2.

The volume content of chalcopyrite is approximately $(0.1/0.345) \times$

(3/4.2) or 0.2 per cent. Hence $x = 0.002$ and similarly $y = 0.0001$. By substituting the values of x and y in Eq. (3.1)

$$n = \frac{0.45 \times 0.002}{(0.0001)^2} = 0.9 \times 10^5 \text{ particles}$$

From Eq. (3.2), the number of particles per gram is

$$n' = \frac{6}{3 \times (0.0104)^3} = 1.8 \times 10^6$$

Therefore, the weight of sample required

$$w = \frac{0.9 \times 10^5}{1.8 \times 10^6} = 0.05 \text{ g.}$$

Above examples make it clear that individual make-up of an ore exerts a great influence on the minimum size of sample required to give results for the required accuracy. Earlier it was assumed that particle size is of great importance in size of sample, which is true only when the particles are free. If the particles are locked, the true yardstick is not particle size but grain size. For ore of larger size, the sample required is much more than in case of fine ore sample.

In sampling of precious metal ores, special difficulties are faced owing to the small value of y i.e. for the given accuracy the sample required is sufficiently large.

3.9 Sampling of Mill Products

Samples of feed, tailing, concentrate or other valuable products are taken during operating intervals to permit effective control of operation. In general, this operating interval is the shift, but sometimes the sample periods may be as short as 30 min. The samples are taken at definite schedules either automatically or by hand methods. The interval between cuts depends on the material to be sampled; in general it is maximum for tailing and minimum for concentrate.

3.9.1 HEAD SAMPLING

The practice of head sampling varies according to the ore, the method of treatment and the capacity of the mill. Head sampling usually is expensive and difficult. If the method of treatment does not involve concentration at relatively coarse sizes, head sampling is postponed until the sample can be taken from tumbling mill or classifier overflow, because at this time the ore would be more or less fine and well mixed, and simple automatic sampler can be used. When a flow of material comprises of a number of parallel or different circuits, samples are generally taken before the point of division. In case of sampling at coarse sizes, a separate sampling mill is used, which usually follows the coarse crushing plant and delivers reject to the fine ore bins.

Large capacity mills make the provision for careful sampling of heads, using automatic samplers taking frequent cuts. Hand sampling of heads is usually confined to small mills only and may vary from grab sampling of mine cars to grab sampling of crusher product. In hand sampling, the cut intervals are generally long.

3.9.2 CONCENTRATE SAMPLING

A satisfactory practice for taking samples of concentrate requires the use of automatic samplers with short sampling intervals. Usually automatic wet-pulp samplers are most suitable. However, in case of thick concentrate (flowing sluggishly) automatic samplers are not satisfactory and a hand cutter must be employed at the discharge of the machine or some other suitable method of sampling concentrate in bins or cars after draining must be employed. Hand cutters are widely used in general cases. Pipe or gun samplers and augers are employed for taking the samples from bins or cars. Sampling of concentrate aids the control on the performance of the smelter.

3.9.3 TAILING SAMPLING

Usually the final tailing from a mill is low-grade, finely crushed, and well mixed. There are no sudden changes in value of tailings, and thus small cuts at long intervals may be permissible. The methods employed for sampling are same as employed for sampling of concentrates. Hand sampling of tailing is likely to be more accurate than in sampling of heads or concentrates. Sampling of tailing is carried out to check the recovery of valuables. Daily samples are composited on a tonnage basis over a monthly period. Assay value of the composite should check within reasonable limits with the daily assay value.

CHAPTER 4

Mineral Liberation, Laboratory Sizing and Industrial Screening

The shape and size of mineral particles (ratio of surface to volume) influence mineral processing to a great extent. Higher the surface–volume ratio, faster will be the rate of reaction. A sphere possesses the minimum surface area for its volume, whereas a thin flat plate possesses a minimum volume for its surface. The most important functions of laboratory sizing control are to find the correct liberation size, and the size of grind for optimum unlocking of the desired mineral from the gangue. Another purpose of sizing, may be the grading of the product. However, comminution control is generally concerned with economic liberation of values from the ore.

4.1 Liberation Size (Mesh)

The ore consists at least of two minerals intimately interlocked. In order to concentrate the seggregated mineral species, the particles of the mineral must be adequately detached from other constituents by suitably controlled comminution. It is not usually possible to achieve the complete liberation by comminuting the ore down to the grain size of the desired mineral particles. Figure 4.1 represents a stylised cross-section, through a cube of ore containing 6 per cent valuables and 94 per cent gangue. If the rectangle shown gets divided into 100 square subdivisions, it may be observed that 86 squares represent clean gangue and 14 partly liberated material in which gangue and value are still closely bound. The incompletely separated squares known as middlings contain varying amounts of valuables. It shows that even with heavy overgrind (100 : 1) instead of 20 : 1, the valuables cannot be liberated completely.

In some cases, the ore may have a weak boundary between gangue and valuables, and the breakage may occur preferentially along the boundary. In such instances, satisfactory liberation may be achieved practically at mineral-grain size. Generally, the adherence of the minerals in the crystalline matrix of the ore is strong and the various constituents cleave across during crushing and grinding. As a result, crushed particles are obtained in the form of locked middlings, composed partly of gangue and partly of valu-

43

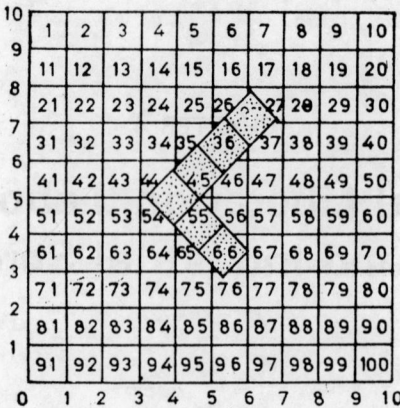

Fig. 4.1. Section showing the phase liberation.

ables. Since more abundant constituent (gangue) is freed at much coarser mesh than the valuables, grinding of ore completely to finer mesh (at which values are fully liberated) may be uneconomical. Therefore, primary coarse grinding should be followed by a stage of treatment to remove some of the free gangue.

Loosely bonded gravels can be disintegrated easily. For example, a clay having nodules of manganese or iron can be disintegrated to liberate the valuables with much less comminution work. When the constituents of the ore differ greatly in hardness or toughness, differential grinding may yield the required liberation by applying forces sufficient to break-up more friable constituent and leaving the tougher one unaffected.

In general, the ore should be comminuted down to an economic liberation mesh size, which can be determined by laboratory test work.

4.2 Screening and Factors Affecting the Screening Efficiency

Screening or sieving is one of the simplest and direct sizing process. In this, each particle is subjected to pass through an aperture of particular size and shape. The material can thus be separated into two groups, i.e. an oversize and an undersize. A size distribution can be obtained by using a number of differently sized apertures. A large number of factors affect the screening. The important factors are described below.

4.2.1 Particle Shape

In general the particles are assumed to be spherical or cubic for relating their surface to volume. In practice, the particles are rarely spherical or cubic. If one gram of ground mineral just passes the apertures of a square-meshed sieve having a linear distance (A) between successive parallel wires, then the surface (S) in square cm can be represented as following:

$$S = \frac{K}{A \cdot \rho}$$

where K is a constant having a value of 60 for cubic, 50 for oblongs, 240 for laminar ones (having a thickness of about 10 per cent of length and breadth), and 82 for the 'average' mineral particle, respectively, and ρ is the specific gravity of the mineral.

For 100 per cent screening efficiency, the particle should be presented to the screen aperture with its smallest cross-section and in the most probable way of penetration. If the particle is small enough, it will pass through. A flaky particle may be presented diagonally to screen aperture, so that it may pass through even being wider than the distance between screen wires. The longest dimension of the particle will not affect the screening in such cases. Similarly, a needle like particle will pass very easily through the screen aperture if it is presented end-on.

4.2.2 SIZE OF SCREEN OPENINGS

The particles of different sizes have similar configurations, and their passage will be inversely proportional to the cube of screen aperture. At the same time, the number of openings per unit area of surface of the screen varies inversely as the square of the screen aperture. Therefore, the number of passages to be affected per opening will vary inversely to the aperture size of the screen. This leads to the conclusion that the capacity of a screen is proportional to the screen aperture, when all other things are equal. The capacity of screens can also be given in terms of tonnes per square metre per day per mm screen aperture, instead of tonnes per square metre per day.

4.2.3 RELATIVE SIZES OF PARTICLES AND OPENINGS

The relative sizes of particles and apertures controls the passage or non-passage of the particles. A relatively small particle can fall easily through the opening but as it approaches near to the size of openings, it will have more difficulty in passing through an unoccupied opening with appropriate presentation.

4.2.4 ANGLE OF INCIDENCE OF PARTICLES ON SCREENS

The angle at which the particle strikes the screening surface is of great importance. For the best efficiency of screening, the particle should fall with its minimum cross-section normal to the aperture and at low speed. However, in practice, the particle competes with a crowd of other particles of random shape and size and thus falls at different angles. With increasing deviation from the normal, the chance of passage decreases and vanishes rapidly at increased obliquity.

4.2.5. PERCENTAGE OF SCREEN AREA OPENINGS

For different systems of screens, different ratios of opening are available for a given mesh. More the area of openings, faster will be the rate of screening.

4.2.6 Friction of the Sieve

If a sample is fully dried, friction on the screen may lead the formation of some electrostatic *bunching* of the particles. In such a case, several truly undersized particles will report as part of the oversize fraction. Such bunching action increases with temperature and thus to minimise it, screening should be performed in a cool room.

4.2.7 Moisture Content

Either bone-dry or wet pulp can be screened easily, whereas, even a small amount of water in a dry feed increases the difficulty of screening. The moisture content (dampness) adversely affects the screening. It leads to the sticking of particles together or their clinging to the screen surface.

4.2.8 Soluble Salts

Care should be taken that no dried salts are adhering to the particles coming from a pulp sample. These dried salts if adhered, would increase the effective cross-section of the particles.

4.2.9 Spreading of Feed over Screen Area

This is an important factor in industrial screening. The wider the feed is spread, easier it will be for a particle to pass through the opening unobstructed.

4.2.10 Blinding of Screen Openings

This arises mainly due to near-mesh material and is worst in closed circuit work. When the return particles are close to *release mesh*, these are retained on the screen and prevent the passage of true undersize and results into reduced capacity of the screen.

4.2.11 Corrosion of Screen Material

Corrosion introduces roughness to the wire which increases the proneness to *blind* and to resist clearance of the blinded apertures. This problem can be overcome by (i) using lime to neutralise the ore and (ii) using a resistant alloy for screen wires.

4.2.12 Rate of Feed, Thickness of Layer and Tautness of Screen

Rate of feed and material thickness influence the resistance of upper particles in penetrating the feed bed, whereas, the shaking and tossing action of the agitating mechanism transmitted to the bed depends on the tautness of the screen.

4.2.13 Motion Given to Particles by Screen Vibration

On mechanical screens the vibrating motion can be from counter-current to co-current. In case of electrically vibrated screens, the upward movement of the screen (vibrating normal to the direction of flow) can be terminated suddenly resulting into *unblinding* of screen at each stroke. This action is not possible in mechanically shaken screens, where opening is assisted by the sloping a action of rubber cords stretched below.

4.3 Laboratory Sizing

The laboratory sizing may be carried out for the following purposes:
a) To check the quality of the comminuted product.
b) To find the extent of liberated values from gangue at different particle sizes.
c) To assist some specific examination of ore constituent.

Size analysis of the product obtained from the crushing section and sent as feed to grinding section shows the extent of work done in comminution by crusher and how much work still remains to be done in grinding section. Samples taken from the mill discharge of wet grinding section and their size analysis will give the size distribution of discharged particles. Size analysis of the samples taken from the closed circuit return will show the sizes of material being returned back for further grinding, whereas the size analysis of closed circuit discharge will check whether a properly liberated material is being sent to the concentrating section. Further, sizing tests at various selected points aid in checking (i) the progress of the material at various stages of treatment, (ii) the efficiency of power and (iii) the effect of grinding on recovery of values from the ore.

A size variable of ore may be removed by screening (as required in gravity concentration methods such as jigging) to exploit the density variable in separation of valuables from gangue. In other concentrating processes, size control is necessary to determine the optimum size of comminution and the point at which pulp is ready for further treatment.

Laboratory screening can be carried down to about 40 microns, below which fabrication of screens to a reasonable accuracy becomes difficult. The routine size tests generally finish at about 70 microns. The apertures (meshes) of sieve form a rigid reference frame-work 'arresting or passing a particle depending on its cross-sectional area. A screen is an assemblage of various apertures designed to test many particles simultaneously.

Microscopic examination may be used to check the aggregation of particles, caused by impact during grinding. Dispersal and free movement of particles can be aided by limiting the size range of the sample. This can be easily accomplished by removing the sub-sieve particles (finer than 70 microns) before performing the screen analysis.

For laboratory sizing, both wet and dry methods can be employed. The wet screening removes a substantial amount of interfering under-size, redissolves the dried-on salts and disperses caked clay. For wet screening test, the dried and weighed sample is dispersed in water with little Na_2SiO_3 (about 0.02 g/litre) and washed gently on the finest screen of the series. The undersize is collected in a bowl below the screen, settled (sometimes a flocculating agent may be added), dried, and retained. The oversize is also dried and then screened over the desired screens. The undersize of this screen is mixed along with the previous one and weighed. Usually mechanical shakers are used for screening, but hand screening can also be used for rough works. Other methods of laboratory sizing include elutriation, sedimentation, infrasizing, microscopic, turbidimetry, permeability, gas or liquid sorption, etc.

4.3.1 HAND SCREENING

When mechanical appliances are not available for shaking the sieves, hand screening can serve the purpose to some extent. In hand screening, a *nest* of screens (usually not more than three sieves) is assembled with the coarsest above and the finest below. In the bottom a tight fitting pan is provided to receive the final undersize. The sieves should fit closely to prevent the dust losses. Before assembling the sieves, they are examined for any displaced wires, ruptures, etc. The apertures may be unblinded by gentle brushing with camel hair brush from below the sieve.

A sample of 200 g for rough work or less (also depending on specific gravity and mesh size) for good work is placed on the uppermost sieve. For good screening, the ideal load is one particle deep, but usually up to four particles deep can be sieved with good accuracy. The nest of sieves is shaken and jarred gently. The direction of shaking is changed every half-minute to 60° round the rim. Shaking with jarring is continued for a certain standard period. The time of sieving depends on the nature of material. For example, a friable material should not be worked with violence and for a long time, otherwise the sample is likely to be degraded.

4.3.2 MECHANICAL SHAKER

Mechanically operated appliances (Fig. 4.2) are commonly used to perform the laboratory work. Up to six standard screens can be nested on a

Fig. 4.2. Mechanical laboratory shaker.

framework of mechanical device. Vertical reciprocation to the sieves is imparted by a rotatory and a tapping motion. The screening period can be controlled by an automatic timer. The amplitude of the vibrating stroke can be varied to suit the material. There is a large number of proprietary makes of screens, which mainly differ in the shaking mechanism. Some commonly known are Rotop Shaker, Sherven Shaker, Genson Shaker, etc. In all the cases some standard sieves are used which are different in different countries.

4.3.3 VARIETY OF SIEVES

Screens may be constructed of sheet metal with either round or square holes, but in general they are constructed from rods or wires either welded or woven to give square apertures. Sieves used in laboratory sizing are of square apertures. The wires of the warp are woven successively over and under those of the woof (Fig. 4.3). The mesh may be defined as the number of apertures (holes) per unit length (inch or cm) measured along either warp of woof. The main systems used in the different countries differ in the thickness of wires used, and thus the apertures framed by these meshes also vary considerably. Different countries employ different standards, e.g. Tyler and ASTM (American Standard Testing of Materials) in U.S.A., BS (British Standard) and IMM (Institution of Mining and Metallurgy, London) in U.K., AFNOR in France, DIN in Germany, ISA in India, etc. The BS series is

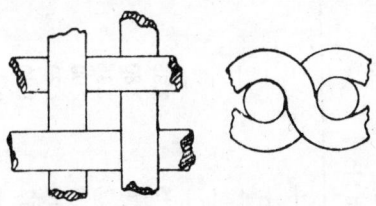

Fig. 4.3. Mesh of woven screen.

similar to the Tyler series, with slight variations in the wire diameter. The smallest wire-woven sieve is 400 mesh or 38 microns. The earlier DIN system (DIN 1171) was designated by the number of apertures per cm^2, which has been superseded by the new standard (DIN 4188). The present system in most of the countries is to include the aperture size. The mesh dimensions for some important systems are given in Table 4.1.

The different series differ with respect to the size range covered, number of sizes in the series, relationship between mesh number, wire gauge and aperture. In IMM specifications, the wire diameter is equal to the width of aperture, giving only 25 per cent screening area. In ASTM system, wires are slightly thinner giving about 30 per cent screening area, while BS system uses still thinner wires giving about 40 per cent screening area. The greater screening area gives higher screening rate but the screens are less robust. The number of apertures in a screen is inversely proportional to the square of

Table 4.1. Mesh dimensions of important standard sieve series[8,9,30]

Tyler (U.S.A.)		ASTM E-11-70 (U.S.A.)		BS 410-62 (U.K.)		NFX-11-501:1938 AFNUR X-11-501 (French)		DIN 4188-1957 (German)		IS 460-1962 (Indian)	
Sieve designation Mesh No.	Width of aperture mm	Sieve designation Mesh No.	Width of aperture mm	Sieve designation Mesh No.	Width of aperture mm	Sieve designation Mesh No.	Width of aperture mm	Opening mm	Opening μm	Sieve designation	Width of aperture mm
3	6.7	3	6.7	—	—	—	—	—	—	—	—
3½	5.6	3½	5.6	—	—	—	—	—	—	5.6 mm	5.60
—	—	—	—	—	—	38	5.00	5.00	—	—	—
4	4.75	4	4.75	—	—	—	—	—	—	4.75 mm	4.75
5	4.00	5	4.00	—	—	37	4.00	4.00	—	4.00 mm	4.00
6	3.35	6	3.35	5	3.35	—	—	—	—	3.35 mm	3.35
7	2.80	7	2.80	6	2.80	—	—	—	—	2.80 mm	2.80
—	—	—	—	—	—	35	2.50	2.50	—	—	—
8	2.36	8	2.36	7	2.36	—	—	—	—	2.36 mm	2.36
9	2.00	10	2.00	8	2.00	34	2.00	2.00	—	2.00 mm	2.00
10	1.70	12	1.70	10	1.70	—	—	—	—	1.70 mm	1.70
—	—	—	—	—	—	33	1.60	1.60	—	—	—
12	1.40	14	1.40	12	1.40	—	—	—	—	1.40 mm	1.40
—	—	—	—	—	—	32	1.25	1.25	—	—	—
14	1.18	16	1.18	14	1.18	—	—	—	—	1.18 mm	1.18
16	1.00	18	1.00	16	1.00	31	1.00	1.00	—	1.00 mm	1.00
20	0.850	20	0.850	18	0.850	—	—	—	—	850 μm	0.850
—	—	—	—	—	—	30	0.800	0.800	800	—	—
24	0.710	25	0.710	22	0.710	—	—	—	—	710 μm	0.710
—	—	—	—	—	—	29	0.630	0.630	630	—	—
28	0.600	30	0.600	25	0.600	—	—	—	—	600 μm	0.600
32	0.500	35	0.500	30	0.500	28	0.500	0.500	500	500 μm	0.500
35	0.425	40	0.425	36	0.420	—	—	—	—	425 μm	0.425

Mesh	Mesh	mm	Mesh	mm	No.	mm	mm	mm	No.	µm	mm
42	45	—	44	—	27	0.400	0.400	0.400	400	—	—
48	50	0.355	52	0.355	26	0.315	0.315	0.315	315	355 µm	0.355
60	60	0.300	60	0.300	25	0.250	0.250	0.250	250	300 µm	0.300
65	70	0.250	72	0.250	24	0.200	0.200	0.200	200	250 µm	0.250
80	80	0.212	85	0.212	23	0.160	0.160	0.160	160	212 µm	0.212
100	100	0.180	100	0.180	22	0.125	0.125	0.125	125	180 µm	0.180
115	120	0.150	120	0.150	21	0.100	0.100	0.100	100	150 µm	0.150
150	140	0.125	150	0.125			0.090	0.090	90	125 µm	0.125
170	170	0.106	170	0.105	20	0.080	0.080	0.080	80	106 µm	1.106
200	200	0.090	200	0.090			0.071	0.071	71	90 µm	0.090
250	230	0.075	240	0.075	19	0.063	0.063	0.063	63	75 µm	0.075
270	270	0.063	300	0.063			0.056	0.056	56	63 µm	0.063
325	325	0.053	350	0.053	18	0.050	0.050	0.050	50	53 µm	0.053
400	400	0.045		0.045			0.045	0.045	45	45 µm	0.045
		0.038			17	0.040	0.040	0.040	40		

the aperture. BS system has also been brought identical to European system (BS 410—1969), where all sizes are designated in mm down to 1 mm and in microns at smaller sizes down to 38 microns (equivalent to old 400 mesh).

The aperture sizes in the standard mesh vary from one to the next, usually in geometrical progression with a common ratio of $\sqrt{2}$ (1.4114), which is convenient in determining the size distribution. For example, the 200 mesh Tyler screen has an aperture of 0.074 mm or 74 microns. The larger screen in the series will have an aperture of 104 microns (i.e. 74×1.414), and the next smaller will have 53 microns (74/1.414). For close sizing a ratio of $\sqrt[4]{2}$ may be used.

The screens of finer sizes should be handled carefully, since they are very delicate. When used for wet screening, they should be dried soon after use. For periodic checking of the work, a master set of sieves should be used, which will reveal any wear and tear, and the damaged sieves can be discarded.

A series of screens given in Table 4.1 is adequate for most of the laboratory tests. If particles are more closely sized, prolonged shaking on the screen may be required, since many particles will be near the aperture size and may cause wedging or fail to find a way through the aperture. An extra care will be required in sieving the ores containing metallic flakes.

Screens of apertures smaller than 1 cm are woven from wire (steel, brass, or bronze) to give square apertures. In coarser sieves, steel wire may be employed, whereas for finer sizes, brass or bronze is used (for the finest sizes bronze is preferred). It is also possible to obtain finer sieves with square apertures by electroplating method. Apertures can be made flaring out and downward to facilitate the clear fall of particles.

4.3.4 PRESENTATION OF SCREEN ANALYSIS

In industries the terminology of mesh size still exists and it also appears in technical literature. In control of mill operations, the screening is universally employed, where a precise knowledge of the state of the material down to 74 microns is required. Below this size, routine control preferably employs other methods based on the rate of fall of particles through fluid. The change-over from a screening to a hydrodynamic is not smooth, and therefore, a good practice will require an overlapping of two systems to adjust them in a continuous record. The screening can be carried down to 60 microns on an unknown ore, and sedimentation may commence at 100 microns. The overlapped portion of the test may be reconciled by using graphical method. In general, fine gravel sizes are considered in between 5 and 2 mm, sand 2 mm down to 60 microns, and silt 70 to 5 microns.

The data of screen analysis can be presented as shown in Table 4.2. A graphic presentation is usually added to these results which illustrates the particle size distribution more clearly. The graphical methods usually employed are direct plot (weight–frequency plot) and the integral plot (cumulative weight plot). The typical plots are shown in Fig. 4.4. The direct plot

Table 4.2. Screen analysis of a sample ore

Mesh aperture in microns	Direct percentage weight retained	Cumulative percentage weight finer
+ 800	7.0	93.0
−800 + 600	10.4	82.6
−600 + 420	14.3	68.3
−420 + 300	13.5	54.8
−300 + 210	9.1	45.7
−210 + 150	8.2	37.5
−150 + 100	8.2	29.3
−100 + 75	5.1	24.2
− 75	24.2	
	100	

confines to the derivative or slope of the corresponding cumulative weight curve, whereas the integral plot is a summation of the area under the direct plot curve.

In order to avoid distortion and misinterpretation of the direct plot curve,

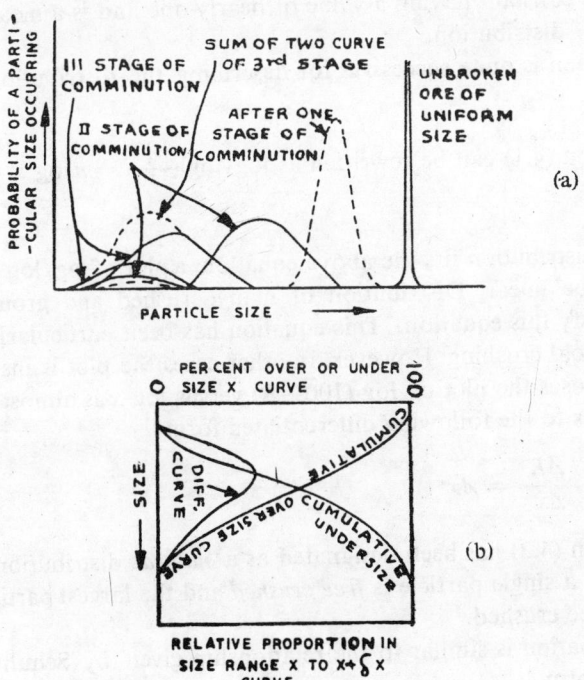

Fig. 4.4. Graphical representation of sieve analysis:
(a) direct plot (weight-frequency plot), and
(b) integral plot (cumulative weight plot).

values of weight increments should be plotted at equal geometrical intervals on the size scale. The intervals taken generally correspond to a $\sqrt{2}$-sieve series and a log-size scale may be used to give equal prominence to all size ranges. If the plot is developed as histogram, the areas will represent the relative proportions of the material in each size interval.

The cumulative weight plot has the advantage that a constant geometrical relationship is obtained even in the absence of $\sqrt{2}$ succession of screens, since the plot is made on logarithmic scale of apertures in microns.

Size Distribution

The curves obtained as in Fig. 4.4 are not amenable to mathematical analysis and thus the distributions obtained by various methods are compared with these types of curves. An equation developed by Rosin–Rammler and modified by Bennett is given below.

$$R = 100 \exp -(x/\bar{x})^n \tag{4.1}$$

where,

R = weight percentage over size x,

$\bar{x}$ = mean particle size corresponding to $R = \dfrac{100}{e}$ i.e. 36.8 per cent,

and n = a constant having a value of nearly one and is a measure of *width* of distribution.

This equation is quite successful for describing the distribution of broken solids.

Equation (4.1) can be rewritten as $\log \cdot \log \dfrac{100}{R} = n \log x +$ a constant

$$\tag{4.2}$$

If the distribution fits the above equation, a plot of $\log \cdot \log 100/R$ versus $\log x$ will be linear. Distribution of many crushed and ground products nearly satisfy this equation. This equation has been particularly found useful in the coal crushing. However, in other cases the plot is insensitive. For many purposes the plot of $\log (100 - R)$ versus $\log x$ is almost linear. This corresponds to the following differentiated form

$$\frac{\Delta R}{\Delta x} = Ax^n \tag{4.3}$$

Equation (4.3) has been designated as a *natural* distribution and it will occur when a single particle is *free crushed* and the largest particles are successively free crushed.

This equation is similar to the relationship given by Schuhmann in the following form

$$y = 100 \left(\frac{x}{K} \right)^m \tag{4.4}$$

where, y = percentage weight finer than x,

m = an arbitrary constant known as distribution modulus.

and K = a constant known as size modulus.

The Schuhmann equation is related to the following equation used by Gaudin in 1926,

$$(100-R) = (x/D)^n \qquad (4.5)$$

where, D = sieve size of the largest particle and

x = another characteristic size more in the nature of a statistically derived mean.

Equation (4.5) ignores the bimodal nature of the size distribution observed in free crushing of single particles.

The Schuhmann equation can be represented by a cumulative weight plot using log scales on both axes of the graph (percentage weight finer than a given size as ordinate, and particle size as abscissa). The distribution in Schuhmann plot is almost a straight line over a major part of its length. The value of m gives the measure of the size distribution of material. The greater value of m represents narrower distribution of the material over the entire size range. The value of m for unclassified material is about $1/\sqrt{2}$. The value of K is equal to x, when $y = 100$ (determined by extending the straight portion until it intercepts $y = 100$). The value of K represents the theoretical upper limiting size of the material.

The size distribution of crushed and ground products tends to follow certain empirical relationships. Each formula should be treated separately. One may fit better in some cases than in others. In all cases, the equations usually fit best in mid-range of sizes.

In general, the graphs plotted for various stages of crushing and grinding yield informations regarding comminution characteristics and mineralogical composition (obtained by analytical tests on each screened fraction) of the ore. This is important, particularly in tailings control, since it shows the amount of loss in each fraction.

4.3.5 SUB-SIEVE SIZING

Routine laboratory sizing is usually carried down to about 70 micron size, which leaves a considerable fraction of the ground ore-sample unsized. The term sub-sieve sizing is applied to the methods employed to grade the particles too small to be easily sized by finest laboratory screens, i.e. -70 microns. The measurements made under microscope reveal the exact sizes of the particles, whereas other methods, such as sedimentation, elutriation, infrasizing, centrifuging, etc., sort out equal settling particles. Therefore, the term sub-sieve sizing is somewhat incorrect, since the shape and specific gravity of the particles are the main factors affecting the settling velocity.

a) Elutriation Method of Sizing

In elutriation, the fall of a particle under the influence of its mass is retarded by the friction between the particles and rising column of fluid, which depends on the total surface of the particle. The elutriation measures the combination of size, shape, and density, and thus it is a sorting or classifying method. However in case of homogeneous materials, volume and shape are responsible for settling behaviour.

When a large number of spheres having same specific gravity, move in a rising column of water, the smaller ones rise, the medium ones dance (teeter) about in a diluted layer (called a teetering zone) and the larger ones fall through the water. The behaviour of individual spheres under these conditions is random and depends on the chances of collision, eddying of the water, and kinetic energy gained during rising or falling. The behaviour of particles in rising currents of water permits a simple and convenient method of sorting. The laboratory elutriator (Fig. 4.5) consists of the following elements:

i) water supply at constant head.
ii) A metering arrangement to control the flow rate of water rising through the elutriator.
iii) A sorting tube having a smooth and parallel sides.
iv) Separate vessels (receptacles) to catch the overflowing and underflowing particles.

The sorting tubes D_1 to D_6 have a progressive increase in internal cross-sectional area by $\sqrt{2}$. The elutriating fluid (usually water), is recirculated with the help of a small pump and is syphoned from the beaker below each tube. The valves A_1 to A_6 are used to control the flow. The overflow from sorting tubes is delivered by the syphon legs B_1 to B_6. The sample is introduced in the first beaker after the liquid flow is adjusted, and stirred by glass stirrers.

The elutriator may also incorporate the cyclone action. One of such equipments is known as Kelsall cyclosizer. It accelerates the movement of fine particles and can sort about 100 g material in -50 to $+8$ micron size range in 30 min.

b) Sedimentation

Sedimentation is based on the falling rate of small particles through static liquids. Sedimentation method can be practised in a simple way by using only some beakers and a laboratory stirrer. Some specially designed tubes are also employed for the same. In sedimentation methods, care should be taken to disperse flocculated particles. Sedimentation method can be used to sort down to 5 micron sizes and in some cases it may be even as low as 2 microns.

c) Infrasizing

This employs a current of air sent with minimum turbulence through a series of tubes of increasing diameter. The sample of ore is introduced into the air stream. The heaviest particles fall out first and the lightest stay in tubes of largest diameter. The Haultain Infrasizer is shown in Fig. 4.6. A

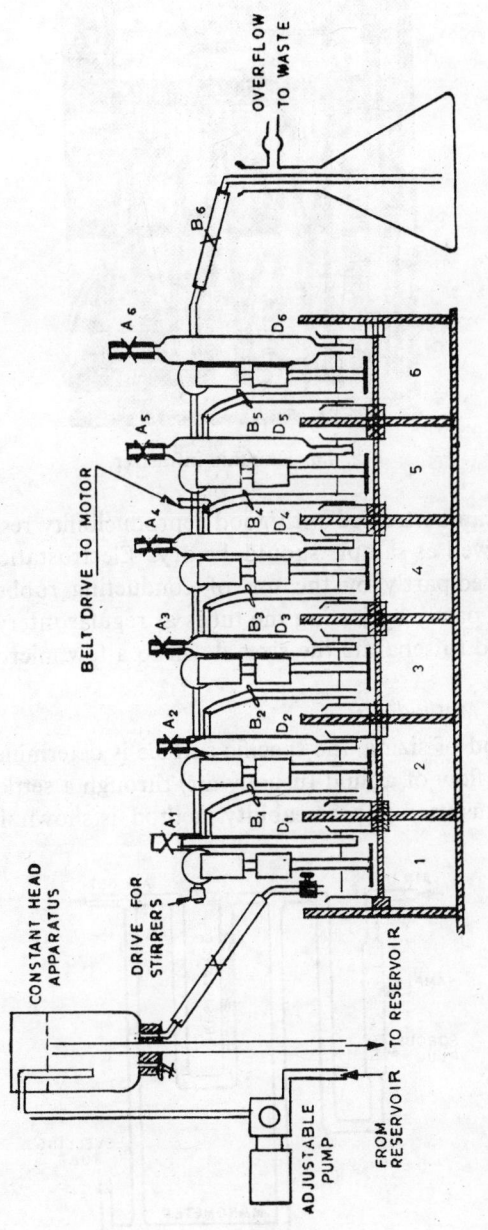

Fig. 4.5. Blyth elutriator.

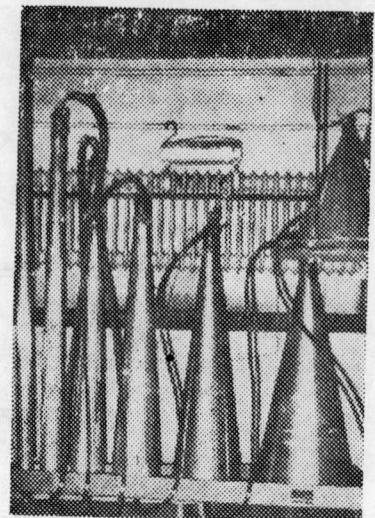

Fig. 4.6. Haultain infrasizer.

sample of 50 g can be treated with good reproducibility results. For infrasizer, the air as well as sample should be dry. Electrostatic seizure of the particles is avoided partly by the use of conducting rubber in the flexible connections and partly by jarring the tubes at regular intervals. This apparatus can be used to separate the sizes down to a few microns.

d) Permeability Method

In this method of sizing, the specific surface is determined by measuring the resistance to flow of a fluid (usually air) through a settled bed of particles. An apparatus used in permeability method is shown in Fig. 4.7. The

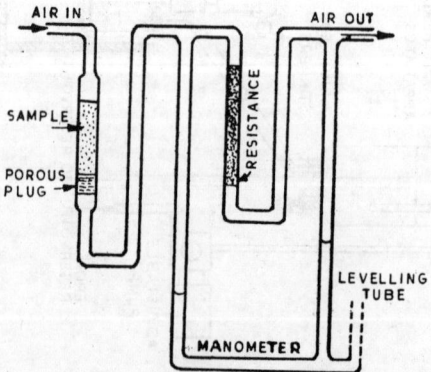

Fig. 4.7. Permeability apparatus.

following formula developed by Royal Institute of Technology, Stockholm[8], may be used in determining the specific surface of particles,

$$Q_m = \frac{\epsilon^3}{k_0\, k_{.\eta}\, s^2 (1-\epsilon)^2} \frac{A\,\Delta p}{L} \frac{8\sqrt{2}}{3\sqrt{\pi}} \frac{1}{k_1} \sqrt{\frac{RT}{MS}} \frac{\epsilon^2}{1-\epsilon} \frac{A}{L} \frac{\Delta p}{p_m} \qquad (4.6)$$

where

Q_m = air volume flowing through the bed measured at the pressure p_m in cm³/sec,

p_m = gas pressure at the middle of the bed in dynes/cm²,

Δp = pressure drop across the bed in dynes/cm²,

ϵ = porosity of the bed as fractional void space,

η = viscosity of gas,

S = specific surface of the powder,

A = cross-sectional area of the bed in cm²,

L = depth of bed in cm,

R = gas constant per mole i.e. 0.8315×10^8 erg/deg-mole,

T = absolute temperature of the gas (K),

M = molecular weight of gas and k_0 and k_1 are numerical constraints.

The constant k_0 is supposed to be real and is equal to 2, while k_1 is supposed to be dependent on the shape of the particles, the value of k_1 varies in between 2.25 and 3.0.

The above formula is based on the flow rate of gas through a compacted bed of powder and gives the reproducible values within $\pm$ 2 per cent on powders with specific surface above 500 cm²/cm³.

e) *Measurement with Microscope*

A metallurgical microscope can be employed to observe and measure directly the particles down to less than 1 micron. Measurement can be carried out with the help of a linear scale engraved on glass which is placed in the eyepiece. The microscope is then focussed on a stage micrometer. This micrometer is a transparent engraved scale which has the squares or other figures of known areas. For example, 1 mm may be divided linearly into tenths and hundredths (1000; 100, and 10 microns). The micrometer is placed on the microscope stage and the two scales can then be correlated for eyepiece–objective system. The stage micrometer is then taken out and the particles are examined and measured. When the particles are placed loose on the glass slide, they will settle in their most stable position presenting the two major dimensions for measurement. In routine size analysis, the particles should be dispersed by adding a drop of suitable reagent such as calgon. The dispersion helps the display of smaller particles which may otherwise be masked by the larger ones. Use of various aids, such as projection facilities reduces the tiring work in counting of particles. Particles dispersed in liquid can be made to flow through a small aperture and the counting is based on the amount of interruption caused to a beam of light.

4.4 Industrial Screening and Its Purpose

The screens employed for industrial work are large in sizes and they should be built strong in structure (made of woven wires or rods) to bear the weight of material. The rods may be made of tapered section to facilitate the passage of material. Wire screens are usually made by laying the wires equally spaced in two directions at right angles and welded together to give either square apertures or slots of equal width. In very large-size screens, the rods or bars are used lying in one direction only.

In order to facilitate the passage of particles, some mechanism is required to turn them many times to have reasonably good probability of their passage. Further, some means should be provided to transport the oversize away from the screen to make way for new feed. Industrial screening may be carried out for any of the following purposes:

a) To retain oversize in a given section or circuit and thus preventing the feed of oversize to the machine not suitable to handle it.

b) To separate undersize from the feed to a crushing machine provided to crush larger pieces.

c) To separate the material into specified sizes.

d) To prepare a correctly sized feed for a given concentrating process.

4.5 Types of Industrial Screens

The choice of a screen depends on the purpose, mesh size, and ore to be treated. The general types of industrial screens are given in Fig. 4.8. The important screens are described below:

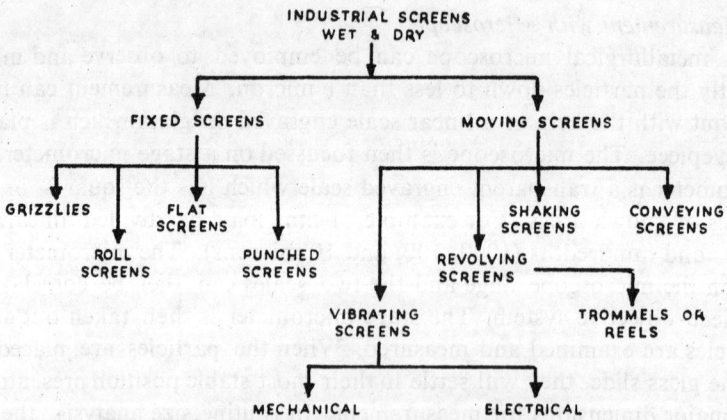

Fig. 4.8. General classification of industrial screens.

4.5.1 Grizzlies or Bar Screens

These screens are sturdy in construction and consist simply of an assembly of bars, rails, or girders having a tapered cross-section as shown in Fig. 4.9. These are held at appropriate spacings and set with thick end of the taper on the upper side. This arrangement provides the clear passage of pieces. These screens are used for largest sizes. The spacing of rods/bars may be as large as 2 metres. The largest sizes of *grizzlies* are basically employed to

Fig. 4.9. Grizzlies or bar screen.

safeguard primary crushers against overloading. The stationary grizzly, is set at a slope of 25–50° (depending on the stickiness of the ore) to ensure the free slide of the ore.

In a cantilever construction, sliding of ore is helped and loss of head is reduced due to vibration action. Sometimes an arresting grid (flat and strong) is provided above the grizzly or in between grizzly and crusher to prevent the passage of large pieces in the crusher. Moving grizzlies consist of alternate bars sliding or moving. One popular arrangement is ring-roll grizzly, in which the ore rotates the rollers. Another one is roll grizzly, which consists of series of grooved rollers driven unidirectionally in a supporting frame. The rolls facilitate the removal of oversize and provide better grip of lumps. Grizzlies can also be vibrated mechanically, electrically or by impact of falling rock.

4.5.2 ROLL SCREENS

These screens consist of a bank of rolls arranged in the form of a cascade having a $\Lambda\Lambda\Lambda$ profile and giving square apertures between adjacent rolls. These screens are used for smaller sizes (10–100 mm) and heavy hard material, where ordinary screens would wear out quickly. The roll screen is quite costly and presents a small area of apertures.

4.5.3 FLAT SCREENS

The flat screens are usually inclined to facilitate the flow of material and are subjected to a periodic motion to facilitate the rolling or bouncing of particles on the deck. Therefore, these screens provide many opportunities to particles to fall through the apertures, which depend on (a) length of screen (increased length gives more chances), (b) slope of the screen (increased slope reduces the chances), (c) amplitude of vibration (should be just sufficient to clear the screen), (d) frequency of vibration (should be just low enough to permit the particles to return within the cycle) and (e) depth of particle bed on the deck (a bed usually five or six particles deep gives the optimum screening rate).

The major problem in flat screens is the blinding of screens, caused by hard, near size particles wedging in apertures or by soft sticky material filling up the apertures. This problem can be overcome by the use of (a) roller screens, (b) drying equipments, (c) light loading of screens, (d) occasional severe jolting, (e) mechanical cleaning with brushes, (f) special types of vibrators, (g) slots flared particularly to the undersize or (h) rubber balls to bounce on the undersize of the screen. If cloth is used, it should be mounted taut to give sufficient springiness to minimise blinding.

The flat screens can be used in series of more than one screen (use of two or three screens at a time is quite common) which give graded products of various sizes in one single operation. The screens may be rectangular (Fig. 4.10 a) where a slope is given or may be circular (Fig. 4.10 b) where slope is not needed.

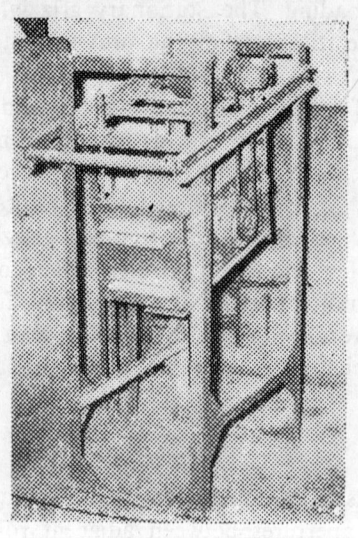 

a   b

Fig. 4.10. Flat screens:
(a) rectangular screen, and (b) circular screen.

4.5.4 PUNCHED SCREENS

These screens are made by punching the round or square holes in metal sheets (usually mild steel and stainless steel) and are quite cheap. The screens having square apertures are usually preferred due to more screening area available compared to that with round holes, but circular openings are better for coarse work. The surface of such screens is rather smooth which aids the sliding of particles about rather than roll on them. The punched plates have a longer life, particularly with abrasive materials. In the field of ore processing, these punched screens are made with large apertures (usually more than 40 mm) and fitted into trommels. However, the screens can be used for quite small apertures down to 70 microns for metal powders.

4.5.5 TROMMELS OR REELS

These are slightly inclined rotating cylinders having the walls of screen material. The material is fed at the top which rolls about while moving towards the discharge end. In practice, compound trommels comprising of several sizes of screen are employed. Two systems can be adopted in screening operation, i.e. (i) cylinders may be used in line (Fig. 4.11 a) or (ii) cylinders may be used in concentric arrangement (Fig. 4.11 b). In the former case the screens are in line having finer size first, and coarse size last, whereas in the latter case the inner one is of coarse size and the outer one of fine size. In concentric cylinders, the material is fed on to the inner screen. It is difficult to replace the worn out screen in case of concentric system. The trommels are usually employed in screening the stone ballast to get three or four fractions and tin and gold dredges (to remove boulders and clay from the gravels).

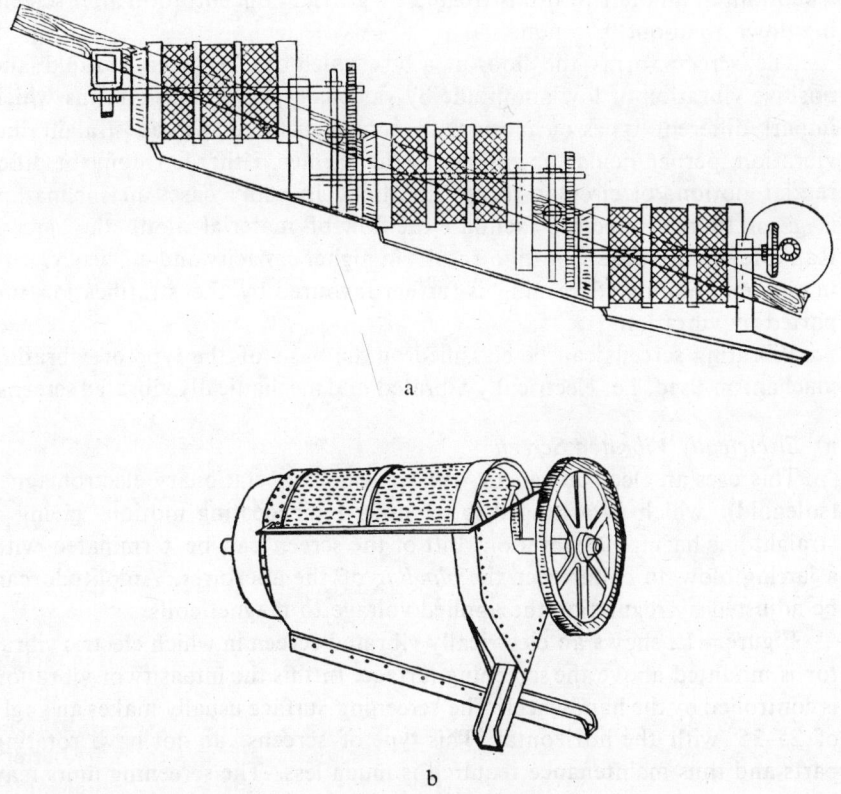

Fig. 4.11. Trommel screens:
(a) cylinders in line, and (b) concentric cylinders.

4.5.6 WOVEN WIRE CLOTH SCREENS

These screens are widely used in the range of 2 cm and 70 microns. Various shapes of apertures, crimps, and weaves are available. Depending on service conditions, the chief materials used for wire cloth are steel, stainless steel, monel metal, copper, bronze, brass, nichrome, galvanised and tinned steel, and silk or rayon (for dry and fine material). For specific purposes, the screens can be made by electroforming instead of weaving. When oblong openings are used, the maximum capacity can be obtained by setting their long side across the flow of feed. For various apertures, different wire diameters are used (Table 4.3) which give varying percentage open area for different apertures.

4.5.7 VIBRATING SCREENS

These screens consist of one or more decks and work at low slopes. These are the most standard sizing equipments in modern mills for medium-coarse and fine screening (250 mm and 0.2 mm). In most mineral processing operations, screening stops at the point of crushing section delivering the fine ore for wet grinding, i.e. usually between, 20 and 6 mm. However, a substantial amount of ore is treated by gravity concentration after screening down to about 0.85 mm.

The screen forms the floor of a box which can be given a rapid and positive vibration of low amplitude by various types of mechanisms which impart different types of motions to screen surface, such as straight line vibration, perpendicular or oblique to the surface, either harmonic or differential motion, or circular throw vibration. In many cases an inclination is given to the screen to facilitate the flow of material along the screen. Rapid low-amplitude vibration results in higher capacity and efficiency, with minimum blinding. Screening is further favoured by the stratification imparted by vibration.

Vibrating screens can be classified on the basis of the type of vibrating machanism used, i.e. electrically vibrated and mechanically vibrated screens.

a) *Electrically Vibrated Screen*

This uses an electromagnetic device, usually a stationary electromagnet (solenoid), which is arranged to set up a reciprocating motion, giving a straight line harmonic vibration. Lift of the screen can be terminated with a jarring blow to counteract the *blinding* of the apertures. Amplitude can be adjusted by regulating the applied voltage to magnet coils.

Figure 4.12 shows an electrically vibrated screen in which electric vibrator is mounted above the screening surface. In this the intensity of vibration is controlled by the handwheel. The screening surface usually makes an angle of 25–35° with the horizontal. This type of screens, do not have rotating parts and thus maintenance required is much less. The screening units may have one or two screening surfaces vibrating by the same mechanism. However, more vibrators can be used for heavy work on the same screen. The screens are usually employed for −12.5 mm feed.

b) *Mechanically Vibrated Screens*

In this case the motion imparted to particles is not necessarily a simple

Table 4.3. Standard coarse screen specifications[8] (Wire diameter and percentage open area for different apertures)

Opening cm	Wire diameters							
	Medium light		Medium		Medium heavy		Heavy	
	Wire dia. cm	Percentage open area	Wire dia. cm	Percentage open area	Wire dia. cm	Percentage open area	Wire dia. cm	Percentage open area
10	1.25	79.0	1.56	74.8	1.875	70.9	2.5	64.0
8.75	1.1	79.0	1.25	76.6	1.56	72.0	1.875	67.8
7.5	1.1	76.0	1.25	73.5	1.56	68.5	1.875	64.0
6.25	0.94	75.6	1.1	72.4	1.25	69.4	1.56	64.0
5.0	0.78	74.8	0.94	71.0	1.1	67.3	1.25	64.0
3.75	0.625	73.4	0.78	68.5	0.94	64.0	1.1	59.9
2.5	0.56	66.6	0.625	64.0	0.78	58.0	0.94	52.9
2.2	0.54	65.3	0.56	63.3	0.62	60.5	0.78	54.3
1.9	0.49	63.4	0.54	61.4	0.62	56.3	0.78	49.8
1.6	0.45	60.7	0.49	58.5	0.56	54.0	0.62	51.0
1.3	0.42	57.1	0.45	54.5	0.49	52.2	0.54	49.8
1.0	0.36	55.6	0.40	51.4	0.41	49.9	0.47	46.4
0.8	0.31	53.1	0.36	48.8	0.38	46.0	0.41	43.6
0.6	0.25	49.6	0.31	45.6	0.35	42.0	0.36	39.6
0.4	0.17	49.1	0.24	42.1	0.30	35.4	0.31	31.6

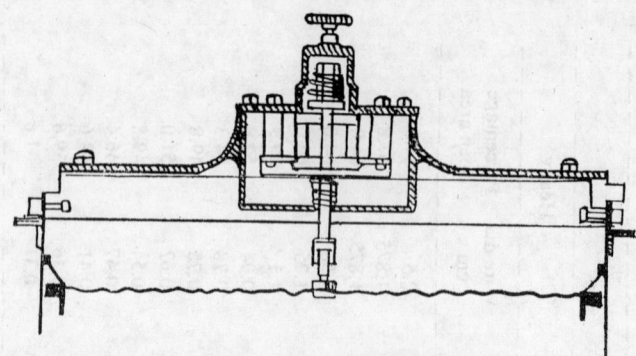

Fig. 4.12. Electrically vibrated screen.

straight-line one. Theoretically, any motion between a straight line and a circular orbit spinning is possible. In a case it is possible to have counter-current ellipse at the feed end, a reciprocation normal to the screen at the centre, and a concurrent ellipse toward the discharge. Thus the entering feed is checked and searched during stratification, then coveyed to an area of screen free from blinding, and finally accelerated off the screen. The variations in the vibrations can be produced by balancing the forces of vib-ration, movement of the tension springs, yield of the screen cloth, the inertia of the framing, and the weight of the passing load. These screens are mainly used for coarser sizing. Vibrations can be produced by three methods, i.e. cam and spring type, eccentric, and unbalanced-pulley type.

In cam and spring type, the arrangement of screening surface and vibra-tor is same as in case of electrically vibrated screens. The vibration is im-parted to the screen through the tappet and connecting rod. The impact and differential vibration imparted to the screen prevent the blinding. Since, many improved mechanisms are developed, this type of vibrating screens is not in practice at present.

In eccentric motion types, a horizontal cross-shaft runs in fixed bearings. The shaft carries the eccentrics which impart an encircling motion. The screen vibration is counterbalanced by some means to insure against the excessive vibration of the screen supporting structures and drive mechanism. The screen boxes can be tilted during running at an angle up to 30° with horizontal. In one type of these screens (shown in Fig. 4.13) a pair of heavy balance wheels (not allowed to rotate about the shaft) is mounted on the shaft. The screen frame is gyrated in a circular path by the eccentrics on the shaft. The screen is either suspended or supported at a slope of 15–30° with the horizontal. Another type of screen is shown in Fig. 4.14 which gives a gyratory motion generated by an unbalanced weight. The screen is operated in the horizontal position.

In unbalanced-pulley type, surface and frame of screens are vibrated by a mechanism consisting of rapidly revolving unbalanced pulleys. The un-balanced pulleys on the two shafts operate to balance each other. The vib-

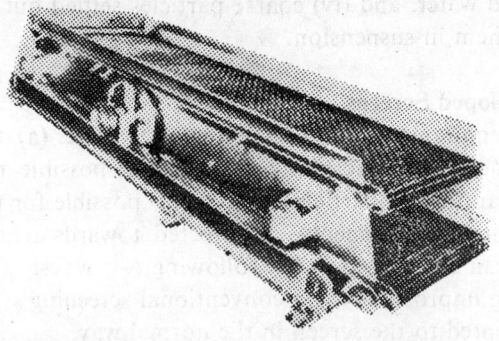

Fig. 4.13. Tyler niagra circle throw screen.

Fig. 4.14. Symons screen.

rator is connected to the screen frame, which is spring suspended, to transmit the vibration at an angle oblique to the screen surface, to allow the setting of screen in horizontal.

4.5.8 SHAKING SCREENS

These screens have plane surfaces and work usually in sorting of dry and soft materials. A typical arrangement consists of a rectangular box having the screen as bottom. The screens hung by chains, links or supported from beneath can be shaken sidewise or endwise. The screens should be provided some slope when driven from eccentrics, whereas it may be set horizontal when driven by a jerking mechanism. When the deck is pushed forward, the material gets lifted and thrown forward, whereas on release, the deck comes backward to its stop point, tossing the material in the air. This loosening of material causes the stratification of material, leaving the largest particles on top and smaller ones pass through the apertures. It is important to have a balanced mechanism for shaking screens, otherwise the supporting structure may suffer serious vibrations.

4.5.9 WET SCREENS

In general, wet screens are similar to the screens discussed in the preceding sections, except that these employ water during screening to overcome the problem of blinding. The wet screens are not in common use due to (i) corrosion of screens, (ii) high power required to vibrate the screens, (iii) difficulty in maintaining the character of pulp during the removal of fine particles and water, and (iv) coarse particles settled out need extra agitation to keep them in suspension.

4.6 Newly Developed Screens

The main requirements of a screening system are: (a) to produce the required size fraction or fractions at the highest possible rate, and (b) to produce the separation of particles as sharp as possible for the lowest cost. Therefore, the new developments are directed towards achieving these objectives, which can be realised in the following two ways:

i) By making improvements in conventional screening system, where the material is presented to the screen in the normal way.

ii) By adopting the probability principle of screening. In this the use is made of the fact that from a stream of particles moving rapidly, almost at right angles to a screening surface, the particles having about $1/2\ d$ size ($d =$ distance between the screening wires) will only pass through the screen.

Some important screens recently developed are sieve bend, Bartles–CTS screen, Mogensen sizer, and HCC–Burstlein screening system.

4.6.1 SIEVE BEND

This is a wet screen and has been developed at Dutch State Mines and is known as DSM sieve. This consists of screen surface composed of wedge wires or rods curved into a quarter circle. The wires or rods lie across the direction of flow of material in the form of a slurry. The wires are equally spaced to act as a screen. These wires or rods can be set as close as 50 microns, but somewhat wider setting (150 microns) works better. The sieve bend can be used for removal of heavy media from products, closing coarse size grinding circuits and screening the final products. The sieve bend suffers the following disadvantages:

a) The wedge wires tend to become blocked, specifically when a large proportion of angular particles is present in the feed which are similar in size to that of screen spacing.

b) Wear of the profile of the wedge wires may also affect the performance.

In order to overcome the blocking problem, vibrating devices may be added to the screens. The vibrating action helps to remove the particles caught in the wedge wires and to detach the *through screen* material tending to cling to the underside of the screen.

4.6.2 BARTLES–CTS SCREEN[80]

This is also a wet screen which consists of a short-radius curved screen cloth fitted with crimps on its under surface as shown in Fig. 4.15. This screen deals the problem of blinding and detachment of through-screen material. The screen is constructed in two stages. The feed is introduced tangentially in a thin laminar film to the upper section, through which about

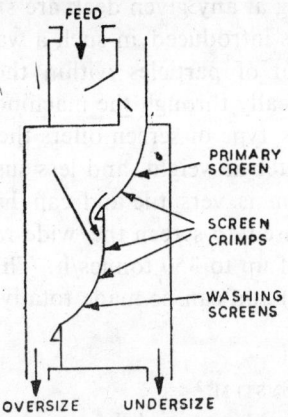

Fig. 4.15. Bartles-CTS screen.

half the undersize passes. The second section then retreats the oversize from the first section, where the remaining undersize is removed.

An alternative configuration is a quadruplex arrangement, having two pairs of screens. The upper pair of screens has coarser aperture than the lower pair. Thus the coarse material is removed first, which reduces the work of lower finer screens making the final separation. Three products can be obtained by such a system. These screens find their application in (i) removal of primary fines in crushing plants, (ii) dense media separations, (iii) closed circuit grinding, (iv) rejection of tramp oversize in flotation circuits, and (v) wet sizing of final products.

4.6.3 MOGENSEN SIZER[81]

This consists of a number of screen decks, usually five, which are set in the form of a fan as shown in Fig. 4.16. The apertures are finer at the bottom and coarser at the top (feed). The apertures of screens are so chosen

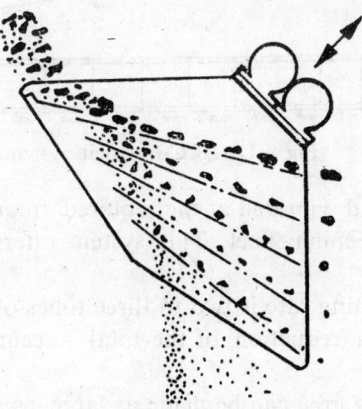

Fig. 4.16. Mogensen sizer.

that the particles arriving at any given deck are smaller than the apertures in that deck. The feed is introduced in such a way as to ensure the individual and free movement of particles within the machine. The principal method of travel is vertically through the machine rather than horizontally across the screens. This type of screen offers the advantages such as less power consumption, lighter in weight, and less susceptible to blinding. This type of screening system is versatile and can handle dry, wet, or moist materials. This can be used to screen the wide range of material (50 to 1 mm) with a capacity of up to 350 tonnes/h. This can also deal with hot materials (up to 800°C) and can be made totally enclosed for dry sizing applications.

4.6.4 HCC–BURSTLEIN SYSTEM[82]

In this system a different approach has been used. The thickness of the screen bed is kept constant or even increasing slightly from feed to discharge end by the slope adjustment, whereas in the usual system of screening, the bed depth usually decreases from the feed end to the discharge end with removal of fines. In conventional screens, the potential capacity is not utilised due to (a) greater thickness of bed to the feed end which prevents the efficient presentation of finer particles (larger in number) to the screening surface, and (b) thin bed at the discharge end, where sufficient number of fine particles are not present. The system (Fig. 4.17) consists of breaking up the screen surface into a number of elements. Each element is inclined at

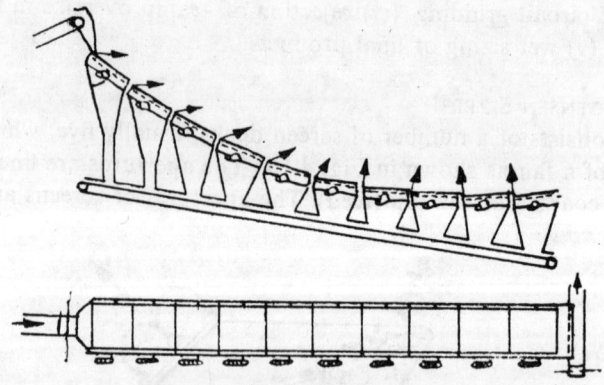

Fig. 4.17. HCC-Burstlein system.

a suitable angle and vibrated at the required frequency and direction to obtain the best screening effect. This system offers the following advantages:

i) Specific screening rate is two to three times of that for conventional screens, and thus a reduction of the total screening surface for a given capacity.

ii) The screening area can be made as large as 100 m² compared with the normal 20 m² of conventional screen installations.

iii) Installation, operating, and maintenance costs are much lower (about 50 per cent of those of conventional screening installations).

These screens have been used successfully in screening iron ore and blast furnace charges.

4.7 Test Sieving Versus Industrial Screening

There is a great difference between test sieving and industrial screening. In the former case, the time allowed for operation may be as long as required to reduce the rate of passage of material to an extremely low value. Theoretically the passage may never stop due to the attrition of particles against the screen. However, a stage may reach at which no measurable increase in any fraction sieved is observed. On the other hand, in industrial screening, each particle may be presented to the screen surface a maximum of 10–12 times before the removal as oversize. Therefore, an average size in any fraction of industrially sized material will be considerably smaller than in the sample obtained by laboratory sieving.

4.8 Operating Characteristics of Screens

4.8.1 SCREENING CAPACITY

Screening capacity is directly dependent on the area of screening surface and screen aperture, and thus it is customary to represent capacity in terms of tonnes/m²/mm screen aperture per hour. However, the comparisons should be made very carefully, since various factors such as specific gravity of ore, size distribution of the feed, moisture content, etc., have a marked effect on the capacity. Capacity of various screens is given in Table 4.4.

Table 4.4. Capacity of different screens[8]

Type of screen	Capacity range (ore) in tonnes/sq m area/mm aperture/day
Grizzlies	10-50
Trommels	3-20
Shaking screens	20-80
Vibrating screens	50-200

There is no single formula which can be used to represent the screening capacity, but fundamentally the capacity decreases with increase of oversize fraction. Further, capacity is more related to width of the screen than length. The oblong apertures give more capacity than square apertures, though it may be at the expense of accuracy. The choice of aperture shape must take into account three inter-related factors, i.e. (i) precision of sizing, (ii) permissible tolerance of wrong sizes, and (iii) effect on overall operating cost. Further the screening capacity is higher when oversize is not recirculated, but sent to different milling system.

4.8.2 SCREENING EFFICIENCY

Screening efficiency is a measure of the effectiveness of the screening operation compared to perfect screening operation. With respect to separation of material, the efficiency is regarded as the percentage recovery of true undersize material in the actual undersize product. The percentage efficiency (E) can thus be expressed as

$$E = \frac{10,000\ U}{uF}$$

where U = tonnage passing through the screen for each F tonnes of feed, and

u = percentage of undersize in the feed as determined by test screening.

In some cases, the screening efficiency may be represented by giving the per cent of true undersize left in the oversize product, which should be zero for perfect screening.

Though this formula is quite convenient, it fails to take into account the particles of near mesh size, which are difficult to screen out. On the other hand, the particles much finer than the screen aperture are readily screened out. Therefore, the same screen on different feeds may give quite different efficiencies. In order to overcome this difficulty, screening efficiency may also be defined as the ratio of difficult grains that are separated by the screen to the difficult grains in the feed, but it is almost impossible to agree on what constitutes a difficult grain. Thus, the screening efficiency is related to the dwelling time of passing material and the openings available during its passage. When the screen is in close crushing circuit, there is a tendency for a near-mesh circulating load to build-up resulting in a progressive decrease in screen efficiency. The efficiency of screening usually varies from 60 to 80 per cent. The efficiency of screening is favoured by the following:

a) Increased percentage of the screen openings to passage of undersize.
b) Smoothness and freedom from pitting of the screen wires.
c) Suitability of the aperture shape to the average particle shape.
d) Increased transit time.

The screening efficiency is adversely affected by:

i) Increased rate of feed.
ii) Increased percentage of near mesh grains.
iii) Increased thickness of bed due to poor presentation of particles.
iv) Increased moisture in the feed.
v) Lack of response of screening surface to vibrating impulses.

4.8.3 OPERATING COST OF SCREENING

The operating cost in screening is quite small compared to the cost involved in other operations of mineral processing. In case of stationary screens, there is no cost of power, the labour cost is low and the cost of supplies is only that of replacing screen cloth or other parts of the screen when worm out. In case of moving screens, in addition to the above costs, power cost is involved. However, the requirement of power is very small, particularly in coarse screening. The cost added by screens in treatment of ore is mainly due to (i) initial cost of screens, since larger building is required to accommodate the screens in the flow sheet, (ii) due to loss in gravity head requiring elevation of ore to compensate the head loss, and (iii) requirement of arrangement for dust control in screening dry and dusty materials.

4.8.4 PRECAUTIONS FOR AN EFFICIENT OPERATION

a) There should be good tension of screen cloth in the frame to give efficient transfer of the vibrating strokes from mechanism, via cloth to load.

b) The combined effect of vibration, speed and amplitude, together with slope of screen should be adjusted in such a way that the material remains well stirred and running freely. If the amplitude is too great, stratification will be disturbed and near-mesh particles will not get sufficient chance to contact the apertures in the suitable position. On the other hand, if the amplitude is too small, the apertures will suffer from blinding.

c) The moisture content of the feed may vary reasonably and thus in such instances several screen cloths of varied apertures should be kept ready and necessary changes in aperture should be made to suit the altered conditions of the feed.

d) The vibrating strokes should be distributed quite evenly over the whole area to ensure sufficient tossing of the feed and to avoid overstress at a point, line or node.

e) When oversize in the product is intolerable, precautions should be taken against delivery of oversize material caused due to unnoticed rupture of screen. This may be achieved by duplicating the same mesh on a double-decked machine.

f) Feed should be delivered across the entire width of the screen with adequate gentleness to avoid wear.

g) In case of wet sizing, the feed should be slurried at a liquid–solid ratio of 2 : 1 and flushed gently and uniformly.

h) In installation, consideration should be given to convenience of maintenance and replacements.

CHAPTER 5

Comminution of Ores—Part I:
Crushing

In industries where raw materials are processed in the solid state, reduction of solid particles is invariably required. The reduction operation is an essential step in ore processing, since the run-of-mine comes in the sizes as large as 3 m. The reduction of material to a smaller size is known as *comminution* which may be defined as the *whole operation of reducing the raw ore to the size required for mechanical separation or metallurgical processing.* The large lumps may get weakened by cracks introduced during mining or by the existence of bedding planes, as a result the big lumps may be broken more easily at the coarse size reduction stage (crushing) than during fine size reduction stage (grinding). However, the mechanism of fracture remains the same.

5.1 Mechanism of Comminution

The fracturing or shattering of the larger pieces is caused by the application of pressure. If a prism of isotropic brittle material is compressed between two plane surfaces (Fig. 5.1), the material will first undergo an elastic deformation and then fracture suddenly along the planes of principal shear stress. If the pressure is released soon after the fracture has occurred, the prism will be found to have broken in a small number (three or four) of major fragments and a much larger number of small pieces, but the weight of small pieces will amount to only a few per cent of the original weight. However, most of the ores are crystalline in nature having well defined cleavage planes and thus the fracture will occur preferentially along such planes which are nearest to the planes of principal stress. If the pressure causes deformation without rupture, the only change will be in dimensions of the crystal, and this is known as plastic deformation.

In general, the best method of causing rupture in solid materials is the application of shearing loads. However, the orientation of crystals in ores is generally so irregular that the compressive load is equally effective as shearing loads. Thus, most of the size reduction equipments use compression, or shear, or both as fracturing forces.

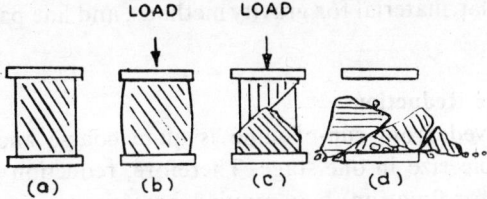

Fig. 5.1. Fracture of ore prism under compressive loading:
(a) unloaded, (b) under elastic strain, (c) fracture
takes place, and (d) fragments.

Ore lumps are usually of irregular shape and thus forces cannot be applied in a symmetrical manner as shown in Fig. 5.1 and the forces may concentrate at edges, corners and projections shearing off small fragments before major fractures occur. This will result in the production of more middle-sized particles in the initial stages than obtainable from the cylindrical model.

Another mechanism of comminution is referred as *attrition* in which shearing-off will take place on some of the high spots of two flat surfaces in contact with each other, resulting into fine dust. If the surfaces are not very flat, larger fragments may be sheared off. This tangential rubbing (attrition) occurs in a majority of comminution processes along with simple compression or squeezing. Many processes may be designed to include both mechanisms.

A third mechanism of fracturing may be found in impact breaking, in which unconstrained particle is struck by a fast moving object (such as swinging hammer) and the particle will fracture in a brittle manner. This fracture is likely to begin at cracks or other surface defects already present. It has been suggested that the impact is unlikely to produce fresh cracks, and thus the impact product tends to resist further breakdown more than the material broken by nipping. Further, it has been suggested that there is a great probability of crack propagation round the grain boundaries by impact fracturing, as a result of which early release of ore particles may occur.

5.2 Objective of Size Reduction
The objectives of size reduction may be summarised as the following:
a) Reduction of large lumps into small pieces.
b) Production of solids of desired size ranges or specific surfaces for direct metallurgical treatment, e.g. powdered coal for boilers and 3–8 cm iron ore for blast furnace.
c) Breaking apart valuable minerals from intimately associated gangue material, i.e. liberation of valuables from gangue.
d) Convenience in handling and transportation.
e) Preparation of feed material for different ore processing methods,

e.g. granular material for gravity methods, and fine particles for froth flotation.

5.3 Stages of Size Reduction

The ore received from run-of-mine is quite coarse and cannot be reduced to the usable size in one stage. Therefore, reduction of run-of-mine size to fine sizes (for flotation) may require three or more stages. The comminution stages are usually divided arbitrarily into convenient stages. In most operations of size reduction, usually two crushing stages, i.e. primary and secondary are employed. In some instances the crushing stages may be more than two. Hukki has proposed the following classification of basic reduction steps:

Explosive shattering:	from infinite size to -1 m.
Primary crushing:	from -1 m to -100 mm.
Secondary crushing:	from -100 mm to -10 mm.
Coarse grinding:	from -10 mm to -1 mm.
Fine grinding:	from -1 mm to -100 micron.
Very-fine grinding:	from -100 micron to -10 micron.
Superfine grinding:	from -10 micron to -1 micron.

However, these steps can be modified depending upon the machines employed. In general, the comminution may be divided into three stages, i.e. coarse size reduction carried out by primary crushing, intermediate size reduction carried out by secondary crushers, and fine size reduction carried out by grinders. One distinction between crushing and grinding processes is that, in the former case the size reduction takes place preferentially on the large fragments, whereas in the latter case the size reduction is less selective.

5.4 Operating Variables

The comminution of an ore is greatly affected by the factors such as moisture content of the material, reduction ratio to be achieved, free crushing, choke feeding, etc. Their effect is described as below:

a) MOISTURE CONTENT OF SOLIDS

If the moisture content is less than 3 to 4 per cent by weight, no specific difficulties in size reduction are faced. When moisture content exceeds 4 per cent, most ores become sticky with a tendency to clog the size reduction machines. Whereas, a large excess of water (50 per cent or more) facilitates the size reduction operation by washing the feed into and the product out of the zone of action. Further, it furnishes the means of transportation of solids about the mill as a suspension or slurry. However, wet comminution is confined to fine reduction only, i.e. grinding.

b) REDUCTION RATIO

The reduction ratio may be referred as the ratio of the average diameter of the feed and the average diameter of the product. In primary crushers, it is the ratio of gape (the maximum size acceptable by the machine) and set (maximum size passing through the discharge end). The reduction ratio of most of the machines is about 3–7, working in coarser range of crushing, whereas reduction ratio in fine grinding may be as high as 100. The energy required during equal reduction ratios increases with increased fineness of the ore being treated, e.g. energy consumption is 0.4,0.6, 1.6 and 10 KWH/tonne in primary crushing, secondary crushing, coarse grinding and fine grinding respectively.

c) FREE CRUSHING

In free crushing (individual particle being crushed freely), the crushed product is quickly removed from the crushing zone. Free crushing operation avoids the formation of an excessive amount of fines by limiting the number of contacts.

d) CHOKE FEEDING

This is the reverse of free crushing, in which the crusher is equipped with a feed hopper and kept filled (or choked) to prevent the complete discharge of the crushed product. This type of operation increases the amount of fines produced and decreases the capacity. In some cases, choke feeding may be preferred as it may result in economy of the operation, eliminating one or more reduction stages due to large amount of fines produced.

e) CLOSED-CIRCUIT OR OPEN-CIRCUIT OPERATION

Usually each stage of size reduction is followed by size separating unit. If the oversize material is returned to the size reduction unit, the operation is referred as closed-circuit, whereas if no material is returned for further size reduction, the operation is referred as open-circuit. The closed-circuit operations are usually preferred owing to being economical in crushing power, smaller units required for a given capacity, and production of more uniform product.

5.5 Factors Affecting the Selection of Comminution Machines

Though the size of feed and product is the most important factor, other factors such as hardness and structure of the ore should be considered while selecting the size reduction equipment. The machines required for coarse crushing of hard ores (magnetite, hematite, etc.) should be sturdy in construction, and elaborate in design, whereas the crushing of soft material such as coals, does not require the machines much sturdily constructed and elaborate in design. For reduction of material in finer size ranges, similar machines can be employed for soft as well as hard materials. The comminution of fibrous material, such as asbestos, requires the machines, which use a tearing action and are called disintegrators.

5.6 Crushing Theories

The structure of a crystal is bound together by its interatomic forces of attraction and the disruption of this bonding will require the stress equivalent to the theoretical strength of crystal which is considerably high, i.e. 7×10^4 Kg/cm². When the applied tension force to a crystal exceeds the elastic limit, a flaw (usually in the form of a crack) is produced which becomes a centre of incipient fracture. The stress which a crystal can withstand then will depend on the length of crack and according to Griffiths it is inversely proportional to the square root of the length of crack, i.e. the fracture will occur at much lower stress loading after a crack is developed.

During the creation of cracks or imperfection by stressing, the work is done to overcome the mutual bonding of the interatomic forces. This work is stored as an elastic energy, which is released as the atoms return to their original positions. If the released energy is sufficient to overcome the weakened interatomic bonds at the tip of the cracks, the crack will propagate at a very fast rate (5000 m/sec). Such a crack may initiate as a scratch or a surfacial discontinuity, as a minute fissure in crystalline structure, or as a defect in the atomic lattice of a crystal grain. Most of the ores and minerals are associated with such points of weaknesses, as a result of which the practical strength falls much below the theoretical. For example, an ideal glass should withstand a stress up to 7×10^4 kg/cm², but in practice a stress of 1.5×10^2 kg/cm² is enough to cause fracture.

When a crystal contains dislocations, the plasticity (plastic deformation by sliding of one plane freely over another) will be blocked at each such dislocation. The stress continues to build-up at this point and causes the rupture of surrounding atomic bonds. It results in a minute crack which becomes a complete fracture on relatively light further stressing. Once crack propagation starts, it proceeds with almost the speed of sound.

From the above, it is clear that in comminution two kinds of stresses may operate, i.e. (i) the reversible below the elastic limit, and (ii) the irreversible, by which surface discontinuities are formed or plastic deformation takes place.

The situation in naturally occurring ores and minerals is far more irregular than that in the perfect and pure crystals due to the following reasons: (a) ores contain a random focal point, (b) explosive shocks during blasting of ore cause further weakening of ore, (c) chemical oxidation between mining and milling, and (d) in a piece of ore at least two mineral constituents are inter-crystallised in different patterns. In order to make use of crushing theory in practice, statistical methods of testing and empirical approach should be added. The crushing may be considered in terms of varieties of disruptive forces, such as compression, tension, shear, torsion, abrasion, and shatter.

In comminution studies, much efforts have been made in measuring the efficiency of energy utilisation, which is the ratio of the energy theoretically needed for a particular degree of size reduction to that actually consumed.

The actual energy consumed can be readily measured as the power input to the mill, which is the sum of (a) energy required to move the working parts of the mill with suitable load, (b) energy required to overcome the friction in the bearings, (c) energy required to grind away the metal from the working faces, and (d) energy required for size reduction. A major part of this energy will be dissipated as heat, and it is extremely difficult to evaluate the various items separately, and thus, obtaining an accurate energy balance on a comminution process is a rare achievement. There has been a great controversy, over the most suitable measure of the minimum energy, required to effect reduction.

5.6.1 Different Principles of Crushing

a) *Rittinger's Law*

In 1867, Rittinger proposed that the energy required for reduction of particle size, should be directly proportional to the new surface formed (increase in surface). In other words, it may be said that the minimum energy required is the surface energy of the new surfaces created. The Rittinger's law can be written as

$$E_r = \frac{a}{d_0} \tag{5.1}$$

where, a is a constant and d_0 refers to the original size.

b) *Kick's Law*

In 1885, Kick suggested that the energy required for producing a specified reduction ratio is proportional to the log of reduction ratio. The Kick's law can be written as

$$E_r = b \tag{5.2}$$

where, E_r is the energy required to obtain a specified reduction ratio, and b is a constant.

The Kick's law was based on the assumption, that geometrically similar particles would always break in geometrically similar manner, irrespective of their size. Thus, if a hypothetical crystal of a mineral initially 1 cm³ in volume and 1 cm in side, is broken in eight cubes, each of 0.5 cm side, the energy required would be exactly the same as that required to reduce the cube of 1 mm side down to 0.5 mm side or even from 1 micron to 0.5 micron. The energy required for unit volume, would thus, be proportional to the number of reduction stages, n. In the above case, the reduction ratio would be $(1/2)^n$.

c) *Dobie's Equation*

The two approaches given by Rittinger and Kick, are quite different. Rittinger has considered the energy in the product of a particular size distribution, whereas Kick has considered the energy required to break the original particle, irrespective of number of pieces produced. In Rittinger's theory, the net free energy of the change of state is measured, whereas, in Kick's theory, the activation energy of the process is measured. Rittinger's energy is an irreducible minimum, while Kick's energy may be modified by changing the mechanism of breaking. Dobie has reconciled the hypotheses of Rittinger and Kick. The equation given, can be written as

$$E_r = b + a/d_p$$

(5.3)

where d_p is the product size.

d) *Bond Theory*

It has been confirmed that the fracture usually occurs by the extension of cracks, already present in the surface of the particle, and such cracks lower the activation energy of fracture to a great extent. Therefore, the energy required should not be proportional to the volume of the particle, rather to the length of crack produced with energy. Following this line of argument, a new approach was given by F.C. Bond. His theory is based on the following three considerations, which were not considered by Rittinger and Kick:

i) The work previously done on feed, cannot be neglected.

ii) In milling, irregularly shaped particles are handled, rather than cubes.

iii) The energy released as heat, cannot be neglected.

Bond has proposed a theory intending to give consistent results over all size reduction ranges for all materials and equipments. Bond theory postulates as following from experimental and mathematical development.

The bulk of the input work is used to deform particles and is released as heat through internal friction. If local deformation exceeds critical strain, a crack tip forms, the surrounding energy flows to it and a breakage will occur. Though, a new surface may finally represent conversion of input power into new surface energy, the main purpose is to form crack tips.

Based on the above, Bond suggested that the theoretical net energy input for a specified amount of reduction, can be given by total crack length multiplied by the net energy used to form a crack of unit length. This energy can be expressed as the fraction of the total energy input, giving the mechanical efficiency of the work.

If Z_1 is the characteristic size of particles that 80 per cent of feed will pass through, and Z_2 is that 80 per cent of product, the energy, W to reduce a material from size Z_1 to Z_2 can be given as following:

$$W = K (Z_2^{-1/2} - Z_1^{-1/2})$$

(5.4)

since crack length is inversely proportional to the square root of that area.

In this equation, K is a constant for any material, which may vary over a small range for a wide variety of materials.

However, the index $-\frac{1}{2}$ is too specific and Holmes proposes to replace it with a variable, r having a value between 0.25 and 0.75 depending on the material.

e) *Holmes' Equation*

Holmes proposed a modified form of Kick's law, suited to ores, in the following equation:

$$B = k D^{1/r} \tag{5.5}$$

where B is the theoretical elastic energy absorbed in deformation to produce unit area of fracture in a cube of side D, and r and k are constants. For a reduction ratio R and product size P, this equation can be written as

$$W = W_i \left[1 - \left(\frac{1}{R} \right)^r \right] \left[\frac{100}{P} \right]^r \tag{5.6}$$

where, W_i is the work index and r is the Kick's law deviation exponent, expressing the degree of variation of particle strength with variation in size, specific material, and the mode of stress application.

5.6.2 VALIDITY OF VARIOUS CRUSHING LAWS IN PRACTICE

Any of the theories discussed above can be confirmed experimentally. The total energy required in any machine far exceeds the fracture energy. The fracture energy can be estimated by comparing the power needed to run the mill empty and loaded. In order to relate energy used with size or surface area, an extremely accurate determination of size distribution will be needed, particularly in the range of fine sizes which make a contribution to their surface area out of proportion to their weight. The specific surface energies of minerals are also not always correctly known and these values depend on the medium used for immersing the mineral (surface energy is different in air and water and different again when surface active agents are in solution with water).

Inspite of the above difficulties, considerable evidence has been produced in support of one or other of these theories. Basically three theories, i.e. Rittinger's, Kick's, and Bond's are considered for evaluation of any crushing process. The others have suggested only modifications. The Rittinger's law deals with measurement of surface areas, the Kick's theory deals with volumes of product's particles and Bond's theory deals with length of cracks formed.

Kick's theory has been supported by test carried out on prepared cubes under compression in the manner shown in Fig. 5.1. This is substantially valid where fracture is essentially of this type, as in primary crushing with jaw crushers and crushing between rolls. The Rittinger's law has been found to be in approximate agreement in the intermediate stage. The Bond's theory becomes increasingly in good agreement at the finer end of comminution, i.e. grinding. In the coarse size range, the difference between the Kick's and Bond's theories is only marginal. In the Bond's model the amount of energy stored per unit volume would be less than in Kick's model. An example may be illustrated as follows.

If a material is subjected to primary crushing from 1 m to 6 cm, then secondary crushing to 4 mm, and finally ground down to 8 microns, the reduction ratios will be 2^4, 2^4, and 2^8, respectively. According to Kick's law, the energy requirements should be in the ratios of $4 : 4 : 8$ or $1 : 1 : 2$ in three different stages. The first two stages may require similar amounts of energy but the third stage of fine size may need 10–30 times as much in ball mill as that of predicted value. This may be explained on the basis that smaller particles have progressively fewer cracks present and the model reverts back from a Bond/Kick mixed to a true Kick's model. However, it may be more likely due to mechanical inefficiency of ball mill. A particle entering a jaw crusher gets crushed in the first three or four bites, and passes through after being crushed, whereas in ball mills there is a high probability of escaping a particle from being crushed. This probability increases with reduction in size. Therefore, a ball mill must work longer as compared to the calculated work for a particular size reduction.

In impact fracturing, such as hammer mill, the activation energy of fracture is probably low. The total strain energy rarely approaches the level envisaged by Kick or Bond. The energy consumed by hammer mills is usually lower than that required by other mills for obtaining the same reduction ratio.

The Bond's theory and its work index have been developed from much test work, and thus provide an excellent basis for finding out the relative performance of crushing systems. However, the work index values are not constant for fixed conditions of reduction, and theory has certain limits of empiricism.

The comminution is the most expensive step of ore treatment and also influences critically both concentration procedure as well as recovery. Therefore, test procedures have been developed for use in each particular problem and for each specific ore. The basic interest in a particular case depends on two factors, i.e. (a) manner in which the association of different mineral-grains in a piece of ore or rock can best be disrupted, and (b) effect of any new development in comminution method on the subsequent treatment.

The efficiencies of different mills have been reported to be in a wide range, i.e. 0.02 to 50 per cent. The highest value can be presumed to be based on Kick's theory, however, it may also be unrealistically optimistic. In fact, the problem of efficiency is an economic one. The cost of energy required for fracturing is very small compared to the other costs, such as costs of power, capital, lining, grinding media, etc. The efficiency of a process can be improved by modifying dimensions and operating conditions, and is best measured in terms of total operational costs per tonne of ore.

5.7 Primary or Coarse Crushing Machines

Primary crushing is referred to the reduction of ore as mined (usually 1 m in size) to 100 mm. Sometimes the feed may be of 2–3 m size. The basic equipments used for this stage of crushing are jaw crushers and gyratory crushers. Sometimes roll crushers may be used for specific purposes. Jaw crushers and gyratory crushers are discussed below.

5.7.1 JAW CRUSHER

Jaw crusher consists of two crushing faces or *jaws*, one of which is stationary and mounted rigidly in the crusher frame, while the other one moves alternately toward the stationary jaw and away from it by a small throw. In jaw crushers, the ore is squeezed until it breaks and the fragments move down to a narrower part of the wedge (space between fixed and stationary jaws), to be squeezed again repeatedly, until the final escape of the crushed product through the minimum gap at the bottom. Jaw crushers can be classified depending on the point of minimum amplitude of motion on the moving face and depending on the way of transmitting the movement to the movable face. The jaw crushers can be made in different designs which combine shear with compression. Blake jaw crushers and Dodge jaw crushers are universal and employ two strokes per revolution. Another one is Telsmith jaw crusher. Industrially only Blake jaw crusher is being employed, whereas the other two crushers are of mainly laboratory importance. These crushers are discussed below:

a) *Blake Jaw Crusher*

This crusher (Fig. 5.2) was patented by E.W. Blake in 1858 and soon attained the final form of use. The Blake jaw crusher consists essentially of cast steel frame supporting two jaws (curved) made of cast steel lined with tough abrasion resistant metal, usually 13 per cent manganese Hadfield steel. The movable jaw is pivoted at the top and operated by the eccentric, pitman, and toggles. The pitman is given almost a vertical motion by the eccentric. One of the toggles is mounted in rigid journals at one end of the crusher frame and as a result the reciprocating motion of the pitman makes the other toggle to move the jaw back and forth. The movable jaw is held against the toggle by a tension link and spring. Crushing is effected only when the movable jaw moves toward the fixed jaw, and during the backward motion of the movable jaw, the crushed product gets discharged. Thus

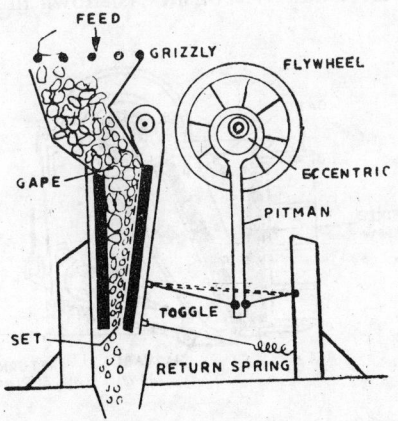

Fig. 5.2. A simplified diagram showing the principle of Blake jaw crusher.

the power requirement is only intermittent (for crushing part only). To equalise this intermittent requirement of power, one or two heavy fly-wheels are mounted on the main shaft of the crusher. The crusher may be driven by flat or V-belts.

The operation of a Blake jaw crusher is influenced by the following factors and can be adjusted to give the best results by a proper interrelation of these factors.

(i) The reduction ratio (gape to set ratio).
(ii) Rate of change of vertical cross-section of crusher throat in respect of fall of ore between strokes.
(iii) Speed and amplitude of strokes of movable jaw.
(iv) Size distribution of feed.
(v) Crushing properties of ore.

A weak point is usually built in the crusher to minimise the damage from uncrushable material with the feed. This weak point can be quickly and cheaply replaced. This weak point may be introduced by (i) weak belt fastener on the drive, (ii) driving pulley weakly bolted to the heavy flywheel of the crusher, (iii) holding down the eccentric by weak bolts which may break and allow the whole pitman to rise, or (iv) scarf jointing one toggle plate by a line of weak rivets.

Nowadays various designs of Blake jaw crushers are available. The aspects in which they differ include (i) variations in toggle plate arrangement to give different kinds of jaw movement, (ii) some designs may include a slight gyratory motion which improves the production rates, and (iii) adding some scrubbing action to simple compression resulting into increased proportion of fines. Blake jaw crushers are manufactured in a wide range of sizes from laboratory models to units accepting the feed size as large as 3 m across. In operation, the *throw* of the movable jaw is only a few centimetres and the frequency is between 60 to 360 strokes per minute. The angle between the jaws is about 20°.

One of the latest models of Blake jaw crusher manufactured by Rexnord Process Machinery Division, Wisconsin[36] is shown in Fig. 5.3. This model is

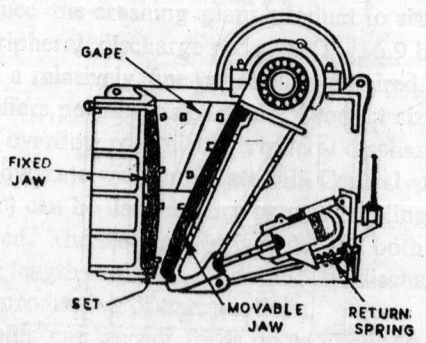

Fig. 5.3. A modern model of Blake jaw crusher.

claimed to have the following advantages and special features:

(i) Lowest down time and maintenance cost.

(ii) Production rates are consistently high.

(iii) Jaws are reversible for long life.

(iv) Toggles are specially designed to reduce friction for long life and quiet operation.

(v) A convenient hydraulic adjustment to reduce the manpower cost.

(vi) The crusher can be made as split bed crusher (two or three piece split beds) to facilitate the transportation.

Capacity of jaw crusher: Taggort has suggested the following empirical formula to calculate the capacity of the jaw crusher:

$$T = 3.75 \, LS \tag{5.7}$$

where T = capacity in tonnes per hour and

L and S are the length of the feed opening and width of the discharge opening (set) respectively, both in cm.

The Blake jaw crushers are capable of processing up to 1,000 tonnes of ore per hour with a reduction ratio of 4 to 10. The capacity variation for different jaw widths[36] is shown in Fig. 5.4. The capacity mainly depends on the discharge setting and the width of the jaw.

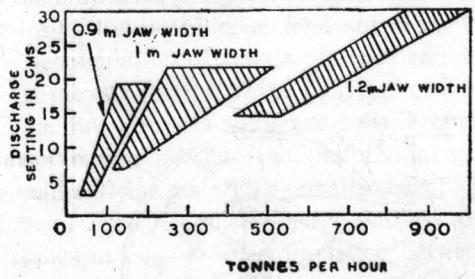

Fig. 5.4. Variation of capacity of jaw crushers-with width of the jaw.

Energy requirements in jaw crusher: The energy consumed in a jaw crusher varies considerably and depends on the following factors:

i) Size of feed.

ii) Size of product.

iii) Capacity of crusher.

iv) Characteristics of the ore/rock.

v) Percentage of idling time.

In general, the energy consumption decreases with increasing size of feed (at constant reduction ratio), with increasing size of product (at constant feed size), and with increasing capacity of the crusher. If the idling time is kept minimum, the energy consumed is usually 0.3–1.5 KWH per tonne of ore crushed.

b) *Telsmith Jaw Crusher*

This resembles the Blake jaw crusher in respect of the motion of movable jaw, i.e. it is largest at the bottom. But it differs in the way of transmitting the motion. In this case the movement is transmitted to the movable plate directly from an eccentric on the main shaft, and not by means of toggles and pitman. It avoids the need for imparting a reciprocating movement to heavy pitman.

The Telsmith jaw crusher is not made in relatively large size as standard Blake jaw crusher, which has met with considerable favour in practice.

c) *Dodge Jaw Crusher*

This crusher (Fig. 5.5) reverses the jaw action of the Blake jaw crusher. In this case, the maximum movement is applied to the largest piece and the minimum to the smallest. The fulcrum is at the lower end of the jaw, and only a slight variation of set occurs on the advancing and receding action of the moving jaw. This crusher does not require toggles and the jaw is operated through the pitman through an eccentric. The power is applied through a long lever. If the crusher becomes clogged, enormous stresses are developed in the members, which become excessive with gapes above 28 cm. This crusher can be employed for the laboratory purposes, where throughput is less important than close control of rock feed sizes. The design of Dodge crusher gives uneven stresses and thus this crusher cannot be built for heavy work and is rarely used in industrial crushing.

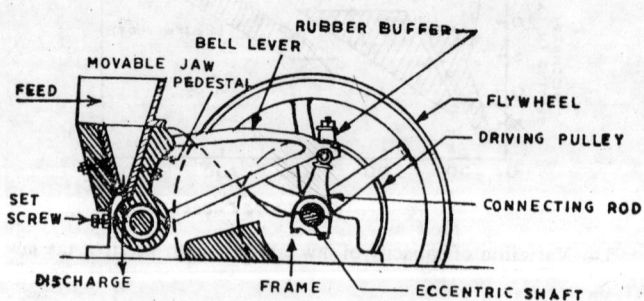

Fig. 5.5. Dodge jaw crusher.

5.7.2 GYRATORY CRUSHERS

Gyratory crushers were developed after jaw crushers to supply the crushing machines of higher capacities. The crushing action of gyratory crushers is similar to that of jaw crusher, since in this case also the moving crushing element approaches and recedes from a fixed crushing element. But the relative motion of the crushing faces is due to the gyration of the eccentrically mounted cone. Some tangential force and simple squeeze are also incorporated. Interaction between particles takes place and as a result the ore is crushed against ore producing some fines reducing the wear of the expensive hard metal (Hadfield manganese steel) facings. There are three types of gyratories, i.e. suspended spindle type, supported or fixed spindle type, and parallel pinch crushers. The first one is the most common in use and is marketed in two designs, i.e. long-shaft and short-shaft, whereas the second one is obsolete.

a) *Suspended Spindle Type Gyratory Crusher*

This crusher (Fig. 5.6) essentially consists of three elements, i.e. suspended spindle, conical crushing head, and the fixed crushing throat. The spindle is suspended from a suitable bearing in the upper portion of the machine. The spindle is free to turn axially. The upper end of the spindle carries the crushing head, whereas the lower end is a circular shaft which is free to rotate in an eccentric sleeve. When the sleeve revolves, the spindle sweeps out a conical path and the crushing head attached to the spindle gyrates with it. The combined system of spindle and crushing head is held in low friction bearings at top and bottom. The eccentric sleeve is driven from a rotating main shaft through a set of bevel gears and rotates within a fixed cylindrical housing. The fixed crushing throat is an inverted conical surface known as *concaves*.

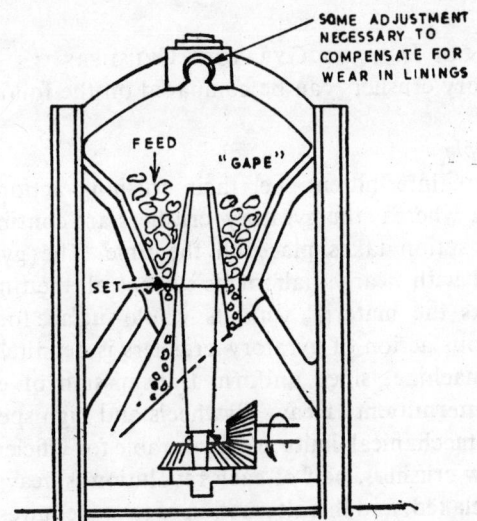

Fig. 5.6. Suspended spindle type gyratory crusher.

When the gyratory crusher runs unloaded, the spindle is free to rotate, but as soon as the ore is fed, it is seized between the head and concave, consequently, the rotation ceases and gyration becomes the only motion causing the head to approach and recede from the concave surfaces. As the head rolls round, the gap between fixed and moving surface becomes steadily restricted and then again relieved, and the pressure is exerted on the material in gap. The feed is crushed as it passes down through the crusher. The differential dilation of cross-section favours relief of choking and makes the crusher a good-free crushing unit.

The gyratory crushers are continuous in action and thus run more smoothly and with larger capacities. One of the largest models can receive a rock as big as 2 m in cross-section and reduce it to 30 cm or less. These gyratories may be 4–12 m in height and weight up to 1,000 tonnes requiring 500 KW power and can crush up to 5,000 tonnes/hr when running at 200 rpm.

The capacity of a gyratory crusher can also be given by the Taggort formula as given in Eq. (5.7), in which L is the perimeter of a circle whose diameter is the arithmetic average of the diameters of the two cones in cm and S is the greatest width of discharge opening in cm.

b) *Telsmith Parallel Pinch Crusher*
In this type the eccentric sleeve is inserted between the fixed vertical shaft and movable vertical cone. The rotation of eccentric sleeve gives a cylindrical motion to the axis of the cone, thus applying the same amount of lateral movement all the way down the crushing head. In fact this is not a gyratory crusher as the crushing head rotates eccentrically instead of gyrating.

5.7.3 COMPARISON OF JAW AND GYRATORY CRUSHERS
Jaw and gyratory crushers can be compared on the following points:

a) *Working Principle*
Jaw crushers are intermittent, i.e. their crushing action is slightly less than half the time, whereas the gyratory crushers are continuous in operation, i.e. crushing action takes place all the time. The gyratory crushers break the material with nearly half of its surface all the time, whereas the jaw crusher breaks the material with its whole surface for about half the time. The continuous action of gyratory crushers is definitely a mechanical advantage to the machine, since uniform transmission of energy is more economical than intermittent. Heavy flywheels and high speeds of rotation within reasonable mechanical limits are favourable for efficient operation of jaw crusher. In jaw crushers, half of each revolution is heavily stressed and the other half is relaxed, and this alternation of stress requires rugged design, fatigue resistant metals for construction, and sturdy mill foundations.

b) *Taper*
The taper (decrease of width between the movable and fixed crushing surfaces per metre or per centimetre of depth) should be small enough to hold the rock properly and not allow its snapping out. The less taper, however, demands the slow running of crusher to allow the rock to fall the necessary distance during each back stroke, and the jaw should also be deeper to give the usual reduction ratio of 4 : 1. The deeper jaw needs greater movement at the end of the greatest movement to prevent packing. The working action of jaw crushers requires less taper than gyratory crushers. The taper given

to gyratory crushers is usually 35–40 cm/m. A taper of 40 cm/m on a gyratory crusher is as good as 35 cm on a jaw crusher.

c) *Capacity*

In both the cases, the capacity is determined by the rate of oscillation of the moving crushing surface, other conditions being same. Further, the upper limit of capacity with certain rate of oscillation is determined by the rate of fall of rock due to gravity between the crushing faces. For large capacities, gyratory crushers are preferred. A gyratory crusher having the same width of mouth as of jaw crusher, will have much larger capacities (due to larger length of mouth and much larger effective crushing space). When the tonnage required is not large, the jaw crusher is preferred. The gape of jaw crusher permits handling of more awkward oversize than the gyratory. For example, a jaw crusher with an opening of 2 m × 3 m taking a 1.8 m rock will have much less tonnage than gyratory crusher capable of handling the same size. Gyratories are built to handle up to 5,000 tonnes of ore per hour which is about 10 times of a big jaw crusher.

d) *Reduction Ratio*

In both cases the reduction ratio varies in between 4 : 1 to 7 : 1. Higher sides are achieved by improved engineering skill, better lubrication and higher working stresses.

e) *Costs of Installation, Maintenance and Power Required*

For a given width of mouth a jaw crusher will be smaller in size, less costly, less excess capacity and a relatively small tonnage can be treated more efficiently compared to gyratory crusher. The gyratory crusher operated at maximum capacity will show lower power consumption. The maintenance of gyratory crusher is heavier than that of jaw crusher. In the places where transport is difficult, gyratory crusher is difficult to transport as it is not easy to sectionise. For a given capacity, gyratories are much more compact than jaw crushers.

f) *Type of Material to be Crushed*

The larger amplitude of motion of jaw crusher is advantageous for crushing soft and clayey materials, whereas gyratories work best on hard and brittle materials.

g) *Type of Feeding*

The crushing machines can be fed either with free feed or with choke feed. Gyratory crushers work better under choke-feed conditions than jaw crushers, and will clear themselves much better after being completely filled.

h) *Product Size*

Gyratory crushers can be adjusted to produce a range of product sizes with allowance for wear on lining.

i) *Compactness of Unit*

Gyratory crushers are considered to be compact and economical in head room. These are usually run with the head burried in ore delivered from the wagons through grizzlies. The top of the gyratory crusher is usually an integral part with a large bin.

5.8 Secondary or Intermediate Crushing Machines

In secondary crushing, the usual size of feed does not exceed 150 mm and the product is usually 12 mm in average diameter. The crushing machines need not have much wider gape and a very sturdy construction is not essential. Since the large pieces of rock have already been reduced in primary crushing to easily manageable fragments, the secondary crushers are generally arranged in series with primary crushers to handle similar tonnage. The main objective of secondary crushing is to reduce the ore to a size suitable for the grinding. The important machines employed in secondary or intermediate crushing are usually cone crushers and modified forms of gyratory crushers. Other machines such as rolls, beater mills, gravity stamps, etc., are also employed. For crushing ores of small quantities, jaw crushers and hammer mills can also be employed. The important secondary crushing machines are described below:

5.8.1 Cone Crusher (Symon Cone Crusher)

This crusher was developed in 1920's and since then it has gained wide acceptance and now it is regarded as the standard crushing appliance in the intermediate range. A standard cone crusher is shown in Fig. 5.7. The essential parts of a cone crusher are spindle, an inner cone or *crushing head*, and an inverted truncated cone known as bowl. In this the spindle is not hung from its upper end (as in case of gyratory crusher) but is supported in a universal bearing below the crushing head. The crushing head is supported by the tapered eccentric journal which is rotated by the bevel gears driven by the main shaft. The entire weight of crushing head and spindle is supported on a bearing plate supplied with oil under pressure. The crushing head gyrates inside the bowl and flares outward.

The principle of operation is identical to that of gyratory crusher, as the crushing results from interaction between three essential parts, i.e. spindle, crushing head and bowl. However, two important differences can be noted.

a) The outer stationary crushing surface flares outward instead of flaring in from top to bottom as in gyratory crusher. This flaring out surface provides an increased area of discharge for the crushed material (a condition for higher capacities), which helps in quick clearing off the machine of the reduced product.

b) The outer stationary crushing surface is held in position by a nest of heavy coiled springs arranged in circular form around the crusher. When an uncrushable object enters the machine, the outer crushing plate gets lifted from the lower surface and prevents the fracture of the plate and damage to the machine.

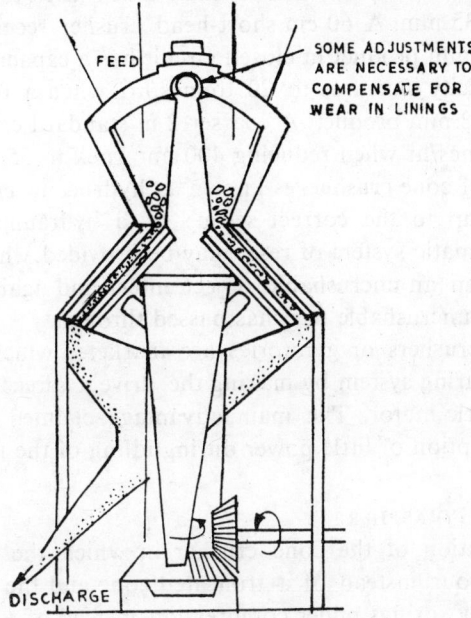

FEED

SOME ADJUSTMENTS
ARE NECESSARY TO
COMPENSATE FOR
WEAR IN LININGS

DISCHARGE

Fig. 5.7. Symon cone crusher.

The cone crushers are made in two types, i.e. (a) standard, and (b) short-head. These differ mainly in the shape of the crushing cavities. The standard crushers are used for the reduction of coarser feeds, and the product is usually in the range of 6 and 60 mm. These are usually operated in open circuit and can be fitted with fine, medium, coarse, or extra-coarse crushing cavities. The short-head crushers have a steeper head angle, a longer parallel section between cone and bowl, and a narrower feed opening. These are used for the reduction of fine feeds and the crushed product ranges from 3 to 20 mm. The short-head crushers generally work in closed circuit.

The *throw* of the gyrating cone is greater than would be practicable with primary crushers and thus the machine should withstand heavier working stresses. In cone crushers, higher speeds (450–700 rpm) are also employed.

The reduction ratio is usually held between 3 : 1 to 7 : 1, which is higher than others. Higher reduction ratios can be worked with some ores. The reduction ratio is controlled by screwing the bowl up or down by means of its capstan and chain mechanism. The capacity of cone crusher is very large compared to rolls or even gyratory crushers for a product of comparable fineness. In order to have the best efficiency of cone crusher, the feed should be dry and free from fine particles, otherwise, the crusher may clog. This requirement makes it essential to use efficient screens in closed circuit with cone crusher.

A 60 cm standard crusher receives a feed of 70 mm and delivers the crushed product at the rate of 15 tonnes/hr of −6 mm size in open circuit.

The capacity is 60 tonnes/hr, when the same unit is used to reduce a feed of −100 mm to −35 mm. A 60 cm short-head crusher receiving 35 mm feed and delivering 3 mm product in closed circuit has a capacity of 6 tonnes/hr, whereas the capacity increases to 20 tonnes/hr, when a feed of 50 mm is reduced to a −12 mm product. A coarser 2 m standard crusher has a capacity of 1,000 tonnes/hr when reducing 400 mm rock to 35 mm product.

One variety of cone crushers is known as hydrocone crusher, in which the cone is held up to the correct setting by a hydraulic jack instead of springs. An automatic system of reset may be provided, which will allow the cone to yield when an uncrushable object enters and again returns to correct setting after uncrushable part has passed through.

Many cone crushers or gyratories are marketed which avoid the complicated bevel-gearing system by making the drive a direct extension of the rotor of the electric motor. The main advantage claimed for this gearless system is consumption of little power during idling of the machine.

5.8.2 Telsmith Gyrasphere

This is a variation of the cone crusher in which the crushing head is spherical in contour instead of a truncated cone and the crushing plate is held in position by springs under compression instead of tension as in cone crusher. The drive and oiling system are identical to that of cone crusher. The spherical contour of the crushing head facilitates the discharge of the crushed product.

5.8.3 Roll Crushers

The rolls were developed in 1806 in Cornwall. It consists of two horizontally mounted heavy cylinders which revolve towards each other, i.e. inward and downward. The feed is nipped and pulled downward through the rolls by friction. In early crushers, one of the rolls was driven whereas the other was driven by friction, but in modern crushers (Fig. 5.8), both the rolls are driven positively at much higher speeds by separate motors. The breakage of the crusher is prevented by mounting the bearing of one of the rolls against a nest of heavy compression coil springs. In roll crushing, there is considerable wear of the rolls and thus the crushing surface consists of a tough steel shell (suitably heat treated) which is shrunk on the main cylindrical castings. These shells can be replaced when worn out.

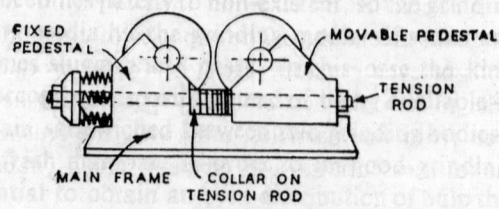

Fig. 5.8. Roll crusher.

The reduction ratio of roll crushers is comparatively small, i.e. 3 : 1 to 4 : 1. The small reduction ratio is a result of the fact that it is difficult to make the rolls of large diameter without unduly increasing the weight of the machine and also due to difficulties in construction. The reduction ratio decreases with increasing size of feed (fine feed can be reduced with higher reduction ratio) until it becomes unity (no crushing) for the coarsest permissible feed. However, due to the springing of the rolls, the actual reduction ratio is still smaller.

The main advantage associated with roll crusher is, small proportion of fines produced and more uniform size of product than other crushers. This may be due to the limited time for which material of a size fine enough to fall through the rolls remains in the machine and thus further crushing does not take place on smaller particles. The diameter and spacing of rolls can be varied over a wide range to permit adequate variation in the feed size and product size. This flexibility is a favourable characteristic of crushing rolls, and this combined with low initial cost has encouraged its application in moderate size reduction in a wide range.

Rolls draw the ore lumps down into the gap and crush between the approaching roller faces by nipping. For a given roll diameter and *set*, there is a maximum size of ore lump which can be drawn in. In selecting rolls for certain work, it is necessary to know the size of feed and product, and tonnage to be crushed. The suitable diameter of rolls, tonnage in tonnes/hr and the required power for crushing rolls can be computed on the basis of coefficient of friction between the material crushed and the rolls. The calculations can be made as follows.

Relation between Roll Diameter, Particle Size, Gap between Rolls and Angle of Nip

Figure 5.9 is a line diagram representing the outlines of a spherical particle, in position to be crushed between a pair of rolls. If gravity is neglected, the forces acting on the particle at the point of contact with roll are a nor-

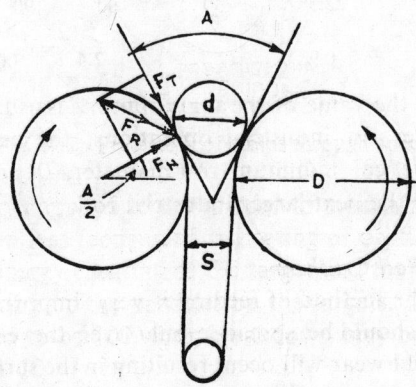

Fig. 5.9. Diagram of rolls for calculation of maximum feed size.

mal force F_N and a tangential force F_T, the resultant of which may be represented by F_R. If F_R is at a negative angle with the horizontal (directed downward), the particle will be nipped and crushed, whereas, if F_R is at a positive angle with the horizontal, the particle will ride in the trough formed by the rolls and will not be crushed.

In Fig. 5.9, A is the angle of nip or wedge angle below which a particle is seized and above which it skids, D is the diameter of each roll, d is the diameter of particle, which can be gripped at zero setting, and S is the set or distance apart of the roll faces at their nearest point of approach. The angle A between the two tangents at the points of contact of the particle with rolls will determine the possibility of nipping the particle. The coefficient of friction is the ratio of the force tangent to the surface to the force normal to the surface, i.e. F_T/F_N. The vertical components of these forces are F_N sin $A/2$ and F_T cos $A/2$, respectively. Under the limiting condition, these two vertical components are equal, i.e.

$$F_N \sin A/2 = F_T \cos A/2 \tag{5.8}$$

or $$\tan A/2 = F_T/F_N \tag{5.9}$$

Under these limiting conditions, angle A is termed as angle of nip.

Since $F_T/F_N = \phi$ (coefficient of friction), the condition required to ensure nip is

$$\tan A/2 \leqslant \phi \tag{5.10}$$

D, d, S, and angle A can be related as follows:

$$\cos A/2 = \frac{D/2 + S/2}{D/2 + d/2} \tag{5.11}$$

or $$\cos A/2 = \frac{D + S}{D + d} \tag{5.12}$$

Maximum sizes of ore lump in relation to roll diameter (calculated theoretically) are as follows, with a coefficient of friction of 0.3.

Roll diameter (cm)	22.5	30	60	90	120	180
Maximum size of ore gripped (cm)	0.9	1.2	2.4	3.0	4.8	7.2

For smooth rolls the value of the angle of nip, A is usually approximately $32°$ for ordinary ores. In industrial operations, the usual practice is to determine the theoretical minimum roll diameter, D, add 2.5 cm to allow for wear and choose the next larger industrial roll.

Method of Feeding Roll Crushers

In roll crusher, the method of feeding is very important for an efficient operation. The feed should be spread evenly over the entire width of the rolls, otherwise partial wear will occur resulting in the surface getting grooved or flanged. A fleeting mechanism may be incorporated to heavy duty crushing rolls, which causes one roll to move to and fro on its axis and thereby reduces the wear. A good practice is to arrange the feed in such a way that some ore falls outside the crushing area at each end, which prevents the uneven wear. Another method is to raise the feeding device, so that the ore arrives on the rotating surfaces at their peripheral speed, giving the best conditions for their nipping. Under lightly fed condition, the rolls can work as arrested (free) crushers, whereas if heavily fed, the rolls are choke crushers (ore crushing ore). The choke feeding is preferable to utilise a machine more effectively. However, with choke feeding there is no individual movement of particles and the classical theory of angle of nip is not applicable. Choke feeding results into the following advantages:

a) A large circulating load (up to 400 per cent) can be used.

b) Wear is reduced due to grinding 'ore on ore' instead of 'ore on metal'.

c) Higher crushing capacity is obtained.

Unless the rolls are of very large diameter, the reduction ratio is limited by the angle of nip, and therefore, crushing rolls may be followed by fine crushing rolls. Whenever, control of maximum particle size in the product is needed, rolls should be worked in closed circuit with screens.

Capacity of Roll Crusher

The theoretical capacity of a roll crusher is the weight of ribbon of feed having a width equal to the width of rolls, a thickness equal to the distance between the rolls, and a length equal to peripheral speed of the rolls in linear units per unit time. This can be expressed as follows:

$$C = 60\, VLS\rho \tag{5.13}$$

where C = capacity in tonnes per hour.

V = peripheral speed in metres/minute.

L = width of rolls in metre.

S = distance between rolls in metre.

ρ = Specific gravity of ore.

The actual capacity is considerably less and usually ranges in between 10 and 35 per cent of theoretical. When the set of rolls is zero, i.e. no clearance between roll faces, the theoretical capacity is zero. But in actual practice, the capacity may be large as the choked feeding results in the roll faces being held apart.

The roll crushers are universally used for road ballast due to uniform size of product and less fines obtained. The roll crushers are often used in series, having each pair of rolls set closer than the previous pair. If the reduction ratios are low and undersize is removed at each stage by screening between rolls, a very close size distribution can be obtained. In ore processing plants, the roll crusher is most suitably used on ores which are sticky or clayey in nature, since roll crusher is self-cleaning, whereas jaw and cone crushers become choked-up with such ores.

5.8.4 Toothed or Slugger Roll

The reduction ratio with smooth rolls is only about 3 : 1. This reduction ratio can be increased above 4 : 1 by providing the ridges or knobs on the roll faces. In some machines, a single roll can be used as shown in Fig. 5.10 (a), whereas in others a pair of rolls (Fig. 5.10 (b)) is used. These toothed or slugger roll crushers are successfully employed for soft and friable materials such as coal, gypsum, iron ore, etc. These 'crushers produce the crushed product with minimum fines, since the sharp teeth or slugs apply a great pressure against the larger lumps of the material on a specific point. Hard rocks, however, will result in an excessive wear on toothed or slugger rolls.

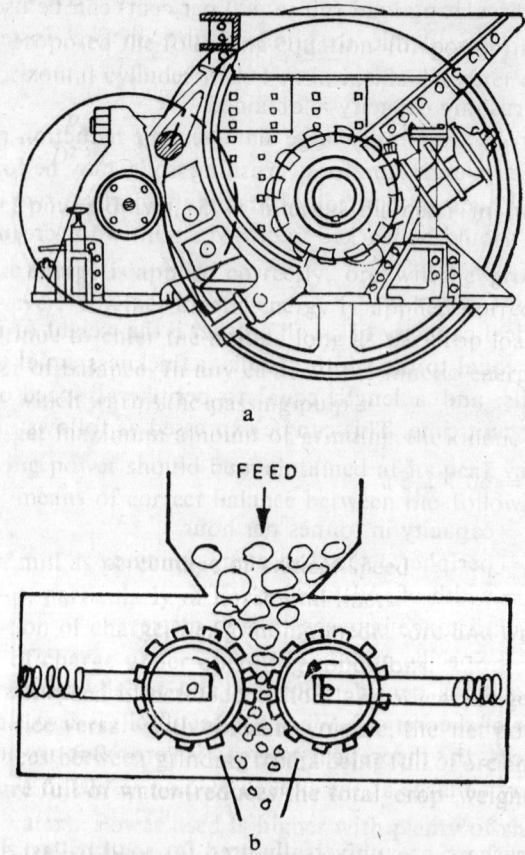

Fig. 5.10. Toothed or slugger rolls:
(a) single roll, and (b) double rolls.

5.9 Fine Crushing Machines

Fine crushing usually refers to the reduction of 25 mm size to −1 mm size, depending on the material and machine. For example, coal can be crushed by hammer mills from a size of even 100 mm down to −1 mm or so. For hard materials such as crushing of quartz, lime stone, etc. to −1 mm size, gyradisc is quite popular. Some of the important fine crushers are described below.

5.9.1 GYRADISC CRUSHER

This has been exclusively designed for the fine product crushing. A sectioned diagram of a gyradisc crusher is shown in Fig. 5.11 (a). In the gyradisc crusher, a fine product is produced by a combination of impact and attrition of a multilayered mass of particles. This process of crushing is termed as *interparticle comminution*. The feed consists of a homogeneous mixture of coarse and fine particles which is introduced to the crusher continuously, so that a head of material is developed and maintained (Fig. 5.11 (b)). At the centre of the crushing cavity, the feed is spread radially outward to the perimeter of the surge chamber, by a motorised rotating feed distributor which reduces the velocity of incoming feed. The feed is deposited

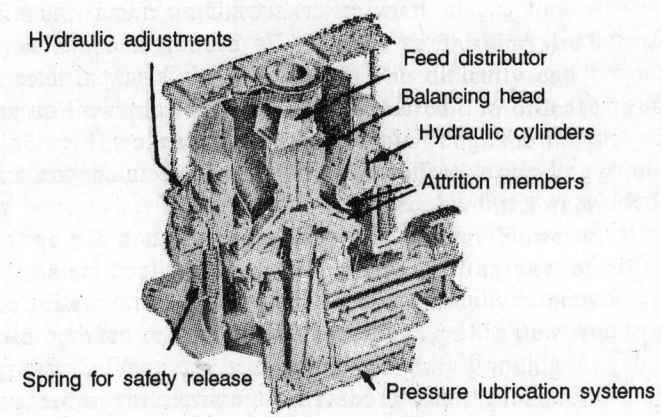

a

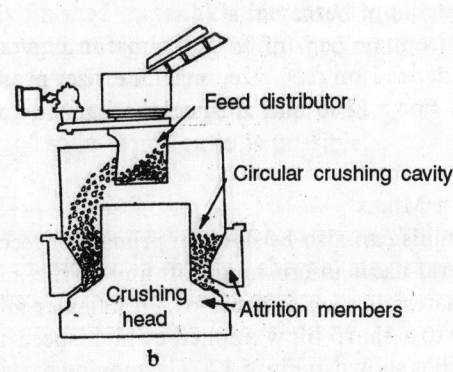

b

Fig. 5.11. Gyradisc crusher:
(a) a sectioned diagram of a gyradisc
crusher, (b) action of crushing.

lightly in a circular path, layer upon layer. Within the feed mass, the particles are separated by a large number of voids created by the loose layering of feed particles. The powered feed distributor working with controlled rate of feed results in the following four conditions, which are necessary for an efficient operation:

a) Sufficient volume of voids to prevent packing.

b) A homogeneous mixture of coarse and fine particles.

c) A uniform distribution of feed around the cavity.

d) A constant maintenance of a head of material in the surge chamber.

The motion of the gyrating head and the pressure due to the height of the surge capacity in the cavity results in the reduced material to move progressively outward through the attrition chamber in a series of short steps. The passage of material through the attrition chamber is not continuous, but intermittent. A variation in number of gyrations and cavity design provides the reduction in oversize production and highly efficient crushing.

When reduction begins, the material is picked up by the lower liner, moved outwardly and caught between the crushing members. The impact of the head results in reduction of material by grinding and by fracture of particle-upon-particle caused by forceful relocation of the particles within the mass. During the upward thrust of the rotating head, two distinct processes combine to crush the feed into a uniform fine product, i.e.:

a) The impact of coarse particle against fine particle similar to the action of balls or pebbles in a grinding mill.

b) The attrition resulting from the high surface friction during forceful relocation of the particles into smaller voids within the feed mass.

Therefore, the combination of impact and attrition in the gyradisc crusher offers the most power utilisation as compared to other fine crushing machines employed today. The gyradisc crushers have been successfully employed in the field of producing sand-sized product from quartz, lime stone, granite, gravel, dolomite, etc. This crusher can also be used to prepare the feed for ball mill to increase the capacity of ball mills, e.g. crushing of moly-sulphide ores, iron ores, copper ore, etc.

The feed size may vary up to 100 mm for a gyradise of 2100 mm. The capacity will depend on feed size, machine size, product size and characteristics of ore. For a 2100 mm crusher, the capacity may be as high as 300 tonnes/hr.

5.9.2 HAMMER MILLS

Hammer mills can also be used for primary or secondary crushing. However, the general use is in crushing soft and brittle materials such as coals, and fibrous materials such as asbestos. In hammer mills, the breaking force is mainly due to a sharp blow applied at high speed to free falling material. A hammer mill is shown in Fig. 5.12. The moving parts are beaters (hammers, rectangular plates, hanging bars, or metal rings), which are fastened by pins to a revolving disc inside a crushing chamber. The beaters move in more

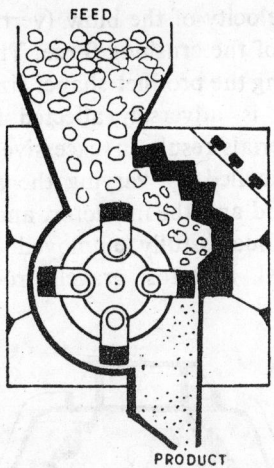

FEED

PRODUCT

Fig. 5.12. Hammer mill.

or less a fixed circle of swift rotation and their centrifugal force delivers the blow to·reduce the rock. The beaters are installed in a robust stationary casing having the openings in the bottom, through which, reduced material leaves the mill.

The feed is introduced through a hopper into the path of revolving hammers (at 500–3,000 rpm), which are freely swinging in smaller machines, but fixed rigid with the rotor in the largest sizes. In hammer mills, the comminution results by the impact fracture mechanism, when hammers hit the ore, or due to shatter of particle against particle or casting plate. All the material remains in the working space until fine enough to pass through the narrow gap between the hammers and the final breaking plate. In some cases, a grid is provided through which material escapes, and the oversize to the grid is swept round the mill for further reduction, until small enough to pass.

The wear on the hard facings is very heavy, and the hammers should be readily reversible and removable. The wearing parts are usually made of high carbon steel or tough wearing alloy steels containing Mn, Cr, or Mo.

The largest hammer mills can accept rocks of about 1m in average diameter and the capacities in such cases can be obtained up to 2,000 tonnes/hr. The main advantage of hammer mills is the high reduction ratio (40:1). Another advantage is, that the product obtained, is strain free, which is beneficial for the rock to be used in road building. Hammer mill is also very useful in grinding ceramic and metal powders.

A modification of hammer mill is known as impactor, in which stoppage is required for hammer adjustment. The rotor is reversible, and the end-wear can also be adjusted while running, by the movement of anvil blocks, which regulate the set of the mill. In this machine (Fig. 5.13), the retaining grid is eliminated, which facilitates the comminution of frozen or sticky feed.

In impact crushing, the velocity of the blow (vertical distance of free fall) is the main determinant of the crushing force. The dropping height can be varied, and thus,influencing the product shape, size, and production of fines. The efficiency of crushing is adversely affected by the moisture content, whereas the abrasive materials result in excessive wear of the beaters. The product size can be controlled by varying the apertures of escape grid, clearance between the grid and the impactor, and the velocity of hammer. This type of mill can be successfully employed in breaking of limestone, gypsum, clays, shale, coal, asbestos, gravels, rock required for concrete aggregates, etc.

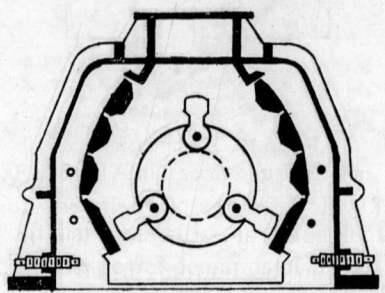

Fig. 5.13. Impactor mill.

5.10 Special Crushers

Some crushers specially designed for specific ores or materials ,can be installed and operated economically for that particular material. Most of these special crushers , have been employed in the past, and now very few industries make use of these, since the requirements have been met by making changes in designs of universal machines. Among this category, the most commonly employed in the past,is the gravity stamps used in gold ore milling, whereas some others such as Bradford breaker, disk crushers, etc., have been used for soft materials. Some of these machines can be used for coarse crushing, whereas some can be employed for fine crushing.

Gravity stamp mills are one of the oldest methods for size reduction in the intermediate and fine size ranges,but are obsolete now. The stamp is an impact crusher and operates on the principle of mortar and pestle. Gravity stamp mill consists of the mortar box on a concrete block,in which crushing is accompanied by 5 (or 10) stamps (heavy blocks). The stamps are raised by cam operating under collars fastened to the upper part of the shaft and drop by their own weight on a hard metal slab. The ore is introduced continuously in the mortar,which falls on the block. The crushing operation is carried out in aqueous pulp.

Its reduction ratio is quite high, i.e. up to 150 (8 cm piece can be reduced to 0.05 cm),and thus,in the past, stamp mills have done a great service in crushing of gold ores for amalgamation.

Steam stamps resemble to the gravity stamp mill,as these also employ a large weight striking the ore. Steam stamps have been used in the treatment of native copper ores.

Squirrel-cage disintegrator: It consists of two or more concentric cages rotating in opposite directions. The feed is introduced into the inner cage. The centrifugal force pushes the material into the spaces between the rotating cages where it is broken apart, and then throws to the outer casing from which the crushed material is discharged to a conveyor or storage bin. This machine is useful in tearing apart fibrous material such as asbestos, wood block, etc.

Disk crusher: It consists of a pair of saucer-shaped disks revolving at almost the same speed and in the same direction. One of these disks has a slight gyrating motion. The feed enters at the centre of the top disk. The centrifugal force throws out the material toward the periphery, where it is nipped and crushed.

5.11 Sequence of Crushing Machines

Generally, the sequence of crushing machines is a jaw crusher, followed by a gyratory which sends its product via by-passing screen to cones, gyra-disc, rolls or any other suitable crusher to yield a final product anywhere between 1 to 20 mm size as required for further grinding operations. The usual trend is to obtain finer product from secondary crushing to reduce the work of the grinding circuit. Below this size, the further reduction is obtained in grinding section. When properly fed, jaw crusher, gyratory crusher and rolls, all crush under free crushing condition, but if the discharge arrangements become obstructed or choke feeding is preferred, then rock grinds upon rock. This results in waste of power, reduced capacity, and also some troubles in the later sections of the flow line.

5.12 Mobile Crushing Units

The mobile crushing units are finding more applications in exploitation of small ore deposits and temporary production of local crushed and graded stone. The mobile crushing unit consists of a train of machines incorporating diesel power unit, and various crushing machines with screens and transfer bins. In open-cast mining, a mobile crusher can be fed directly by a digging shovel and the product is discharged via a bridge conveyor to a more permanent conveying system. In this case, preliminary crushing is necessary, since large rocks should not be carried on belts. The units having capacity up to 100 tonnes/hr using jaw or gyratory crushers are common in use. However, for easily broken ores/rocks as quarried in cement industry, the equipment can handle more than 500 tonnes/hr.

CHAPTER 6

Comminution of Ores—Part II: Grinding

Grinding may be referred to breaking down the relatively coarse material (produced by crushing) to the ultimate fineness. Apart from the grinding of ores, for their subsequent beneficiation, the grinding is also done for preparing the materials for industrial applications, such as grinding of quartz and felspar to fine mesh below 70 microns, grinding of talc to prepare body powder, grinding of iron ore for preparation of pellets and many others.

The size to which the ore/material should be reduced is govened by the requirements of the subsequent processes. For the purpose of beneficiation, theoretically the ore should be broken down until every particle is either fully mineral or fully gangue. If fractures can be made along the interfaces between grains of different kinds and not across them, the valuables can be released by reducing the ore to a size similar to that of the valuable particles. However, in practice, the fractures run indiscriminately across the various crystals resulting into mixed particles even at much smaller sizes. This can be demonstrated with the help of Fig. 6.1. If an ore contains x per cent valuable by volume of an ore entirely in the form of cubes of side a and these are broken by slicing, making parallel cuts with a distance of a/n apart in all the three directions parallel to the cube sides, with the provision that plane of no-slice coincides the cube face, in such a case the volume of each ore particle would be a^3. On slicing it may be observed that each particle of ore mineral remains in the centre of a slightly large cube having a volume equal to $(n + 1)^3 (a/n)^3$ comprising $(n-1)^3$ small cubes of ore mineral each of $(a/n)^3$ volume, and $6 n^2 + 2$ particles equally sized and made up partly of ore mineral and partly gangue mineral, which result from the faces, edges, and corners of the original ore mineral particle under consideration. The gangue not associated with the mineral will be broken into cubes of side a/n.

If all the cubes containing even small amounts of mineral are collected and all others rejected in a concentration process, the volume percentage of valuable mineral would be equal to $100 \, n^3/(n + 1)^3$ and this is independent of the initial percentage. With different values of n, the concentration

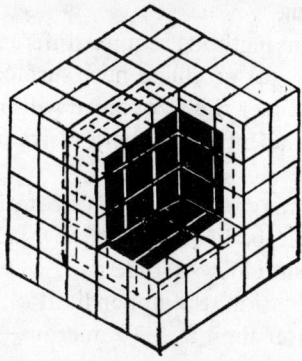

Fig. 6.1. Three dimensional representation of a cubic particle of ore
minerals (dark) associated with gangue mineral (light).

levels obtainable will be different as shown in Table 6.1. It may be observed
that a satisfactory level of concentration can be obtained only when the ore
is broken down to sufficiently small size by grinding. However, since grinding costs are quite high, the suitable level of size reduction is decided on
economic grounds.

Table 6.1. Possible volume concentration at different values
of reduction factor 'n'

n	Volume concentration $n^3/(n+1)^3 \times 100$
0.5	1.2
1.0	12.5
2.0	29.6
3.0	42.1
4.0	51.1
5.0	57.9
10.0	75.1
20.0	86.5

The particles obtained by grinding consist of a wide range of sizes and
therefore a material ground to a predetermined maximum size, would have
most of its portion in much finer size. Therefore, some compromise is made
to grind until, say, 75 per cent of the material is ground to the required
size and the remaining fraction may be assumed to be nearing that size
and not much over that size.

For liberation of the mineral, it is necessary to reduce the size of the
particles until individual particles are sufficiently rich or poor in desired
value to be concentrated, and also until the size is obtained which responds
to the subsequent concentration process. The grinding mills are usually employed for comminution of materials ranging from 5 to 40 mm down to
varying product sizes.

6.1 Objectives of Grinding

Different concentration methods require different sizes of feed. Grinding serves two purposes, i.e. (a) produces new surfaces for reaction, and (b) provides specified sizes. Finer the size, more will be the new surfaces produced. In general, the concentration methods require any of the following grinding preparations:

i) In case of gravity, magnetic, and electrostatic separation, the desired minerals of the ore should be liberated at the coarsest practicable size and overgrinding or *sliming* should be avoided.

ii) In case of froth flotation, the upper limit of particle size is 200–300 microns, whereas the lower limit is 5–10 microns, with the desired mineral exposed at part of the particle's surface.

iii) For chemical treatment such as leaching, the constituents to be dissolved should be sufficiently exposed at the surface of each particle. In this case, overgrinding improves the recovery, but excessive cost involved in overgrinding should be avoided.

Thus the grinding practice depends on the method of ore treatment. When mixed methods are employed, grinding may be carried out in stages and different grinding methods are adopted to suit different stages. The grinding operation is carried out in different branches of industry such as ore dressing, cement, lime, porcelain and chemical industries.

6.2 Types of Grinding

The type of grinding employed depends on the type of ore, quantity of ore to be ground, the use of the ground product, etc. The mills can be designed to carry out following types of grinding.

6.2.1 BATCH OR CONTINUOUS GRINDING

The batch grinding confines to the grinding in a batch by putting a load into the mill and grinding for a determined period or condition to reduce all the particles to the desired size. The mill is then opened and the contents are discharged. The grinding media may be held back by a grate during the discharging operation. Batch grinding is quite inefficient from an ore dressing view point, since most of the product is overground to obtain a definite maximum size. This practice is usually employed in laboratory testing, preparation of paint mixtures, pharmaceutical products and ceramic materials, where quantity involved is less and energy consumption is of less importance. Tumbling mills are commonly employed for batch grinding. However, batch grinding mills are not limited in design, and shapes which can be fed at one end and the product is discharged on the other end. A variety of shapes, such as cylindrical, cylindroconical, oval, polygonal, and even cubic can be used.

In continuous grinding, ore is fed in the mill at a controlled rate and the product is discharged partly from the mill after a suitable dwelling time. Continuous grinding is employed in standard mineral processing,

6.2.2 DIFFERENTIAL GRINDING

Ores and minerals vary in their relative grindabilities. Accordingly, when an ore is ground in a mill, the softer constituents are ground finer than hard ones. This action in grinding is known as *differential grinding*. This differential action can be increased in closed-circuit grinding.

6.2.3 DRY AND WET GRINDING

The materials should be ground when either wet enough to form a slurry (wet grinding), or completely dry (dry grinding), since the grinding in the moist and sticky state is difficult and requires extremely high energy. Dry grinding is performed where the subsequent concentration process is dry or in the areas where water is scarce. Dry grinding is employed in cement industry due to the nature of the material. Other materials treated by dry grinding are ores of Cr, Au, Pb, Mn, Mo, phosphate, Pt, carborundum, coal, talc, etc. Wet grinding is most widely used in mineral processing industry since most subsequent processes such as flotation, magnetic separation, leaching, etc., are done wet. However, both dry and wet processes may be employed for grinding the various materials, depending on the requirements and other considerations. Dry grinding requires the consideration of following points:

a) Feed material should be low in moisture content (less than 1 per cent). A moist and sticky material will require artificial drying.

b) Temperature of circulating air rises as the ore is milled and its humidity increases due to pick-up of moisture from the passing feed. Therefore, enough air should be bled off to prevent saturation with moisture.

c) Less consumption of liner and grinding media per tonne of material ground compared to wet grinding.

d) Costly filtering and drying equipments are not required.

e) A simple storage system is required. In some cases cooling of product may be required before further handling.

f) It requires dust control system.

Whereas in case of wet grinding, following points should be considered:

a) It requires less power per tonne of material than dry grinding.

b) It requires less space than for dry grinding.

c) Generally, the installation costs are less for a closed-circuit operation.

d) It does not require dust control equipment.

e) It requires large quantities of water.

f) Pump maintenance is high.

6.2.4 OPEN-CIRCUIT OR CLOSED-CIRCUIT GRINDING

In open-circuit grinding, the mill receives feed and grinds to the desired product size in one pass. Depending upon sizes of feed and product, open-circuit grinding requires more power than closed-circuit grinding. Open-circuit grinding is usually employed in the following cases:

a) For coarse grinding.

b) To grind the wet cement-raw mix.

c) Where sizing costs make a closed-circuit grinding uneconomical.

d) When tramp oversize and extreme fines can be tolerated.

In continuous open-circuit operation the discharge from one mill goes to the next operation.

In closed-circuit grinding (Fig. 6.2) the mill discharge is fed to a sizing device to separate out the oversized material which is recycled to the grinding mill. The particles may pass several times through the mill before being ground to a final size. The recycled material is termed as circulating load and is referred in terms of per cent of new feed. In general, the closed-circuit system requires minimum power except for rod mill applications. The product size is controlled by the sizing device, and this system lends to instrumentation and automation. Closed-circuit grinding requires less skilled operators to give a constant product size analysis. Closed-circuit grinding is used in most mineral processing industries.

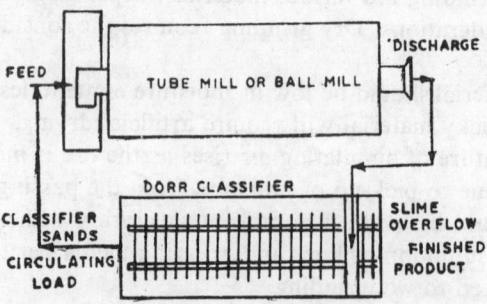

Fig. 6.2. A schematic digaram showing closed-circuit grinding.

6.2.5 PRIMARY AND SECONDARY GRINDING (OR STAGE GRINDING)

Grinding may also be carried out in stages, i.e. primary and secondary as in case of crushing. In primary grinding the milling speeds, types of liner, and size and shape of crushing bodies are chosen to develop shattering or impact milling, whereas, in secondary grinding a more gentle abrasive action is obtained. Usually, the primary grinding mills work vigorously on a fast passing stream of ore, whereas the secondary mills work more gently and with longer retention time of pulp.

Various types of mills employed for grinding, can be classified on the basis of the shape of mill, grinding media used, and the action (force) of grinding. Among the various mills, the most widely employed mills in ore processing are tumbling mills, whereas other mills are designed and used for specific purpose and particular type of materials.

6.3 Some Old Methods of Grinding

The older devices for grinding consisted of two main parts, a stationary surface and a surface rubbed against the stationary surface. The mill stones used for grinding flour from grains are typical, which cause the fracture mainly by the application of shear loads. In between the old style of shear grinding devices and the commonly used ball mills and rod mills, several grinding devices appeared, in which the material was reduced between rollers, or heavy balls rolling against a crushing ring. Some of these mills include pendulum or roller mill, disc grinder, edge runner mill, ring ball mills, vibration mills, hammer mills, rolls, etc. Most of these mills are obsolete. Only hammer mills and rolls are still in common use, whereas pan mill or edge runner is used at Andhra Pradesh Mining Corporation for the recovery of asbestos fibre.

Pan mill or edge-runner mill consists of rotating disc on which spring loaded stationary rollers (made of iron or ceramic) are mounted. Hammer mills and rolls have been described under Sec. 5.9.2 and 5.8.3. These can also be used for fine grinding.

6.4 Tumbling Mills

The tumbling mills are cylindrical or cylindro-conical in shape and rotate about a horizontal axis. A load of crushing bodies (rods, balls or pebbles) is called the grinding media which forms a part of the mill load. In these mills, the particle of ore in the tumbling mixture meets the abrasive and/or impacting force which reduces the ore to particles of desired size. Rod mills, ball mills, pebble mills, cascade mills, etc., fall under this category.

The tumbling mill is nearly half filled with crushing bodies. The ore is fed in the mill at one end and discharged continuously at the other end. At the feed end, water is added (in wet grinding) to flush the feed through the mill. When the tumbling mill is rotated, the mixture of ore, crushing bodies, and water is churned with flying, crushing bodies depending on the speed of the mill. The kinetic energy of the tumbling load is dissipated in the form of heat, noise, and crushing of ore, grinding media and linings. The useful energy (kinetic energy) can be applied to particle by the following actions:

a) Collision between ore pieces.

b) Pressure loading of particle pinned between grinding media or between grinding media and liner.

c) Shear and abrasion caused due to dragging of particle between moving grinding media.

d) Impact of falling grinding media.

e) Shock wave transmitted through crop load by falling grinding media.

The conditions are quite different than those in dry crushing. The inside of the mill is slippery with water and fine particles. The particles undergoing comminution are generally too small to be firmly gripped. In tumbling mills the fixed surface in movement is the lining inside the cylindrical shell, which is buried under the mill load during the part of each revolution and then rises clear. Some control on the average size of particles discharging from the mill can be achieved by adjusting the balance between various factors.

6.4.1 Liners of Tumbling Mill

The mill liners perform two major functions, i.e., (a) protect the cylindrical shell of the mill against abrasion by the tumbling mixture of ore, crushing bodies and water and thus prevent the wear of the mill, and (b) help in tumbling and rotating the mill load. The liners are fixed or bolted to mill shell and thus revolve with the peripheral speed of mill without slipping. The crop load (mill load consisting of crushing media, ore and water) rests on liners, which is displaced by frictional drag against the rising side of the revolving mill. The energy consumed in grinding is mainly that used in unbalancing the crop load. The farther the load is lifted out of balance, the greater is the amount of energy used during comminution. The grip of liners transfers this grinding force into churning action of crop load, which is carried upward by the rising shell-liners during the revolution of mill. The forces causing the mill load to rise are (a) centrifugal force of the rotating mass, (b) collective pseudo-viscosity, and (c) interlock produced by mutual adjustment of crushing bodies, liners and ore.

If the liners are smooth, the crop load will loose the contact soon after it starts to rise, causing the churning of load quietly. If the liner surfaces are made with projecting ribs, a positive lifting force can be applied to the load, resulting into higher climb of the load. Thus in the later case, the load falls with considerable violence, adding a shattering action to the abrasion characteristic of churning motion. The liners are subjected to grinding action as well as to corrosion effect of chemicals (either in mine water or acid produced by decomposition of sulphides). The wear of liners is much more in the central part than of liners at the end. Therefore, the materials employed for liners should be hard, tough, and resistant to corrosion. Further, the material chosen should not introduce any element in the ground product which may present problems in subsequent concentrating process. The usual materials employed for liners is alloy steel and alloy cast irons. The consumption of liners depends on type of ore and mesh of grind, and usually varies from 0.015 to 0.15 kg/tonne of ore.

Mill liners can be made in many shapes but in general, they can be classified into the following three groups:

a) Smooth liners having simple frictional contact with the crop load.

b) Grate liners (having pocket, grid, etc.) in which the ore and crushing media tend to wedge in the openings which form the wearing surface.

c) Lifting liners provided with longitudinal waves, ribs, lifters, wedge bars, etc., which grip the crop load to some extent on the rising side.

The primary grinding mills are usually provided with liners having ribs, steps, bars, or similar projections, whereas secondary grinding mills have wave-type or smooth liners. However, the choice of liners depends on the following considerations:

a) Capital cost of liners per tonne of ground ore between fitting and scrapping of liner.

b) Grinding time lost in replacing the liner.

c) The mesh of grind obtained (coarse, fine or slimes) and its effect on recovery and grade of concentrate.

d) Power requirement per tonne of ore with different liners.

e) Influence of liner action on general economics of the process.

The wearing of liners causes the increase in diameter and capacity of tumbling mills. The wear is concentrated on projections and bolt heads. This results in the following effects:

a) Slow change of product fineness and increase in proportion of very finely ground material.

b) Loosening of liners due to wear of bolts which allows the leakage of pulp through the liners to the shell, which is highly undesirable.

c) As the mill diameter increases with wear, an increased weight of crushing bodies will be required together with increasing amount of ore in crop load.

d) Power consumption and capacity rise accordingly.

e) A definite increase in peripheral speed is obtained which increases the centrifugal force available to lift the crop load on the rising portion of the mill.

In case of heavy duty mills, liner backings are used to reduce severity of shock and danger to liner fracture. Such backings may be about 2.5 cm thick, made out of rubber sheet, old belts, or zinc metal. In general, the cost of liners involved is relatively small, and thus best liners should be chosen for the work, even if costly. The time lost in replacing quickly worn out liners would offset more than the extra cost of high quality liners. Usually, alloy steels are preferred for heavy work in primary mills, whereas in case of secondary mills where the impact is low, cast iron or self-renewing liners are used.

The recent trend is towards the use of rubber in linings and end plates. This gives the advantages of higher productivity and faster and cheaper replacement and repair due to the possibility of vulcanising in worn out small areas. The rubber liners require substantial lifters closely set together to minimise slip between the load and the liners. However, rubber cannot be used in larger mills or where the balls are bigger than 85 mm in diameter. In any case, the thickness of rubber must be sufficient to withstand damage by the action of grinding media (however, less than that of steel liner). Rubber also withstands both abrasion and corrosion and has the additional advantage of being quieter in action than steel.

6.4.2. Capacity of Tumbling Mills

The mill capacity is governed by the specifications of pulp (size of solid particles) entering the mill, and mesh of grind. Therefore, the capacity should be represented in terms of tonnage to be ground with mesh of grind. The grinding work is usually carried out in stages, which should be mutually adjusted to keep the final stage running at full load. Since the final stage should grind the material completely below the optimum mesh size, capacity can be increased by adjusting the work along the line till each machine becomes well loaded and the last stage of feed has been made to such a condition that a required volume can be handled efficiently. Efficient grinding requires to apply the force only to oversized particles. In most of the concentrating operations, this condition improves the recovery of values. However, this is not true in case of extra-fine grinding as required to expose minute particles of gold for chemical attack of cyanide solution.

In order to obtain the maximum capacity of a mill, several variables should be adjusted and balanced. These variables may be divided into three groups as follows:

Group I: Ratio of mill length to diameter, speed of mill, percentage of mill volume charged with crop load, and number of grinding stages.

Group II: Grinding media size ratio, sizes of new feed, and solid–liquid ratio.

Group III: Circulating load, dwelling time of material in mill, and feed rate.

The variables of group I will change with wear of liners and grinding media. The variables under Group II can be maintained fairly constant during grinding. Any change in feed rate of group III will immediately affect the other two variables influencing the capacity. Other factors influencing the capacity are grindability (depending on hardness, brittleness, toughness, cleavage, etc., of ore), flow capacity (new feed + return feed + water passing through the mill), and speed of mill in terms of percentage of critical speed (theoretical speed at which the outer layer of grinding media will just centrifuge with no slippage between them and the shell).

The capacity of a grinding mill is related to the mill volume and the portion of the mill occupied by crop-load. Capacity of a mill can be increased by (a) increasing the crop load till unbalance reaches its maximum, (b) increasing the mill speed, (c) increasing the circulating load (if classifier and mill trunnions are able to handle more tonnage) which decreases the amount of overgrind resulting into increased throughput, (d) giving a liner contour such as to give higher lift and greater impact grinding, and (e) increasing the diameter by using thin alloy steel liners. However, in practice, the grinding capacity of a mill is increased by assigning the maximum comminution work to the crushing section. Sometimes, it may be justifiable to introduce a further stage of fine crushing by roll mill or rod mill when a grinding plant is overloaded.

6.4.3 Grinding Bodies in Tumbling Mill

In tumbling mills the comminution is effected by the impact and abrasion. In primary mills, larger or heavier grinding bodies capable of delivering more powerful blows will be needed than in the secondary grinding mills. Though some grinding would result due to the kinetic energy of colliding and abrading components, an efficient grinding would not be produced, unless there is an adequate percentage of large pieces of ores or heavy grinding media alongwith ore.

The various physical properties of crushing bodies, such as density, size, shape, hardness, and toughness influence the quality of grind. The crushing medium in a ball mill is steel balls of assorted size (25 to 125 mm). The balls are usually made of high manganese steel, but nowadays balls of alloy steel (chrome steel, chrome-moly steel, etc.) are used as grinding medium. In autogenous grinding, large pieces of ore are used, whereas in pebble mills, flint pebbles are used. In some special cases tungsten carbide balls, and ceramic balls are also used, their high cost gets compensated by their long life.

6.4.4 TYPES OF TUMBLING MILLS

a) Ball Mill

The ball mills date back to 1876 and are characterised by the use of balls (made of iron, steel or tungsten carbide) as grinding medium. These mills are horizontal, rotating cylindrical or cylindroconical steel shells, usually working as continuous machines. The size reduction is accomplished by the impact of these balls as they fall back after being lifted to a certain height by the rotating shell. The length of the cylinder is normally equal to the diameter. The feed enters at one end and the product is discharged either through the opposite end or through the periphery. The ball mills may be operated dry or wet. The product may also be discharged by overflow through a hollow trunnion and the finer particles are carried away by the circulating fluid (air or water). The ball mills can be classified according to the shape of mill (cylindrical or cylindroconical), the method of discharging the product (overflow discharge, or grate/diaphragm discharge mill), discharge rate (high discharge mill, low discharge mill), and whether operated dry or wet. In dry grinding, the load is kept lower (below 40 per cent) than in wet grinding to avoid over-carry at cataracting speeds.

Theory of ball mills: In action the balls are lifted up the rising portion of the shell and rolled down or thrown over into the pool, nipping ore particles against other balls, the lining, or other pieces of ores. The effectiveness of grinding will depend on the number of opportunities for each particle to fracture. As the balls are lifted, their path is circular, whereas the path of dropping balls is parabolic or near parabolic as shown in Fig. 6.3 (a). If the mill is revolving at a speed of N rpm, a ball at a distance r will leave the circular path when the centripetal component of gravity exceeds the centrifugal component of angular acceleration (Fig. 6.3 (b)), i.e.,

$$\frac{mV^2}{r} = mg \cos \alpha \qquad (6.1)$$

where, m is the mass of a ball, V the linear velocity of the ball, r the radius of mill, and g the acceleration due to gravity. Since $V = 2\pi r N$,

$$\cos \alpha = \frac{4 \pi^2}{g} N^2 r \tag{6.2}$$

The locus of points X represents the beginning of the parabolic path (Fig. 6.3 (a)) for different positions of balls from the centre of.mill to periphery, which is the curve OEA of Fig. 6.3 (c). This curve can be evaluated graphically from calculated values of α in function of r. ,

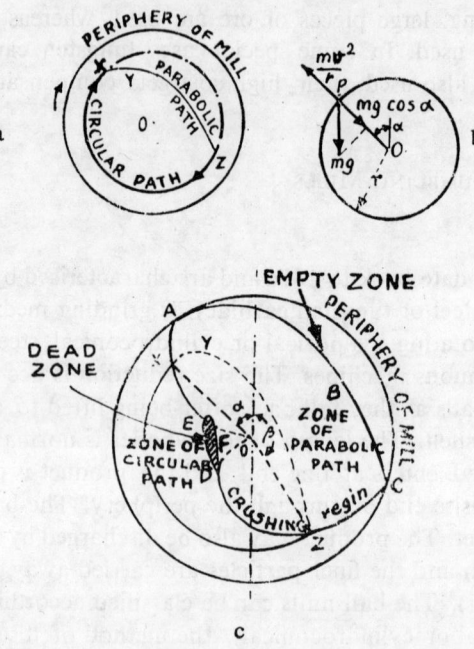

Fig. 6.3. Theoretical principles in ball milling:
(a) Path of ball in ball mill, (b) Forces acting on a ball
at distances from the centre of mill, and (c) Various
zones in a ball mill.

Centrifuging of the outermost layer of balls commence as $\cos \alpha$ exceeds unity [according to Eq. (6.2)]. From Fig. 6.3 (c), according to Davis

$$\beta = 3 \alpha \tag{6.3}$$

From the relationship of Eq. (6.3), the locus of the points Z.represents the end of the parabolic path, which can be drawn in an identical manner as the locus of points X. This gives the curve CDO of Fig. 6.3 (c). Further, Davis has shown that the arcs DO and EO correspond to unstable equilibrium and the zone EFD is a dead zone having no effective motion. Thus, the inside of a ball mill can be considered to consist of four zones, i.e. (a) an empty zone, (b) a dead zone, (c) a zone of circular path, and (d) a zone of parabolic path, which are shown in Fig. 6.3 (c).

According to Davis, crushing action should take place along DC and nowhere else, which is not true, since a substantial crushing takes place within the zone of circular path arising from the rolling of balls on each other by slippage between ball layers. The slippage is a function of many variables, such as pulp density, size and type of ores, pore space in ball charge, speed of mill, etc. Secondly, the path of each ball is not compounded of two segments, i.e. one circular and one parabolic, but is compounded of three segments, i.e., one circular, one parabolic, and one near parabolic (*XY* in Fig. 6.3 (a)).

In ball mills attrition also occurs between balls as they are half dragged and half rolled through the pulp under continuous impact from above. The speed of the mill should be adequate to raise the balls to about two-thirds of the height of the inside of the mill and throw them across to the bed of the load having coarser material accumulated in it. If the mill is about half-filled with balls and pulp, the speed of mill should be about 75 per cent of the critical speed N_C, at which the load will centrifuge (at this condition, no grinding takes place). The critical speed may be given by

$$N_C = 42.3/D \qquad (6.4)$$

where, D = diameter of mill in cm. However with smooth linings, slip occurs between load and liner over a wide range of speeds and centrifuging does not occur even at the speed several times of critical speed (N_C). This is true particularly under low loadings.

At these super-critical speeds, there is much more attrition between balls, ore, and linings. This results in improved efficiencies, but the wear on lining is excessive. The super-critical grinding may not be favoured with metal liners but can be effectively used with autogenous linings, i.e. lining having ore lumps fixed to or embedded in the lining to protect the steel lining.

The cross-section of a ball mill showing the ball movement is shown in Fig. 6.4, which represents the action in the various parts of the crop load. In the segment *X–X* to *Y–Y*, the load rises quietly with certain amount of ball spinning due to slip, causing little grinding. The segment between *YY* to *ZZ* corresponds to free fall, during which the balls convert their potential energy into kinetic energy. If the fall is cascading, most of their energy is used in abrasive grinding during the descent of balls. At the toe of the mill,

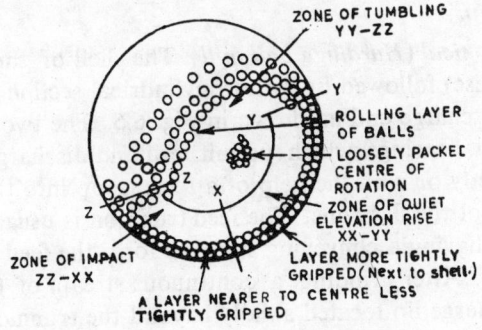

Fig. 6.4. Cross-section of ball mill showing ball movement.

churning of mass takes place, resulting into most of the grinding work. In addition to the direct contact between falling balls and pieces of ores (lying above balls and liners), the ore packed between balls receives the shock transmitted through the segment ZZ to XX by the continuous battering. At the extreme toe of the crop load, the churning mass is folded continuously and is carried up to the breakaway point.

The material just adjacent to the liners is held more firmly compared to any other central part of the crop load. The larger balls and ore lumps are less able to penetrate the charge near the toe and thus there is a tendency of balls and ore lumps to align themselves concentrically with respect to the centre of rotation of the crop load, i.e., largest remain in outermost layer and are most stable during rise, whereas the smallest ones are most loose at the centre of charge. As a result of this alignment, the largest balls are lifted at full peripheral speed and thrown outward from the breakaway point with the maximum force. Most of the large balls and ore pieces travel outside at a speed little less than the peripheral speed, and these are the last to leave the shell at breakaway point because (a) they are held to the shell by the inside layers, and (b) extra centrifugal force works on the balls in this position of maximum mill radius and minimum load slip.

The necessary cataracting effect can be obtained by employing largest balls having maximum individual weight, velocity and inertia. They should not miss the proper target (line ZZ) and carry on to hit the liners. Then smaller and smaller balls and pieces of ores come inward through, to the approximately oval-shaped part of the cross-section. These rotate more slowly and decrease in their peripheral speed with increasing slip. Thus there is less centrifugal force available to maintain the inside of the crop load in its climb which causes the sagging away of material from the rising portion of the mill earlier than the outer part of the load. Between XX and YY, the entire load is in most compact condition and thus most of its weight and centrifugal force work on the ore, being abraded during slip or spin of balls in this section. Above YY as well as the centre of turn, the texture of the load opens outward and upward and becomes continuously less dense as it gets away from the centre. This also favours flight of largest balls and ore pieces.

Various Ball Mills

i) *Cylindroconical (Hardinge ball mill)*: The shell of this mill consists of a flat cone (obtuse) followed by a short cylindrical section and a steep cone (acute) at the discharge end as shown in Fig. 6.5. The two ends carry two hollow trunnion bearings which permit feed and discharge. Feed may be introduced directly or with the help of a feed scoop into the feed end trunnion at the end of the flat cone. The feed trunnion is usually provided with a wearing plate having a conveying spiral to force the feed forward into the body of the mill. After grinding, a continuous stream of thick pulp comes out from a discharge lip located axially beyond the trunnion at the end of the steep cone. The discharge end may be provided with a retaining grid

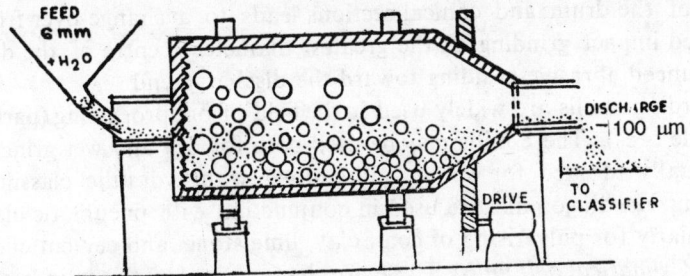

Fig. 6.5. Simplified diagram of a cylindroconical ball mill.

having punched holes) to avoid the risk of discharging balls. The mill may be driven by a crown-wheel bolted round the steep cone, by a belt, or through a speed reducer. The interior of the shell is lined with cast iron or steel plates which may be backed by shock absorbing material such as rubber sheet or old belting.

These mills are designated according to the diameter and width of the cylindrical section. The power required per tonne of ball load increases with the diameter of drum and is usually 7–12 kW/tonne of balls at 75 per cent of critical speed. In these mills, maximum centrifugal effect is obtained in the drum section due to maximum peripheral speed, and therefore, the ability of the mill to lift the crop load is highest in this section. At the same time, the load present inside the acute cone tends to work back down-slope, which causes the accumulation of coarse particles and large balls in the cylindrical section of the mill and the size of balls and ores tapers off towards the discharge end, i.e. fine particles and small balls in conical section. This type of arrangement makes possible the preferential grinding of coarse material by large balls and fine material by smaller balls. The effect of conical shape on peripheral speed and kinetic energy on the crop load at various cross-sections is shown in Fig. 6.6. The difference between the peripheral

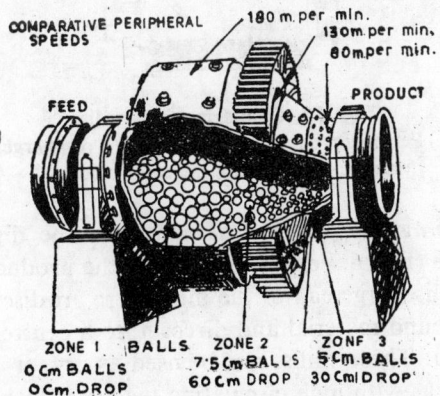

Fig. 6.6. Hardinge conical ball mill—showing the kinetic energy and speed in different sections.

speed of the drum and conical sections leads to a change-over from pronounced impact grinding at the greatest diameter (centre of the drum) to pronounced abrasive grinding toward the discharge end.

Hardinge mills are widely used in the field of ore processing (particularly sulphide ores). These mills may be used both for dry and wet grinding, but are usually adopted for wet grinding along with hydraulic classifier. Dry grinding Hardinge mills are used in conjunction with pneumatic classifiers, particularly for pulverising of coal, clay, lime stone, and cement clinker.

ii) *Cylindrical ball mills:* These are characterised by the cylindrical shape of the shell having uniform diameter throughout. The cylindrical mills can principally be classified into following categories depending on the mode of discharge of the ground product.

1) *Peripheral discharge ball mills* have the discharge through screens along the cylindrical shell. Though this type of mills is less common due to the expensive screen wear, these mills (Fig. 6.7 a) can be employed for grinding non-abrasive materials either wet or dry.

2) *Overflow ball mills* have the discharge as overflow from the exit of the mill through the trunnion (Fig. 6.7 b) and this is the most common discharge arrangement. This type of mill operates only wet. Its major advantages are simplicity of design and construction, easy access for inspection, and easy replacement of liners. This type of mill provides relatively fixed or constant pulp level and thus the effectiveness of the mill can be controlled only by the size and quantity of balls or the rate of feed.

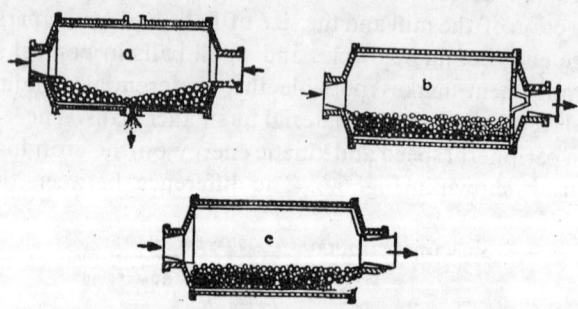

Fig. 6.7. Cylindrical ball mills:
(a) Peripheral discharge, (b) overflow discharge,
and (c) grate (diaphragm) discharge.

3) *Grate (diaphragm) discharge ball mills* have the discharge end fitted with vertical grates (Fig. 6.7 c) through which the product passes. Between the grate and the discharge end of the mill, there are discharge lifters (pans) which raise the ground material and direct it to a centre cone and into the discharge trunnion. These mills can be used as dry or wet. This type of mills generally works with high circulating load, producing very little fines. This has the advantages of less power consumption than overflow mill, and higher capacity per unit volume (15–25 per cent higher than overflow mill).

In this case the pulp level can be independently controlled at any desired level by making the grate/diaphragm solid for the desired distance from the periphery.

iii) *Multiple compartment ball mills:* This consists of one very long cylindrical mill (rather than multiple units) divided into two or more compartments. Each successive compartment is of smaller width and contains sized balls for most efficient grinding. Such a mill is essentially an equivalent of a series of mills operating continuously. These mills are used when high reduction ratios are to be achieved. A simplified diagram of a three-compartment mill is shown in Fig. 6.8. This mill is also referred as tube ball mill owing to its higher length to diameter ratio (5 : 3).

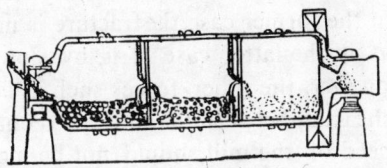

Fig. 6.8. Multiple compartment ball mill.

The ball mills are extremely flexible in geometry and operating speeds. Mills having lengths three to five times their diameter are usually employed for applications where surface area of the product is critical and a high recycling load is undesirable. The mills having lengths one to two times of their diameter are usually selected for applications where it is desirable to have the product size mainly in a narrow intermediate range, e.g. required in liberation of mineral grains from gangue. In this instance, the classifier is expected to remove finished product as early as possible and a recycling load of 200–300 per cent of new feed rate is desirable. The ball mills have been successfully operated at speeds of 60–90 per cent of critical speed, but in general most of the mills operate between 65 and 75 per cent of critical speed.

b) *Rod Mills*

These may be defined as rotating cylindrical shells using rods (75–100 mm in diameter) as grinding media, which are laid parallel with the axis. The rods should span over the whole length of the mill. The rod mills are more difficult to operate than a ball mill, since rods wear thin and break and the broken pieces interfere with the performance of the mill (the rods shorter than the mill diameter may become wedged against the lining). This calls for removal of the thin rods. The rod mills are made of length greater than their diameter (length/diameter ratio should be at least 1.4) in order to prevent jamming of the rods in the mill. Cylindroconical shells can also be used in rod mills, provided the cylindrical section is relatively longer compared to the diameter.

The basic principle of grinding in rod mills differs from ball mill and in this case the reduction is by line contact between rods extending the full length of the mill. In grinding action, the rods are kept apart by the coarsest particles and as a result selective grinding is carried out on the largest particle sizes at all stages. Thus the rod mills produce minimum of extreme fines or slimes and the grinding work is more effective as compared with a ball mill. The selective grinding is also aided by means of water flowing along the spaces between the rods which carries the undersize particles toward the discharge end. This mill minimises attrition of fines, overgrinding and heavy oxidation of exposed minerals.

The rod mills can be run at high rotational speeds with deep lifters for crushing coarse feed (from about 40 mm), or slowly with shallow lifters for controlled grinding. In the former case the fracture is mainly by fast squeezing (nipping), whereas in the latter case it is by slower squeezing (more tangential action). However, the other forces such as compression between lifters and rods, and shearing between the rods also contribute to the reduction. The operating speed of the mill should not be more than 70 per cent of its critical speed and preferably in the range 60–68 per cent. Most rod mills do not run under free falling conditions (cataracting), but run under cascading conditions (the emerging rods rolling down the slope of the crop load to the bottom of the mill) and the rods act as multiple rolls. As the rod mills are usually operated in open-circuit, the largest bridging particles will be near the feed end of the mill, and the smallest near the discharge end.

Rods wear down in use and this affects the nipping of particles. The preferred material for rods is high-carbon steel. Highest efficiency of the mill can be obtained by periodic removal of worn out rods and charging the fresh rods.

The rod mills may be classified according to the discharge arrangement. As shown in Fig. 6.9, the rod mills can be made with (i) overflow discharge, (ii) end peripheral discharge, and (iii) centre peripheral discharge.

Overflow discharge rod mill (Fig. 6.9 a) is the most common and is used extensively in ore processing applications. The grinding is done wet and the mill is used to reduce the crushing plant product to size required for ball mill feed. End peripheral discharge rod mill (Fig. 6.9 b) is mainly used for dry grinding when a relatively fine product is required. It can be used for wet grinding but offers no advantage as the product size will be almost the same as in case of overflow rod mill. Peripheral discharge gives a high gradient and good flow rate with dry material. Central peripheral discharge rod mill (Fig. 6.9 c) can be used for dry or wet grinding when a coarse product (sand) is desired. The feed is introduced from both ends, which results in a short grinding length and high gradient. The discharge rate is very high which reduces the production of extreme fines.

Though rod mills can accept feeds up to about 50 mm depending on hardness of material, an ideal feed is considered to be in the size range of

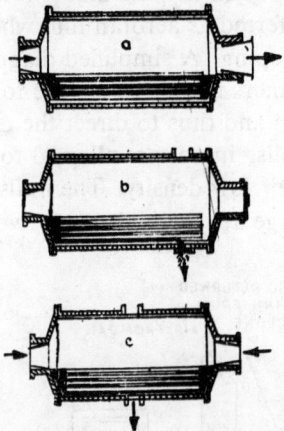

Fig. 6.9. Rod mills:
(a) overflow discharge, (b) end peripheral discharge,
and (c) central peripheral discharge.

12 to 20 mm. The principal field of rod mills is the preparation of coarse product (5 to 0.4 mm), for which a rod mill is usually superior to ball mill. Rod mills can be used for preparation of feed for gravity concentration, magnetic concentration, and to produce suitable feed for the fine grinding stage.

c) *Pebble Mills/Autogenous Grinding Mills*

These are cylindrical shells provided with porcelain or rubber linings, and filled with porcelain or suitable flint pebbles. When the pebbles of the same ore are used as grinding media, the mill is known as autogenous grinding mill.

Pebble mills or autogenous mills are mainly used for secondary grinding, where low media costs and linear wear are of prime consideration, or iron contamination with charge is to be avoided, e.g. grinding of ceramics and gold ores. These mills are similar to grate/diaphragm discharge ball mills in nearly all aspects of design. The autogenous grinding possesses many advantages, such as gentle liberation action, less power consumption, less wear and elimination of preliminary crushing. These mills are mainly of two types, i.e. (i) tube mill, and (ii) cascade or aerofall mills.

i) *Tube mill:* These mills are characterised by their more length and less diameter (narrower than ball mills). These mills are made narrow to prevent breakage of pebbles and are made longer to compensate for the reduced diameter of the mill and lower density of the pebbles. These are usually run at speeds between 75 and 85 per cent of critical speed. These mills are mainly used in the ceramic industry and grinding of gold ores.

ii) *Cascade and aerofall mills:* These mills are large in diameter and less in width (with D/L ratio of more than three) and are comparatively a recent addition to tumbling mills. This is quite different in shape from other tumbling mills. The mill is referred as aerofall mill when used for dry grinding in conjuction with air sweeping. A simplified diagram of a cascade mill is shown in Fig. 6.10. The liners used are concave to give maximum diameter at the centre of the drum and thus to direct the crop load away from the vertical sides. In these mills, lumps are allowed to fall through a large distance to make-up for their low density. The mills may be operated dry or wet. The feed and discharge are made through two support trunnions.

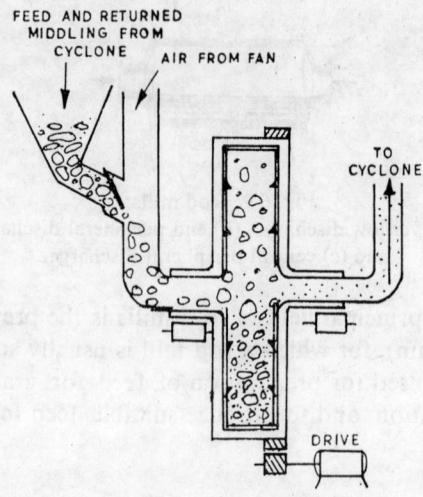

Fig. 6.10. Aerofall mill with air sweeping.

The reduction ratio obtainable in these mills is very high. This requires maintenance of large proportion of heavy particles (about 25 cm across) in the load to act as a grinding medium. The product is usually less than 1 mm. However, every material will not respond to this type of milling. Usually, the brittle materials with a granular texture can be ground economically and efficiently. These mills possess the advantages, such as, economy of space, cheap to build and operate, and a good control over the size distribution of the product. The mills are made with diameters as large as 14 m, running at a peripheral speed of 3–8 m/sec and require a power up to 5,000 KW. Under favourable circumstances, these mills can be used as very large mills for autogenous grinding in conjuction with hydrocyclones (in wet grinding) or pneumatic cyclones (dry grinding). A successful application of these mills is reported in grinding of asbestos, iron ore, and gold ore. These mills are in operation for the last 20 years.

This type of mill can also be made to give peripheral discharge. The openings are guarded by a screen.

6.5 Special Methods of Grinding

Many grinding devices have been developed for fine grinding of specific materials and grinding in submicron size range. The important ones are discussed below:

6.5.1 JET PULVERISER OR MICRONIZER

In this, the feed of -3 mm material is carried in a stream of air or steam at a pressure of about 8 kg/cm^2 through suitable circular expanding chambers. The material is streamed out from the chambers tangentially. This results in extremely fine grinding, partly due to mutual rubbing between the solid particles, and partly by contact with the chamber walls and by pressure release. The product is discharged from the centre at sizes of 1–2 microns. The capacity of the mill is quite good but wear is heavy due to which this method is limited to specialised work.

6.5.2 FLUID ENERGY MILLS

In these mills, high velocity jets of air or steam are used in which fine solids are suspended. Two or more jets are made to collide in a high pressure chamber. The particles in collision break by impact fracturing. A more sophisticated system uses the jets only to induce violent turbulence. The product is circulated through a cyclone to remove undersize. This method can successfully be employed for fine grinding of weak materials. The feed is typically of -150 micron size and the product is 1–2 micron in size. The energy consumption is quite high. The same principle has also been applied to coarser materials, where the solids are fed into a magazine with compressed air or steam at about 30 kg/cm^2 pressure. This pressure is released through a special fast-release valve, as a result the solids emerge out explosively at high speed to collide either with impact plates or with similar particles released simultaneously in opposite direction from another gun. The fracture caused may be partly due to the sudden expansion of gas within the pores of the particles, and partly due to impact fracturing as the particles collide.

6.5.3 ELECTRICAL METHODS

These include induction shock heating and the production of shock waves by the discharge of condensers within a fluid, in which solids are suspended. The solids are fractured as the waves pass through them, identical to impact fracturing in a hammer mill. Many devices of this kind have been developed. However, they cannot be adopted for commercial work due to high operating costs.

6.5.4 CENTRIFUGAL CRUSHERS (GRINDERS)

A centrifugal crusher is shown in Fig. 6.11. A rotor is driven by an electric motor with a vertical spindle. The feed is introduced at the centre, which is accelerated along radial blades and gets ground on targets made of some hard material, built into the casing. The shaft is enclosed in a stanch joint and the rotor turns in vacuum. The rotor has only three blades of 15–30 cm diameter. Its maximum speed of 20,000 rpm corresponds to an impact velo-

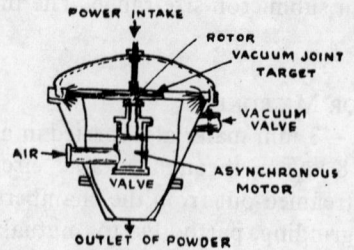

Fig. 6.11. Principle of centrifugal crusher (grinder).

city on targets of 200–400 m/sec depending on the diameter of rotor blade. This action produces fine powders. The energy consumed increases and the output decreases with increasing fineness of product.

In order to have efficient grinding, the grains must be separated from each other when they reach the targets, and hence a thin layer of material should be maintained in the rotor. Though it appears that the output may be limited by maintaining a layer of such a small quantity of material, but this is not the case. For example, a small 30 cm rotor having six blades of 5 mm height and having a radial velocity on the blades of 200 m/sec has an output of about 7 tonnes/hr.

Mechanism of Working

The centrifugal force produces a radial movement, whose acceleration increases with the distance from the centre. The combination of this relative displacement outwards with the speed at which the rotatory movement of the rotor that is driving it, determines the exact path of a grain through space. If the friction on the blade is negligible, the radial movement of a particle can be given by the following equation:

$$\frac{d^2r}{dt^2} = \omega r_0 \tag{6.5}$$

where r = distance of the grain from the axis at any time t, and
ω = angular velocity of the rotor, taken to be constant.

The general form of the above equation on integration will be

$$r = K_1 e^{\omega t} + K_2 e^{-\omega t} \tag{6.6}$$

where K_1 and K_2 are two arbitrary constants.
When

$$t = 0, r = r_0, \frac{dr_0}{dt} = \omega r_0$$

These initial terms represent the fact that at the outset ($t = 0$) the radial velocity of particles along the blade is equal to its speed (ωr_0) as carried by the rotor. Therefore, the particular integral governing the movement is

$$r = r_0 e^{\omega t}$$

The equality of carrying speed and radial velocity is maintained throughout the movement. Therefore, the actual path of the particle is a logarithmic spiral with its asymptotic point at the centre and intersecting the radii at a constant angle of 45°. This type of curve will move away from the axis at such a high speed that the particles complete only a fraction of·a turn between its injection at the base of the blade and its leaving the end.

During the movement, a particle is subjected to two perpendicular accelerations as follows:

i) Radial acceleration $v_r = \omega^2 r$.

ii) Acceleration comprising normal Coriolis (double the centrifugal force) at the blade, which is equal to double the vectorial product of the angular-velocity by the radial velocity, i.e.

$$v_0 = 2 \, \omega \, \Lambda \, V_r \qquad (6.7)$$

In a simple case when friction is negligible,

$$V_r = \omega r \quad \text{and}$$

thus

$$v_0 = 2 \, \omega^2 r \qquad (6.8)$$

In practice, the coefficient of friction of the particle on the blade is approximately 20 per cent, which decreases the radial velocity. The angle at which the logarithmic spiral intersects, the vector radii increases from 45° to 53°. When the particle reaches the extremity of the blade, it leaves the rotor and its acceleration becomes zero, but its movement continues at a constant velocity along a straight tangent out of the spiral, at the point where the latter reaches the edge of the rotor as shown in Fig. 6.12. Thus, the true path followed by each particle in the space is the logarithmic spiral extended by a straight line.

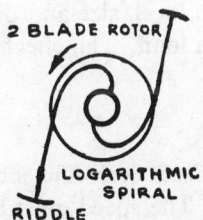

Fig. 6.12. Trajectory of·grains through space.

The total energy supplied to the grinder to maintain the rotation is divided in the friction loss (consumed in heating of particle and rotor) in the rotor and kinetic energy (heating of the target and the heat remaining in the material) for breaking up the particles on the targets. If the initial driving power is considered to be 100, their breakdown may be as follows:

Heating of grain 10

Heating of the rotor 10

Heating of the target 32
Heat remaining in the
material after impacting
the target 48.

When the material passes through the grinder, the layer of particles on the blade is so thin that the particles reach on the target in singles. At the target, they are crushed by a single impact lasting only for 10^{-3}–10^{-8} sec or less. As soon as the foremost part of the particle is halted abruptly by the hard surface of the target, inertia develops great stresses inside the particle which quickly exceeds its mechanical resistance and the front portion of the particle gets crushed into finer particles. The remaining part of the particle becomes quite weak when it approaches the target and thus gets crushed.

The impact velocity should be maintained between certain limits for a given size of material. If the impact velocity is too high, excessive fines will be produced, and if the velocity is too low, the particles rebound back to the targets without creating any new surfaces and their kinetic energy is completely washed.

Applications

The centrifuge crusher/grinder can be used to break up lumps of material of any size to finer sizes directly. However, the best size of feed material to be introduced is between 1 and 10 mm, depending on the requirements. These grinders can thus be installed in direct sequence after the conventional crushers. Compared to ball mills, these grinders give about 30 per cent savings in power.

6.6 Factors Affecting the Operation of Grinding Mills

The various factors affecting the grinding are speed of mill, load of grinding media and material, pulp level, size and distribution of grinding media, solid–liquid ratio, and return load. The effect of these factors is discussed below:

6.6.1 Speed of Mill

The speed of the mills affects both efficiency of action of grinding media as well as grinding capacity. The speed must be as high as possible without centrifuging the charge. During the rising of grinding media, there is very little grinding. As the speed of the mill is increased, work input increases at first in proportion to the speed and then slippage increases with a slow increase of work input compared to speed. The work input increases until a critical speed [according to Eq. (6.4)] is reached, beyond which the power input decreases rapidly to the vanishing point. Under this condition, the solids are centrifuged on the shell of the mill and no work is done by the mill. Whereas at low speeds, the grinding media will roll and slide over one another in a mass near the bottom of the mill, giving an action called *cascading action*. At intermediate speeds, the grinding media will be carried up to a certain extent and then fall on the material in the bottom of the mill, and this action is called *cataracting action.*

The speeds resulting into cataracting action accompanied by free fall of grinding media are favourable for impact grinding, giving higher capacity and efficiency. On the other hand, slower speeds result into cascading action giving attrition grinding, resulting into lower capacity and efficiency.

In case of ball mills, the actual operating speeds range from 55 to 85 per cent of critical speed and within these limits, capacity and power consumption both increase with speed. Higher speeds are used for high capacity of fine product, whereas low speeds may be better for most economical power and steel consumption for moderate capacities. The pebble mills are operated at much higher speeds than ball mills, sometimes even several times of critical speed (super-critical speeds). Rod mills are worked at somewhat lower speeds, i.e. about 50-55 per cent of critical speed.

6.6.2 LOAD OF GRINDING MEDIA AND MATERIAL

Increasing load of grinding media and material will require more power until the maximum value is reached and then power requirement decreases with increasing load as the centre of gravity of the load approaches the axis of rotation. In case of ball mill, the mill should be slightly more than half-full with material and balls. However, the volume contributed by water and material will be comparatively very small, since they occupy the interstices between the balls or rods.

6.6.3 PULP LEVEL

In wet grinding, the pulp level or quantity of material being ground is an important factor in the operation of the mill, since higher pulp level or quantity of material (density being high) requires more power. The lower pulp levels result in greater freedom of movement of grinding media with consequent more effective grinding. In a simple overflow mill, the grinding media loose their kinetic energy when falling into the dense media, as a result grinding is less effective. Low levels of pulp and decreased time in the mill reduce the chances of overgrinding.

6.6.4 SIZE DISTRIBUTION OF GRINDING MEDIA

The size distribution of grinding media should be near to a close packing grading, but in practical replacement, grinding media are fed in at a single size (in case of balls and rods) or nearing the same size (pebbles and rock) and allowed to degrade to extinction. In case of ball mills, a proper ball rationing is important for grinding control. The ball rationing may be referred to the ratio between balls of various sizes, proportioned to the feed and discharge products. This proportion can be maintained by two methods, i.e. (i) by ascertaining the wear rate by observation, and (ii) by trial and error method combined with sizing analysis. The control is exercised by charging a suitable amount of new balls of various sizes into the mill daily in appropriate proportion. However, no general rule can be given as the ores and mills vary widely.

Davis has made a study of the size distribution of balls under equilibrium. If the wear rate varies proportional to the weight of the ball, then

$$W = \frac{d_a^3 - d_b^3}{d_f - d_r}$$ (6.9)

where, W = per cent weight of balls in size range d_a–d_b, and d_f and d_r are the sizes of balls fed to and rejected from the mill. Though this relationship is not very accurate, it provides a good working guide for replacement of the crop load. When an open-circuit rod mill is used between the secondary crushers and the ball mills, the ball ratio is of less importance, since the discharge of rod mill is diverted at about 1 mm size and the size disparity between particles is a minor factor.

The grinding media balls, pebbles, rods, etc., should be of a size proportionate to the feed size, otherwise nipping of particles will not occur. If the size of grinding media is smaller, the particles will only become rounded and superficially abraded, but not ground.

6.6.5 SOLID-LIQUID RATIO (PULP DILUTION)

The pulp dilution also influences the grinding efficiency and power input. The energy input is maximum for a certain critical range of solid–liquid ratio of the pulp, which is usually obtained in a pulp consisting of 60–75 per cent solids. However, this range depends on the chemical composition, particle size, and specific gravity of ore.

Finely ground dry ore in the grinding mill acts as a fluid and possesses good transporting power. With increase in moisture, first a stiff mud is produced beyond about 8 per cent and up to about 15 per cent moisture. Beyond 15 per cent moisture, the fluidity again begins to return and from 20 per cent onward, efficient movement begins to appear. After about 40 per cent water, the pulp becomes watery and chances of coating the grinding media get reduced. However, this general character of pulp may be modified by average particle size and density of ore.

For an efficient use of energy in grinding, the ore particles should form a coating on liners and grinding media. At the same time, the pulp should be sufficiently fluid to flow steadily through the mill. With too thin a pulp, the solids tend to settle and centrifuge outward, and the coatings on the grinding media becomes patchy to non-existent, so the grinding force is washed as the grinding media hit the grinding media. If water used is too little, the charge becomes sluggish and pasty. In this case the kinetic energy will be wasted in overcoming viscosity instead of being available for grinding and the ore will remain sandwiched between two grinding bodies instead of giving a way for a fresh material. In order to get good grinding and high capacity, it is essential to obtain an even distribution of pulp through the crop, and a clinging layer of particles on the metal surfaces everywhere in the mill.

In general, for continuous ore treatment, an optimum solid–liquid ratio should be established, which depends on the coarseness of grind and density of material. The pulp should give enough frictional grip to the crop load, to give the desired lift during rotation and proper coating on the grinding media. Within the limits imposed by the above considerations, thicker pulp results in less wear of steel, greater lifting friction and more throughput.

A rod mill usually operates in circuit and grinds at a size coarse enough. The grinding action is thus quite different in rod mill than in ball mill. Rods remain apart by relatively large pieces of ore, and do not get coated, since the ore is not pulped but is coarse enough to permit rapid settlement of solids till an appropriate solid–liquid ratio is reached. The supernatant water transports the ore through the mill. Therefore, the usual solid–liquid ratio employed in ball mills ranges from 70 : 30 to 80 : 20, whereas the figure is 30 : 70 as solid–liquid ratio, at discharge of rod mills. However, if the rod mill is stopped, the apparent density may be much higher.

6.6.6 RETURN LOAD OR CIRCULATING LOAD

Any attempt to reduce all the feed to finishing size in one operation would result in wasteful overgrinding. This difficulty can be overcome by building up the circulating load in closed circuit. In order to have minimum overgrinding, the ore should move rapidly through the mill (passage or dwelling time may be as short as 1 min). More effective grinding in terms of output and least overgrinding, can be obtained by faster rate of ore passing through the mill and higher capacity of classifier. However, there is a limit to the practical application of this concept, as transport of heavy circulating load costs money. The practical limit is usually in between 5 : 1 to 6 : 1, as a ratio between return (circulating) sand and the new feed in case of primary grinding. The ratio is lower in secondary grinding due to the limited sorting speed of the classifier. Increased circulating load results in the following effects:

i) Mean size of entering feed becomes lower.
ii) Amount of nearly finished material is increased in circulating load.
iii) The retention time and overgrinding of finished material are reduced.
iv) Range of ball sizes is reduced.
v) Interstitial loading of the ball charge is improved.
vi) Close adjustment of solid–liquid ratio is possible.

6.6.7 PRESENCE OF FINE MATERIAL

The presence of excessive fine material such as primary slime, clay, overground friable sulphide or gangue in the mill charge may create problems, since the surfaces of grinding media become coated with a film of fine material, which acts as a lubricant resulting into slip and reduction in impact grinding.

6.7 Control of Wet Grinding Mill

The charge of a grinding mill consists of grinding media (adjusted to weight, composition, and proportion of mill volume occupied), ore in ratio to the charge (limited by the size of new feed and circulating load), and water in ratio of the ore (limited by giving a good fluidity and coating to metal). The various controls include prevention of undue amount of tramp oversize entering the mill, grinding media ratio, solid–liquid ratio, dwelling time, speed of mill, etc.

In small mill the control to prevent the entry of undue oversize can be exercised visually, but methodical cutting of head sample is better. Cutting of head sample also yields a sample for assay of entering feed.

The grinding media ratio is controlled to obtain the right proportion of abrasion and shatter which is judged by the finished product. If other conditions remain constant, increase in percentage of heavy grinding media results in increased impact grinding and reduced abrasive grinding. The grinding media size, and proportion of the sizes depend on the size of finished product. Fine size of end-product requires more abrasive grinding. Bond has developed a formula which relates the size of grinding media with work index, which can be given as follows:

$$B = \left[\frac{F}{K} \right]^{1/2} \left[\frac{SW_i}{C_s \sqrt{D}} \right]^{1/4} \tag{6.10}$$

where,

B = diameter of grinding media (ball, rod, pebble) in inches,
F = size in microns passing by 80 per cent of new feed,
K = proportionality constant depending on media and circuits,
S = specific gravity of ore,
W_i = work index at feed size F,
C_s = percentage of critical speed of mill,
and D = diameter of mill inside liners.

The above formula can be approximately modified for dimensions in cm (for B and D in cm) as follows:

$$B = 3.9 \left[\frac{F}{K} \right]^{1/2} \left[\frac{SW_i}{C_s \sqrt{D}} \right]^{1/4} \tag{6.11}$$

The proportion of ball-size to ore-size is best worked out by trial and error, since there is a great difference in grindability of various ores and no generalisation to the ratio can be made. The end-product should be sampled and sizes of solids should be determined by screen analysis. If the analysis indicates more impact, the daily addition of new balls to make up for wear should include a larger proportion of bigger balls.

Altering of mill speed or liner contour is generally not convenient except at the time of replacing liners. However, it is possible to vary the dwelling time of ore in the mill, solid–liquid ratio, and the kind of blow struck. Dwelling time is a function of speed of feed (feed displaces a similar volume at the discharge end) and thus it can be altered by varying the feed rate. Solid–liquid ratio can be changed by altering the setting of water cock in the feed launder. The blow struck depends on the size or weight of grinding media.

The change in one factor requires adjustment of the other factors affected and thus it becomes to some extent by trial and error. Thus it will need some patience and skill to rebalance the crop load after each change. Since overall grinding effect is presented in the form of new surfaces produced, which can be determined by sizing analysis of the sample, the control of grinding efficiency is carried out by laboratory sizing of the products. For this purpose, the samples are taken at regular intervals and sent for testing. Similarly, solid–liquid ratio can be maintained at the required setting by determining the pulp density.

If all the mechanical conditions are good and the crop load is constituted correctly, the maximum production will correspond to maximum intake of power. This is markedly observed in the case of primary grinding low-discharge mills. Changes in the grindability of the ore can be compensated by varying the feed rate.

6.8 Energy Requirements in Grinding

The power consumption in a grinding mill varies slightly between its fully loaded and underfed conditions, good and bad ratio of grinding media to ore and water and good or bad operation. In fact, most of the power required in grinding is used in overcoming mechanical friction and only a fraction is used in size reduction. Theoretically, the energy used in size reduction is proportional to the new surface produced, since there is no change in material except size and creation of new surface. The values vary for different ores/materials depending on elastic constant and its relation with ultimate strength and manner or rate of application of grinding force (this differs for different machines and operating conditions).

About 10 per cent of the power applied is lost in motor, and 10–15 per cent is lost in gears and mechanical friction of the mill. The balance of about 75 per cent is available as useful power, which is converted into kinetic energy in the tumbling load. If the crop load is not suitably constituted, a part of this kinetic energy will be wasted by conversion to avoidable heat, sound, and wear of metal. When a mill is at rest, much extra power will be required to obtain the system in motion. The crop load is displaced in the direction of the rising side and the kinetic energy is built up in the components. Soon after the normal speed is reached, the requirement of the power falls to a constant level. The load is gripped by the liners and carried upward on the rising side and toward the limit of rise, the upper part of the load breaks away and cascades or cataracts to the bottom of the mill. The intake of useful power is balanced by the displacement of load, which is continuously converted to kinetic energy and finally into heat, sound, and newly developed surfaces (of ore, balls, and liners). The useful power will be nil at zero speed and also at critical speed.

Davis has proposed the following equation for best operating speed (N)

in terms of internal stability of crop-load and frictional grip from the shell:

$$N = \frac{0.8158}{\sqrt{r} \cdot \sqrt{1 + K^2}} \quad \text{rev/min.} \tag{6.12}$$

where $K = r_c/r$ (r_c is the inner radius of charge, r being the mill's radius). This equation can be modified to include the influence of slippage.

The rate of converting power into kinetic energy is quite steady in case of ball mills, but varies abruptly in a rod mill due to hold-ups, entanglements, and momentary seizures of the rods during running.

In general, the power input (P) can be related to diameter (D) of mill as following:

$$P = KD^{2.5} \tag{6.13}$$

Hukki has proposed the following equation for power index (P_i) for a base mill (a horizontal cylinder with 1 m in inside diameter and in length)

$$P_i = \frac{P_n}{D^{2.5} L} \quad \text{kW} \tag{6.14}$$

where P_n = net power drawn, D = internal diameter in metres and L = length in metres.

If the kinetic energy is applied correctly, ore will be ground with best efficiency. However, whether kinetic energy is applied correctly or not, the power will continue to enter the mill as long as the crop load is being held dynamically out of balance. In any case, all the kinetic energy is finally dissipated as heat which warms the passing pulp.

In order to get maximum amount of grinding, the kinetic energy generated by the driving power should be maintained at its peak value, which can be obtained by means of correct balance between the following four main factors:

a) Speed of mill as percentage of critical speed.
b) Liner grip, particularly of horizontal liners.
c) Constitution of charge, i.e. grinding media, ore and water.
d) Volume of charge under operating conditions.

When the mill speed is increased without any other change, the crop load is reduced and vice versa. With constant volume, the net power is highest with the interstices between grinding media being full of ore and lowest when the interstices are full of water (reduces the total crop weight due to much less density of water). Power used is higher with plenty of sharp sand in the charge than with slimy sand only. The extra friction of sharp sand helps the liners to grip the load more firmly and thus raises the power used. The power used is higher in case of low-discharge mill as the pulp rises centrifugally on the rising side, but can escape near the periphery of the diaphragm/grate. Whereas, in the case of high-discharge mill the pulp can only overflow from the trunnion, causing the larger volume of pulp to be retained on the falling side of the mill, which reduces the useful power.

At constant mill speed, the larger diameter gives higher centrifugal force and stronger liner grip. This results in increase of power used and grinding capacity, but with increased wear of liners, since the power used is more with greater lifting grip of the liners. With decrease in wave contours of the liners with wear, the grip is reduced, which neutralises the lifting grip of the liners. The useful power is higher for ores of higher density. If the water in the charge is less, it may lead to the formation of paste which lubricates the liners and reduces the power consumption due to increased slip.

In practice, the power required to crush large rocks down to gravel size (8–10 mm) is far less than that required to grind the same amount of gravel size to fine sand. The finer the mesh size required of the finished product, more will be the cost of grinding.

6.9 Calculation of Charge Volume of a Grinding Mill

The charge volume (level) of a grinding mill is generally expressed as a percentage of volume within liners that is filled with grinding media. When the mill is at rest, the charge volume can be obtained by measuring the mill diameter inside the liners (D) and the distance from the top of the mill inside the liners to the top of the charge (H). The percentage loading or charge volume can then be calculated from the following equation:

$$\text{Percentage loading} = 113 - 126\ \frac{H}{D} \tag{6.15}$$

Measurements should be made at several points down the mill length to obtain the average charge volume. The weight of the grinding media in the mill can be calculated from the bulk density of the media.

When the charge occupies 50 per cent of the mill volume, maximum power draw is obtained. However, with the increase of charge volume from 45 to 50 per cent, the power draw increases to only 1 per cent. Therefore, in most cases, the increase in consumption of grinding media would be much greater than the increase in production. As a result, the mills are usually run with the charge levels of less than 45 per cent.

In rod mills, the charge is expanded by feed particles which separate the rods. If the mill is shut down just after the feed is stopped, the charge level will be greater than if the mill had been run for some time before shutdown. Due to this reason, the rod mills are usually operated with 32–40 per cent charge by volume. In operation this becomes a 40–50 per cent charge, since the bulk density of the charge is considerably lower than that of stacked rods.

6.10 Calculation of Motor Size

The power required to grind a material from a given feed-size to a given product-size can be calculated from the following equation:

$$W = 11.15\ W_i \left[\frac{1}{\sqrt{P}} - \frac{1}{\sqrt{F}} \right] \tag{6.16}$$

where, W = power consumption in KWH/metric tonne for wet grinding, closed-circuit operation with P more than 70 microns. Multiplying W by the new feed rate (in metric tonnes/hr) will give the power requirement at the mill pinion shaft including bearing and gear losses. Motor losses, and other drive component losses, such as reducer and clutches are not considered. W_i is work index, which is the power in KWH/metric tonne required to reduce a material from theoretically infinite size to 80 per cent passing 100 microns. Table 6.2 represents the average work indices for various materials.

P = size in microns of the screen opening, through which 80 per cent of the circuit product will pass.

F = size in microns of the screen opening through which 80 per cent of the feed will pass.

Table 6.2. Average work indices for different materials

Material	Work index W_i	Specific gravity S	Material	Work index W_i	Specific gravity S
Bauxite	105	2.4	Lead ore	124	3 4
Cement clinker	144	3.1	Lead-zinc ore	125	3.4
Coal	124	1.6	Lignite (coal)	147	1.4
Coke	231	1.5	Lime stone	113	2.7
Copper ore	143	3.0	Molybdenum ore	143	6.0
Fluorspar	110	3.0	Nickel ore	132	3.3
Gold ore	163	2.9	Phosphate rock	116	2.7
Hematite	142	3.8	Tin ore	123	3.9
Magnetite	126	3.9	Titanium ore	131	4.2
Taconite	167	3.5	Uranium ore	187	2.7
			Zinc ore	136	3.7

Based on data from Ref. 8, 16-18.

In making use of Eq. (6.16), the following points should be considered:

a) The power consumption per metric tonne, W, will only be correct for the specified size reduction when wet grinding in closed circuit. If the grinding circuit is changed, then the power consumption will also change. Letting G = gross KWH/metric tonne, the power consumption for other circuit becomes.

  i) For wet grinding, open circuit, product top size not limited, $G = W$ (applicable to most rod mill applications).

  ii) For wet grinding, open circuit, product top size limited, $G = W$ to 1.25 W.

  iii) For dry grinding, closed circuit, $G = 1.30\ W$.

  iv) For dry grinding, open circuit, product top size not limited, $G = 1.30\ W$.

  v) For dry grinding, open circuit, product top size limited, $G = 1.30\ W$ to 2.0 W.

vi) If P is less than 80 per cent passing 70 microns, $G = \dfrac{W(P + 10.3)}{1.145\,P}$.

b) The values of P and F should be based on materials having a natural particle size distribution.

c) The work index, W_i, should be obtained from plant data or tests, where the feed and product size distributions are as close as possible to those under consideration.

The most reliable work index values are those obtained from long-term operating data. However, if this is not available, standard grindability tests can be carried out to provide approximate values. Further, the value of work index varies considerably for similar materials and also across one ore body or deposit. The grindability results obtained for rod and ball mills should only be applied to the respective methods of grinding.

EXAMPLE FOR CALCULATION OF MOTOR SIZE

A dry grinding ball mill in closed circuit is to be fed 100 metric tonnes/ hr of a material with a work index of 15 and a size distribution of 80 per cent passing 0.8 cm. The feed was prepared in a cone crusher. The required product size is 80 per cent passing 60 microns. In this case the power required can be determined as follows:

The power consumption W can be calculated from the formula (6.16), i.e.

$$W = 11.15 \times 15 \left(\frac{1}{\sqrt{60}} - \frac{1}{\sqrt{8,000}} \right)$$

or $\qquad W = 21.61 - 1.87 = 19.74 \quad$ KWH/metric tonne.

Since W is for wet grinding in closed circuit with P greater than 70 microns, the gross power consumption G can be determined as follows:

$$G = 19.74 \times 1.30 \, \frac{(60 + 10.3)}{1.145 \times 60} = 26.26 \quad \text{KWH/metric tonne.}$$

Now multiplying the new feed rate by G, i.e.

$$26.26 \times 100 = 2,626 \quad \text{KW}$$

Therefore, the power required is 2,626 KW.

6.11 Calculation of Mill Size Matching to Required Power

In Sec. (6.10) the calculated power is the power that must be applied at the mill drive in order to grind the tonnage of feed from one size distribution to a finer size distribution. The size of the mill required to draw this power can be calculated as the following.

Figure 6.13 represents a mill section in operation. The power input required to maintain this condition is theoretically

$$\text{KW} = \frac{M \times C \times \sin \alpha \, (2\pi N)}{12,400} \qquad (6.17)$$

where M = weight of charge in metric tonnes.

$\qquad C$ = distance of centre of gravity of charge from centre of mill in metres.

$\qquad \alpha$ = dynamic angle of repose of the charge.

$\qquad N$ = mill speed in rpm.

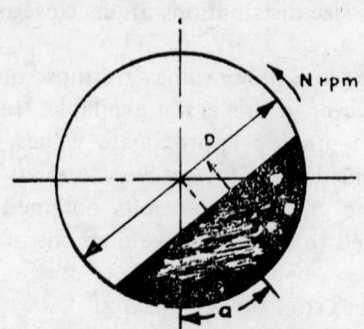

Fig. 6.13. A mill section in operation.

If the data from a similar installation is available, the value of the angle α can be determined and the power demands of mills with various diameters of the same speed calculated. The value varies with the type of discharge, per cent of critical speed, and grinding condition.

Thus, direct comparison can only be made between mills with similar types of discharge. If various types of discharge are to be used, the following factors should be applied for mills of the same size and speed:

a) For dry diaphragm 1.0

b) For wet diaphragm 0.9

c) For wet overflow   0.8.

In order to use Eq. (6.17), considerable data are required on existing installations. Therefore, this approach has been simplified as follows:

The basic conditions determining the power drawn by a mill include diameter, length, per cent loading, speed, and type of mill. These conditions have been built into factors which are given in Table 6.3. The approximate KW of a mill at the pinion shaft can be calculated from the following equation:

$$KW = 2.5 \times A \times B \times C \times L \qquad (6.18)$$

where, A = factor for diameter inside the shell liners,

$\qquad B$ = factor for mill type and charge volume (per cent loading).

$\qquad C$ = factor for mill speed expressed as a percentage of critical speed of mill.

$\qquad L$ = Length in metres of grinding chamber measured between head liners at the junction of the shell and head liners.

The factors shown in Table 6.3 are given for steel grinding media. For other types of media or pebbles, the factor B should be adjusted by the ratio of actual charge density in g/cm³ to 1.6 or charge density/1.6.

Table 6.3. Factors for power calculation of rod and ball mills[38]

Diameter		Speed		Per cent loading	Mill type and loading — Factor B					
					Ball Mills			Rod Mills		
Dia. inside shell liners in m	Factor A	Per cent critical speed	Factor C		Dry diaphragm	Wet diaphragm	Wet overflowing	Dry peripheral	Wet peripheral	Wet overflow
2.40	32.0	60	0.1340	20	4.30	3.87	3.44	4.73	4.25	3.78
2.70	43.4	62	0.1400	22	4.57	4.12	3.66	5.04	4.54	4.03
3.00	56.2	64	0.1460	24	4.80	4.32	3.84	5.27	4.75	4.22
3.30	71.6	66	0.1521	26	5.07	4.57	4.06	5.58	5.02	4.47
3.60	88.6	68	0.1583	28	5.31	4.77	4.24	5.82	5.24	4.66
3.90	109.0	70	0.1657	30	5.53	4.97	4.42	6.08	5.47	4.86
4.20	131.0	72	0.1724	32	5.71	5.14	4.57	6.28	5.65	5.02
4.50	156.0	74	0.1798	34	5.90	5.32	4.72	6.48	5.83	5.19
4.80	182.5	76	0.1878	36	6.05	5.45	4.84	6.67	6.00	5.33
5.10	213.0	78	0.1958	38	6.16	5.55	4.93	6.78	6.10	5.42
5.40	245.0	80	0.2040	40	6.27	5.65	5.02	6.90	6.21	5.52
5.70	281.0	82	0.2124	42	6.34	5.70	5.08	—	—	—
6.00	320	84	0.2208	44	6.41	5.77	5.13	—	—	—
		86	0.2294	46	6.46	5.82	5.17	—	—	—
				48	6.49	5.84	5.19	—	—	—
				50	6.50	5.85	5.20	—	—	—

6.12 Selection of Optimum Size of Grinding Media

In any grinding mill (rod, ball or pebble mill), sizing of the grinding media is influenced by two major opposing factors:

a) With increase in size of grinding media, the pressure between the surfaces increases, which causes the breaking of larger particles.

b) With decrease in size of grinding media, the surface area available for grinding small particles increases, which results in an increased grinding capacity.

However, there are several other secondary factors which influence the selection of grinding media size. Some of the important ones are:

i) Harder rock requires larger grinding media to break the particle of given size.

ii) The mills of larger diameter require smaller grinding media to break a particle of given size.

iii) At faster mill speeds, smaller grinding media are required to break up the particle of given size.

All these factors are taken into consideration in the following equation, which can be used to determine the maximum size of grinding media required in a given mill to grind a given material:

$$X = \sqrt{\frac{FW_i}{KC_s}} \sqrt{\frac{S}{\sqrt{D}}} \qquad (6.19)$$

where, X = diameter of top size of grinding media in cm,

$\quad\quad F$ = size in microns of the screen opening through which 80 per cent of the new feed passes,

$\quad W_i$ = work index,

$\quad C_s$ = per cent of critical speed,

$\quad\quad S$ = specific gravity of feed,

$\quad\quad D$ = diameter of mill inside liners in cm,

$\quad\quad K$ = constant (for ball mills $K = 13$, and for rod mills $K = 20$).

Equation (6.19) gives the maximum size of grinding media required for a given application. In most of the cases, this will also be the size of grinding media added to the mill to make up for wear during grinding operation. However, in order to have the most efficient grinding, the initial charge of grinding media should have the size distribution in a complete range from maximum size down to very small size. This distribution of sizes is referred to as seasoned or equilibrium charge. If only maximum size of grinding media is used, this equilibrium charge can be obtained only after a very long period of operation. In order to have a seasoned charge as soon as the mill starts, the initial charge should be graded with respect to diameter. A list of graded charges for various top sizes of grinding media is given in Tables 6.4 and 6.5 for rods and balls, respectively.

6.12.1 OPTIMUM SIZE OF RODS

The rods used as grinding media are hot rolled high carbon or alloy steel.

Table 6.4. Rod size distribution in the charge

Rod dia. in cm	Rod size distribution for initial charge (per cent wt.) Rod top size in cm (X)				
	12.50	11.25	10.00	8.75	7.50
12.50	19	—	—	—	—
11.25	17	21	—	—	—
10.00	16	19	24	—	—
8.75	15	18	23	30	—
7.50	13	17	20	26	38
6.25	10	15	18	24	33
5.00	10	10	15	20	29

Based on data from Ref. 38.

Table 6.5. Ball size distribution in the charge

Ball diameter in cm	Ball size distribution for initial charge (per cent weight) Ball top size in cm (X)								
	12.50	11.25	10.00	8.75	7.50	6.25	5.00	3.75	2.50
12.50	17.0	—	—	—	—	—	—	—	—
11.25	25.0	16.0	—	—	—	—	—	—	—
10.00	20.0	30.0	20.0	—	—	—	—	—	—
8.75	15.0	21.5	32.0	22.0	—	—	—	—	—
7.50	10.0	14.0	21.0	35.0	26.0	—	—	—	—
6.25	6.4	9.5	12.8	19.8	37.0	32.0	—	—	—
5.00	3.8	5.8	8.9	15.3	23.4	42.2	38.0	—	—
3.75	2.8	2.8	3.7	6.0	10.6	19.7	45.3	56.0	—
2.50	—	0.4	1.6	1.9	3.0	6.1	16.7	44.0	10.0

Based on data from Ref. 38.

The rods should be machined and cut at the ends to obtain straight rods and free from cracks at the ends. The rods should be sufficiently hard to give good wear properties but not so hard as to cause excessive breakage. The sizes used vary from 13 cm down to 4 cm diameter. The length of rod used in a given mill will be 15–23 cm shorter than the distance between new head liners measured at the juncture of the head and shell liners. In larger mills, the rods are charged mechanically. In order to maintain the optimum power draw, the rods should be charged on a regular basis.

When new rods are charged, they will be in line contact and the voids will be about 22 per cent and the new stacked rods will weigh about 6 tonnes/m³. With ore particles, the rod charge becomes expanded and the bulk density of the total charge may become much lower if the proper procedures are not followed. Broken rods also cause increase in void space and decreased efficiency. Table 6.6 represents the data for grinding rods.

Table 6.6. Data for grinding rods

Rod size in cm	Weight/m kg	m/tonne	Surface area (m²/tonne)
3.75	8.95	115.34	13.59
5.00	16.11	64.88	10.19
6.25	25.05	41.52	8.15
7.50	35.79	28.83	6.79
8.75	48.91	21.18	5.82
10.00	63.83	16.22	5.10
12.50	99.92	10.38	4.08

Based on data from Ref. 38.

6.12.2 OPTIMUM BALL SIZE

The balls employed as grinding media are generally of either forged steel or cast Ni-hard or cast Mn-steel. The choice of composition depends on the material to be ground, the ball size required, and the proximity of availability. Balls should be as nearly spherical as possible with no excessive forging or casting ridges. The charging of balls should be made as frequently as possible to maintain optimum power demand. When new balls are charged in a mill, they will have point contacts with one another and the voids will be about 45 per cent. The weight of new balls is approximately 4.5 tonnes/m³. Table 6.7 represents the data for grinding balls.

Table 6.7. Data for grinding balls

Ball diameter in cm	Weight each in kg	Number of balls per tonne	Surface area m²/tonne
1.88	0.029	36,614.9	40.66
2.50	0.067	15,570.8	30.57
3.12	0.132	8,010.7	24.50
3.75	0.228	4,613.6	20.38
5.00	0.540	1,946.4	15.29
6.25	1.054	996.53	12.23
7.50	1.821	576.7	10.19
8.75	2.891	363.2	8.74
10.00	4.316	243.3	7.64
12.50	8.429	124.6	6.12

Based on data from Ref. 38.

6.12.3 OPTIMUM PEBBLE SIZE

If rock, flint, or ceramic pebbles, etc., are to be employed as grinding media, the size of these should be such that a rounded pebble would have a similar weight to the steel ball required for the same feed and mill conditions. This gives the first approximation to the pebble size, and rock pebble grinding should be thoroughly tested on pilot plant scale for any given material. However, there is no need of using graded charge initially in most of the pebble mill operations, since the media consumption is relatively rapid and the equilibrium size distribution is achieved relatively in a short period.

6.13 Consumption of Grinding Media

In wet grinding, the consumption of media results from two major sources, i.e. (a) due to abrasion of the media surface by contact with the material being ground, and (b) due to corrosion of the freshly exposed media surface. The latter source contributes the major proportion of media consumption. In dry grinding, the consumption of media is only from abrasion and is, therefore, much less than in wet grinding. In some cases the consumption of media is reduced to a very low level due to formation of a coating of grinding material in dry grinding. Table 6.8 shows the expected levels of media consumption in ball and rod mills.

Table 6.8. Expected media consumption

Mill	Consumption of media kg/tonne of feed			
	Dry grinding		Wet grinding	
	Range	Average	Range	Average
Rod mills	0.05–0.2	0.15	0.25–1.0	0.45
Primary ball mills	0.05–0.35	0.15	0.25–1.0	0.45
Secondary ball mills	0.05–0.25	0.07	0.15–0.7	0.40

Based on data from Ref. 38.

6.14 Recent Trends in Crushing and Grinding

The processes of crushing and grinding are the most important links within a flowsheet, due to the following facts:

a) Partial liberation of mineral constituents or their overgrinding will result in losses of valuables.

b) Crushing and grinding operations consume 40–65 per cent of the total power required in a concentrator.

c) These two operations account for about 50 per cent of total capital cost and 60 per cent of total operating costs.

d) In processing of low-grade ores, the proper liberation of ore mineral constituents is a decisive factor.

The necessity of improving the art of crushing and grinding is not only important for better grain liberation and minimum overgrinding, but also for reduction in power consumption, capital costs and operation costs.

The requirement of handling higher tonnage of ore, and increasing amounts of low-grade and finely disseminated ores have brought about a wide range of problems to be solved within the existing framework. In installing the new mills, the main problem is usually to increase the capacity of crushing and grinding equipment along with reducing the size of ground particles.

During the sixties, primary autogenous grinding was successfully employed in the iron ore processing; later on in the seventies, application of autogenous and semi-autogenous grinding in non-ferrous ore processing plants became common. However, due to energy crises, a return to the grinding process with steel grinding media has been observed (autogenous grinding consumes more energy).

In the last two decades, the most important features for increasing the unit productivity for both conventional and autogenous mills has been a constant scaling-up of crushing and grinding unit dimensions. But now, the simple scaling-up of equipment dimensions appears to be exhausted. For example, the diameter of a cone crusher cannot be increased to more than 3 m without having problems in the manufacture and operation of the cone crusher. Rod mills have also been made to their maximum dimensions (4.6 m diameter $\times$ 7.3 m length) and further scaling-up will lead to operational problems due to inefficient motion of rods. Similarly, for an efficient and trouble-free operation, the ball mills should not be made in more than 5 m diameter. In case of autogenous mills, a diameter beyond 11–12 m causes the overgrinding of the material.

In recent times, there has been a strong tendency towards increased capacity of mills by optimisation of equipment operation and process pattern. In case of crushers, increased throughput can be achieved by:
a) Providing feed distributors to fill the crusher cavities with ore.
b) Proper profiling of the working chamber.
c) Mechanical and automatic control of the discharge opening.
d) Incorporating the devices for maintenance and mechanisation.
e) Choosing the optimum liner profiles.

Optimisation of crushing and grinding is carried out by rational distribution of different kinds and values of breaking loads applied to the material at different stages in process. In case of conventional crushing and grinding, the general trend is to transfer comminution as far as possible from grinding to crushing circuit (i.e. reduced size of crushed product). To achieve this objective, fine and ultra-fine cone crushers have been developed with closed-circuit operation. A similar objective can be achieved by optimum distribution of load in grinding stages.

In autogenous grinding, the hardest ingredients of the ore may not be broken, which will result into accumulation of initial-size lumps and consequently a reduction in the output will be caused. This problem can be overcome by introducing crushers for handling the circulating loads of autogenous mills, or by directing the discharge of autogenous mill to the ball mills for further grinding.

The output can also be increased by introducing the automatic process control systems in the circuit.

Some new developments have also been made in the field of crushing and grinding. The inertia cone crusher developed in Soviet Union is an important development. In this case, the number of cone oscillations has been reduced to 425–485 and the drive power is 500 KW. The important feature of this cone crusher is that the crushing force is due to rapid rotation of an unbalanced load and not due to an eccentric drive as in the conventional crushers. Planetary mills offer the most promising developments in drum mills for grinding, which are operated with grinding media or used as autogenous mills. These mills intensify the process greatly by using gravity and centrifugal force simultaneouly.

Another new equipment developed for crushing and grinding is the centrifugal crusher which can reduce the size from 1 cm or so to very fine size directly. This is also a very compact unit and needs a small space for equivalent production as for conventional units.

CHAPTER 7

Movement of Solids in Fluids and Classification

The movement of solids in fluids plays an important role in a large number of mineral processing methods, such as thickening, filtration, gravity concentration, and classification. Therefore, a clear understanding about the movement of solids in fluids under various conditions is extremely important in order to evaluate the conditions required for separation/concentration of minerals. Movement of solids in fluids is governed mainly by two conditions, i.e. *free settling* conditions, and *hindered settling conditions*, which are independent of each other. *Free settling* refers to the falling of particles freely in still water or against an opposing upward current, without the interference of other particles. This condition is observed in settling tanks, thickeners, and classifiers. Whereas, *hindered settling* refers to falling of particles of mixed sizes, shapes and densities in a crowded mass. The velocity of these particles is much less than the free falling velocity of the particles, but fast enough to keep the particles in motion. The examples of hindered settling are jig beds, hydrotators, heavy media separation, etc.

7.1 Factors Affecting the Settling of Particles

The rate of falling particles under free settling conditions depend upon the following:

a) *Specific gravity of particles*: If two particles have the same size but different specific gravities, the particles having higher specific gravity will fall faster.

b) *Shape of particles*: Spherical grains will fall faster than the long, narrow grains and the latter will fall faster than the flat grains.

c) *Size of particles*: If the particles have the same specific gravity, larger particles will fall faster.

d) *Density of fluids*: Particles will fall faster in lighter fluids.

e) *Viscosity of fluids*: With increasing viscosity of the fluid, the rate of fall becomes slower, or the resistance to fall increases.

7.2 Laminar and Turbulent Flow

When a particle falls through a fluid, the fluid flows around the particle.

This flow may be either *laminar* (streamline flow round a particle) or *turbulent*. In laminar flow the fluid flows around the particle without forming any eddies or swirls (Fig. 7.1(a)). In turbulent flow the fluid breaks into eddies and swirls, around the particle (Fig. 7.1(b)). With increased rate of flow, eddies become larger and more complex and the flow becomes more turbulent. If the particle is small, the flow is usually viscous or laminar and the resistance to the fall of the particle mainly depends on the viscosity of the fluid. This laminar flow is limited to conditions where Reynolds number, R_e is less than 0.2. If the particle is large, the flow is usually turbulent and accompanied by the formation of eddies and vortices in the fluid. These eddies give large resistance to the fall of particle and the viscosity of the fluid becomes less important and can be neglected in determining the resistance under fully turbulent conditions. Fully turbulent flow is limited to Reynolds number, R_e of more than 800. Between the Reynolds number 0.2 and 800, the flow is in between true laminar and true turbulent.

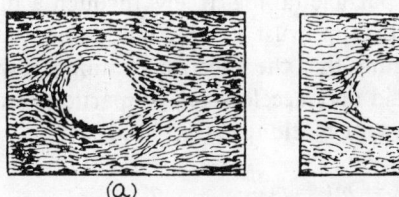

(a.)   (b)

Fig. 7.1. Flow of fluids:
(a) laminar flow, and (b) turbulent flow.

Reynolds number (R_e) was derived by Osborne Reynolds and is a dimensionless ratio given by

$$R_e = vl\,\rho_l/\eta \qquad (7.1)$$

where l is the parameter characteristic of a linear dimension of the particle. For spheres this is diameter of the particle, and thus l can be replaced by d, i.e.

$$R_e = v\,d\,\rho_l/\eta \qquad (7.2)$$

ρ_l = density of medium, for water $\rho_l = 1$ and
η = viscosity of fluid, for water $\eta = 0.01$.

Therefore, the laminar flow around particle is possible when the particles are quite small, whereas turbulent flow is possible only with comparatively larger ones. However, in practice, most of the cases involve the intermediate range of R_e between 0.2 and 800.

7.3 Fluid Resistance and Terminal Velocity

All the fluids (liquid or gas) exert a resistance to the motion of a particle falling through it. The resistance, R is a complex function of the velocity v of the particle, i.e.

$$R = f(v) \tag{7.3}$$

Solution of this function for all possible cases is the great problem of hydrodynamics and aerodynamics. Particles falling through a fluid start with zero velocity, which increases until the resistance offered by the fluid, buoyancy of the fluid and the effects of eddying, balance the downward forces acting on the particle. The velocity reaches a maximum and remains constant thereafter. It is referred as maximum or *terminal velocity*. In mineral dressing, maximum or terminal velocity is of particular importance.

The terminal velocity may be evaluated in terms of physical characteristics of the solid (such as specific gravity, size, and shape), the physical properties of the fluid (viscosity and specific gravity) and the forces in the system. Therefore, terminal velocity for different cases should be evaluated independently. The various cases are discussed below:

7.3.1 SETTLING OF FINE SPHERES (FREE SETTLING UNDER STOKE'S LAW)

The forces acting on a particle falling freely through a fluid are its net weight (its weight minus the upthrust of the fluid) and the resistance, R offered by the fluid. Assuming that the effect of container's walls is negligible, the resistance, R of fluid and acceleration of particle, f can be written as (following the second law of motion; mass × acceleration $= \sum$ forces).

$$(m - m^1) g - R = mf = m \frac{dv}{dt} \tag{7.4}$$

where, $m = $ mass of the particle, $m^1 = $ mass of the fluid displaced by the particle of mass m, $g = $ acceleration due to gravity, v velocity, and $t = $ time.

When the velocity v is low enough to cause viscous or laminar flow, the resistance is due to shear of the liquid, since the particle surface along with water attached to it moves through the stationary fluid. Under these limiting conditions, Stokes has shown that the resistance to the motion of a shpherical body can be given as follows:

$$R_s = 6\pi \, \eta \, r \, v \tag{7.5}$$

Putting the values of $m = 4/3 \, \pi \, \rho_s \, r^3$ and $m^1 = 4/3 \, \pi \, \rho_l \, r^3$ and inserting the value of R_s from Eq. (7.5),

$$\frac{4}{3} \, \pi \, r^3 \, \rho_s \, \frac{dv}{dt} = 4/3 \, \pi \, r^3 \, (\rho_s - \rho_l) \, g - 6 \, \pi \, \eta r v \tag{7.6}$$

in which ρ_s and ρ_l are the specific gravities of the solid and the fluid, respectively.

Dividing Eq. (7.6) by $4/3 \, \pi r^3 \, \rho_s$

$$\frac{dv}{dt} = \frac{\rho_s - \rho_l}{\rho_s} \, g - 9/2 \, \frac{\eta v}{\rho_s \, r^2} \tag{7.7}$$

when $\frac{dv}{dt} = 0$, v becomes maximum. The terminal velocity thus can be obtained by putting $\frac{dv}{dt} = 0$ in (7.7), i.e.

$$v_m = 2/9 \frac{(\rho_s - \rho_l)}{\eta} r^2 g \qquad (7.8)$$

Therefore, the velocity of fall in case of fine spheres is proportional to the square of the particle radius (or diameter). Equation (7.5) is the form in which Stoke's law is usually represented. The quartz particles having diameters 5 to 100 microns follow the Stoke's law. Therefore, thickening and classification of fine flotation pulps are mainly based or Stoke's law, whereas, gravity concentration processes usually dealing with coarser particles (coarser than 100 microns) are not based on Stoke's law.

7.3.2 SETTLING OF COARSE PARTICLES (TERMINAL VELOCITY) UNDER NEWTONIAN RESISTANCE

Settling of coarse particles results into turbulent flow conditions for which Newton's law is applicable. The Newtonian resistance to motion of a coarse sphere in a fluid is given by

$$R_n = \frac{\pi}{2} \cdot \rho_l r^2 v^2 \qquad (7.9)$$

However, it is found that the above Newtonian relationship (Eq. 7.9) is not satisfactory. The facts can be made to fit in the relationship by inserting a coefficient Q known as the *coefficient of resistance* in Eq. (7.9). Equation (7.9) then represents the facts accurately. However, Q cannot be considered as even approximately constant except within narrow limits.

$$R_n = Q \frac{\pi}{2} \rho_l r^2 v^2 \qquad (7.10)$$

For spheres of minerals in water, Q can be taken as about 0.4 if r is larger than 2 mm.

The fundamental equation for turbulent resistance obtained according to second law of motion is the same as Eq. (7.4). Now putting the values of m and m^1 and inserting the value of R_n in Eq. (7.4) for Newtonian resistance,

$$4/3 \ \pi \ r^3 \ \rho_s \frac{dv}{dt} = 4/3 \ \pi \ r^3 \ (\rho_s - \rho_l) \ g - Q \frac{\pi}{2} \rho_l \ r^2 \ v^2 \qquad (7.11)$$

or

$$\frac{dv}{dt} = \frac{\rho_s - \rho_l}{\rho_s} g - \frac{3Q}{8} \frac{\rho_l}{\rho_s} \frac{v^2}{r} \qquad (7.12)$$

when $\frac{dv}{dt} = 0$, v becomes maximum.

The terminal velocity for coarse particles thus can be obtained by putting $\frac{dv}{dt} = 0$, i.e.

$$v_m = \sqrt{\frac{8}{3Q} \, g \, \frac{\rho_s - \rho_l}{\rho_l} \, r} \tag{7.13}$$

The relationship (7.13) is also referred as Rittinger modified equation, where value of Q is about 2.5 under typical ore-dressing condition.

From Eqs. (7.8) and (7.13), it can be noted that in viscous flow, the terminal velocity varies as the square of the particle diameter, whereas in turbulent flow it varies as the square root of the particle diameter.

In commercial operations, Newton's law is most applicable because the particles are relatively coarse in most of the gravity concentration processes. The value of Q in Newton's law may be taken as 0.4 for all practical purposes.

7.3.3 SETTLING OF PARTICLES OF INTERMEDIATE RANGE

Bulk of gravity concentration processes are conducted in the size range in which neither Stokes nor Newton relationship applies. Many attempts have been made to propose a formula taking care of facts. Oseen formula has been widely used which can be expressed as follows:

$$R_0 = 6 \pi \eta \, R \, v_m \left(1 + 3/8 \, \frac{\rho_l v_m}{\eta} \right) \tag{7.14}$$

where R_0 is the Oseen resistance.

Equation (7.14) implies that the resistance experienced by the sphere in the intermediate size range is the sum of two resistances, one Stokesion $6 \pi \eta \, r \, v_m$ and the other Newtonian $9/4 \, \pi \rho_s \, r^2 \, v_m^2$.

The Oseen formula may be considered as an approximation of a more general formula due to Goldstein, which comes closer to the facts than Oseen's, but again it fails at moderate and high speeds.

The idea of fluid resistance of two resistances, one due to viscous flow, and the other due to eddy current, is expressed very simply by Budryk, i.e.

$$R = 6 \pi \eta \, r v_m + Q \, \frac{\pi}{2} \, \rho_l \, r^2 \, v_m^2 \tag{7.15}$$

However, it also deviates from the facts. For particles of intermediate size range, theoretical treatment has not been successful, except that the settling velocity could be shown to be a variable function of the diameter.

7.3.4 FACTORS AFFECTING THE THEORETICAL TREATMENT OF FALLING PARTICLES

The above theoretical considerations ignore many factors in the settling of mineral particles. These are discussed below:

a) The particles are never spherical and thus the derived equations are not correct, with respect to the numerical coefficient. A shape factor k should be incorporated, which would be constant for all sizes of any particular mineral, but may be different for different minerals. The diameter of the particle should be replaced by some convenient statistical size parameter l

equivalent to the diameter of the sphere having the same volume, or the diameter of the circle having the same projected area in the direction of motion. Equations (7.5) and (7.9) then become $R_s = k\,l\,\eta\,v$ and $R_n = k^1\,\rho_l\,l^2\,v^2$, respectively. Shape of the particles affects rate of fall, largely through the flow pattern. The rate of fall of sphere is faster than most other shapes. The rate of fall of flat plates like mica flakes is slowest.

b) Free settling conditions do not approach in practice. Interference caused due to the walls and bottom of the container as well as other particles is appreciable. These effects can be incorporated by applying the empirical corrections to Eq. (7.8). For example, if an experiment is carried out in a cylinder of radius x and the velocity is measured at a distance of h from the bottom end, Eq. (7.8) can be represented as follows:

$$v_m = 2/9\; r^2\; \frac{(\rho_s - \rho_l)}{\eta}\, g \cdot \frac{1}{\left(1 + 2.4\,\dfrac{r}{x}\right)} \cdot \frac{1}{\left(1 + 1.10\,\dfrac{x}{h}\right)} \qquad (7.16)$$

From Eq. (7.16), it may be observed that wall effect alone will contribute to about 1 per cent when $x = 200\,r$ and 30 per cent when $x = 6\,r$. Similarly, the end effect alone contributes to about 5 per cent when the end is about 20 r distant. The effects of the container walls and other particles thus cannot be ignored.

c) When large particles fall through a suspension of fine particles, the effective density and viscosity will be equivalent to those of the suspension as a whole, instead of the true fluid medium.

d) When there is a wide range of size and densities of particles, settling is a complex one and the fall of any particle will depend on not only its own size but also on the distribution of sizes and velocities of the particles around it. Particles having similar size and weight will tend to fall together through a suspension of the finest particles. Particles of any specific size will tend to be dragged down by larger ones, and impeded by smaller ones. The effective density and viscosity will be different for different particle sizes. A dense particle will fall faster than a lighter one of the same size.

7.3.5 EVALUATION OF REYNOLDS NUMBER (R_e), COEFFICIENT OF RESISTANCE (Q) AND VELOCITY (v) IN TERMS OF r, ρ_s, ρ_l, η AND g

It has been shown that Stoke's law is valid up to $R_e = 0.6$ and the Newton–Rittinger relationship is valid approximately from $R_e = 800$ to 2,00,000 with Q ranging from 0.35 to 0.48. In the range of $R_e = 0.6$ to 800, no theoretical formula is valid. There is an extraordinary drop in the value of Q from about 0.40 to about 0.15 when the value of R_e becomes from 2,00,000 to 3,00,000. The values of R_e, Q, and v can be determined graphically in terms of r, ρ_s, ρ_l, η and g with the help of Eq. (7.2) and (7.13).

By taking logarithms of Q and R_e and eliminating v,

$$\log Q + 2 \log R_e = \log \frac{8r\,(\rho_s - \rho_l)\,g}{3\rho_l} + 2 \log \frac{2r\,\rho_l}{\eta} \qquad (7.17)$$

For given values of r, g, ρ_s and ρ_l, the Eq. (7.17) becomes

$$\log Q + 2 \log R_e = C \qquad (7.18)$$

where C is a constant.

The locus of the above equation is a straight line having a slope value of -2, and the intersection of this line with $\log Q$ versus $\log R_e$ curves defines settling velocity, v.

7.4. Equal Settling Particles

The particles having the same terminal velocities in the same fluid and in the same field force, are called *equal settling* particles. Spheres of one material and of the same size are equal settling. Spheres of different materials may be equal settling, having a proper size ratio. Thus, spheres of specific gravities ρ_1 and ρ_2 are equal settling if

$$(\rho_1 - \rho_l)\,r_1^2 = (\rho_2 - \rho_l)\,r_2^2 \qquad (7.19)$$

(under Stoke's law)

and

$$(\rho_1 - \rho_l)\,r_1 = (\rho_2 - \rho_l)\,r_2 \qquad (7.20)$$

(under Newton's law)

In general it can be represented as

$$r_1\,(\rho_1 - \rho_l)^{m_1} = r_2\,(\rho_2 - \rho_l)^{m_2} \qquad (7.21)$$

In Eq. (7.21), the values of exponents (m_1 and m_2) range from 0.5 under laminar conditions to 1.0 under turbulent conditions.

7.5 Free Settling Ratio

Free settling ratio of spheres of different specific gravities, ρ_1 and ρ_2 in the fluid of specific gravity ρ_l can be represented as follows:

$$p_f = \frac{(\rho_1 - \rho_l)^{m_1}}{(\rho_2 - \rho_l)^{m_2}} \qquad (7.22)$$

Generally, m_1 and m_2 are very close to each other, and thus free settling ratio p_f can be represented as

$$p_f = \left(\frac{\rho_1 - \rho_l}{\rho_2 - \rho_l}\right)^{m} \qquad (7.23)$$

7.6 Settling of Large Spheres in a Suspension of Fine Spheres

When large spheres fall through a suspension of fine spheres, the effective fluid density will be the specific gravity of resulting suspension. The velocity of large spheres of a mineral in a suspension of fine spheres of the same mineral can be determined from the following equation:

$$v = \sqrt{\frac{8}{3}\frac{g}{Q}\left(\frac{\rho_s - \rho^1}{\rho^1}\right)r} \qquad (7.24)$$

where ρ_s is the specific gravity of mineral and ρ^1 is the specific gravity of suspension.

7.7 Hindered Settling

When a crowd of particles is present in a fluid, the motion of the particles is affected by their mutual interference. The rate of settling is appreciably less than that calculated, according to the equations derived for the free motion of particles. The particles in fact fall through the suspension of particles in the fluid. Therefore, the density of the clear fluid in Eqs. (7.8) and (7.13) should be replaced by the bulk density of the suspension (slurry). The *hindered-settling ratio*, P_h of two particles is given as

$$P_h = \left(\frac{\rho_1 - \rho^1}{\rho_2 - \rho^1}\right)^m \qquad (7.25)$$

where ρ_1 and ρ_2 are the specific gravities of heavier and lighter particles, respectively, ρ^1 is the specific gravity of suspension, and $m = 0.5$ for Newton's relation, and 1.0 for Stoke's relation. An example of sand suspension in water will show the importance of hindered settling ratio in separation. A suspension consisting of 35 per cent solids and 65 per cent water by volume will have the apparent specific gravity ρ^1 equal to

$$0.35 \times 2.65 + (1 - 0.35) \times 1.00 = 1.58 \text{ (specific gravity of sand}$$
$$\text{is 2.65 and that of water is}$$
$$1.00).$$

In this suspension the settling ratio of galena (specific gravity 7.5) and quartz (specific gravity 2.7) can be obtained by substituting the value of $\rho^1 = 1.58$ in Eq. (7.25)

$$P_h = \left(\frac{7.5 - 1.58}{2.7 - 1.58}\right)^{1/2} = 2.29$$

Whereas the settling in water will be

$$\left(\frac{7.5 - 1}{2.7 - 1}\right)^{1/2} = 1.96$$

Thus, the hindered settling ratio indicates that the sand-water mixture will perform a better sorting work than plain water.

7.8 Classification

Classification is generally restricted to processes in which particles of various sizes, shapes and specific gravities are separated by their different rates of travel through a fluid, usually water or air. The coarser, heavier, and rounder particles settle faster than the finer, lighter, and more angular grains. The fall of each particle through a fluid medium is made to take place under controlled conditions. The feed is sorted into oversize and undersize fractions. The overflowing liquid carries away the slow settling (undersize) grains while the sediment of fast settling grains (heavy grains) is removed simultaneously from the classifier through its discharge gate. The particles of intermediate size are held in suspension (teeter column).

7.8.1 MECHANISM OF CLASSIFICATION

Basically classification depends on Stoke's law of settling. The suspension of fine particles results in an increased specific gravity and increased viscosity of the fluid. These result in a decreased settling velocity and decreased importance of eddying resistance.

In classifier, the particles fall under a vertically acting hydraulic force provided by the velocity of rising fluid (water or air). The magnitude of this force depends on the velocity (or volume) of water passing upward through the horizontal cross-section of the classifier at a given point. If a particle has to fall against these forces, it should overcome frictional drag and collision in the teeter zone. Under these conditions the particles will be separated by *hindered* settling. If the classifier also imposes horizontal flow on its contents, the falling particles are also displaced horizontally to a distance proportional to the time taken in passing through the current. In mineral processing, a pulp of water and fine particles is the medium normally used as classifying medium, whereas fine and dry powders can be classified in vertical or horizontal currents of air.

Classification deals with small particles in movement varying from slight drift in some areas to turbulence in other areas. The fall of an individual particle is affected by the packing density (number of particles per unit volume of pulp) in its immediate neighbourhood. There is a certain critical packing density in the pulp, below which unhindered movement of individual grains occurs and above which increasing intergranular interference is encountered. The various factors affecting the movement of particles in a fluid are already discussed under Sec. 7.1.

Since classification depends partly on frictional retardation, coarse material cannot be treated efficiently. The usual range of application is 1 mm to 50 microns. The range can be extended further to lower sizes by the use of centrifugal force. In hydraulic classifier (using water) the mixture of fine particles and water acts as a fluid medium, density of which depends on the specific gravity of the fine particles and the solid-liquid ratio. When relatively coarse particle falls through this fluid, its potential energy gets converted to kinetic energy. The motive power is mainly due to the difference between the weight of particle and the weight of an equal volume of classifying fluid. Therefore, higher the ratio of solid to liquid (finer particles) the smaller becomes the gravitational effect.

The generated kinetic energy of the particle is used to overcome the viscous resistance of the fluid and to start vortexes and displace other particles during collision or frictional contact. In classification, there is no sharp cut-off point as can be had in careful screening, since at the separating cut-off point some particles will be diverted by drifting vortexes into the wrong stream. However, at a given separation point, bulk of the particles will respond in the desired manner. In milling schemes the cut-off point is kept sufficiently elastic.

Classification can be carried out under *hindered settling* conditions (separating fluids carrying 40–70 per cent solids) or *free settling* conditions (separating fluid carrying 3–35 per cent solids) depending on the various factors.

7.8.2 SETTLING VELOCITIES IN CLASSIFIERS

Classification is mainly applicable to fine particles. The settling of fine particles is given by Eq. (7.7). For determining the settling velocity in a classifier, two factors should be introduced in Eq. (7.7) to take care of solid-liquid ratio, and shape of the particles, since the effect of these two factors is not taken into account, while deriving Eq. (7.7).

a) *Correction Factor for Solid-Liquid Ratio*

Since classification is carried out in a relatively thick suspension, the settling velocity is reduced to some extent. The proportion of this can be represented by a correcting factor f_1 expressed as

$$f_1 = (1-x^{2/3})(1-x)(1-2.5\,x) \qquad (7.26)$$

where, x is the fraction of the volume of suspension occupied by the solid.

b) *Correction Factor for Shape of Particles*

A relationship between shape of particles and settling velocity can be expressed as the ratio of the size (diameter of a sphere of equal volume) of equal settling particles, where one is irregular in shape, and the other is spherical. When compared with spheres (for spheres value of this ratio is 1.0) this ratio is 1.19 for cubes, and 1.28 for very thin disks. In classification, particles are fine enough to settle by viscous resistance and thus little variation occurs in the shape–correction factor. A ratio of 1.24 may be taken as a good average for irregular particles having the degree of roundness intermediate between that of cubes and of very thin disks. Since the velocity of particles is proportional to the square of size, the correction factor f_2 is approximately

$$f_2 = \frac{1}{(1.24)^2} = 0.65 \qquad (7.27)$$

If the term $\dfrac{2}{9}\dfrac{g}{\eta}$ is considered as a third factor, f_3, then Eq. (7.8) can be expressed as

$$v = F\,(\rho_s - \rho_l)\,r^2 \qquad (7.28)$$

where $\qquad F = f_1 \cdot f_2 \cdot f_3$

7.8.3 Conditions Required for Efficient Classification

The conditions employed in classification depend on the use made on classifier products, i.e. whether the objective is sorting or sizing. When the classifier products are to be subjected to gravity concentration (such as tabling), emphasis should be laid on the effect of difference in specific gravities of minerals. In such a case, *hindered* settling conditions should be exploited to maximum possible extent, i.e. suspension should be as dense as possible. This operation of classifier is called *sorting* and is not strictly *sizing*. Sorting classifiers require a solid content as high as possible (usually 40–70 per cent by weight) depending on specific gravity and size of the minerals. On the other hand, when a classifier is used only as a sizing device to assist a grinding mill, the effect of difference in specific gravities should be exploited to the minimum possible extent. This can be accomplished by providing free settling conditions, i.e. by using dilute suspensions.

In practice, classifiers employed for sizing require dilute pulps with solid content of 3–4 per cent by weight ($x = 1$–2 per cent) for extremely fine end of the practical range of classifiers and up to 30–35 per cent by weight ($x = 12$–18 per cent) for extreme coarse end.

7.9 Classifiers

Classifiers can be divided into various categories depending on their function, mechanism, fluid used, etc. A general classification is shown in Fig. 7.2.

7.9.1 Sorting Classifiers

In sorting classifiers, more or less hindered settling conditions are used. In this, sizing is modified by specific gravity and shape of particles. It is usually employed for relatively coarse products. Commonly known sorting classifiers are launder classifier, cylindrical, trapezoidal tank type, and hydrotator classifier. Presently, a little use is made of sorting classifiers.

Launder classifier (Evans classifier) consists of a sloping launder having attached to it several rectangular boxes which open into launder and are capable of discharging through a spigot. Water is introduced in controlled amounts through pipes having valves. The faster settling particles discharge through the spigot, whereas the slower settling particles overflow.

Cylindrical type classifiers (Anaconda classifier) use sorting columns in place of rectangular boxes (used in launder classifiers). The hydraulic water is introduced from below into the cylindrical sorting column, and from there the water is allowed to pass into an inner conical column through tangential and radial parts.

Trapezoidal tank classifier (Fahrenwald sizer) consists of five sets of rectangular and cylindrical pockets which yield a spigot product (heavy product) whose size decreases gradually from the first rectangular pocket to the cylindrical pocket. When the pressure exerted by the teetering column exceeds a certain predetermined value, the spigot is allowed to discharge.

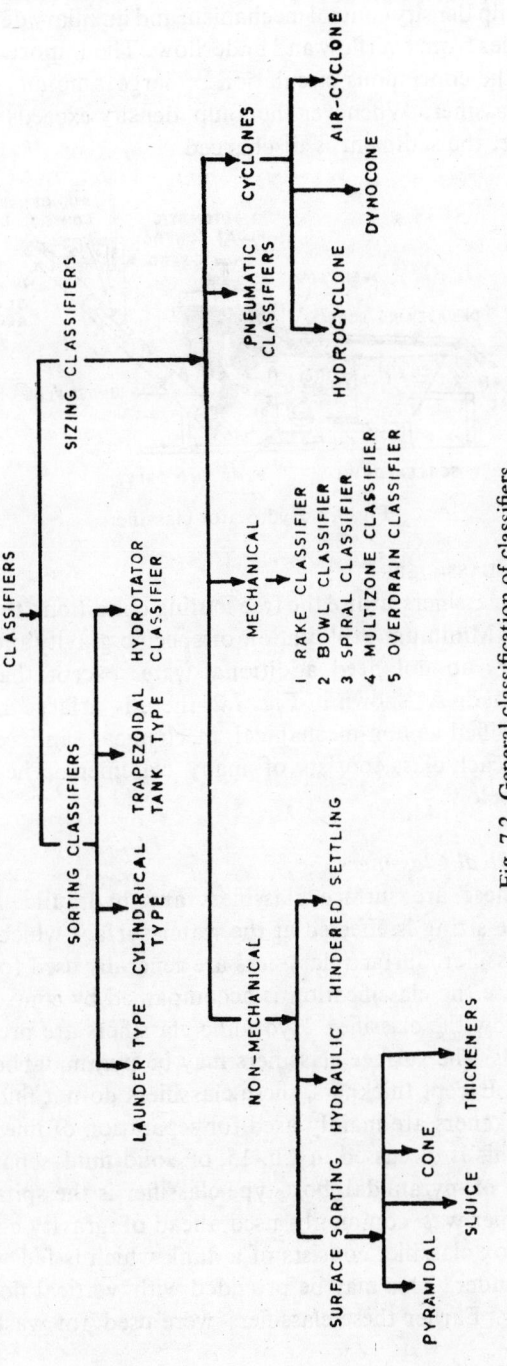

Fig. 7.2. General classification of classifiers.

Hydrotator classifier (Fig. 7.3) was used earlier for cleaning of coal. This consists of a pulp density control mechanism and auxiliary devices for removal of slime particles from overflow and underflow. The important feature of the hydrotator is the continuous circulation of large amounts of fine material through the classifier. Whenever the pulp density exceeds a certain predetermined value, the sediment is discharged.

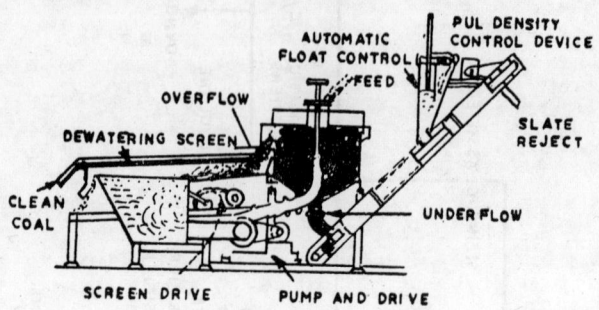

Fig. 7.3 Hydrotator classifier.

7.9.2 SIZING CLASSIFIERS

The sizing classifiers utilise the free settling conditions to effect maximum possible sizing. Minimum exploitation of specific gravity and shape is made. These classifiers do not need additional water except that present in the pulp to be treated. As shown in Fig. 7.2, there is a large number of sizing classifiers, classified as non-mechanical, mechanical, and centrifugal classifiers. Further, each class consists of many classifiers. The important ones are described below:

a) *Non-mechanical Classifiers*

Basically these are surface classifiers and hydraulic classifiers. In the former case the sizing is effected at the water surface which brings the material to the classifier. Surface classifiers are generally used for fine materials. In the latter case the classification is accompanied by controlled passage of water from below the classifier. Hydraulic classifiers are preferably used for coarse materials. The surface classifiers may be pyramidal box, cones, boxes, thickeners, etc. Except thickners, these classifiers do not find any use in present time. Thickeners are mainly used for separation of fine solids from suspensions and this is discussed in Ch. 15 of solid-fluid separation. The elementary form of pyramidal box type classifier is the spitzkasten. At one time this classifier was commonly used ahead of gravity concentration. A simple sluice box classifier consists of a tank which is fed with pulp at one end from a launder. This may be provided with vertical flow of water and conical hutches. Earlier these classifiers were used for washing off the fine gangue.

A cone classifier consists of conical shells of steel or concrete, having the apex at the bottom and a peripheral overflow launder at the top. Though these can be used for desliming or dewatering, but thickeners have replaced them.

A hydraulic classifier (upward current classifier) is of cylindrical shape (or tapering) in which water rises upward at a controlled rate. The feed is introduced centrally near the top. The light particles overflown from one column are passed to another column, in which the velocity of rising water is less in order to obtain a second fraction. Free settling as well as hindered settling conditions can be employed in these classifiers.

In general, hydraulic classifiers are used for sizing of coarse material (above 180 microns). However by employing air columns instead of liquid, much finer particles can be treated. Hydraulic classifiers have been developed in many shapes and varieties.

In a class of hydraulic classifiers known as hydrosizers the teeter bed is controlled depending on the composition of the mineral. The bottom discharge aperture can be opened or closed by a suitable control mechanism actuated by pressure changes, to take care of the load of teetering sands. This type of classifier can deliver a series of graded products.

b) *Mechanical Classifiers*

Mechanical classifier mainly consists of a rectangular or bowl shaped tank. The pulp is fed under such conditions which allow the fall of heavier and coarser solids downward quite freely and flowing away of the lightest particles to a weir discharge. The mechanical part of the classifier generally consists of a drag belt, a set of rakes, or a spiral screw which stir the pool of pulp and remove settling solids. The pulp from the grinding mill with more addition of water (pulp density adjusted to about 30 per cent solids) flows to the classifier through a short launder under gravity (sometimes aided by a centrifugal pump). If the feed contains equal size particles, a clean separation of feed into heavy (sinking) mineral and light (overflowing) mineral can be obtained.

In practice, the feed contains a wide range of sizes with various stages of liberation. As a result some particles drop swiftly, some flow away, while others accumulate in the sorting zone which varies in its density from the uppermost watery layer to the densest part of the pool just above the raking zone. These strata of varying densities are stirred by the rakes or spirals as they move in the pool of the classifier. An undisturbed stratum lies below the rakes which packs the clearance space between the rakes and bottom of the tank. Each layer of the pool continuously receives new particles from above which would be either (i) allowed to fall through the layer, (ii) rejected back to the layer above it, or (iii) retained by the layer. As long as the layer sorts the entering particles according to (i) or (ii), the layer will maintain its integral composition and will carry out the work consistently. However, in practice, capture of more particles is inevitable and thereby its density increases. The continuous rise of pool density is partly offset by the stirring action of the rakes or spirals.

The required turbulence in the pool can be produced by varying the rate of movement of rakes or spirals. For example, separation at −70 microns to yield a fine overflow, the rate may be as low as 9 rake strokes per minute, whereas the rate may vary up to 30–35 strokes per minute for 0.6 mm size rapid settling sands. Another major function of rakes is cleaning of the settled sands up-slope to the mill feed launder. A large number of mechanical classifiers having different mechanisms have been developed. The important mechanical classifiers are described below:

Dorr rake classifier: A section through the rake classifier is shown in Fig. 7.4. It consists of an inclined rectangular settling tank (inclination of the bottom may be about 12 cm/m) provided with rakes moving through an elliptical orbit with the help of an eccentric motion. Mechanically operated rakes start their climb at the lowest settling portion of the tank and gather settled sand which is lifted sharply at the rising end of the stroke. The rakes then return and drop the sand. The cycle is repeated. The rake classifiers may be simplex when the trough has one compartment, duplex (two compartments), triplex (three compartments), or multiple rake unit (many compartments), where the compartments are separately raked. The rakes can also be raised.

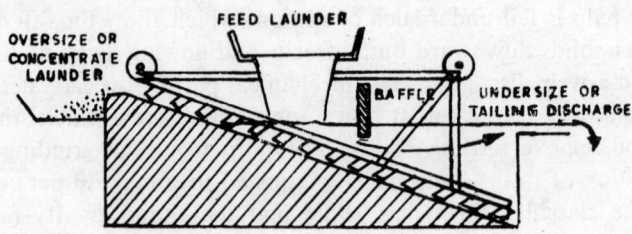

Fig. 7.4. Dorr rake classifier.

The feed is introduced at either end of a transverse trough near the overflow, which is provided with splitter vanes to ensure even distribution of feed across the entire width of the tank. The tanks are usually made of steel, but sometimes can be constructed of wood or concrete.

Dorr bowl classifier: This is a variant of Dorr rake classifier, in which large shallow cylindrical settling tank is provided as shown in Fig. 7.5. The lower part of the tank is in the shape of shallow cone (sloping 18–20 cm/m). The feed from the mill is introduced centrally and allowed to spread radially to provide larger surface area for drifting of particles from the outward streaming pulp. The rakes revolve slowly (2 rpm) in the settling tank to move settled material gently towards the centre, where it falls into the trough of the classifier. From the classifier trough, the settled sand is scraped up to the discharge end with the help of rake mechanism. The slime overflows from the periphery of the bowl.

In bowl classifier there is an appreciable loss of head, due to which it is

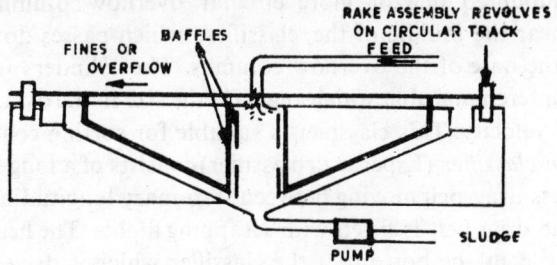

Fig. 7.5. Dorr bowl classifier.

not normally possible to use it in a close circuit without the use of a centrifugal pump to elevate the mill discharge head of the classifier. A modification has been developed to process fine ground products (requiring more settling area), where a bowl is somewhat like a miniature thickener. This is termed as hydroclassifier. The principle of operation is the same as in bowl classifier. In this classifier the required conditions are a long transit time and a minimum agitation.

Spiral classifier (Akins or Hardinge spiral classifier): The spiral classifier (Fig. 7.6) consists of a usual sloping trough (as in rake classifier) in which the pulp is maintained in stirring condition with the help of one or more helix (spiral ribbons) mounted on the shaft. This helix does the same work as the rakes in rake-classifier, i.e. to remove the material from the bottom of the tank. The helix also works as an elevator for the bottom sand. In the classifier, the settled material is turned over and over again, before its final discharge. It offers better opportunity for complete disliming of sand. The helix is run at a speed varying from its maximum diameter. The speed of

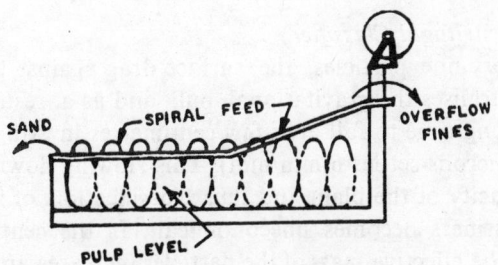

Fig. 7.6. Spiral classifier.

the shaft is usually 3–20 rpm (lower rpm for a large spiral, and higher for a small) depending on the size of the spiral. Steeper slopes (25–35 cm/m) in the classifier tank are possible due to the continuous and gentle action of a spiral.

Multizone classifier makes use of two settling zones, one providing the hindered settling conditions near the rakes and the other providing the free settling conditions. The annular spaces above the hindered settling zone work as sorting zone. The lower portion of the classifier is decked, over which is surmounted one or more circular overflow columns. The pulp is introduced near the middle of the classifier, which passes downward along the deck to the base of the overflow columns. The cylinders in the overflow column are interchangeable which regulate the size in the overflow by changing the fluid velocity. This classifier is suitable for sorting coarse material.

Overdrain classifier (Esperanza classifier) consists of a long sloping trough and an endless drag belt moving between stationary longitudinal walls of the classifier. The drag belt is fitted with scrapping flights. The heavier and coarser grains settle in the bottom of the classifier which is driven up the slope of the trough at the upper end, from where it is discharged continuously. The finer and lighter material overflows through the side openings. Various bottomless compartments may be provided to facilitate the removal of material trapped between successive moving sections.

c) *Pneumatic classifiers* (Hardinge and Gayco pneumatic classifiers)

A pneumatic classifier consists of two conical shells (one outer and one inner) and employs air as a medium. The settling velocities in air are much higher (about 100 times) than in water. The dilute suspension of solids in air is allowed to ascend in the annular space between the two conical shells. Stationary vanes (partly radial and partly curved) are fastened to the outer shell at the top of the inner shell. Coarse and heavy particles descend to the inner shell and pass across the annular flow of air at the bottom of classifier. The material of intermediate sizes remains in considerable circulation. Control over the size of classification is obtained by raising or lowering the inner cone. Pneumatic sizing classifiers were used in conjuction with dust collectors. Presently, these are not much in use, as air cyclones are more convenient.

d) *Cyclones (Centrifugal Classifier)*

In case of very fine particles, the surface drag against the surrounding fluid nearly neutralises the gravitational pull and as a result particles may require a very long time to fall even few centimetres in still water (a silica particle of 10 microns settles 6 mm/min). This slowing down of settling rate reduces the capacity of the plant and thus classification of fine particles by mechanical classifiers becomes uneconomical. If the centrifugal force is superimposed, the effective mass of the particles increases and thereby gravitational pull increases tremendously. The net result is the increase in settling velocity many times by the application of centrifugal force. Therefore, cyclones have replaced mechanical classifiers in most of the grinding mills. The fluid used in cyclones may be either water or air and depending on the fluid used they are termed as *hydrocyclone* (using water) and *aircyclones* (using air), respectively. In mineral processing, hydrocyclones are more common, since grinding is wet. Air cyclones are used in dry grinding circuits.

Hydrocyclone: The hydrocyclone is made of conical shape (Fig. 7.7) having a cylindrical top. The feed is introduced tangentially near the top. A spinning motion is imparted to the suspension in the cylindrical portion and a vortex is generated about the longitudinal axis. The accompanying centri-fugal acceleration increases the settling rate of the particles, and tends to throw them radially.

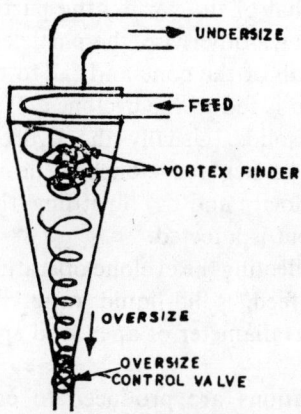

Fig. 7.7. Hydrocyclone.

The coarser particles spiral down to the cone's wall (where a zone of reduced pressure exists) and flow downward to the apex through which they are discharged. The amount of feed joining the coarse product depends on the size of the inlet and vortex finder (provided the underflow does not exceed about 30 per cent of the feed). At the centre of the cyclone, a zone of low pressure and low centrifugal force exists, which surrounds an air-filled vortex. Therefore, a part of the feed carrying finer particles tends to move inward towards the vortex. As a result, some newly entrained feed may be picked up by the vortex finder (a central pipe provided for removal of overflow) and removed along with overflow. Therefore, the vortex finder is so adjust-ed as to project into the conical section of the cyclone, and thus to mini-mise short circuiting of newly arriving pulp. The diminishing cross-section of the cyclone helps in superimposition of flow toward a vortex developed along the axis. The net flow is upward through a vortex finder which stabi-lises the vortex working as a carrier of overflow.

Any particle in the size is subjected to a centrifugal force mv^2/R, charac-teristic of its tangential velocity v and the radius of curvature, R of its path (almost equal to radius of cyclone), and a drag force caused by the local inward flow of the fluid which depends on the relative radial velocity, the viscosity of the fluid, size of particles, and specific gravities of particles. The tangential velocity of a particle v_r can be given by the following equa-tion:

$$v_r = \frac{2u_t^2(\rho_s - \rho_l)r^2}{9\eta R} \tag{7.29}$$

where, u_t = tangential velocity of fluid. Other symbols have the same meaning as in Eq. (7.8).

Each particle follows the path of a particular radius at which these forces are in balance. This radius will be greater for larger and denser particles (having high terminal velocities in free settling). If the radius of this equilibrium path is less than that of the vortex, the particles will be carried in the vortex stream to overflow, otherwise the particles will join the larger and denser particles at the wall of the cone and fall to the apex of the cyclone.

Generally, dilute pulp is fed to the cyclones. The underflow may contain up to about 80 per cent solids. Usually the hydrocyclone is designed for a particular purpose. Any particular cyclone may be used for any work through variations in the feed velocity and by throttling the apex discharger-pipe, but the rate of throughput is affected.

The various factors affecting the cyclone operation are feed inlet diameter, pressure of feed, rate of feed, solid-liquid ratio, position of vortex finder, diameter of vortex finder, diameter of apex, and specific gravity of solids in feed.

Some empirical equations are produced to calculate the diameter of particle in equilibrium in terms of other factors, such as cyclone overflow diameter, feed rate, cyclone inlet diameter, shape, gravity, etc. Following empirical equation has been given by Dahlstrom

$$D_{50} = \frac{152 \, (b \, e)^{0.68}}{Q^{0.53} \sqrt{\rho_s - \rho_l}} \qquad (7.30)$$

Where, D_{50} = 50 per cent particle diameter in microns (this is the equilibrium particle size at which half the solids are discharged as underflow through the apex and the rest through the central overflow pipe),

b = cyclone inlet diameter in cm,

e = cyclone overflow diameter in cm,

Q = feed rate in litres per minute,

ρ_s, ρ_l are specific gravities of solid and liquid respectively.

Another equation for the size of particle revolving in equilibrium at the circumference of the cyclone cylinder is given by Tarjen, i.e.

$$d = \frac{42 \, e^2}{\sqrt{(\rho_s - \rho_l) \, h Q}} \qquad (7.31)$$

where, h is the height of the cylinder in cm and others have the same meanings as in Eq. (7.30).

Usually a hydrocyclone has two types of characteristics, i.e. (i) fixed by construction, i.e. diameter of the cyclone, area of feed entry, length of drum, length of vortex finder, and cone angle, and (ii) operating variables, i.e. pulp density, feed pressure, diameter of overflow pipe and diameter of vortex finder.

Advantages of hydrocyclone: Hydrocyclones offer many advantages over the other classifier such as less floor area for a given capacity, cheap to build, less power consumption, less maintenance, possibility of shutting down the mill under full load, possibility of bringing circuit rapidly into balance, and elimination of cycling surging. However, the wear and tear on the feed pipes can be severe and expensive.

The hydrocyclone is increasingly used for classification in the finer grinding ranges (150 to 5 microns). The cyclone does not work as an effective substitute for the thickener in treating materials finer than about 5 microns in size. However, it can be used to remove bulk of the solids in the underflow and then the overflow will carry a relatively small percentage of the finest solids increasing the capacity of thickeners.

Dynocone: This consists of mechanically spinned separating vessel as shown in Fig. 7.8. This equipment also consists of a revolving conical shell with screw conveyor rotating at a slightly higher speed. This can be employed for classification of much finer materials. The solids settle to the inner wall of the cyclone, which are removed by the screw. The finer fraction overflows at the other end.

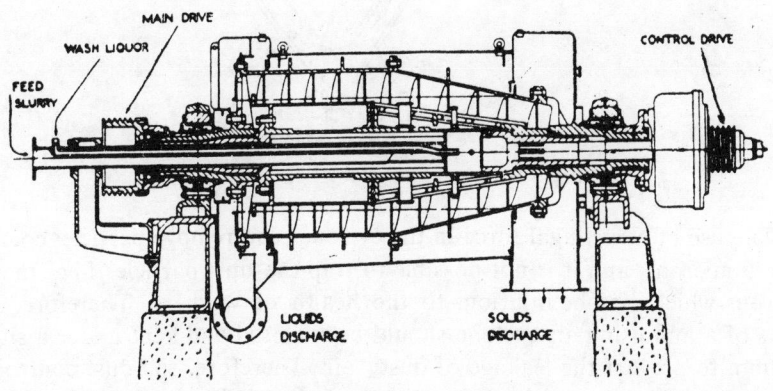

Fig. 7.8. Dynocone.

Air cyclones: Air cyclones are identical in construction as well as in operation to hydrocyclones. These are used for classification in dry grinding. In air cyclones, air currents can be manipulated to collect the dust and its collection in suitable containers. In the first case, larger particles are separated from smaller, whereas in the second case all possible dust is removed from the air in order to reduce the air pollution.

In case of air cyclones, frictional drag upon air and particles touching the cyclone walls is quite pronounced, which will require special precautions/devices for preventing this from becoming serious. The effects produced by the air stream depends on its velocity, humidity, viscosity (pressure effect) and the way in which the air moves/circulates. The behaviour of a given particle in an air stream will depend on its size, shape, density, and liability to collide with other particles.

If the velocity of the transporting stream of air is properly controlled, relatively coarse particles are dropped first, while the finer ones will be carried further. Baffles or deflectors may be in the stream to sort out the particles according to their inertia. In air cyclone, the dust slides to the sides and then falls down, while a vortex of comparatively dust-free air rises at the centre. The air is usually blown by fan. A modified form of air classifier known as *gravitational inertial* classifier is shown in Fig. 7.9. This can be used to remove −150 microns material.

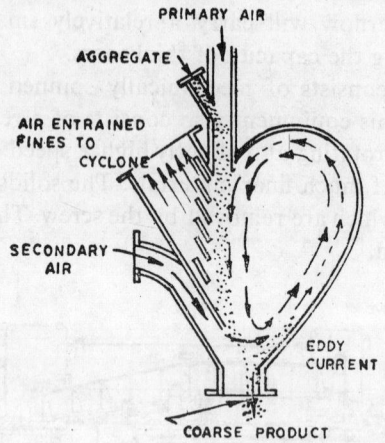

Fig. 7.9. Gravitational inertial classifier.

The use of centrifugal force in the cyclones can remove particles coarser than 5 microns, and it is not possible to trap the dust particles finer than 5 microns which may be injurious to the health of workers. Therefore, the parts of a mill using air cyclone should be enclosed and kept under a slight vacuum to prevent the leakage of dusty air. Therefore, the dust control is an essential feature in use of air cyclones. Industrially, electronic dust precipitation is employed to deal with particles too fine to settle by gravity.

7.10 Performance of Classifiers

The performance (capacity), C of a classifier is directly proportional to four factors, i.e. (a) cross-sectional area of the sorting column, A, (b) the rising velocity, v of the fluid in the sorting column, (c) pulp density (solid content of the feed by volume, γ, and (d) specific gravity, ρ_s of the solids. The capacity of a classifier can be represented as following:

$$C = a\,Av\,\gamma\,\rho_s \qquad (7.32)$$

where a is a constant.

Practical capacity of a classifier for the tank-type machines (such as hydroseparator), is found to be in agreement with Eq. (7.32). Whereas, in classifiers of trough-type (Dorr and Akins classifiers), the practical capacity is quite low, if the full cross-sectional area of the settling trough is considered.

7.11 Efficiency of Classifiers

It is somewhat difficult to quantify the efficiency of a classifier. Usually the separation efficiency is measurable as the percentage of misplaced product in either the overflow or underflow. This is assessed by performing screen analysis of suitably collected samples of feed material, underflow (coarse product), and overflow (fine product), and using these data in calculation. It is customary to use the same formula as employed to calculate the recovery, i.e.

$$E = 100 \frac{c}{f} \left(\frac{f-t}{c-t} \right) \qquad (7.33)$$

where, E is the efficiency expressed in percentage, c, f, and t are the content of $-x$ size material in the overflow, feed, and underflow, respectively, x being any size at which none of the c, f and t is zero.

However, Eq. (7.33) is not accepted by all, since it is likely that some feed may be by-passed into the overflow, which will represent higher efficiency calculated from Eq. (7.33). Another equation expressing efficiency as the ratio on a percentage basis, of the classified material in the overflow to classifiable material in the feed, can be represented as:

$$E = \frac{10,000 \, (c-f)(f-t)}{f(100-f)(c-t)} \qquad (7.34)$$

Equation (7.34) gives lower values than obtained from Eq. (7.33). The efficiency calculated according to this equation, is generally 50–80 per cent. However, it would be considerably greater, if measured by sedimentation analysis instead of screen analysis.

However, this practice of assessing classifier efficiency is not fair, since classifiers do not size material in the same way as screens. It would be more appropriate to assess the classifier efficiencies from careful laboratory classifications of overflow and underflow obtained from industrial classifiers.

7.12 Classifiers Versus Cyclones

The screens fail to handle fine materials efficiently in closed circuits employing wet or dry grinding. Therefore, screens are unsuitable in processing of most of the low-grade and fine grained ores being treated presently. Therefore, the choice lies mainly between mechanical classifiers and cyclones. However, the flowsheets using gravity processes of concentration, may employ hydraulic classifiers. During the past five decades, the growth in the use of cyclones has been rapid. The mechanical classifiers have the following advantages over the cyclones:

a) They can smooth out surges.

b) They can return oversize to the mill launder without using an extra device.

c) They are robust in structure.

d) Their wear is low.

e) They are easy in control.

The phenomenal growth in the use of cyclones,is attributed to the following facts:

i) Use of centrifugal force speeds up the settling rate, due to which either larger tonnages can be handled with an equipment in a small area,or separation can be made at finer sizes than the other classifiers.

ii) Though operating costs are quite comparable, capital and installation costs are much less.

iii) Oversize can be fed directly to the feed trunnions of the grinding mill,and does not require a scoop and feed box.

iv) A small tonnage of material remains in circulation, which reduces the oxidising effects in the grinding circuit (important in flotation of sulphides).

v) The mechanical classifiers put a limitation of fine grinding in closed circuit due to moderate circulating load,limited by the free settling speed of the near release particles, whereas the accelerated settling due to centrifugal force in the cyclone,makes possible the circulation of large circulating load. Overgrinding is thus,reduced due to repeated passage of feed through the secondary mill as with the primary one.

7.13 Classifiers as Concentration Devices

In general, classification is a sizing operation, or a sorting operation adjunct to gravity separation, but it can also be used as a means of concentration in some cases. When two particles having same shape and size, but different specific gravities, are introduced into a classifier, the heavier particle may sink and be returned for further grinding, while the lighter one will overflow. Under these conditions, the classifier becomes a concentrating device. It is usually undesirable that a particle should remain longer in a closed circuit. For example, a metallic sulphide after its almost complete liberation,can be caught more effectively in the concentrating section of the plant when the particle is comparatively coarse. Therefore, a classifier may be used to remove maximum possible of the desired value in the form of a rough concentrate to minimise its over grinding or accumulation in the closed circuit.

Concentration by classifiers may be effected under the following two types of conditions:

a) When the valuable constituents are in one range of sizes and the waste is in another range of sizes.

b) When settling can be crowded sufficiently to cause stratification, the lower stratum will consist of heavy material and the upper stratum will consist of light material.

Following examples illustrate the use of classifiers in concentration:

7.13.1 Separation of Fine Valuable Mineral from Coarse Waste

This can be illustrated by the cleaning of clays (silicate minerals in an extremely fine state of subdivision) contaminated by coarse impurities (mainly quartz, feldspar, micas, and pyrite). In this case, classification yields a sedimented residue (containing coarse impurities) and suspended washed clay. The conditions for classification are those of free settling.

7.13.2 Separation of Coarse Valuable Mineral from Fine Waste

This is the reverse of the above example. In this case, a classifier sediment is the concentrate and a classifier overflow is the waste. When there is no difference in specific gravities of mineral and gangue, free settling conditions are employed, whereas in case of coarse valuable material having higher specific gravity, hindered settling conditions are preferred. The important examples are beneficiation of iron ore, phosphate rock, and washing of gravel.

CHAPTER 8

Dense Media Separation

8.1 Introduction

When an ore or material having particles of different densities is immersed in a liquid having specific gravity in between the specific gravities of associated particles, lighter particles would float and the heavier ones would sink. The floats and sinks can be withdrawn separately. This process is based purely on density of minerals, and does not depend on rate of fall or size of an ore or material. The process is known in the industries by various names, such as *sink and float, dense media separation, heavy media separation*, and *heavy liquid separation*. However, the process is usually referred as dense media separation (DMS) or heavy media separation (HMS). The separation can be carried out in a quiet single bath or with controlled agitation of the bath. Centrifugal action may also be incorporated in some cases. The principle of DMS can be illustrated as follows:

Feed
↓
Minerals ←— Media bath —→ Minerals
sp. gr. −1.5   sp. gr. 1.5   sp. gr. + 1.5
(floats)   (sinks)

The most common example illustrating the above process,is the separation of wood chips from sand and gravel using water as the medium. However, water cannot be employed in separation of minerals, since all minerals are heavier than water. Therefore, some heavy liquids such as organic liquids (specific gravity 1.5 to 3.5), aqueous solutions of soluble salts, and heavy pseudo liquids made by suspending solids in water,should be employed,depending upon the specific gravity for a given ore system.

Though the use of heavy media has been known for laboratory separation,since more than 100 years back, the first commercial success was achieved in 1917 by the Chance process using hydraulically dilated sand. It was followed by the development of various other processes using $CaCl_2$, clay, gypsum, pyrite, galena, barytes-clay, etc.

The DMS process is applicable in separating the minerals having adequate difference in specific gravities and liberated at coarser size (+ 0.3 mm), e.g.

separation of tungsten, uranium, and vanadium minerols from calcite and quartz. The process can also be used to remove light weight waste barren rock at an early stage in crushing to warrant treatment in the concentrator. However, the characteristics of DMS processes and the densities of pseudo liquids,commonly obtainable,make the process most attractive for cleaning/ washing of coals, where a graded end-product is clean coal having very low ash content.

8.2 Mechanism and Working Principle of Dense Media Separation

Separation in dense media is regarded almost a pure gravity separation and the results obtained in quiet bath,are nearly equivalent to such an effect. An essential counter-restraint is viscous-shear. The forces acting on a particle for free falling, can be shown by the following Stoke's equation as discussed in Ch. 7.

$$V = \frac{d^2 g (\rho_s - \rho_l)}{18 \eta} \tag{8.1}$$

and

$$R = \frac{V d \rho_l}{\eta} \tag{8.2}$$

Now, let us consider the behaviour of spherical particles having same size,but different specific gravities. When five spheres equal in volume having specific gravity 1.8, 1.9, 2.0, 2.1 and 2.2, respectively,are placed in a quiet (truly static) bath of heavy liquid of specific gravity 2, the potential energy of spheres of specific gravity 2.0 will be neutralised,and thus,these will remain suspended at the point of their release into the bath. In the absence of frictional restraint, the lighter spheres will float upward and the sphere of lowest specific gravity (1.8) will reach first to the surface. Similarly, the heavier spheres will sink,and the sphere of highest specific gravity (2.2) will reach first to the bottom. Now, if the experiments are repeated with smaller and smaller spheres without changing the other conditions, until their size (mass) is reduced to the point where gravity pull is neutralised by the resistance to viscous shear of the liquid media, spheres having specific gravity 1.9 and 2.1 will be the first to become suspended. If the size is reduced further, all the five spheres become suspended. The size effect is therefore, equally important in dense media separation, since the largest (specific gravity 2.0), the medium (specific gravity 1.9 and 2),and the small (specific gravity 1.8 and 2.2) will remain in suspension together.

The spheres lacking in sufficient gravitational pull or push for their movement (up or down) with respect to the heavy liquid, will join middling. Therefore, for a medium of given viscous resistance, there is a minimum size limit for a sphere of given specific gravity to move up or down.

In industries, dense media separation rarely uses true heavy liquids. The suspensions of suitable heavy materials (such as magnetite, ferro-silicon) are used in water. The materials are finely ground to have a slow rate of sedimentation. The retarding force in such a system is mainly a function of specific surface of heavy materials used for suspensions. However, some influence to this, is also contributed by the slimes, clays, and worn-out materials.

Theoretically, the separation may be considered as a hindered-settling classification, and the following equation (Rittinger's formula) derived previously for hindered settling relations (Ch. 7), may be used.

$$V = C \sqrt{d(\rho_s - \rho_l)} \tag{8.3}$$

If two equal settling particles, A and B of different specific gravities are considered, and their settling velocities are denoted by V_A and V_B, then

$$C \sqrt{d_A(\rho_A - \rho_l)} = V_A = V_B = C \sqrt{d_B(\rho_B - \rho_l)} \tag{8.4}$$

or $\qquad\qquad d_A(\rho_A - \rho_l) = d_B(\rho_B - \rho_l) \tag{8.5}$

or $\qquad\qquad \dfrac{d_A}{d_B} = \dfrac{\rho_B - \rho_l}{\rho_A - \rho_l} \tag{8.6}$

in which $\dfrac{d_A}{d_B}$ is the ratio of diameters of two equal settling particles A and B.

The hindered settling column in dense media separation consists of the following:

a) Closely sized (average size about 0.3 mm) solid material (magnetite, ferro-silicon, etc.) responsible for making heavy media.

b) Ore or coal particles in size range of 1.5 to 150 mm.

c) Refuse (tailing) particles in size range of 1.5 to 150 mm.

For example, in cleaning of coal, specific gravity of heavy media required is about 1.5, which is to be maintained. The equilibrium may be established with respect to suspension material in the cone when rising current of water will be almost equivalent to the average settling velocity of the particles in stable suspension. Under equilibrium condition, Eq. (8.6) may be used to calculate the diameters of coal and tailing (refuse) particles having different specific gravities, which will rise or just fall in the hindered settling zone. If d_s and δ_s represent the diameter and specific gravity, respectively of the medium (solid) particle, and d and δ represent the diameter and specific gravity, respectively of coal or refuse particles which will first remain in suspension with the medium particles, then,

$$\frac{d}{d_s} = \frac{\delta_s - \rho_l}{\delta - \rho_l}$$

or $\qquad\qquad d = \dfrac{d_s(\delta_s - \rho_l)}{\delta - \rho_l}$

If the respective values are put in the above

$$d = \frac{0.3(5.5 - 1.5)}{\delta - 1.5} = \frac{1.2}{\delta - 1.5}$$

If the density of medium is controlled at 1.5, all the particles having specific gravity less than 1.5,will float, irrespective of hindered settling conditions, whereas it is not true that all the particles having specific gravity of more than 1.5,will sink, since it can occur only in an ideal gravity separation.

The efficiency of separation using dense media, largely depends on the differences between specific gravities of the ore constituents and the specific gravity of medium.

The dense media formed is affected for, its viscosity, by the size and shape of the particles used to constitute the dense media, and pH of the media. For a given media density, certain minimum size of particles has to be used which can move sufficiently fast through the bath at that density. pH of the media affects the dispersive and coagulative tendencies of the solids in the bath. The rubbing action,and thus,generation of fine slime is dependent on the shape of the particles, and distance created in the medium is proportional to the cross-section of the particles.

In the above discussion, the separation has been considered in still bath, neglecting the disturbance. In separation of continuously arriving stream of sized dry ore, the conditions would be quite different. Assuming that 100 particles of ore are fed, out of which 45 will sink and be withdrawn, 45 will float and be allowed to overflow, and the remaining 10 will form an equipoised teeter. In the beginning, the separation will be clean, but with growing amount of teetering particles and further arrival of middling, the separation will be adversely affected. The sinking fraction will be obstructed and part of its gravitational force will be lost in collision and pushing through the crowded middlings,and as a result,it will be retained there. If bath is left uncontrolled, choking will occur soon near the middlings,and separation would cease. This problem is overcome by incorporating the special arrangements in design of the cone, and suitable controls in the dense media separation.

8.3 Types of Media Employed in DMS

The media used in DMS can be classified in four classes, i.e. (a) solutions of salts in water, (b) organic heavy liquids, (c) autogenous media provided by the ore pulp, and (d) suspensions of solids in water. Nowadays, operation (d) is most commonly practised in commercial plants.

The solutions of highly soluble salts such as $ZnCl_2$ and $CaCl_2$,can be used, provided it does not react with the material treated. At one time, the Lessing process using solution of $CaCl_2$ in water at specific gravity 1.35 was extensively used in cleaning of coal. Various organic liquids of low viscosity are available in the wide range of specific gravity (1.4–3.0),which can be used for sharper separation even on small feed sizes. Some commercially available heavy liquids include, bromoform ($CHBr_3$, specific gravity 2.89), methylene bromide (CH_2Br_2, specific gravity 2.48), tetra-bromo-ethane ($C_2H_2Br_4$, specific gravity 2.96), ehylene di-bromide ($CH_2Br \cdot CH_2Br$), pentachloro ethane ($CCl_2 \cdot CHCl_2$, specific gravity 1.67), methylene iodide ($CH_2 I_2$, specific gravity 3.31), thallous formate solution (H COOT1, specific gravity 3.39) and many others. However, due to various operation problems and losses of costly liquids, very little commercial use could be made and these are mainly used in laboratory and research work.

Pyrites and magnetites can be used as autogenous media in separation of the respective ores.

The solid materials chosen for producing dense media should possess the following properties:

a) The material should be sufficiently hard and should not break or wear down into slime under operating conditions.

b) The material should not be chemically corrosive, and reactive with the ore minerals being treated. For example, if galena is used, it should not oxidise.

c) The resulting viscosity of the medium should be low. It should form a fairly stable pulp with reasonable coarse particles, since very fine material results in high viscosity of the medium.

d) The specific gravity of material should be high enough to give the required bath density under reasonably non-viscous conditions (usually with less than 30 per cent solids).

e) An easy method should be available to clean the media, as the media is to be cleaned before recycling.

f) A certain amount of media may enter the cracks of cleaned and washed ore lumps, which should not upset the subsequent treatment of the ore.

The substances which can meet these requirements are galena (cleaning by flotation of galena), silica (cleaning by flotation of coal), and magnetic materials, such as magnetite and ferro-silicon (cleaning by magnetic methods).

Initially, galena (specific gravity 7.4–7.6) was used in the treatment of lead-zinc ores. The recent trends are towards the use of magnetic materials rather than galena, due to their easy regeneration and lower cost. Magnetite (specific gravity 5.0–5.2) can be used for bath densities below 2.5 and thus it cannot be used in separation of ores where the density of gangue minerals is more than 2.6. It finds an extensive application in cleaning of coal. Mill scale and flue dust from steel plants can also be used as an alternative to magnetite.

Presently ferro-silicon is most widely used due to its magnetic property, its reluctance to oxidise, brittleness, and high specific gravity (6.7–7.0). By using ferro-silicon, suspensions having a specific gravity of 3.2 or somewhat higher can be obtained. Ferro-silicons containing 15 per cent Si are usually employed, since below 15 per cent Si, ferro-silicon is prone to rust and above 22 per cent Si, the ferro-silicon is only feebly magnetic. Sometimes a certain amount of magnetite (10–20 per cent) can be used along with ferro-silicon to adjust the density of the medium. Ferro-silicon for suspensions can either be prepared by grinding lump material (usually in a wet ball mill) or by atomisation of molten alloy. Atomisation process produces the particles of spherical shape which is beneficial due to (i) higher bath densities obtained due to less specific surface area, (ii) reduced loss through drag out, since the possibility of attaching spheroids to the ore products is less, and (iii) reduced rusting, as spherical ferro-silicon is chemically more stable.

8.4 Cleaning and Recirculation of Media

Efficiency and economics of regeneration and recycling of clean media is the critical control factor in DMS processes. The important steps involved in recovery, cleaning, and recirculation of media are as follows:

a) Draining out of media from products leaving the bath.

b) Washing of products to remove remaining adherent media.

c) Collection of foul media, its thickening by suitable methods such as thickening and its cleaning by suitable methods such as flotation (for galena and barytes mixed with coal), magnetic separation (for ferromagnetic materials), or hydraulic separation, such as tabling, mechanical classification, etc. (for sands).

d) Reconstitution of media and its return to bath.

After cleaning, the media is usually obtained in the form of a thick slurry or sludge, which is collected in surge tanks from where it is continuously returned to the separating bath. In case of magnetic materials, the media is passed through demagnetic and de-flocculation systems. The returned amount is always in excess of that removed for cleaning due to small drag-out loss during removal of medium from the products. The density of medium is adjusted enroute from storage to bath.

8.5 Operation of Dense Media Separation

Dense media separation is generally applicable to a feed ranging from about 8 cm down to 2 mm. However, for larger or smaller sizes, special equipments may be designed. Various steps involved in carrying out the separation by dense media separation include preparation of feed suitable for separation, presentation of prepared feed to separating bath, separation of floats and sinks in dense media, withdrawal of products, removal of dense media adhered to products and cleaning, reconstitution and return of media to bath of clean dense media. All these steps can be merged into one continuous process as shown in Fig. 8.1.

Preparation of ore consists of removing the colloidal material, primary slimes and fine ore. These fines reduce the efficiency of treatment due to increase in specific area of the particles. Sometimes a controlled amount of such fine material may be required to obtain stability of the dense media by decreasing its rate of settlement. After removal of fines, the ore is sufficiently washed and then drained before entry to the bath to avoid undue dilution of the media.

The washed feed and reconstituted media are fed continuously into the

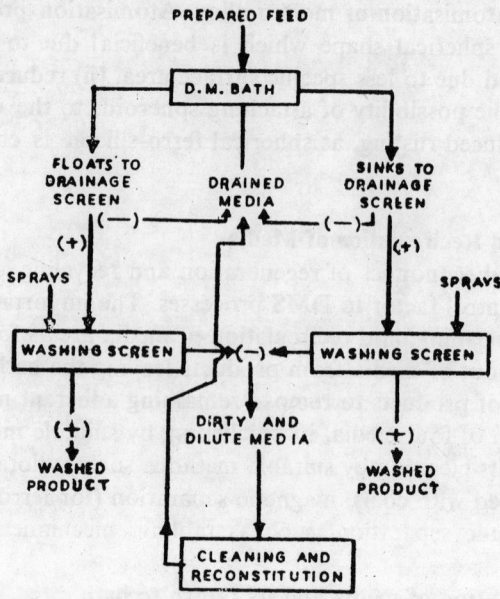

Fig. 8.1. Schematic representation of dense media separation.

bath. Floats and sinks are withdrawn into separate channels/screens and drained off the adherent or drag-out media. The drained media may either be pumped back into the bath direct or given a cleaning treatment before its return. The floats and sinks are washed on screens by water sprays to remove almost all the residual media. The floats and sinks are then transported to separate places. The media removed from the finished products by washing is collected, thickened, cleaned, made up to working density and finally returned to the bath.

8.6 Dense Media Separation Processes

The usual equipments used for these processes are simple in design. These consist of mainly (a) a separator vessel in the shape of cone, pyramid, drum, cyclone, etc., in which the heavy density liquid or pulp is maintained, (b) mechanical or hydraulic means or combination of both to keep the material moving and maintain the pulp in suspension, (c) arrangements for removing adhering particles of medium from the products (usually screens with wash water jets), (d) settling devices to remove excess water from the drained medium, and (e) devices for cleaning of thickened media, e.g. flotation cell for galena, and wet magnetic separator for ferro-silicon and magnetite.

Heavy media separation (HMS) or dense media separation (DMS) is quite economic, simple and an efficient method for gravity separation of coarse and medium sized materials, particularly coal.

Basically the heavy media separation methods can be divided into two classes, i.e. (i) processes using heavy liquids (usually organic liquids and solutions of cheap salts such as $CaCl_2$, $ZnCl_2$, etc.), and (ii) processes employing cheaper solids for suspension, such as galena, magnetite, ferro-silicon, etc.

8.6.1 PROCESSES USING HEAVY LIQUIDS

Earlier processes of heavy media separation employed only heavy liquids. Earlier, extensive use of $CaCl_2$ solution has been made in cleaning of coal, but presently it is not practised. Since, organic liquids such as carbon tetrachloride, are quite costly, their use is restricted to special cases.

Lessing and Bertrand processes employed calcium chloride solution of specific gravity approximately 1.4 to clean coal.

These processes produce extremely clean coal. The major disadvantage of Lessing process was cost for thermal concentration of separating liquor. Bertrand process consisted of five circulating liquors, i.e. hot water, weak solution, medium solution, strong solution, and separating solution and thus it had the advantage over Lessing process that it avoids the costly thermal concentration of the dilute wash liquor. However, the flowsheet of Bertrand process is quite complex.

Du-Pont process employed heavy organic liquids (mixture of several halogenated hydrocarbons) as separating media. Du-Pont process can be used for coarse minerals and is not applicable to fine particles. Consumption of the medium is about 0.5 kg/tonne of coal/ore. This process could not receive much commercial importance due to various working problems and high cost of parting liquid. However, this process can be successfully and economically used for the separation of amber from associated impurities.

8.6.2 PROCESSES USING HEAVY SUSPENSIONS

The use of heavy suspensions in washing of coal has been in practice for the last seven decades, but their application to concentration of ores requiring a specific gravity of about 2.6 is only a recent one. Main advantages of heavy suspensions are, (a) solids used for suspension are cheap, (b) easy and economic regeneration of the medium by flotation or magnetic separation, and (c) possibility of maintaining any desired specific gravity of the medium. A large number of processes based on heavy suspensions and using various materials for medium, have been developed during the last eight decades. Many of them are improved and are still in use, where some new methods have also been developed. The important processes used in industries are discussed below:

i) *Vooys Process*

This process employs the suspension of clay and finely ground baryte (70-100 microns) in water at a specific gravity of about 1.5. This method can be employed for coal of much finer size (about 150 microns) due to the finer solids being used in medium. It is possible to produce coal of high purity containing only 3.5–4 per cent ash with almost the same yield as obtained by other methods. Thickener is employed for regeneration of the medium. The loss of baryte is usually 1 kg/tonne of raw coal.

ii) *Wuensch Process*

This process was adopted for the concentration of ores, where the light constituent (waste) has a specific gravity of 2.7 or higher. Since the separation of ores from gangue will require the specific gravity of the medium to be more than 2.8, and the suspension can have maximum 30 per cent solids for required properties, a medium solid should be of specific gravity 6.7 or higher. Therefore, for this reason either galena or ferro-silicon has been successfully used. Further, these materials of medium can be regenerated easily (galena by flotation and ferro-silicon by wet magnetic separator). This process can be used to separate locked particles from waste and as a preliminary step in treatment of low-grade ores to reject a large tonnage of coarse barren tailing at low cost. This process is not used presently for the technical and economic reasons.

iii) *Huntington–Heberlein Sink–Float Process*

First time this process was used in 1937 at Halkyn Mines, in North Wales for the treatment of ores. This consists of a pyramidal bath and the media employed is galena. The density of media is controlled by the rate of flow of media through the open topped pyramidal bath. The important feature of this process is its quiet bath having no mechanical agitation to stir its contents. This process became less applicable with the depletion of ore grades.

iv) *Differential Density Process*

This is commonly known as Cyanamide Process. In this case, under ordinary working conditions, there is considerable rise in specific gravity of media from the top to bottom of the separating cone (vessel). A schematic flowsheet of the process is given in Fig. 8.2. This process employs a separating tank of conical shape with cylindrical section at the top. The media used is magnetite or ferro-silicon. A cone having a diameter of about 4.5 m can work at a capacity of more than 200 tonnes/day. About 60 tonnes of ferro-silicon charge would be needed for this cone. The *drag out* loss of ferro-silicon is about 0.25 kg/tonne of ore treated. The media of any required operating density can be introduced at any point or set of points along the vertical axis of the cone. A stirring mechanism is employed inside the cone. The upward current of the medium becomes gentler as the horizontal cross-section widens.

For high operating densities, coarse sized ferro-silicon (−200 microns) is favoured and it leads to a higher density differential from top to bottom of the cone and aids in the quick formation of a teeter bed, while for lower densities, finer sized ferro-silicon (−150 microns) is used. A mixture of

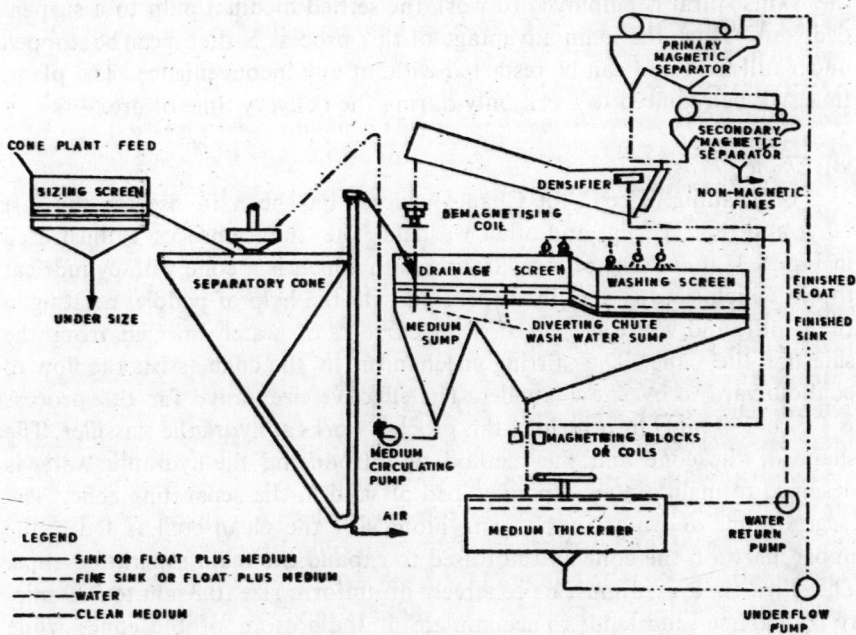

Fig. 8.2. Schematic flowsheet of differential density process.

magnetite and ferro-silicon can also be employed for lower densities.

The ore feed of + 1.7 mm size is charged to the separatory cone. The ore gets separated into floats and sinks, the floats are discharged from the periphery of cone whereas sinks are pumped up by an air-lift. Floats and sinks are conveyed separately by two parallel drainage screens. The drained medium is either pumped back to the separatory cone or a part of this may be sent to wash water which receives all the washings from working screen. Reclamation of drainages is carried out by the use of magnetic field. The magnetite and ferro-silicon particles used as media become flocculated into aggregates due to magnetic changes and thus settle at a faster rate. The thickened sediment is then treated by wet magnetic separators where non-magnetics are separated from the feed. The magnetic flocs are thickened and discharged through a demagnetising coil back to the medium drainage sump. From the sump, the feed is pumped to the separatory cone. The thickener can work as a storage tank during shutting down, as dense medium should not be allowed to settle solidly in the cone, because the restoration of proper fluid condition will be difficult.

This process is capable of treating the ore down to fine gravel size. However, if the viscosity and friction are high, movement of smaller particles will be retarded considerably and thereby reduce the capacity.

v) *Akins Dense Media Separation Process*

This makes use of Akins classifier as a separating vessel. In this system the Akins spiral is employed to work the settled medium pulp to a suspended state. Thus, the main advantage of this process is that it can be stopped under full load and can be restarted without any inconvenience. The plant, thus will be required to work only during the delivery time of ore.

vi) *Chance Process*

For cleaning of coal, the Chance process has been in use for the last 60 yr and replaced jigs and other washers. The Chance process is illustrated in Fig. 8.3. It consists of a separating bath which is a cone with cylindrical top in which sand is kept in suspension with the help of paddles rotating in the middle and with the help of rising currents of water injected from the sides of the cone. The stirring mechanism in the cone assists the flow of coal outward to overflow launder. The effective size range for this process is 1.5 to 250 mm. In principle, this process works as hydraulic classifier. The shape of the cone and the method of introducing the hydraulic water is designed to maintain an expanded bed of sand in the separating zone. The large volume of water overflowing alongwith the clean coal is fed to the upper part of the cone and not used to expand the main separating zone. The silica sand used must be relatively of uniform size (0.3 mm to 200 microns). Coarse sand tends to accumulate in the bottom of the cone, while the fine sand tends to accumulate in the upper layers of the cone.

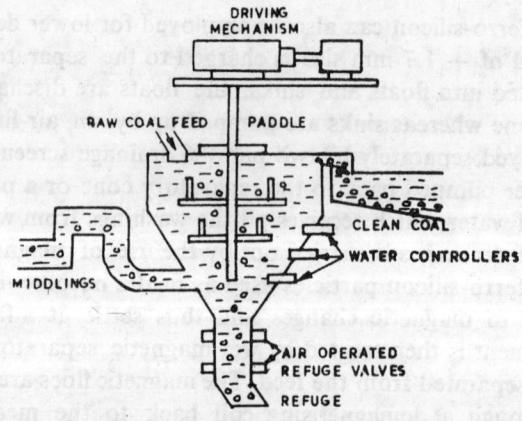

Fig. 8.3. Schematic illustration of Chance process.

The specific gravity of medium is between 1.25 and 1.8, which can be adjusted by varying the proportion of sand and water. Control of gravity is maintained by varying flow rate in the high gravity and low gravity zones. Cleaning of anthracite coal needs higher specific gravity of medium than required for bituminous coals.

The feed is introduced at the top of the cone at one end. The overflow of clean coal and sand passes from the other end over clean coal screens having water sprays which desand and dewater the coal. The heavy shale sinks to the bottom of the cone and is discharged as underflow through an automatically controlled refuse valve on refuse screens, from where the refuse is discarded. The mixture of diluted sand and fine coal is purified in a cone thickener. The sludge coal is wasted and the regenerated sand is returned to the top of the cone. Medium is added continuously to the cone to maintain the medium level up to the overflow lip at the top of the cone. In India this process is being used for coal washing at Jamadoba (Bihar).

A combined process using froth flotation and Chance process may be used, where −1.5 mm material can be removed from the raw feed for froth flotation and the feed of 1.5 to 250 mm coal is treated by Chance process.

vii) *Stripa Process*

This process is used in separation of heavy minerals such as magnetite, employing magnetite suspended in water. Similar to Chance process, in this case also the water rises into the shaking trough through valves. In a typical operation, a fairly coarse magnetite sand of + 0.5 mm size is treated at a medium density of 3.4. The coarse feed entering the separator floats on this teeter or sinks into it. Two fractions are separated by adjustable splitter plate into floats and sinks. The media drains out through washing screens. The washed media is either recirculated to separator or sent for further treatment. Since in this case sufficient media is supplied autogenously, no regenerative treatment is required. The process can also tolerate enough amount of unscreened feed, as there is no problem of media.

viii) *Wemco Drum Separator*

This consists of rotating drum with lifters fixed inside the shell as shown in Fig. 8.4. This employs ferro-silicon as a medium. The feed is charged through a chute at one end of the drum into the medium. The light fraction overflows and is discharged from the other end of the drum. The sink is

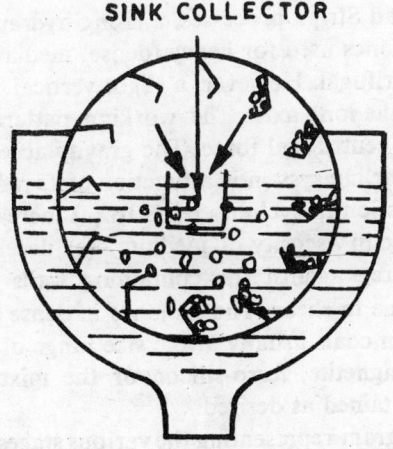

SINK COLLECTOR

Fig. 8.4. Illustration of Wemco drum separator.

lifted by the lifters and removed into a separate launder. The products are conveyed and washed on the screens. Ferro-silicon from the drainage is recovered by magnetic methods and is recirculated. This method can be employed for beneficiation of manganese ore.

ix) *Counter-Current Dense Media Separator*

This is illustrated in Fig. 8.5. This is suitable for large material such as coal, where gentle handling is required. Various substances such as clay, barytes, magnetite, ferro-silicon, etc., can be used in the bath, but finely ground magnetite or ferro-silicon is preferred due to the possibility of their cleaning and reovery by magnetic methods. A continuous flow of media through the separating bath, further reduces the tendency of substances to settle. Separating baths may be designed for two-product division into *floats* and *sinks* or to deliver an intermediate middling in one pass. Alternatively, it can be designed for three products, by operating two baths in series at different densities.

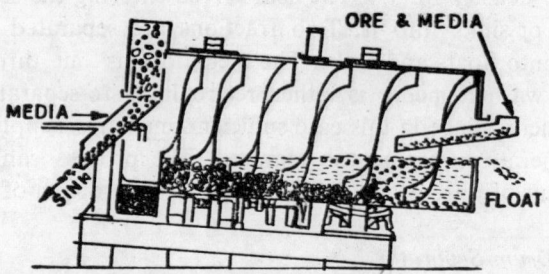

Fig. 8.5. Illustration of counter-current dense media separation.

8.7 Recent Developments

8.7.1 HEAVY MEDIA SEPARATION IN CYCLONES

In the Chance and Stripa processes, a rising hydraulic current is imposed, whereas in cyclones used for heavy (dense) media separation, principal driving force is centrifugal. However, a slight vertical movement is caused roughly parallel to the long axis. The working pattern of cyclones is quite complicated, due to centrifugal force. The gravity acceleration on any particle in an evenly swirling system is a function of its radial distance from the axis. The swirling action in cyclones leads to (a) increase in specific gravity outward, (b) increase in viscosity of medium, (c) deceleration towards the periphery due to drag against the containing walls of the cyclone, and (d) stirring action due to shear. The cyclones in dense media separation can be employed to clean coals usually in the size range of 10 to 0.4 mm. The medium used is magnetite, ferro-silicon or the mixture of the two. The density can be maintained as desired.

A schematic diagram representing the various stages involved in a heavy-media cyclone plant is shown in Fig. 8.6. In operation, six stages are involved, viz. (a) feed preparation, (b) feeding, (c) cycloning, (d) separation, (e) cleaning, and (f) densifying the media and handling of products. The feed comprising of heavy media and coal is introduced at the periphery of the cyclone. The lighter product, i.e. clean coal leaves the cyclone via vortex finder and the heavier, i.e. shale leaves via the apex. Two products are conveyed on separate screens having water sprays. The drainages obtained are thickened and magnetite is recovered by wet magnetic separators. The various factors affecting the separating action are specific gravity of media, solid-liquid ratio, feed rate, constitution of coal (particle shape, size, etc.) and feed velocity.

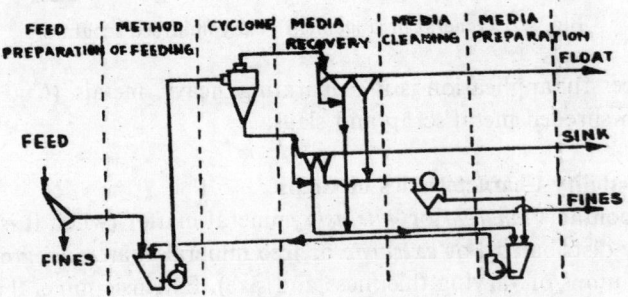

Fig. 8.6. A schematic diagram of heavy media cyclone plant.

In cyclones, the linear velocities of the media and pulp are much higher, which give the better gravity conditions. Though there are certain operational problems, this gives higher throughputs for smaller units. This process finds extensive use in coal cleaning and can be employed successfully for other minerals (separation of diamonds from quartz, uranium and tungsten from gangue, etc.).

A great variety of heavy-media cyclones has been developed, such as Univer, Dyna, Whirlpool, Dutch State Mines Cyclone, Vorgyl separator, etc. The important features of cyclone are high capacity of treatment, low media losses, and high ratio of concentration. In India, most of the washeries in Bihar employ this process.

8.7.2 DRY FLUID-BED SEPARATOR

This uses air fluidised medium of ferro-silicon to effect a sink–float separation similar to that used in heavy-media plants using suspension of ferro-silicon in water. A simplified diagram showing the general arrangement is given in Fig. 8.7. This can be used for separation of material ranging from 5–7.5 cm down to about 0.7 mm. The full-scale separator has about 125 cm wide bed and contains its own magnetic media cleaning circuit. It can treat 5–10 tonnes of ore per hour depending on the density and size-range of the feed. Density of fluid bed is continuously measured and it is linked to the media cleaning circuit to control bed density. Similar separations to those achieved with normal dense media are possible with fluid-bed separator. Its

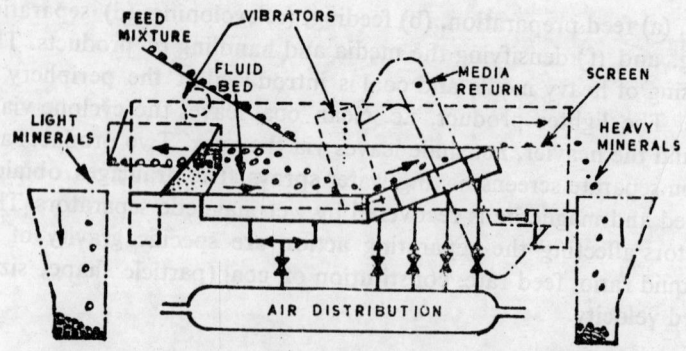

Fig. 8.7. General arrangement of dry fluid-bed separator.

most successful application is in separating heavy metals (Cu, brass, Pb, etc.) from shreded metal scrap and slags.

8.8 Washability Characteristics of Coals

Coal contains *inherent* or *intrinsic* mineral matter (when it is finely and uniformly distributed), or *extrinsic* or free mineral matter (segregated into bends or lumps of varying thickness and size). Extrinsic mineral matter being less intimately associated with the coal substance than intrinsic mineral, can be removed more or less completely by cleaning. An increased ash content of coals or cokes adversely affects the metallurgical operations, and thus upgrading of inferior coals (since better grades of coal are being exhausted at a faster rate) becomes necessary.

Basically coal cleaning can be carried out by wet as well as dry methods. Wet cleaning methods include jig washing, heavy-media separation, trough washing and froth flotation. Dry methods include pneumatic methods using air-currents. The main criteria in coal washing is to obtain the maximum yield of clean coal of the desired ash percentage from run-of-mine coal at an economic price. The most important factor in deciding the choice of the method is the characteristic of the coal, i.e. the ease or difficulty with which it can be separated into clean coal and dirt.

The washability characteristics of coal can be best evaluated by dense-media separation (float and sink test). In this test specific gravity is the only factor affecting the separation. Size and shape of particles have hardly any effect on separation. The moisture in the coal influences the density and thus coal should not be dried before conducting the test.

In most cases, a suitable range of specific gravity of separating fluid is from 1.30 or 1.35 with an increment of 0.05 up to 1.60. Gravity baths can be made from either organic liquids or inorganic salt solutions. Bromoform and carbon tetrachloride are the most common organic liquids. Zinc chloride is the most common salt solution employed for float and sink testing, but calcium chloride can also be used for lower density solutions.

The float and sink test starts at the lowest density bath with the float coal being removed and the sink material is placed in the next higher density liquid. The information required is the weight of float coal at each gravity and the weight of the sink material in the highest gravity bath. Each gravity fraction should be dried before weighing. After sampling, each fraction is analysed for its sulphur and ash content. The washability studies of the coal under consideration are made before building the washing plant. The specifications of the final coal product include the percentage of ash, sulphur, moisture, and ash fusion temperature. Studies of the washability characteristics of coal reveal the following informations:

a) The separating density to produce the specified ash and sulphur content.

b) The per cent recovery and the per cent reject when the coal is washed at the theoretical specific gravity.

c) The percentages of ash and sulphur of the clean coal and the reject.

If a sample of raw coal (to be studied for its washability characteristics) is placed successively in a number of liquids of increasing specific gravities, then a certain proportion will float at each density from the sink in the preceding one. The clean coal floats at the lowest density, whereas the pure dirt sinks at the highest density. If the percentage weights of these float-sink fractions and their ash contents are determined, the results obtained can be plotted as characteristic curves (float curve and a sink curve) which will represent the composition of coal and its washability.

Figure 8.8 represents the washability curves for easy, normal and difficult coals. The characteristic curve for easy coal (curve A) in Fig. 8.8 (a) shows an abrupt flattening at a specific gravity of 1.5 and at this point an effective separation between coal and dirt can be readily obtained. This point represents an yield of 67 per cent of clean coal with an ash of 10 per cent. The yield and cumulative ash percentages of the floats or clean coal are obtained from the float curves *B*, whereas the cumulative ash of the sinks is obtained from the curves *C*, in case of easy coals this value is about 70 per cent. In case of normal coals (Fig. 8.8 (b)), it may be observed that there is no sharply defined natural cut point. The separation can be achieved without great difficulty by employing a specific gravity of 1.5 or 1.6. A

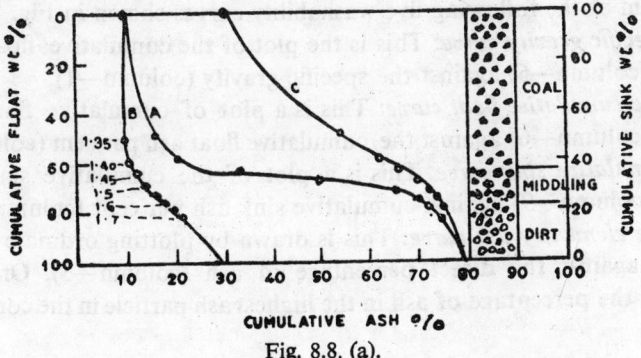

Fig. 8.8. (a).

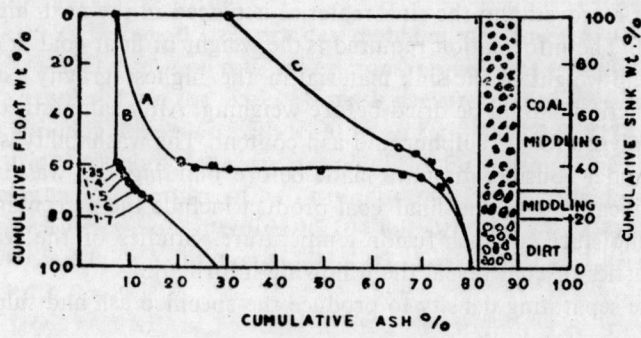

Fig. 8.8. (b)

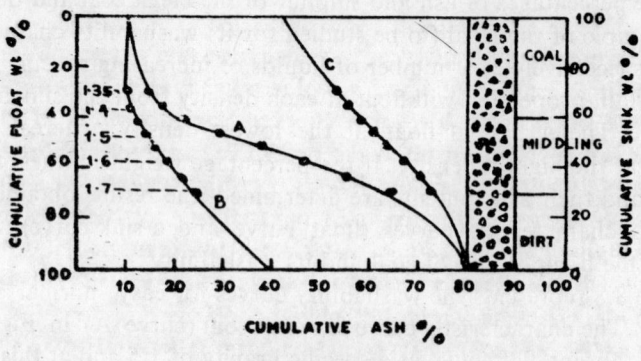

Fig. 8.8. (c)

Fig. 8.8. Washability curves for easy, normal, and difficult coals.

small proportion of middlings would be present in both the products. In case of difficult coals [Fig. 8.8 (c)], the characteristic curve represents a gradual increase in yield and ash from pure coal to pure dirt. In case of difficult coals, the cut point is purely arbitrary, which depends on the maximum yield of marketable product.

An example of washability data is given in Table 8.1, which can be plotted in the form of the following five washability curves shown in Fig. 8.9.

a) *Specific gravity curve*: This is the plot of the cumulative float weight per cent (column—6) against the specific gravity (column—1).

b) *The cumulative float curve*: This is a plot of cumulative float weight per cent (column—6) against the cumulative float ash per cent (column—7).

c) *Cumulative sink curve*: This is a plot of the cumulative sink weight per cent (column—9) against cumulative sink ash per cent (column—10).

d) *The elementry ash curve*: This is drawn by plotting ordinate D (column—12) against the direct percentage of ash (column—3). Ordinate D represents the percentage of ash in the highest ash particle in the correspond-

Table 8.1. Example of washability data

Specific gravity fractions	Wt.% float	Ash %	Ash product (column 2 × column 3)	Cumulative float			Cumulative sink			± 0.1% material	Ordinate D
				Ash (cumulative summation of column 4)	Wt.% (cumulative summation of column 2)	Ash % $\frac{\text{column 5}}{\text{column 6}}$	Ash (cumulative summation of column 4 from bottom)	Wt.% (cumulative summation of column 2 from bottom)	Ash % $\frac{\text{column 8}}{\text{column 9}}$		
1	2	3	4	5	6	7	8	9	10	11	12
F 1.35	44.2	4.3	190	190	44.2	4.3	1,498	100.0	15.0		22.1
S 1.35-F 1.40	24.2	9.6	232	422	68.4	6.2	1,308	55.8	23.4		56.3
S 1.40-F 1.45	11.4	16.2	185	607	79.8	7.6	1,076	31.6	34.1	45.5	74.1
S 1.45-F 1.50	6.0	23.7	142	749	85.8	8.7	891	20.2	44.1	23.3	82.8
S 1.50-F 1.55	3.9	30.3	118	867	89.7	9.7	749	14.2	52.7	13.0	87.8
S 1.55-F 1.60	2.0	37.6	75	942	91.7	10.3	631	10.3	61.3	7.4	90.7
S 1.60-F 1.70	1.5	43.3	65	1,007	93.2	10.8	556	8.3	67.0	1.4	92.5
S 1.70-F 1.90	1.4	53.1	74	1,081	94.6	11.4	491	6.8	72.2		93.9
S 1.90	5.4	77.3	417	1,498	100.0	15.0	417	5.4	77.2		97.3

Based on data from Refs. 6, 9, and 11.

ing specific gravity fraction. The value of D may be determined as follows:

Ord. $D = A + B/2$

where A = cumulative weight per cent of the float material down to the specific gravity fraction being considered, and

B = weight per cent of the material (column—2) in the gravity fraction.

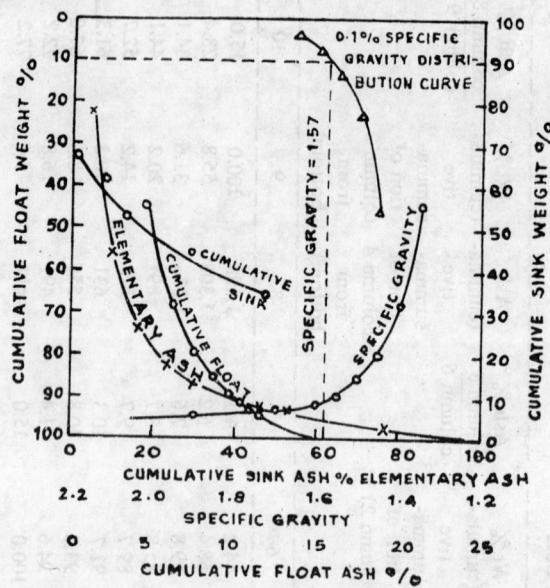

Fig. 8.9. Washability curves of data given in Table 8.1.

e) *A ± 0.10 per cent specific gravity distribution curve*: This is drawn by plotting ± 0.10 per cent values in column—11 using the cumulative float weight per cent scale, against the corresponding specific gravities. The amount of·material in the gravity fractions, 0.10 above and 0.10 below is calculated. For example, if the cumulative float in 1.55 gravity fraction is 89.7 per cent and in 1.35 gravity fraction is 44.2 per cent, then the amount of material in the gravity fractions between 1.55 and 1.35 will be 89.7 per cent −44.2 per cent = 45.5 per cent, which is the ± 0.10 per cent in the 1.45 gravity fraction.

A point at the 10 per cent cumulative float weight per cent line is intersected by ± 0.10 per cent curve, representing the specific gravity, which may be considered as the boundary between a moderately difficult and difficult separations.

CHAPTER 9

Gravity Concentration in Vertical Currents (Jigging)

9.1 Introduction

Jigging may be considered as hindered settling, consisting of stratification of the particles into layers of different densities caused by repeated upward and downward current of fluid to a very thick suspension of the mixed particles to settle or fall for short periods of time. A pulsator forces the water up through the screen with adequate velocity to bring all the particles momentarily into suspension resulting into fluidised condition. This upward movement is called pulsion and the minerals of two or more specific gravities arrange themselves according to the law of hindered settling (Ch. 7). The water is then allowed to drain back through the screen and this downward movement is called suction. During this action, small grains move downward through the interstices between the large grains. The cycle of pulsion and suction is repeated continuously. Finally the stratified layers are discharged into concentrate, tailing, and middlings.

Details of cycle are important and vary from case to case. The cycle normally includes a very sudden upthrust, a period of free fall (in which hindred settling and differential acceleration bring the dense particles below the lighter ones), and finally draining or suction period, during which small particles (particularly small and dense) are drawn low in the bed and even some may pass through the grid. A schematic representation of a jig is given in Fig. 9.1. Further, after the large particles come to rest on a grid, the small ones will continue to fall and may trickle through the interstices between the big ones. This *consolidation trickling* can be exaggerated by allowing the water to drain through the arrested solid bed for a short duration.

Jigging operation yields three products, i.e., (a) gangue tailing skimmed over a weir, (b) concentrate drawn via a gate which excludes the tailing, and (c) a hutch product (usually dense fines) drawn from a spigot at the bottom of the cell. However, the separation is not clean enough and thus jigs are operated in series or in batteries to produce a concentrate and a tailing. The middling produced may be re-ground and returned for further sorting.

In continuous jigs, a surface carrying current (usually caused by mecha-

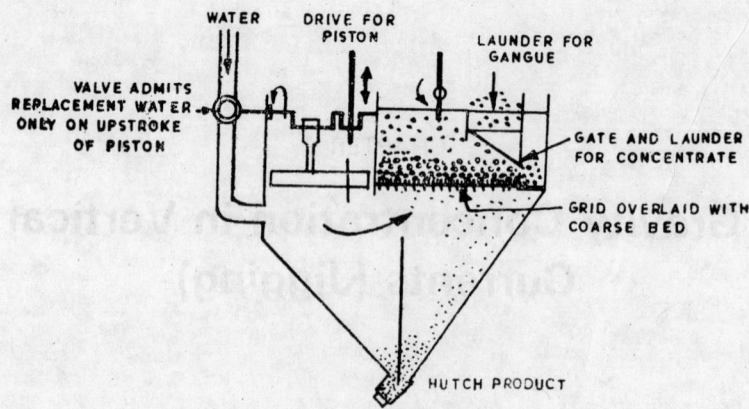

Fig. 9.1 A schematic illustration of jig.

nical devices) helps in transporting of lighter grains forward until they are discharged over the tail.

9.2 Theory of Jigging

The classification is based on the maximum or terminal velocity of a particle, which is a significant characteristic determining the separation of particles. If the particles are given only very short settling periods, they will never attain a maximum or terminal velocity, and the separation will depend on the initial settling velocities of the particles. The velocity of the particles at the outset of settling is extremely low and the resisting forces due to frictional effects are not developed. Equations (7.7) and (7.12) represent the initial acceleration. Since the value of resisting forces due to frictional effects is practically zero, the initial acceleration in both the cases becomes

$$a = \frac{dv}{dt} = \left(\frac{\rho_s - \rho_l}{\rho_s}\right) g = \left(1 - \frac{\rho_l}{\rho_s}\right) g \tag{9.1}$$

This initial acceleration a of settling depends on the force of gravity, and the densities of the particle, pulp and the fluid, and does not depend on the size or shape of the particle. Equation (9.1) shows that the initial acceleration is maximum in case of most dense and largest particles. This situation indicates that separation of two minerals according to their density may be possible by providing extremely short durations of settling. The relative acceleration or relative initial velocity of particles of two different materials X and Y (value and gangue) can be obtained as follows:

$$\frac{a_x}{a_y} = \left(\frac{\rho_A - \rho_l}{\rho_B - \rho_l}\right) \frac{\rho_B}{\rho_A} \tag{9.2}$$

The difference between sorting on the basis of initial velocity and terminal velocity is shown in Fig. 9.2. This figure represents the relative velocity of settling as a function of time for different particles. For example, a particle of heavy mineral (curve-1) having same shape and size as particle of

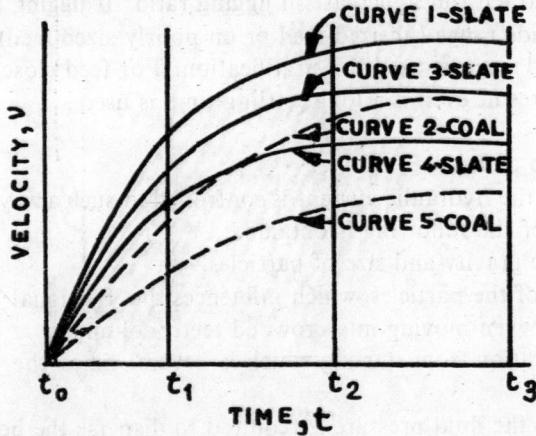

Fig. 9.2. Sorting difference on the basis of initial velocity and terminal velocity.

light mineral (curve-2) will settle faster than the particle of light mineral due to greater density. A smaller particle of heavy mineral may be considered as shown by curve-3, which will have a terminal velocity equal to that of larger particle of light mineral (curve-2). A still smaller particle of heavy mineral (curve-4) will have a still slower terminal velocity. Light mineral of the same size as the smallest heavy particles will always settle much slower (curve-5) than the heavy mineral. However, the initial velocity or acceleration of all heavy particles is greater than that of light particles.

The distance of fall for a particle in time t can be represented by the area under the curve $\left(\int_{0}^{t} v\, dt \right)$ in Fig. 9.2. If the time of fall is quite short, e.g. t_1, all heavy particles will fall to a greater distance than all light particles, and thus sized feed can be completely sorted into clean heavy mineral concentrate at the bottom and clean light fraction (tailing) at the top. If the time of settling is extended to t_2, when areas under curve-2 and 4 are approximately equal, it will not be possible to separate the smallest heavy particles (curve-4) from the largest light particles (curve-2) as they are *equal jigging*, i.e. they fall to an equal distance in a given settling period. The ratio of the diameters of these particles is termed as *jigging ratio*, which is similar to the settling ratio. The jigging ratio represents the maximum size ratio that can be completely sorted under the given conditions.

If settling periods are further extended beyond t_3, the terminal velocity controls the separation, and it would not be possible to make an effective separation between light and heavy particles with the size distribution indicated for this feed.

As shown in Fig. 9.2, the jigging ratio greatly depends on the duration of settling. When the duration of settling is decreased from 0.50 to 0.10, sec. the jigging ratio becomes four times, whereas decreasing the duration to 0.05

sec results into ten times increase in jigging ratio. If jigging is practised on a feed of a wide range (unsized feed or on poorly sized feed), a very short settling period must be used for stratification. For feed closely sized, stratification may result even if a long settling time is used.

9.3 Operation of Jig

In jigging the hydraulic current is controlled in such a way that optimum exploitation of the following is obtained:

a) Specific gravity and size of particles.

b) Shape of the particles, which influences the frictional resistance and cross-section when moving in a crowded teeter column.

c) Acceleration from starting which is several times the whole mass in the bed.

In jigging, the fluid pressure is required to disperse the bed of material into individual teetering particles, as well as to reverse this action to start their settling back into solid mass.

Each pulsion should impart sufficient energy to the separating fluid to lift all the particles except the heaviest ones, to expand the jig box contents to teetering state. Therefore, thrusting action needed should be more powerful for heavier particles. The various components of jigging column effecting separation are:

a) Amount of fluid in motion above screen.

b) Horizontal area of cross-section through which the fluid passes upward and downward in each jigging cycle.

c) The rate of build-up of heavy particles on the screen.

d) The maximum velocity of the fluid attained in each cycle.

e) The point of the half-cycle where the maximum rate is achieved.

The factors determining the volume of water rising and falling through the jig screen in each complete cycle are:

a) Length of stroke of the piston.

b) Leakage around the plunger.

c) Effective area of energing unit, i.e. area of piston, diaphragm, etc.

d) Effective area of jig screen, i.e. open area.

The velocity of the rise and fall varies from zero at the beginning to again zero at the end, passing through a maximum value which depends on the volume and cross-sectional area. This area may be modified by the porosity of the jig bed at any horizon and at any time, which changes continuously. The jigs are run usually at speeds 60–300 rpm. Thus each half-cycle takes only about one-sixth to one-tenth of a second from zero to zero.

The jigs are usually driven by an eccentric or linking motion and the jigging cycle is more complex than an unidirectional teetering column. The most common fluid used in jigging is water, but air has also been used in pneumatic and fluidised bed jigs. The water can be set into motion in several ways, such as by gentle oscillation between the limbs of U-shaped box in which pulsing force is applied by a low pressure air introduced through valves, or by moving the jig box itself up and down in a tank of water.

Since smaller particles pass through the screen during suction, there is a size limit to feed of about + 1 mm for treatment in jigs as the screening of smaller sizes is slow and expensive. If the sizing problem can be overcome, jigs can be used in place of shaking tables for the treatment of sand ores.

9.4 Factors Affecting Stratification in Jigging

As a result of repeated cycles of pulsion and suction, the individual particle finds its position in its most stable horizon in the mass and becomes stratified with other particles. This stratification taking place in jigging is influenced by the following factors:

a) Size range of the total load in the jig box.

b) Size, shape, and specific gravity of different particles.

c) The apparent density of the bed.

d) Average particle shape in the bed.

e) The density of surrounding particles.

f) The pulsion and suction rates of fluid and height of rise.

g) Back pressure exerting on its horizon and resisting its freedom of movement.

h) Volume of voids and their distribution, layer by layer.

The above factors are common to batch loaded jigs as well as continuously loaded jigs. In addition, the following factors affect the stratification when the feed is introduced and the products are withdrawn:

a) Horizontal stratifying influence of cross-streams.

b) Gradual changes in density between the feed-end and discharge-end of the jig box.

c) Rate of lateral transport of the jigged strata.

d) Amount and percentage removed of heavy mineral during passage across jig box.

e) Difference in densities of withdrawn and overflowed (rising) particies.

In addition to the above factors, there are some more, which indicate the complexity of jigging operation, and thus there should be an emperical approach for individual ore.

9.5 Jigging Cycles

The movement of hydraulic current follows the form of a sine-curve as shown in Fig. 9.3 (a). The inertia of the particle is overcome at some point on the upward portion of the curve during its rising and acquires the kinetic energy from the water. In case of light and small particles, the movement begins during the build-up of hydraulic energy, whereas in case of large and heavy particles, the movement starts while the water decelerates towards zero from the top of the pulsion stroke. Therefore, as long as the water moves faster than the particle, the latter receives the energy continuously from the water. During suction, the particle uses the excess energy it has stored in continuing its rise, and then begins to fall, and follows a path of a sine-curve of smaller vertical amplitude compared to that of pulsing water. For a sufficiently heavy particle, a point may be reached where the particle continues to descent even when the water is rising on its next pulsion. The point of reversal of motion, and the direction and strength of water flow at the point depends on the mass of the particle. One particle

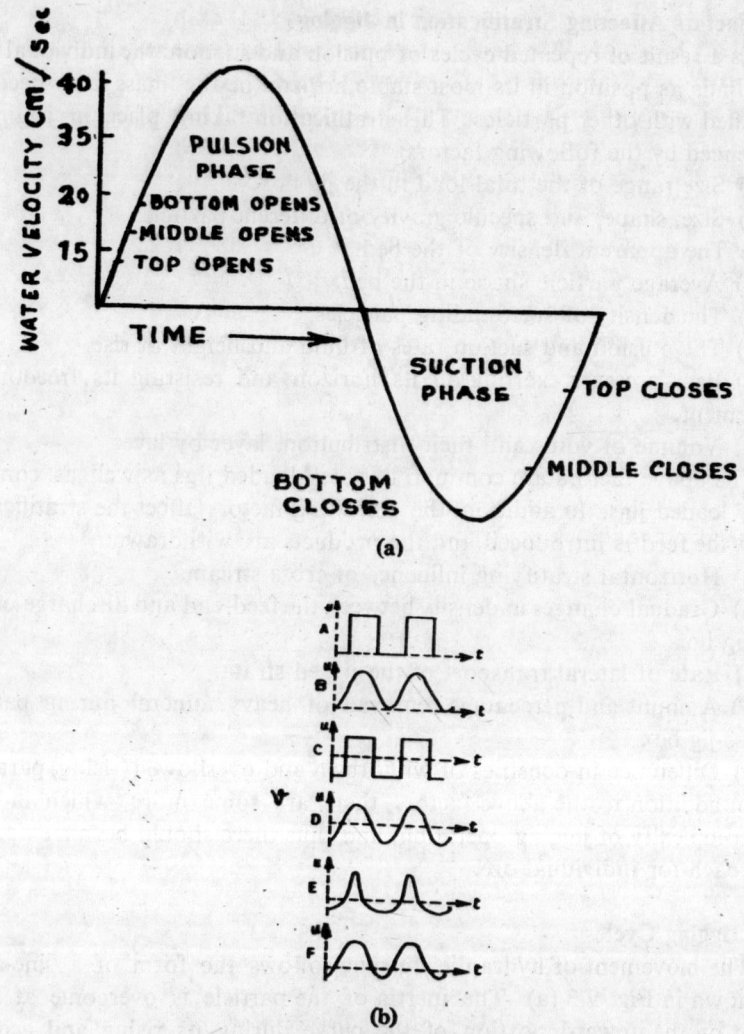

Fig. 9.3. Jigging cycles:
(a) sine-curve as followed by movement of hydraulic
current, and (b) some possible jigging cycles.

may begin to fall during the dying away of pulsion, whereas another may be carried by its stored energy over to the suction stroke.

The jigging cycles consist of pulsion (fluid moves upward) and suction (fluid moves downward). The movement of fluid is with respect to a stationary point (screen). All jigs use pulsion, most jigs suction but the latter may not be used in some cases.

The form of the jigging cycle may be different in different cases, where the position and the amplitude of the curve differs. In some jigs, suction may not be employed and only pulsion stroke is used. Some possible jigging cycles are shown in Fig. 9.3.

9.6 Stratification During Jigging

Stratification during two jigging cycles is shown in Fig. 9.4. One cycle is shown at the beginning of the treatment and one on completion of stratification. In the pulsion stroke (*A*) water is forced through the screen and the mixture of closely sized heavy and light mineral particles is dilated and teetered. Then the pulsion diminishes and dies out, and the suction stroke starts. During suction, heaviest particles are the first to be influenced by the reduced hydraulic lift and thus fall quicker and faster. As soon as the teeter collapses, the bed as shown in *B* (Fig. 9.4) will be formed where the heavier particles begin to concentrate at the bottom and lighter ones at the top. The stratification is further aided by the fact that heavier grains possess a higher kinetic energy and thus they are able to push them through the lighter grains even when the bed is close. (*Y*) and (*Z*) represent the stratification position after a number of cycles, when the strata of particles are formed by their mass and shape.

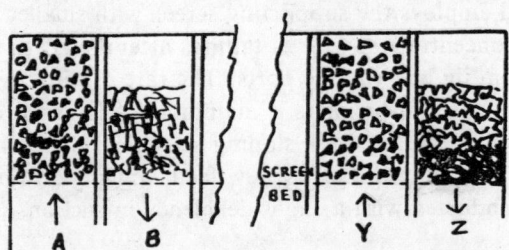

Fig. 9.4. Stratification during two jigging cycles.

If the feed is long ranged, different particles (either of same or of different density) do not travel the same distance during a particular settling period allowed to them, and thus they will come at rest at different instants. The coarse particles may remain in suspension for a very short duration of the cycles (e.g. 0.05 sec out of a cycle of 0.25 sec), whereas the small particles may remain suspended for longer duration (e.g. 0.2 sec out of 0.25 sec). However, a period of time exists, during which the small particles settle on top of a bed of larger particles. The larger particles bridge against each other and become incapable to move further, whereas the fine particles still remain free to move. In addition to the settling of these small particles under the influence of fluid velocity, they settle under the influence of gravity in the voids between the larger particles and this is known as *consolidation trickling*.

The stratification in jigging may thus be summarised as following:

During the stage of bed being open, the stratification is essentially controlled by hindered settling classification modified by differential acceleration. This process brings the coarse heavy grains at the bottom, the fine–light grains at the top, and the coarse–light and small–heavy in the middle. When the bed is close, the stratification is controlled by consolidation trickling which does the reverse of the previous case, i.e. it brings the fine–heavy grains at the bottom, the coarse–light grains at the top, and the coarse–heavy and fine–light grains in the middle. By adjusting the time of these two processes, almost perfect stratification can be obtained according to density alone. The small particles pack the voids between the larger ones and thus increase the bed's resistance to penetration (consolidation trickling). This packing effect is important in determining the size range of feed. The greater the dilation of the bed, easier is the movement of particles. The dilation can be varied by varying both the quantity of water used and the force applied to it.

9.7 Methods of Jigging

Jigging methods are mainly classified in two classes based on arrangements of withdrawing the products, i.e. (a) jigging on screen, and (b) jigging through screen.

9.7.1 Jigging on Screen

This method employs the supporting screen with smaller apertures than the feed size. Concentrate as well as tailings after their desired stratification are removed from jig bed via side ports. The rate of withdrawal is so adjusted that a bed of desired thickness is maintained. The travel of bottom layer may be facilitated by the use of a sloping screen. The top layer may be removed through crowding of bed by new feed. Jigging on the screen should preferably be conducted with a slight deficiency in suction.

9.7.2 Jigging Through the Screen

This method employs the screen of larger apertures than that of feed size. An oversized bed of selected material (broken ore, steel shot, metal discs, steel punchings, etc.) called the bedding or ragging is required on the sieve to prevent the passage of the light product into the hutch and permitting the passage of the heavy product in the hutch. The bed is maintained in a suitable thick layer. The bed evens out the upthrust during pulsion giving better stratification and provides a controlling barrier preventing the passage and mixing of too light particles with concentrate. These particles of the bed act like inefficient valves which open during the violent pulsion stroke and shut down (pack tightly) under the combined force of a suction stroke and the gravity effect of the mass above them.

9.8 Types of Jigs

In general, the jigs are tanks of rectangular cross-section with a screen fitted below the rim or overflow at a short distance in a horizontal or slightly sloping position. The screen or water is given a pulsating or jigging motion which causes an upward and downward motion of fluid through the screen. The feed is introduced over the screen at one side and subjected to a series of short settling periods as it moves across the screen to overflow. The jigs can be classified based on the fluid used, i.e. hydraulic jigs (using water) and pneumatic jigs (using air). The hydraulic jigs can be further classified into hand jigs, fixed screen jigs, and movable screen jigs.

9.8.1 HAND JIG

This is one of the oldest and simplest, and presents the essential features of the jigging process. This consists of a water-tight box or hutch in which a framed sieve (an ore box) can be jigged up and down. The accumulated heavy mineral is removed from time to time after a series of stratifications carried out by moving the screen up and down through the hutch water. Presently hand jigs are not in use since the jig motion can be duplicated with mechanical devices.

9.8.2 FIXED SCREEN JIG

In this type of jigs the currents are produced either by a plunger or by a stream of hydraulic water. The fixed screen jigs are more common. The important fixed screen jigs are described below:

a) *Fixed Screen Plunger Jigs (Harz Jig)*

This is the oldest type of jig in use. This was developed in the Harz Mountains in Germany to treat lead–zinc ores mined there for several hundred years. The Harz jig as shown in Fig. 9.5 is usually made of steel or concrete. It has a fixed screen consisting of an ore box which occupies about half the plan area of the hutch (compartment in a water filled tank). The other half of this compartment consists of a loosely fitting plunger which moves vertically to pulse the water upward and downward. Additional hy-

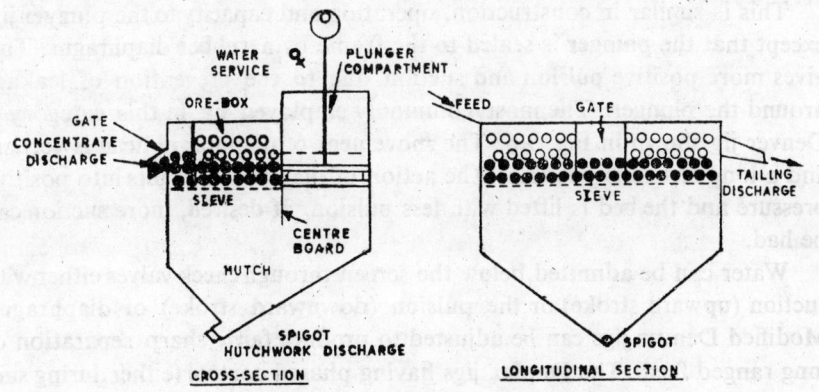

Fig. 9.5. Fixed screen plunger jig (Harz jig).

draulic water is used to aid in controlling the tightness of the closed bed. Several successive compartments (usually four) are placed in series in the hutch. The sized feed is charged gently and evenly at the head of the first compartment without upsetting the stratifying work.

The controls are so manipulated (maximum amplitude of jig in the first compartment and minimum in the last) as to produce high-grade concentrate in the bottom of the first compartment. The lighter fraction stratifies upward and passes over from the far end of the first compartment to the second compartment, and passes further to the next compartment till it is finally discharged as tailing from the last box. The control over the pulsion and suction amplitudes can be obtained by controlling the opening of hutch discharge and by varying the amount of rising water.

The tailings (upper layer) can be discharged over a weir at the side opposite the feed. The concentrate (lower layer) can be discharged either through the gate, (when feed is coarser than aperture of screen) or as hutch product through a valve (when feed is finer than aperture of screen).

The length of the stroke depends on the size of the feed and may range from 0.5 to 8 cm (coarse feed requires long strokes). Jigging cycles may range 100 to 300 strokes per minute. The power required is about 1 KW/ sq. m of the screen area. The capacity of jigs is usually 10–40 tonne/sq. m of screen area per day. Coarse feeds give higher capacity. The amount of water used varies appreciably from one mill to another. Usually 20 tonne of circulating water is required per tonne of ore treated in four compartment Harz jig.

Many variants of Harz jigs having different plunger mechanisms and product discharge systems have been developed. Many modern jigs use flexible diaphragms to produce the stroke in place of plungers. The diaphragm jigs permit higher running speeds and are more economical of water. Harz jigs are generally employed in treating metallic ores in the approximate range of 3.5 to 0.025 cm particle size.

b) Fixed Screen Diaphragm Jig (Denver Jig)

This is similar in construction, operation and capacity to the plunger jig, except that the plunger is sealed to the frame by a rubber diaphragm. This gives more positive pulsion and suction due to the prevention of leakage around the plunger. The most commonly employed jig in this category is Denver jig shown in Fig. 9.6. The movement of diaphragm head is vertical and a long stroke is possible. The action of diaphragm results into positive pressure and the bed is lifted with less pulsion. If desired, more suction can be had.

Water can be admitted below the screen through check valves either with suction (upward stroke) or the pulsion (downward stroke) of diaphragm. Modified Denver jigs can be adjusted to produce fairly sharp separation of long ranged feed. The Duplex jigs having phased control (either during suction or during pulsion) of hydraulic water can be used in open circuits, where-

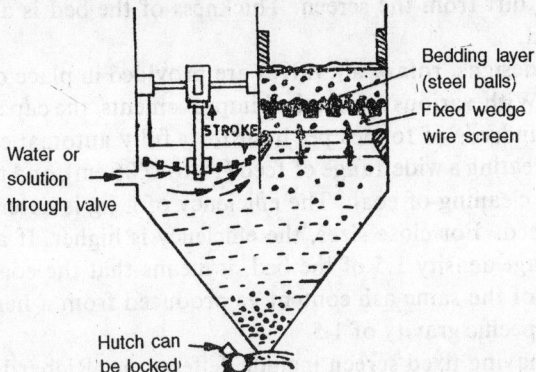

Fig. 9.6. Fixed screen diaphragm jig (Denver jig).

as the ordinary one can be used in closed circuits. One important application of Denver jig is the roughing of diamond gravels.

Another jig of fixed screen diaphragm type is Bendelari jig, in which the diaphragm is reciprocated from below.

c) *Fixed Sieve Air-Pulse Jig (Baum Jig)*

The most important of fixed sieve air-pulse jigs is Baum jig (Fig. 9.7) developed in Europe in 1892. The distinct feature of this jig is that the water is pulsed by means of compressed air and no suction stroke is used. The bed recloses by unassisted gravitation. The jig box has a U-shaped cross-section. Two compartments of such a section forms a complete jig. The sieve of the first compartment slopes backward towards the feed end. The feed is admitted gently below a very shallow baffle to ensure wetting and prevent excessive disturbance to upset the stratification. The air pressure used is approximately 0.1 to 0.15 kg/cm². Most of the dirt falls in the first compartment and a part of it falls through the screen to a spiral conveyor, which takes it out. In the modern Baum jigs, an automatic control is provided to remove

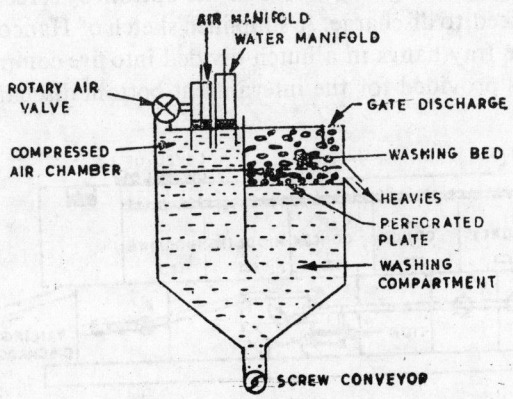

Fig. 9.7. Fixed sieve air-pulse jig (Baum jig).

the buildup of dirt from the screen. Thickness of the bed is also automatically controlled.

In the new designs, rotary air valves are provided in place of older slides and plungers. With various mechanical improvements, the capacity of jig has increased to hundreds of tonnes per hour for a fully automatic jig. Baum jig is capable of treating a wide range of feed (15 to 0.06 cm), and thus it is most widely used in cleaning of coal. The efficiency of a jig is determined by the size range of feed. For close sizes, the efficiency is higher. If a jig is operated on an average density 1.5 of the bed, it means that the coal produced by this should be of the same ash content as produced from a heavy-media separation at a specific gravity of 1.5.

Other jigs having fixed screen include Jefferey jig, Richards pulsater jig, Pan American jig, Simon Craves jig, etc. They are similar in principle and vary in fixing of screen, withdrawal method, mechanism of pulsion of water, etc.

9.8.3 MOVABLE SCREEN JIG

In this type of jigs, the screen box is moved upward and downward with fore and aft motion in a tank of water either by hand or power. Acceleration at one end is more than at the other to cause the bed to move forward. The compound motion can be obtained from a cam through a set of levers, links, and rocker arms.

The jigging is carried out through the screen. The concentrate accumulates in the hutch of the first compartment, whereas the middlings are collected in the hutch of the last compartment. The other compartments make concentrate or tailing, depending on the characteristics of the ore and requirements. The important jigs of this class are Hancock jig, James coal jig, and the Halkyn jig. These jigs can be employed to treat comparatively small amounts of ore, where the mines are pockety and a large mill does not pay.

a) Hancock Jig

This consists of a long tray closed at the bottom by screens increasing in aperture from feed to discharge. A simplified sketch of Hancock jig is shown in Fig. 9.8. The tray hangs in a hutch divided into five compartments. Each compartment is provided for the intermittent bottom discharge for the con-

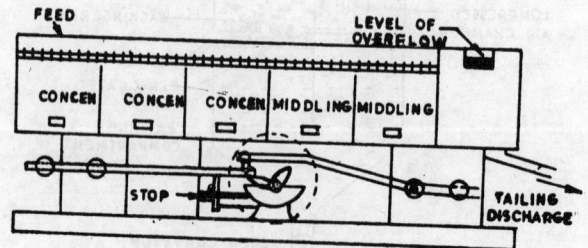

Fig. 9.8. Movable screen type Hancock jig.

centrates, accumulated in the hutch. The tray is lifted by cam-gear to provide the forward and upward throw to the feed. The jigging cycle thus results into stratification as well as a positive push of the bed from feed to discharge. The return stroke can be made mild by introduction of hydraulic water into the hutch below the tray. The Hancock jig is noted for its higher capacity (300–600 tonnes/day) for a jig 8 m × 1.25 m.

b) James Coal Jig

This jig is also known as pan-jig and used for cleaning of coal. In principle, this jig is similar to Hancock jig. The sieve is moved by a crank mechanism and water is introduced to the hutches to reduce the suction. Jigging is carried out on the sieve. A bed of very heavy ragging is used to prevent passage of slate (shale) in the hutch. The cleaned coal is discharged from the end of the jig and the slate is removed from the side passage under a baffle by a gate and dam discharge system.

c) The Halkyn Jig

This was developed to concentrate coarsely crystalline lead–zinc ore occurring in limestone gangue. It has a tank divided into several compartments and a screen box closed with a wedge–wire screen bottom. Similar to Hancock jig, it also receives vertical and horizontal motion which stratifies the feed and helps it to move toward the withdrawal end. Hydraulic water can be introduced below the screen. The fine particles fall through the screen to the hutch. The concentrate is removed through an adjustable gate in the discharge end of each compartment of the screen box.

9.8.4 PNEUMATIC JIGS

Pneumatic jigs (Fig. 9.9) make use of air in stratification of particles. This is particularly suitable when a dry product is to be obtained or water deteriorates the characteristics of the material. The separation achieved is usually inferior, owing to the lower density of the fluid. At one time, al-

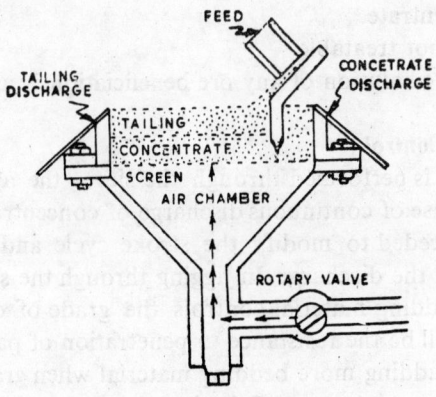

Fig. 9.9. Pneumatic jig.

though this jig treated 30–40 m tonnes of coal annually, now it has become only a technological curiosity. The plumb pneumatic jig (Fig. 9.9) is a pulsator jig which is controlled by a rotary valve using 400–500 pulsations per minute. Capacity for the coarsest feed is about 5 tonnes/sq. m/hr. Earlier this jig was also employed for concentration of sulphides. For effective concentration, the feed is required in closer sizing than required in hydraulic jigs. Specifically, pneumatic jig cannot be employed to very fine material, since it creates a dust problem.

9.9 Advantages and Disadvantages of Jigs

Pneumatic jigs offer the following advantages:

a) A dry finished product is obtained, which is particularly important in treatment of fine coal. Dry coal also gives better appearance and less abrasion losses in transit.

b) Process is adoptable in desert countries.

Some disadvantages of pneumatic jigs are:

a) Necessity to dispose and remove the dust.

b) The separation obtained is somewhat inferior.

c) Though pneumatic jigs can be used to treat as fine as 200 micron size and as coarse as 2.5 to 4 cm size, one single jig cannot treat a wide range of sizes.

Hydraulic jigs offer the following advantages:

a) These can be used to reject waste at a coarser size. The middling (low-grade concentrate) product can be further crushed and ground and then treated by flotation.

b) Hydraulic jigs are cheap and simple to operate.

c) These jigs can be inspected easily.

d) Heavy suspensions may be used in place of water for jigging course material.

Some disadvantages of hydraulic jigs are:

a) Large quantity of water required.

b) Usually incomplete liberated ores are concentrated and thus do not yield clean concentrate.

c) Fines are not treatable.

d) A complete solution of any ore beneficiation is usually not provided.

9.10 Operation Control

When jigging is performed through the sieve, the apertures should not be blinded. In case of continuous discharge of concentrate from the hutch, excess water is needed to modify the stroke cycle and to compensate the amount lost with the discharge. In jigging through the screen, thickness and density of the bedding material controls the grade of concentrate (thicker the bed, more will be the resistance to penetration of particles). The bed can be thickened by adding more bedding material when grade of concentrate is to be raised. Whereas, in case of jigging above the screen, slower the feed rate, more severe is the selective action and higher will be the grade of concentrate.

In both the cases, when other conditions are adjusted properly, and the tailings are high in value, both the feed rate and the quantity of hydraulic water should be reduced. This provides more dwelling time resulting into better grade of concentrate and low-value discharge in tailing.

For an efficient operation, feed should be sized from 5 cm downward. Another control is speed and stroke amplitude. Heavy mineral jigs need speeds varying from 100 rpm with 5 cm stroke amplitude for coarse feed to 300 rpm with 3 mm stroke for fine feed. The coal jigs usually work at lower speeds (50–60 rpm).

The factors controlling the operation and its efficiency can be summarised as below:

    a) The specific gravity of the oversize particles in the separating bed.

    b) Thickness of the bedding material.

    c) The rate and quantity of hydraulic water introduced.

    d) The rate of feed.

    e) The rate of concentrate removal.

    f) Speed and length of stroke amplitude.

In order to obtain the concentrate from the jig of uniform grade, automatic control devices for discharging the concentrate, should be adopted which can be based on the use of one of the following principles:

    a) Pressure exerted by the concentrate (heavy product) on the gate to open it fully or partly.

    b) Pressure exerted by the water below the jig screen to control the gate opening.

    c) Variation in specific gravity of the jig bed with change in concentrate level to raise or lower a float and thus open or close the gate.

9.11 Consideration in Design of a Jig

The basic features of a jig design may be summarised below:

    a) A suitable jigging cycle should be developed. The jigging cycle should be adjustable with respect to the length of stroke, duration, and character of cycle.

    b) There should be an even transmission of jigging motion from the point of application to the point of use of motion.

    c) A suitable bedding material should be used to secure a hutch product.

    d) Provisions should be made for rapid withdrawal of strata and their transport from the jig.

    e) Design should be made for relative tonnage of light and heavy strata.

    f) Devices should be incorporated for automatic discharge of heavy product.

    g) Application of the jig should be borne in mind, i.e. whether for heavy minerals such as sulphides, or light minerals such as coal, since the volume of the jig will depend on this.

9.12 Applications of Jigs

For many centuries, jigs have been the chief concentrating devices. The jigs are totally obsolete for sulphide minerals of non-ferrous metals owing to the establishment of flotation process. Presently, they are mainly used in coal cleaning and to a lesser extent for non-magnetic iron ores and non-metallics. In a closed circuit between the mill discharge and the mechanical classifier, a jig (Denver type) can remove a heavy fraction from the pulp, which would otherwise tend to recirculate and over-ground. Another important area of the application of jig is rough concentration of diamondiferous gravels. Some other applications of jigs are concentration of tin and tungsten ore.

CHAPTER 10

Gravity Concentration in Streaming Currents

Gravity concentration in streaming currents is also known as flowing film concentration. This method may be considered as reverse classification, in which concentration of mineral particles is carried out on flatish surfaces (sluices, shaking tables, etc.) by flowing film of fluid depending on the sizes and density of particles. The transporting force is proportional to the cross-section exposed to flow, which can be employed to size the homogeneous sand, to wash out particles of a light mineral from heavier particles, and to separate particles of similar density but different shape and size.

10.1 Theory of Flowing Film Concentration

Behaviour of a number of spheres composed of two kinds of minerals (heavy and light) and of different sizes introduced into a thick layer of water running gently down a smooth plane can be represented as shown in Fig. 10.1 (a). The relative displacement and thereby their separation is based on the mass effect, according to which largest heavy spheres will fall faster and will be least affected by the water current and therefore, these spheres will collect nearest of downstream to the feed entry point. On the other hand, smallest light spheres will be drifted farthest downstream. The other spheres will overlap to some extent. Therefore, this process will result in clean coarse concentrate upstream, a clean light gangue downstream, and overlapped mixture in between, which can be separated by screening.

The supporting plane gives a new system of forces as shown in Fig. 10.1 (b), i.e. (a) a gravity pull will try to anchor the sphere to the plane, (b) on inclined planes gravity will aid their rolling, and (c) the flowing water will exert a pressure against the cross-section of the sphere aiding the movement downstream. Further, the differential velocity of flowing water (lowest near the plane and maximum at the top of the spheres) will have a slight tendency to overturn and roll down the spheres. If a system is employed to allow the spheres to come to rest (i.e. multiplane surface, flattening out, or a broadening of the stream), the spheres will change their arrangement as shown in Fig. 10.1 (c).

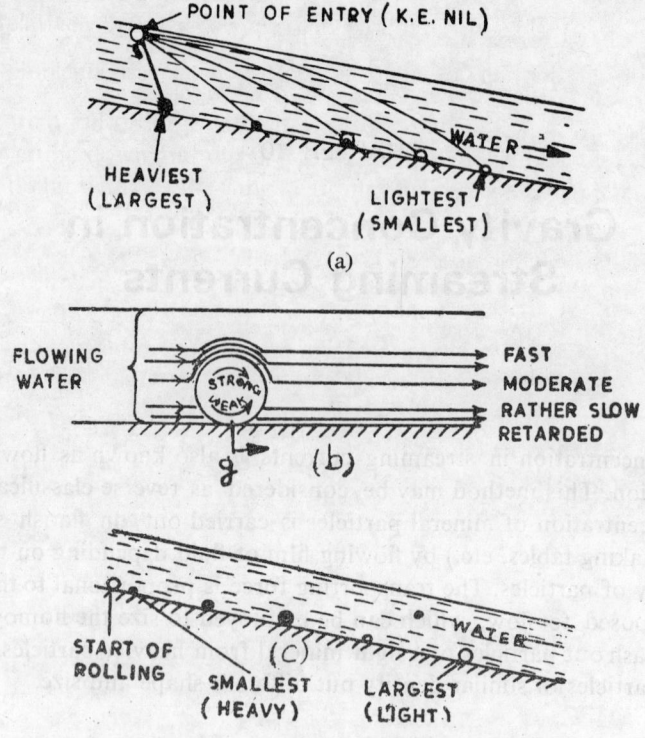

Fig. 10.1.　Schematic illustration of flowing film concentration:
(a) Behaviour of spherical particles of two kinds of
minerals and of different sizes, (b) Effect of various
forces, and (c) Classifying effect on streamed
particles.

In practice, the mineral particles are never spheres and their shape is an important determining factor to the push of the stream and the ease of rolling. The flattest flakes would be the first to stop, the cubes would come next and rounded particles would roll farthest. Further, in practice it is not possible to provide the large surface area required to develop the separating conditions fully and thus for efficient separation, a bed of material having many particles along the depth is necessary, which is provided by the use of riffles across the flow of water.

The riffles introduce turbulence and vortexing in water flow. The lighter particles will climb over the obstacles and would be carried along by the water stream, while the heavier particles find their way in the pockets. Therefore, an increased dynamic pressure along a separating surface (sluice) may be obtained by increasing (a) flow rate of water, (b) slope of the surface, and (c) roughness of surface. The main factors controlling the separation are:

a) Slope of the separating surface.
b) Thickness of water film and rate of flow.

c) Roughness of separating surface or friction coefficients between the various particles and the surface.

d) Relation between maximum and minimum settling rate of feed particles in water.

e) Percentage of solids in pulp (solid–liquid ratio).

f) Relative density of concentrate and tailing particles.

g) Shape and size of particles.

h) Tightness of particle interlock along the bed.

i) Viscosity of the fluid, if water is not used.

In flowing film concentration, it is essential to classify the feed to certain size range. This mainly depends on (i) the relative thickness of fluid film and particles, (ii) the relative time required for the fluid to be discharged from the table, and (iii) the relative time required for the particles to settle on the separating surface. In case of employing deck without riffles, particle size should not be more than the film thickness. If the particles are larger, they would be exposed above the film and fall without the range of their own slope-push. Whereas in case of riffle-decked tables, the maximum particle size should be equal to the riffle depth rather than film depth and also should not be more than one-third of the riffle width, in order to prevent the bridging. Further, the particles should be heavy and sufficiently coarse to facilitate their settling on the table before the suspended particles are washed off the table by flowing water.

10.2 Devices for Flowing Film Concentration

In the past, a large number of concentration devices have been employed for beneficiation of various ores by flowing film concentration principle. Many of them were simple, cheap and crude, whereas the devices developed at a latter date incorporated mechanical devices also. In general, concentration devices used in flowing film concentration can be divided into two broad classes:

a) Devices with stationary separating surfaces.

b) Devices with moving separating surfaces.

10.2.1 DEVICES WITH STATIONARY SEPARATING SURFACES

Concentration devices in this class are usually crude and cheap to construct and operate. These devices do not result in high recovery and had been in use considerably in the past, but have become obsolete by the development of better flowing film concentration methods and flotation processes.

In principle, mineral particles are agitated by a water current flowing over an uneven surface. The heavier particles settle into catch pockets, while the lighter ones pass on. Some of the important devices in this category used in the past include simple sluice systems, inclined stationary tables, fixed round tables and Humphreys spirals. Out of these, only Humphreys spirals are still in use.

In simple sluice system, the surface waters wash away the lighter grains, leaving the grains of heavy minerals concentrated in natural depressions lying in the way of water current. The simplest sluicing system known as pinched sluice consists of a stationary inclined plane (narrowing in width from feed end to discharge end) with a gentle slope downward, just sufficient to permit the gentle flow of the pulp. The light mineral particles roll down, while the heavier ones will stop on a removable cloth fixed on the sloping surface.

Blanket, corduroy and other rough surfaces are some of the oldest devices for recovery of free gold. Even after the development of other processes, they played an important role in recovery of free gold. Corduroy is a blanket material developed in England in 1930s specifically for the recovery of gold from fine pulps.

Inclined stationary tables are the simplest means of flowing film concentration and are also known as buddles. These consist of inclined rectangular tables (made of wood or canvas). The pulp is fed at the upper end, which flows down over the table surface. The slope is so adjusted that the lighter particles get washed off, while the heavier ones settle and accumulate on the surface.

Fixed round tables or circular buddles have obtuse–conical smooth surfaces and permit the operation to be continuous, instead of intermittent. The tables are usually made of cement and can be either convex or concave in the centre. In the former case, the pulp is distributed radially from the centre head. As the pulp spreads, the water film becomes thinner and moves more gently. The lighter particles travel faster and thus get washed away from the periphery of the table, while the heavier particles get arrested near the centre. A light drag of brushes may help in preventing the channelling. Whereas in case of concave buddle, feed is peripheral. The lighter material gets immediately flushed away as the pulp-stream accelerates towards the centre.

Humphreys Spiral

This equipment (Fig. 10.2) consists essentially of five or six turns of troughing, having a cross-section of nearly a quadrant of a circle. The fall per turn depends upon the mean radius of spiral and the material to be concentrated. In general, for a mean radius of 15 cm, a fall of 35 cm is needed. However, for the treatment of lighter material such as coal, the slope may be much less. The wash water runs down a small trough alongside the inner (horizontal) part of the cross-section. The wash water can be deflected as and when needed into the portion carrying the ore pulp. Withdrawal parts are provided at different points on the inner part of the spiral, which are provided with discs having adjustable deflecting wings to remove part of the inner pulp stream to the central concentrate collecting pipe.

In spirals, separation is affected by the combined effect of centrifugal action with multiplane sluicing in a partially controlled sink–float medium,

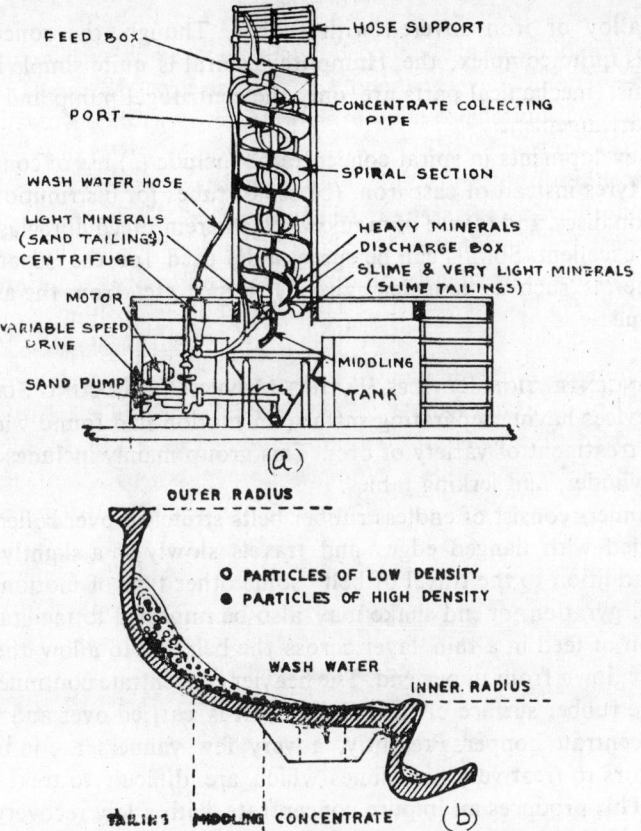

Fig. 10.2. Schematic representation of Humphreys spiral.

which is autogenously provided by the pulp. If the spiral is viewed from the top [Fig. 10.2 (b)], the heaviest and coarsest particles remain nearest to the centre and on the flattest portion of the cross-section, whereas the lightest and finest particles rise to the side walls. The rising action of light and fine particles is further assisted by wash water fed in at an angle. The innermost part of the pulp band is diverted by the deflecting discs at suitable intervals. Since, the outer edges of the heavy fraction concentrated in the innermost part contain middlings and slimes, the concentrated material should be treated more than once.

In general, pulps having 20–30 per cent solids by weight are used for treatment. However, more dilution is used in handling fine sands and more solids (up to 50 per cent) in treating coarse materials. In plant practice, a battery of spirals is employed and the material travels from the highest block of roughers under gravity and gradually gets concentrated. The throughput varies from ore to ore, 1–2 tonnes/hr of feed at 25–50 per cent solids and 850 micron size is quite common.

The spiral can be constructed from Ni-hard cast iron or similar wear resisting alloy or iron covered with rubber. Though, the concentrating principle is quite complex, the Humphreys spiral is quite simple in operation. In this, mechanical parts are only the centrifugal pump and feed distribution arrangement.

New developments in spiral concentrator include (a) use of concrete and old truck tyres instead of cast iron, (b) use of tubes for distribution instead of deflection discs, and (c) use of moulded plastic reinforced fibreglass spirals, which are excellent. Spirals can be successfully used for the separation of heavy minerals such as tungsten, zircon, garnet, etc., from the associated light gangue.

10.2.2 Concentration Devices Having Moving Separating Surfaces

The devices having separating surface in motion had found wide applications in treatment of variety of ores. This group mainly includes vanners, rotating cylinder, and jerking tables.

The vanners consist of endless rubber belts stretched over rollers, which are provided with flanged edges and travels slowly on a slightly inclined plane. In addition to the travel of belt, some other type of motion such as side shake, gyration, or end shake may also be imparted to facilitate gentle distribution of feed in a thin layer across the belt and to allow the flow of wash water down from upper end. The heavier concentrate continues to hold back to the rubber surface of the belt, until it is carried over and removed into a concentrate hopper. Presently, a very few vanners are in use in tin concentrators to treat very fine slimes which are difficult to treat by other methods. This produces an impure concentrate with a low recovery (50 per cent or less).

Cylindrical Concentrator (*Johnson Concentrator*)

It is a rotating (5–8 rpm) steel cylinder (usual size being 1 m dia × 4 m length) lined circumferentially with corrugated rubber. A slight slope (5°) is given to facilitate the flow of pulp. When the feed flows downward, heavy mineral particles collect in the corrugation, whereas the lighter particles are washed away. As the concentrator rises, concentrate is washed off by water spray into an axial launder.

Jerking Table

It consists of a deck (with or without riffles) which is shaken rapidly in the direction of its length. This table is inclined laterally a few degrees with the horizontal to facilitate the water flow across the surface, approximately at right angles to the shaking direction.

Except the shaking tables and some newly developed processes, most of the old devices for flowing film concentration have become obsolete due to the following reasons:

a) Particles finer than 10–20 microns cannot be treated and thus a fraction of valuable minerals is not recovered from slimes, whereas it is possible by flotation process.

b) Compared to flotation, the throughput of flowing film concentration devices is much smaller for the treatment of finer particles.

c) The concentrates obtained are of too low a grade for satisfactory recoveries.

Since shaking tables are still in use, these are dealt separately under Sec. 10.3.

10.3 Shaking Tables

10.3.1 CONSTRUCTION OF SHAKING TABLES

The shaking table (Fig. 10.3) consists of a large flat surface (deck) made of canvas, linoleum, wood or aluminium, which is inclined slightly (about 5°) both front to back and from left to right. Its surface may be roughened or provided with grooves or cleats to form riffles. The cleats are usually made of wood, and are 5 to 15 mm wide and 5 to 10 mm high, spaced at 15 to 50 mm apart. The cleats cover most of the deck as shown in Fig. 10.3. The riffles form a large number of parallel channels (parallel or at an angle to the direction of table motion) usually 2 cm wide × 1 cm deep at the starting point and tapering down to zero depth at the lowest end. Pulp is fed at the top high corner and flows through the channels. A flow of water is maintained across the table. The table is given a reciprocating horizontal motion

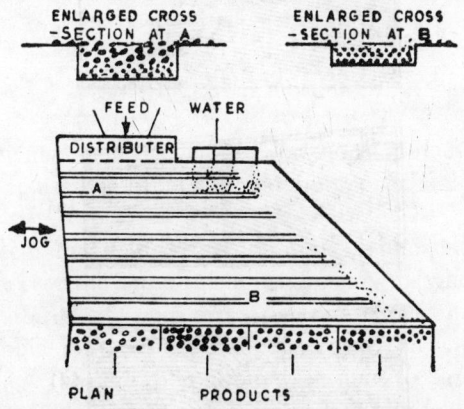

Fig. 10.3. Simplified diagram of a shaking table.

at the rate of about 300 cycles per minute. Each cycle consists of a relatively slow forward motion and a very rapid return which causes the particles to creep forward along the channels.

Riffles may be made of hard wood, plastics, rubber, metals, etc. Riffles on the deck offer the following advantages:

a) Riffles are responsible for the increased capacity of concentrating tables over smooth ones due to the possibility of treating a suspension of material many particles deep.

b) Riffles assist and spread the entering feed.

c) Each channel between successive riffles provides place for hindered settling and consolidation trickling.

d) Shaking action is transmitted more efficiently to the enclosed material.

e) Riffles help in exposing of the top layer of material after it is stratified to the cross-flow of wash water down the table deck.

f) Delivery of feed is promoted to succeeding riffles, wash planes and discharge points.

Various riffling plans may be employed depending on requirements. Usually, a riffle plan providing the movement of feed with the deck and charging the pulp smoothly on the top row of riffle is best. Some of the important riffle plans are shown in Fig. 10.4.

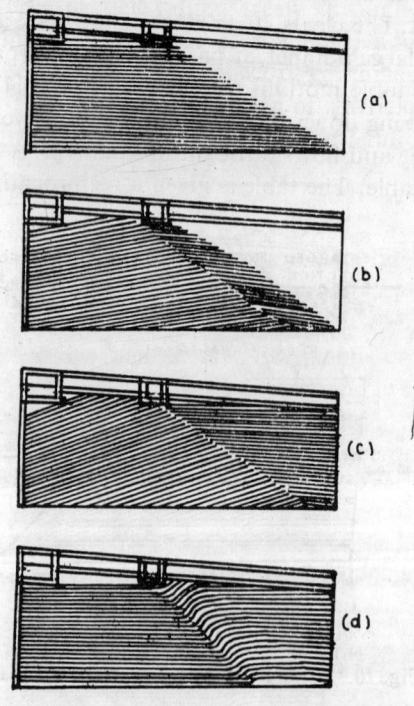

Fig. 10.4. Some important riffle plans.

The shaking motion to the table may be imparted by eccentric toggles, springs, cams, electromagnetic, etc. Provision is made to alter the slope (tilt) of the deck. The drive has the variable speed mechanism which is coupled to the table to adjust the length of jerking stroke (1 to 3 cm).

The tables can be constructed in varying sizes depending upon (i) material to be treated, (ii) quantity of material to be treated per day, and (iii) size of the material. A small table about 40 cm × 90 cm can treat 0.5–2 tonnes of solids per day, whereas a large table about 200 cm × 500 cm can handle 15–60 tonnes of solids per day. Power requirements for each table may vary from 0.5 to 2 KW and the water requirements may vary from 15 to 150 litres per minute per table.

10.3.2 WORKING PRINCIPLE AND OPERATION OF SHAKING TABLES

In shaking tables a number of spheres may be considered as rolling down a slighlty tilted plane under the combined influence of a flowing stream of water and gravitational pull. Under such condition, the movement of various spheres can be presented as in Fig. 10.5. The largest sphere will travel fastest and the smallest one slowest in case of spheres having the same density. On the other hand, if two spheres have the same diameter but densities are different, the lighter spheres will travel faster. If the plane is moved sideways during the downward travel of these spheres, the horizontal displacement of the spheres will vary according to the time taken in rolling down (varying inversely as the square of diameter of particles). Under the combined effects of all the forces, the particles will travel along a diagonal line beginning at the feed box. Figure 10.5 shows that the largest light sphere has undergone the least horizontal displacement due to its maximum velocity, while the smallest heavy one has been carried farthest to one side. Therefore, the feed can be spread into suitable bands according to the size and density of the

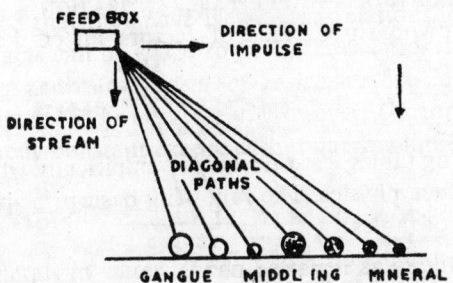

Fig. 10.5. Representation of movement of various spheres under shaking conditions.

particles by applying a displacing movement to the separating plane. The feed thus gets separated into three products as follows, which can be collected separately as they leave the deck.

a) *Fastest moving*: These consist of coarse light grains (usually gangue).

b) *Medium moving*: These consist of fine light and coarse heavy particles (middlings).

c) *Slowest moving*: These consist of fine heavy particles (usually concentrate).

The working principle of shaking table can thus be summarised as follows:

a) The movement of particles is affected by the slope of the table, velocity of water, and thickness of water film.

b) Light particles are carried into suspension with water and heavy particles tend to settle into the grooves or riffles.

c) Heavier particles travel forward toward the left and get discharged over the lower left corner of the table, whereas the lighter ones overflow to the suitable launders.

d) The small particles are brought below the larger ones, and heavy ones below light particles of the same size.

e) The largest and lightest proceed almost straight down the table, while the smallest and heaviest follow a diagonal route via the bottoms of the channels.

The process of concentration on shaking tables is continuous. The tables work best on well classified material, so that concentration can be effected by density rather than on size and density simultaneously. The tables can be used successfully for concentrating materials from about 3 mm to 50 microns but concentration of particles below 0.3 mm (300 microns) can be effected more economically by flotation. The success of tabling mainly depends on the wide difference in densities of the minerals and thus the shaking tables are still in use for separation of heavy density minerals such as gold, garnet, etc.

10.3.3 FACTORS CONTROLLING THE TABLE OPERATION

Basically, following three kinds of factors affect the control of table operation:

a) *Design Features*

Initially shaking tables were developed empirically when much was not known about surface physics. The following basic principles should be considered in design of tables:

i) The type and material of riffles.

ii) The material of deck surface.

iii) The acceleration and retardation during one cycle of the drive.

iv) The mode of feed presentation.

v) The forward part of the stroke must end with an abruptness to promote some skidding of the particles in the direction of discharge.

vi) The return stroke should die away gently.

vii) The return part of the stroke should commence abruptly.

viii) Suitable stroke amplitude and frequency should be incorporated.

The above are the major points to be considered for a good design, which can be varied by reconstruction only. Therefore, these factors are decided by carrying out the tests on the material to be treated under local conditions.

b) *Mechanical Features*

These include running speed and amplitude of stroke. Speed depends on motor and pulley system, whereas, stroke can be adjusted by varying the toggle spread, or by altering the amplitude of the solenoid vibration in electromagnetic motion. The amplitude of strokes can be varied in running position. These adjustments should be made as early as possible in a working plant, otherwise inefficiency of the operation and working trouble may be caused.

Operating Controls
Following items fall under the operating controls:
  i) Slope of the table.
  ii) Solid–liquid ratio in the pulp.
  iii) Amount of wash water.
  iv) Position of the product dividers.

10.3.4 Factors Affecting the Capacity of Tables
The tonnage treated depends on the following factors:

a) *Feed Size*
The tonnage of coarse feed treated is always larger than that of fine feed. The treatable tonnage is at least proportional to the square of the average size.

b) *Difference in Density of Minerals*
More the difference in density of the minerals, higher will be the treating capacity of the table, as this condition permits clean separation without much middlings.

c) *Relative Presence of Locked Particles*
The behaviour of locked particles in tabling will be intermediate between that of constituent minerals and thus it affects the capacity to be treated.

d) *The Average Specific Gravity of the Ore*
A bulky substance having low specific gravity such as coal cannot be treated in the same quantity as a heavy ore, provided other factors remain the same. However, much coarser size of coal used in tabling counter balances the adverse effect of its low specific gravity.

e) *Whether the Operation is Roughing or Cleaning*
A roughing operation is preferably carried out on a fully riffled deck providing higher capacity, whereas, cleaning operation is performed on a partly riffled deck having comparatively small capacity.

10.3.4 Operation of Shaking Tables
The tables can be used in series where the first group handles to produce a low-grade concentrate containing all recoverable values and a tailing for discard. This treatment is called roughing. The low-grade concentrate is then

treated on a smaller group of tables for cleaning. The tailing obtained from the cleaner tables may be given further treatment by returning them to the rougher feed, as shown in Fig. 10.6.

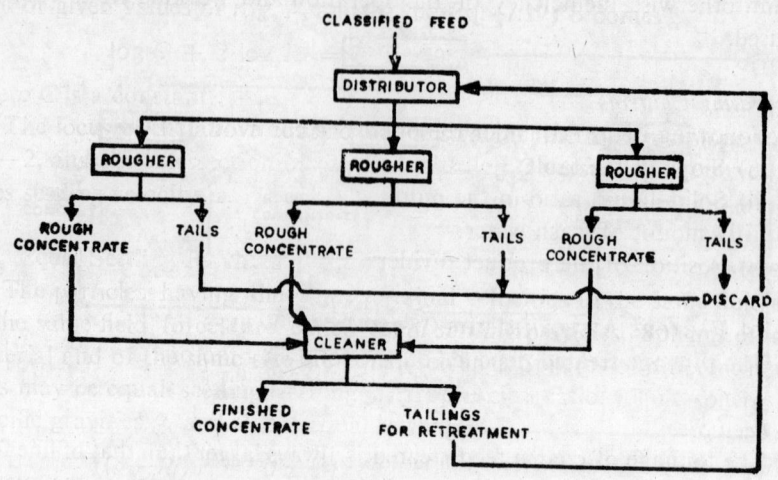

Fig. 10.6. A rougher cleaner flowsheet of tabling.

Since the tabling depends mainly on a good gravity difference between the minerals, unnecessary grinding should be avoided of values which are sufficiently unlocked to respond to tabling treatment. Fine particles unable to fall rapidly through still water result in poor recovery of a low-grade concentrate. Therefore, in tabling, the desired mineral should be removed at the coarsest size of liberation. In tabling, a middling product is invariably obtained which may contain true tails and true concentrates. If this middling product is recirculated, its volume will continue to increase and after some time it may join one of the other products. Therefore, recirculation is not the only solution to handle a true middling. If the middling consists of incompletely unlocked particles, it should be subjected to further grinding before its reprocessing.

The variables required to be controlled for an efficient operation include slope of the table, thickness of flowing film and amount of water, rate of acceleration of deck, coefficient of friction between deck and particles, character of riffling, length of stroke, etc. Suitable adjustment of these variables may be exercised as follows, in order to have good results:

a) Roughing operation requires greater feed rate, more tilt, more water, longer stroke, and more riffling.

b) Cleaning operation requires less water, less feed, less tilt and a shorter stroke.

c) Treatment of coarse feed requires higher rate of feed, more water, slow reciprocation, and longer stroke.

d) Treatment of fine feed requires less water, less feed, faster reciprocation and shorter strokes.

e) The solid–liquid ratio of the pulp must be maintained steady for a particular ore. Auxiliary water service must be suitably and smoothly distributed in order to maintain a correct balance between cross-streaming and down-streaming.

10.3.5 Wilfley Table

Various shaking tables mainly differ in the design of riffles and the way in which motion is imparted. Most widely known shaking table is Wilfley table. Original Wilfley table was made by Arthur R. Wilfley in May 1896 at Colorado. This introduced a big advance in the field of flowing film concentration. At one time over 23,000 Wilfley tables were in use. These tables are still in use in concentration of some heavy minerals. The most important features of Wilfley table were (a) the introduction of riffles which increased the capacity and permitted the treatment of coarser material, and (b) use of effective and rugged head motion.

The Wilfley table consists of a four-sided nearly rectangular deck with the adjustable slope toward one of the long sides. This deck is actuated by a pitman and toggle type adjustable head motion as shown in Fig. 10.7. Other types of head motion (including electromagnetic) may also be used. The riffles in the table run parallel with the long axis and are tapered from a maximum height on the feed side and merge with the deck at the opposite side. The deck is tilted transversely. The corner of the upper side of the table is provided with a feed box which is near the head motion. A washwater distributing box is also provided on the upper end. This wash-water distributor consists of a perforated trough having fingers at the perforations for control of water flow.

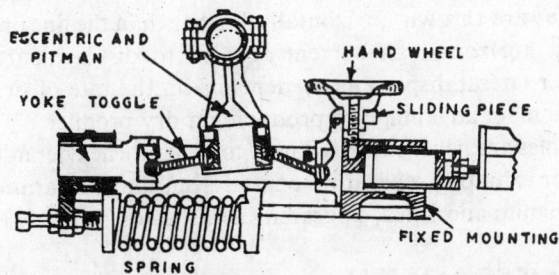

Fig. 10.7. Toggle type adjustable head motion of Wilfley table.

Riffles are usually wooden cleats tacked on the linoleum-covered surface. The cleats usually taper in thickness from 1.5 cm at the feed end to a thin edge near the concentrate end and the width is usually 0.75 cm. The ends of the riffles form a diagonal line in such a way that about two-thirds of the deck is riffled and rest of the portion is plane.

10.4 Dry Tables or Pneumatic Tables

Use of water in wet tabling is associated with the following drawbacks:

a) A large volume of water is required (up to 15 times) in circulation.

b) The water becomes foul with use and if slimy, may reduce the frictional grip of the sand to the deck.

c) Treatment of very fine slimes may require still more water.

d) Wet tabling can be used where large quantities of unfrozen water arc available during the operation season.

e) Sometimes the properties of the product get affected by wetting.

f) Soluble salts in water may also affect the quality of product.

g) Sometimes the cost of drying a wet product may be prohibitive.

To overcome the above problems, dry method or pneumatic table can be employed. Basic principles of pneumatic tables are similar to those of wet tables. A pneumatic table has a finely perforated flattish riffled deck through which a blast of air passes upward to keep particles moving down the table, while a jogging action keeps them moving across it. The feed is introduced near the top of the inclined table. The lighter particles are supported by air above the surface of the table and flow downward to discharge over the lower end. The oscillating motion of the table causes the heavier particles contacting the surface of the table to move upward and discharge over the higher end of the table.

In wet method the particle size increases and the mineral density decreases from the top of the concentrates band toward the tailing, whereas, in case of pneumatic table, particle size as well as mineral density decrease from the top end to the tailing. The middling consists of coarsest particles having the lowest density. Thus, pneumatic tabling gives the sorting effect similar to the hydraulic classification, whereas in wet method, pre-classification of the feed is required. The dry sands get stratified during their descent to the table and fanned out as they fall from the launder. During the fall, coarser particles are thrown horizontally farther than the fine ones. In pneumatic tables a horizontal air-current is used to displace particles falling vertically. Their lateral displacement depends on the rate of drop. The dry tables have the main advantage of producing a dry product.

The dry tables are generally employed to separate heavier minerals from lighter ones, for example, separation of coal from shale, separation of monazite from sillimanite and silica, upgrading of asbestos, etc.

10.5 Treatment of Slimes on Tables

The slimes can be treated on shaking tables by having the following adjustments:

a) Large area for treatment is provided, since the horizontal area required varies inversely to the size of feed.

b) Gentler and slower action should be given to feed.

c) Distribution of feed must be even and channelling should be avoided.

Separation of slimes can be effected more efficiently if there is large difference between the specific gravity of lighter and heavy particles. For example, the concentration criterion for a mixture of quartz and cassiterite at less than 50 micron size is 3.5, which is not sufficient to give effective separation, whereas for a quartz–gold mixture, the concentration criterion is nearly 9, and good separation can be obtained at the same size, provided gold does not become flaky during grinding.

10.6 Applications of Flowing Film Concentration

Flowing film concentration method was widely used up to about 1920s and its use declined gradually. This process lost its field in the treatment of sulphides of Cu, Pb, and Zn and to some extent in treatment of oxidised ores of Au, Cu, and Pb, since flotation and hydrometallurgical processes steadily gained importance. Presently, among the different flowing film concentration processes, shaking tables are still used in the following cases:

a) Recovery of a part of the galena and sphalerite in coarsely aggregated lead and zinc ores.

b) Concentration of cassiterite, barytes, and fluorspar.

c) Concentration of some free-milling gold ores, and auriferous ores.

d) Beneficiation of many non-metallics, such as silica sand, chromite, tungsten ores, etc.

e) Recovery of relatively coarse pellet-like metal from the by-product such as foundry dross.

f) Beneficiation of some iron ores.

10.7 New Developments in Flowing Film Concentration

10.7.1 BARTLES–MOZLEY CONCENTRATOR

This is shown in Fig. 10.8. The concentrator has 40 thin, flat fibreglass decks, which are stacked over one another at a spacing of 1.3 cm. Each deck is 1.5 m long and 1.22 m in width. The decks are placed in a frame in two packs of 20 each. The two packs are separated by a space in which an out-of-balance weight rotates in a horizontal plane at about 230–250 rpm, as a result an oscillating motion is imparted to the packs. The frame is freely

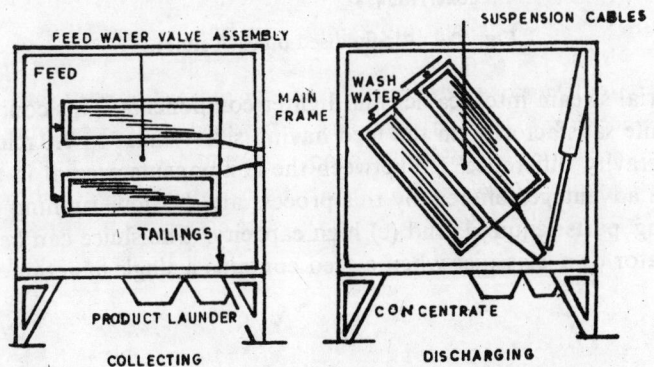

Fig. 10.8. Simplified diagram of Bartles-Mozley concentrator.

suspended on wires. In running condition, the frame is inclined at an angle of 1–3° downwards from feed-end to discharge-end. Pulp (10 per cent solids) is distributed to all the 40 decks at a total flow rate of about 400 litres/min. The tailings are discharged continuously, while the concentrate fraction is held on the deck surfaces. After the decks are loaded, the operation is stopped, the decks are tilted to about 45°, and the concentrate is removed into separate launder. The sequence of feeding and discharging is controlled by an automatic timer.

This concentrator can be used to treat most materials in the size range 5–100 microns. However, heavier materials such as gold, platinum, silver, etc., can be recovered at much finer sizes. The concentrator is particularly suitable for the processing of fine cassiterite, wolframite and tantalite ores.

10.7.2 AIR-FLUIDIZED PINCHED SLUICE

A schematic diagram of air-fluidized pinched sluice is shown in Fig. 10.9. This sluice is a high capacity device for the treatment of sand-sized feeds. The sluice is in the form of a shallow, tapered trough which is fitted with a porous plastic deck. An air box is attached to the trough which supplies low pressure fluidising air to pass upwards through the deck. An adjustable splitter unit is provided at the end of the trough, which divides

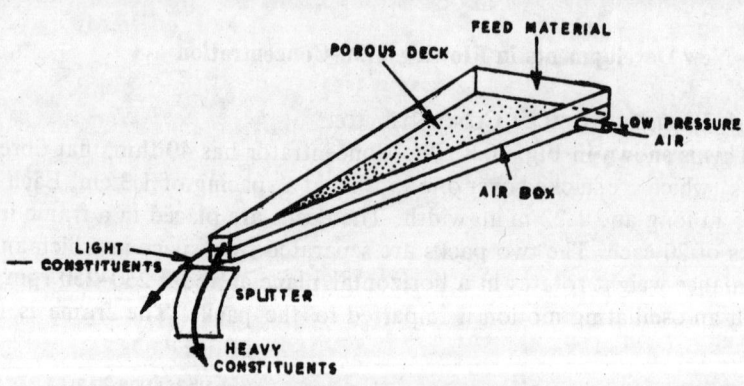

Fig. 10.9. Air-fluidised pinched sluice.

the material stream into heavier and lighter components. This concentrator works quite satisfactorily on the feed having size 1.5 mm to 7.0 micron and specific gravity difference 1.0 between the components present in the feed. The main advantages offered by this process are: (a) low running cost, (b) no moving parts required, and (c) high capacity. The sluice can be used as concentrator or a classifier when a feed contains a single mineral.

CHAPTER 11

Flotation History and Theory

11.1 Introduction

Flotation is a process of separating fine particles of different minerals from each other by floating certain minerals at or on a water surface. In flotation the mineral particles heavier than water are maintained in suspension by the action of surface tension forces. The separation is effected from the adhesion of some species of solids to gas bubbles which are introduced in the pulp and the simultaneous adhesion of other species of solids to water. Wettable particles cover themselves completely with water and thus sink into a water phase, whereas the particles adhering to air will float on the top of the pulp in the form of froth due to effective specific gravity of mineral—air association being lower than that of pulp.

The floatability of any given mineral is the tendency of its particles to adhere to air bubbles. This is primarily dependent on the properties of the mineral surface and specific gravity, whereas pure physical properties of the bulk mineral hardly play any role in flotation separations. The natures of the other surfaces or interfaces in the flotation system, such as air–water interface also have a direct influence on the effectiveness of the separation. In flotation processes, surface properties can be modified by the use of some specific chemicals to achieve a variety of separations, which mainly depends on the ore and number of commercial products to be obtained.

At present, flotation process is the best and almost universally employed for the concentration of finely ground ores (less than 150 microns). A majority of the ores mined presently require fine grinding for high degree of liberation of valuable minerals, and thus flotation becomes the only possible means for the efficient treatment to recover the values in the form of high-grade concentrates.

Originally the flotation was applied for the concentration of metallic sulphide minerals and even presently this technique is predominant in this field. Flotation technique has also become useful in the treatment of a great variety of oxide ores, native ores, and non-metallic minerals. The facts used in flotation can be summarised as follows:

a) Most minerals adhere to water in preference to air.

b) Some mineral substances adhere to air due to some inherent property of the minerals.

c) Most hydrocarbons and paraffins adhere to air in preference to water.

d) Various minerals can be made to adhere to air by adding a small quantity of suitable agents to the pulp.

e) Practically all the minerals can be made air adherent or water adherent by the use of a suitable agent or a combination of agents.

f) The flotability of a mineral is greatly affected by the changes in the character of the surfaces of mineral because of oxidation or other processes.

11.2 History

There is a good evidence that the ancient people used oil and pitch for collecting minerals, and undoubtedly phenomenon underlying flotation were observed accidentally around mining camps with little or no recognition of their possible practical application until relatively recent times. The method based on the use of conditioned surface for selective attraction of mineral particles was used in the remote past. Herrodotus (5th century B.C.) records separation of gold grains from silt by means of goose feathers covered with bitumen or grease. In the 15th century A.D., Arabs separated azurite by a process in which the mineral adhered to softened resin and was then floated out with it.

Flotation as a physico-chemical concentration method dates only from the 19th century, particularly its last three decades, when extensive efforts were made in search of new surface-active substances. In 1860, William Haynes (in England) patented the use of selective clinging of particles at an oil–water interface and separated the sulphide mineral particles by washing the oil–ore mixture with water. Later this process was used to some extent for concentration of copper ore containing precious metals. In 1877 the Bessels Brothers floated graphite from an aqueous pulp by employing a small quantity of mineral oil using steam or CO_2 gas. Flotation of the same ores today uses air to render the selected particles buoyant. In the beginning of flotation in 1886, an American woman Carrie Everson patented the use of some reagents (fats, oils, acids, and salts) in the flotation of oiled sulphides from gangue by suitable means. In 1894 Rubson (England) patened the separation of sulphide minerals and rocks with a mixture of oils in a small amount of water.

Until 1898 no successful commercial process was worked out, when F.E. Elmore patented a *bulk oil flotation* process and applied it to the concentration of a gold bearing chalcopyrite. The first breakthrough in applied flotation was at the beginning of the 20th century, when in 1902 C.V. Potter, G. Delprat (Australia) and Froment recognised the action of gas in lifting oiled particles of sulphide to the surface of a pulp and about 6 tonnes of concentrate containing about 42 per cent Zn was produced. Froment's patent was acquired by Minerals Separation Limited, an English organisation in the subsequent development of oil flotation.

In 1904, Elmore patented methods of introducing the gas either by electrolysis or by the use of vacuum. This method was used in several concentration plants. A great stride forward was made by Sulman and Picard, the two metallurgists of Mineral Separation Limited in 1905, when the development was made for the introduction of air into the pulp by suction effect of a rotating mixer. This led to the evolution of the agitation in froth process. In the following year, true nature of surface conditioning was recognised and a patent was taken for the use of strong agitation of the pulp and very small quantities of oil. However, the amount of oil used was as high as 5 per cent of the pulp volume.

The development of selective separation of sulphide minerals was the most important stage in flotation. It was established in 1912 that bichromate depresses galena in flotation. In 1913 Bradford established that copper sulphate could be used as an activator for sphalerite. In 1912, soluble organic compound containing trivalent nitrogen or bivalent sulphur were used as collectors. The first collector not based on oil (x-naphthylene) was patented by Corliss in 1917. Another important achievement was the introduction of selective depressants in selective flotation. In 1922, Sheridan and Griswold patented the use of cyanides and $ZnSO_4$ in an alkaline solution for the separation of galena, sphalerite and pyrite. Up to that time, flotation of sulphide ores resulted only in a bulk concentrate and the further separation of various metals was made by expensive smelting methods. Alkaline xanthogenates and dithiophosphoric acid esters (aerofloat collectors) were employed as collectors in flotation in 1925 and 1926, respectively. These are still the most widely used collectors in flotation of sulphide ores.

Further developments in flotation included the studies of reagent–mineral action and the introduction of new reagents. Attention was also given to study the physical and chemical aspects of surface phenomena, in order to have the correct understanding of the flotation process. Table 11.1 summarises the important steps in development of flotation process.

11.3 Types of Flotation Processes

a) FROTH FLOTATION

The flotation process relies primarily on the fact that hydrophilic particles are wetted by water, whereas hydrophobic particles are wetted by oils and air bubbles; therefore, if air bubbles are introduced into an aqueous slurry, the bubbles adhere to the hydrophobic solid particles. As a result, air-solid aggregates are carried to the surface, forming a froth layer; this explains the name froth flotation. The froth layer can be removed manually or mechanically; the result is the separation of hydrophobic from hydrophilic particles.

This is the most commonly employed technique in concentration of minerals/ores. This process is based on making use of differences in the physico-chemical surface properties of various minerals. The difference can also be enhanced by treatment of minerals with specific reagents. In the froth flotation, valuable part of mineral usually passes into the froth product (sometimes gangue is also floated) and becomes the float fraction, whereas, the other part remains in the pulp as an unfloated fraction (chamber product). Both the fractions can be removed separately from flotation machine by suitable mechanisms. When the desired minerals are separated as froth product, the process is referred as *direct flotation*, whereas, if the

Table 11.1. Important steps in the evolution of flotation

Year	Contributor	Work in flotation
1491	Mansur	Beneficiation of ultramarine and/or azurite based on differences in wettability of minerals by oils and water.
1731	Petit	To float submerged solids by air–solid adhesion.
1860	Haynes	Bulk-oil process.
1877	Bessel	Graphite beneficiation by boiling process.
1885	Bessel	Graphite beneficiation by the process based on chemically generated gas.
1886	Eversen	Use of acidified pulps.
1902	Froment, Potter and Deprat	Use of gas to float sulphide ores.
1905	Schwarz	Recovery of oxidised base metal minerals by Na_2S.
1906	Sulman, Pickard, and Ballot	Reduced oil consumption and introduction of gas by violent agitation.
1909	Greenway, Sulman, and Higgins	Use of soluble frothing agents.
1913	Bradford	Use of $CuSO_4$ as activator for sphalerite and SO_2 to depress sphalerite.
1921	Perkins and Sayre	Use of specific organic collectors and alkaline circuits.
1922	Sheridan and Griswold	Use of cyanides to depress sphalerite and pyrite.
1924	Sulman and Edsen	Flotation of oxides with soaps.
1925	Keller	Use of xanthates as collectors.
1929	Gaudin	Use of pH control.
1929	Jeanprost	Flotation of highly soluble salines.
1933	Nessler	Separation of water-soluble chemical salt mixture by flotation.
1934	Chapman & Littleford	Agglomeration.
1934	Chapman & Littleford	Use of alkyl sulphates as collectors.
1935	Chapman & Littleford	Use of cationic collectors.
1950	Chapman & Littleford	Use of dithio-phosphates (aerofloats).
1960	Chapman & Littleford	Increased size of flotation cells.
1970	Chapman & Littleford	Automatic control and optimisation.

gangue is drawn into the froth product, the process is known as *reverse flotation.*

In froth flotation, air is blown through the pulp containing flotation reagents. The particles having water-repellent surfaces (not wetted easily) attach the air bubbles to them and rise to the surface through the pulp in the form of mineralised froth. On the other hand, particles having water-avid surfaces (wetted readily) do not attach air bubbles to them and thus remain in suspension of the pulp.

b) FILM FLOTATION

In this process, the ore to be concentrated is ground and sprinkled into the surface of water. The water repellent (unwettable) particles remain on the water surface which are separated as the flotation product, while the water-avid (wettable) particles pass into the water. Presently film flotation does not find much application.

c) OIL FLOTATION

It is based on the selective wetting of mineral particles by some oil such as kerosene dispersed in water. The particles covered with oil collect and float to the pulp surface. Presently oil flotation is rarely used in concentration of ores except for graphite.

d) COMBINED FLOTATION–GRAVITATIONAL METHOD

When the lifting power of an air bubble is inadequate, particle–bubble agglomeration may be employed for the selective separation of mineral particles. For example, sulphide particles forming a particle–bubble complex (granules with air bubbles) of reduced specific gravity, can be removed by gravitational separation. This combined process can be employed mainly in the concentration of coarse grain minerals.

e) ION FLOTATION

This is a new flotation technique and differs from the collection of already existing particles by flotation. This process is employed when the desired concentrate is at first in aqueous solution, either in the form of ions or colloidal suspension. This process is based on the conversion to a product having hydrophobic sites, which attach themselves to a bubble and collapse on reaching the surface. This results into a scum rather than a froth as produced in froth flotation. The aquated ions to be stripped from the aqueous phase and floated are conditioned with a non-micellar collector and floated in a fragile froth. The ion flotation can successfully be employed for the recovery of sols, gels, or polynuclear ions. The ions possessing attraction to collector must become insoluble in water before they are captured by a rising bubble.

It has been suggested by various researchers that ion flotation has great potentiality to treat leach solutions, mineralised water, process water, etc. With the help of ion flotation it may be possible to extract dissolved mineral wealth from sea water.

11.4 Physical Chemistry of Flotation

The real surface property of minerals of interest is the chemical and crystallographic capability of interacting with particular organic ions resulting in a very thin layer (probably a monomolecular layer) of these ions attached to the surface. This condition results in a high effective surface tension and is recognised as *wettability*, which is measured in terms of *contact angle* at an intersection with an air–water interface. Physical chemistry of flotation process includes various aspects such as surface modification, ions and changed lattice points on surface, activation theory, double electrical-layer theory, etc. These are discussed as follows:

11.4.1 PARTICLE SURFACE, AND ITS MODIFICATION

Surface modification in the pulp is usually considered as a rebalancing between aquated ions and charged lattice points. If the polar strength of water molecules aided by a suitable pH is more than the attraction binding of a mineral ion into its crystal lattice, the ion will hydrate and pass into the aqueous phase. Ions resisting aquation remain bound electrostatically in the crystal.

The particle surface is not homogeneous and thus due to various defects and cracks resulting from explosive shattering during mining and shearing during comminution, it includes both hydration-prone and water-repellent charged points. In the crevices and eroded pits of the particles, dissolved gases from the aqueous phase may precipitate and prevent these from chemical attack. Research shows that only 2–5 per cent of the particle surface needs a coating of collector agent to ensure flotation in a number of cases. The surface area of the ore considerably affects the flotation efficiency.

Most minerals are readily wetted with water in preference to air when their surfaces are clean (freshly cleaned), but become water-repellent and flotable when coated with hydrocarbons such as grease, oil, paraffins, alcoholtates, etc., sorbed from the air or lubricants. However, this generalised coating serves no purpose in selective separation. Therefore, flotation should be preceded by conditioning treatment, during which only the desired mineral is rendered aerophilic by the reaction with a specific reagent on the particle surfaces. The quantity of chemical reagent required to condition a clean surface is very small, e.g., in case of chalcopyrite 0.05 kg/tonne of collector agent ensures the efficient flotation of 2 per cent copper ore of −100 micron particles.

The surface of mineral must be sufficiently clean in order to react with the chemicals introduced into the pulp. If surfaces are contaminated with oil, tightly adherent slimes, iron stains, etc., the treatment would be interfered and thus special remedial measures such as use of dispersing agents and/or pH control would be needed. Some minerals exhibit inherent or native flotability which is a fundamental characteristic of crystals having no ions at their surface.

11.4.2 SURFACE ENERGY OF MINERAL PARTICLES

The adsorptive properties of a mineral are affected by their crystal structure and consequently their response to the hetropolar surfactants used in the flotation. In a crystal lattice of a mineral there are energy differences between the atoms on the surface and those in the body of the lattice. The atoms in the body of the mineral spend all their energy in reacting with their neighbouring atoms which surround them on all sides. The particles having greater lattice energy are more resistant to chemically induced changes. Lattice energies of silica and alumina are about 3,700 kcal/mole and thus their surfaces are resistant (or poorly responsive) to activation by flotation reagents. The lattice energy of sphalerite is 800 kcal/mole, and this mineral is moderately responsive to suitable surfactants. For NaCl the lattice energy is only 180 kcal/mole, making it readily ionised and highly soluble in water.

Thermodynamic values of floatable surfaces differ considerably from those of the other system. The system having free energy at a minimum, attains the equilibrium faster and therefore leading to a change considered thermodynamically in terms of decrease in free energy.

However, the reaction energy of the surface atoms being in the extreme outside layer, is exerted only in relation to the atom layers beneath them and to the neighbouring atoms in the surface layers. As a result, the energy of some of the surface-layer atoms remains free, which is called *free surface energy* and usually surface energy. The surface energy is usually quantified in erg/cm^2. For liquids, an equivalent term known as surface tension is used instead of surface energy. Surface tension is expressed in dynes/cm or in erg/cm^2. The surface energy for any two adjacent phases is measured on the interfacial surface (e.g. the liquid–gas interface) and is designated by the appropriate symbols (σ_{l-g}).

The free unsaturated bonds of the surface layer atoms are due to the existence of free surface energy. The magnitude of this surface energy determines the nature and ability of any mineral surface to react with water and the substances dissolved in it.

In flotation the phases in contact are virtually immiscible and mutually insoluble. The solid particle is reduced to such a small mass that its gravitational drag is much less and its surface–volume ratio quite high. The behaviour of the particle (whether aerophilic or hydrophobic) depends initially on adsorption to this surface from the ambient aqueous phase. The force determining the ability to attract external ions is the available surface energy of a particle, which is at its highest and most characteristic value when the particle is just broken during comminution, with its crystal surface bonds newly sheared.

11.4.3 SURFACE ACTIVATION

In flotation reaction technology, activation refers to causing of selective reaction on the surface of a desired mineral, usually resulting in attraction into the air–water interface. Due to activation, particle becomes eager to be drawn into a rising bubble, rises, and persists as part of a selectively mineralised froth. The reagents used to induce such property are called activators, whereas the chemicals used for the opposite purpose are called wetting agents or depressants. Both types are surface active agents or surfactants which influence the conditions in the air–water–solid triphase. These reagents usually consist of inorganic ions ($+$ or $-$) which are chosen for their ability to increase or decrease the difference between the absorptive powers of the various species in the pulp.

The tenacity and density of fixation of most commonly employed collector (xanthate), depend on the structure of the crystal lattice and the cleavage features developed during grinding. The minerals possess heteregeneous surfaces, which lead to preferential attachment to air and preferential adsorption of a surfactant or localised patches of oxidation. Some sulphide surface may contain both anodic and cathodic areas. In the case of non-sulphide minerals (e.g. fluorite, calcite) the crystal structure greatly affects the action of gases from the pulp as well as surface interaction with reagents. In dealing with surface activation, following points may be remembered:

a) The fresh surface of a sulphide mineral is wettable to some extent in the absence of air or oxygen.

b) Oxygen is adsorbed from water in preference to other gases present in the system.

c) Adsorption takes place in stages.

d) Surfactant is fixed after the adsorption of oxygen.

Adsorptive forces in case of non-sulphides such as metal oxide, silicates, carbonates. etc., are much weaker than the chemisorptive bonds between surfactant and sulphide. Most commonly floated minerals are metal sulphides, which differ considerably in physical properties (such as lustre, colour, solubility in water, oxidation tendency, etc.) from each other. Despite this difference of common physical properties, these can be floated easily with a suitable surfactant (collector). The reason responsible for flotation is not found with low solubility which is a common factor of sulphides. In most sulphide minerals the metal–sulphide lattice is strongly bound together by internal ionic forces, as a result of which the minerals are immune from attack across the interphase by solvating counter ions at normal pH values (near neutral) in the pulp. The unstable sulphides such as pyrrhotite and marcasite readily take up external oxygen and produce chemisorbed surface layers of sulphite and thiosulphate raising the acidity of the pulp to some extent.

An ion-exchange process has also been considered as the attracting mechanism in which incipient oxidation is required and considerable evidence is found for the necessity of oxygen in the pulp as a prerequisite of collector coating.

The function of the pH (H-ions) has been recognised to promote an attracting potential at the surface of each specific mineral within a certain pH range or below a specific pH value. Activation is also facilitated in accordance with the inverse solubilities of the metal sulphides. Compounds of Hg, Ag, and Cu effectively replace Zn at the surface of sphalerite.

Maximum adsorption of collector molecules is found at the iso-electric point of a mineral, rendering it floatable with that collector. The electrical character of the surface is determined by both electrochemical and electrokinetic potentials with excessive transfer into solution of one type of ion (+ve or −ve) by which the electric equilibrium is disturbed and the surface becomes charged. Ions of opposite charge now find it difficult to escape from the crystal lattice into the liquid phase. Ions already present in solution are drawn nearer to surface to balance its charge and due to which an electrical double layer is formed. Its inner component of this layer is due to the charged surface on the mineral while the outer counter ions are more mobile due to stirring effect by thermal vibration in the liquid. Counter ions next to the surface move with the mineral while the distant ions diffuse away setting up the electrokinetic potential known as *zeta potential* (ζ). If a change in the magnitude of ζ is accompanied by a change in its sign, then the reagent ions at work in the pulp determines the potential and can penetrate to the inner layer of the zone of shear and thus adsorption to the mineral surface is affected. Xanthates have been reported to affect the electrochemical potentials of conducting minerals.

The structure and stability of hydrated surface layers are dependent on the nature and interplay of the phases in contact, whereas, the process of hydration influences the solubility of minerals, attachment of particles to bubbles, and coagulation.

Flotation properties of minerals may be tested in terms of surface hydration or contact-angle test (discussed in subsequent sections). Attraction of water dipoles to surface lattice points is due to free atomic or molecular force and surface hydration increases with the adsorption of hydrated ions. If during grinding strong electrostatic forces are set up, the mineral surface becomes strongly hydrated due to rupturing of bonds and uncompensated atomic and ionic charges. Hydration of minerals possessing these charges is slight (e.g. graphite, sulphur, MoS_2, talc, etc.). These minerals possess weakly adherent bonds across the cleavage slip planes and are characterised by their native floatability. Gas adsorption results in lowering the hydration of minerals surface considerably. Multivalent cations may be hydrated with six to eight molecules of water, whereas anions attract a much weaker hydrating layer. Dissolution of a mineral constituent occurs when the energy of hydration exceeds the lattice energy. In hydrated layer, the water molecules just in contact with the true ion appear strongly oriented and the effect decreases outwardly. Polar molecules become adequately hydrated, whereas non-polar ones only slightly. The reaction taking place in hydration is exothermic in nature. The amount of heat produced indicates the work done in surface wetting and strength of water attachment to the mineral. Minerals having the same anions will show that the work done in wetting is inversely proportional to the radius of cations.

The possibility of chemical action in modification of surfaces is considered to be an essential energising activity for many minerals, whereas ion exchange at surface monolayer is the main factor in the action of collecting and modifying agents.

11.4.4 REACTION BETWEEN WATER AND MINERAL

The wetting of mineral surfaces is the most important reaction of mineral with water. The wetting may be defined as the adsorption of water ions and molecules on the mineral–water interface. The final effects of mineral–water reactions will depend on both, the mineral surface energy and the energy of reaction of water ions and molecules among themselves. In case of reaction between molecules of the same substance (e.g., water), the mutual tension (energy of reaction) is called *cohesion* which is characterised by the *work of cohesion* (in erg/cm^2). This is the work to be done to break a column of liquid 1 cm in cross-section into two columns of the same cross-section. When the molecules react in two phases (e.g. water and mineral) making their appearance in the interface, the mutual tension (energy of reaction) is called *adhesion*. The tension exerted by one phase on another through their interface requires the application of work to separate these phases. The work, which is related to the unit of interface area, is known as the work of adhesion (in erg/cm^2).

The first stage in a molecular reaction between water and a mineral surface is the wetting of the mineral by the water occurring at the boundary line of three phases (solid, liquid and gas, or solid and two immiscible liquids). In wetting, liquid drop flows fully or partly over the surface of a mineral, depending on the wettability of that surface by the liquid. A thin hydrated film forms on a dry mineral surface due to adsorption of water vapour. This film always assists in formation of a hydrated skin and thus wetting the surface by water. Thickness of the hydrated skin may reach 0.1 micron corresponding to many thousands of water molecules. The hydrated skins are formed when the bond energy between the water dipoles and the mineral surface exceeds the bond strength between water dipoles themselves (cohesion force).

The greater the water-repellent forces of a mineral, the easier is the flotation due to detachment of water from the mineral surface and attachment of air bubble to the mineral. The extent of water-repellence of a mineral surface is represented by the contact angle (discussed in the next section), which increases with increase of water-repellent properties. In order to render the minerals water-repellent, special flotation reagents are added.

The reaction between a mineral surface and water depends considerably on the nature of the free (unsaturated) bonds present on the crystal surface. The atoms, ions, ion groups or molecules in solid bodies are interconnected by forces of different strengths. Ionic and covalent (atomic) bonds, and the universal polar bond (intermediate between the two) are characteristic of polar mineral surfaces. All these types of bond are very strong, which give a high free energy value on a polar surface, whereas a non-polar mineral surface is characterised by molecular bond forces (Van der Waals forces) being very much weaker than ionic and covalent bonds.

11.4.5 WATER-AVID (HYDROPHILIC) AND WATER-REPELLENT (HYDROPHOBIC) SOLIDS

A *water-avid* mineral surface is one on which unsaturated ionic or atomic bonds predominate and due to the presence of these bonds, the surface reacts with water molecules. On the other hand, a *water-repellent* mineral surface is one on which unsaturated molecular bonds predominate, making the conditions unfavourable for attachment of water molecules. For example, graphite can be readily split along the cleavage planes due to weak molecular forces between the layers, and the surface corresponding to cleavage plane is water-repellent. Similarly, water-repellent properties of sulphur can also be explained.

The ionic bond in minerals having oxygen bearing anions is stronger than in case of sulphides. As a result, the water dipoles on sulphide surfaces are less firmly attached and it is easier to render the surface of sulphides water-repellent, by treating it with the suitable reagent than the oxide minerals.

A mineral surface may also react with gases, particularly oxygen. Strong atomic bonds of freshly broken sulphide mineral crystals are exposed and thus a freshly broken sulphide mineral is completely water-avid. However, after a short time, oxygen is adsorbed on the mineral surface which is always present in solution in water. The primary action of oxygen on a mineral surface is to render the mineral water-repellent to some extent. The further action of oxygen on sulphide minerals leads to their oxidation, resulting into the development of water-repellent properties. Thus the natural water-repellent and water-avid properties represent the extent to which a mineral surface is active in relation to water and gases (particularly oxygen).

Although most naturally occurring minerals are hydrophilic, some are hydrophobic. Examples of the latter include graphite, sulfur, antimonite (Sb_2S_3), molybdenite (MoS_2), talc, and high-rank coals such as anthracite. Many polymers, such as Teflon and Nylon, are also hydrophobic.

Hydrophilic materials can be made hydrophobic by the adsorption of chemicals. For example, calcite $(CaCO_3)$ can be readily made hydrophobic by treating it with low concentrations of sodium oleate $(C_{17}H_{33}COONa)$, Silica (SiO_2) can be made hydrophobic by treatment with dodecylamine $(C_{12}H_{25}NH_2)$. A solid made hydrophobic by chemical treatment can revert to the hydrophilic state on further chemical change.

11.4.6 SOLUBILITY OF MINERALS IN WATER

The dissolution of mineral in water may result from the reaction between water and a mineral. A certain amount of molecules and ions (complex and varied in nature) from the compounds of minerals is normally present in the liquid phase of a flotation pulp. The direct reaction of water molecules on mineral surfaces in a pulp may lead to breakdown of their crystal lattice resulting into separate hydrated ions (Fig. 11.1), which form an ionic or dispersed molecular solution.

The mineral ions (or molecules) passing into solution may enter into reactions with mineral particle surfaces as well as with the flotation reagents fed into the pulp. Some ions known as inevitable ions are always present in some quantity in the flotation pulp. The concentration of these ions sometimes exceeds the concentration of flotation reagents introduced into the pulp and thus these influence the flotation practice considerably.

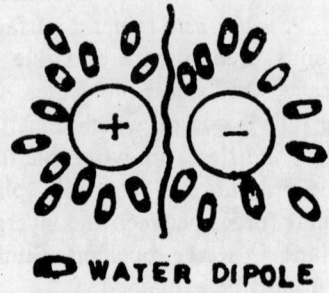

WATER DIPOLE

Fig. 11.1. Hydrated ions, forming an ionic or dispersed molecular solution.

On dissolution of mineral in water, crystal lattice energy of the mineral is absorbed by the solution and at the same time ion hydration energy is given off. The difference between the two energy values is the heat of solution of that mineral in water. Minerals dissolve in water when the ion hydration energy (H) is more than the crystal lattice energy (L). Ion hydration energy (H) increases with the increase of valency of the ion, the decrease of ionic radius, and increase of lattice energy. However, hydration energy increases much more slowly with increase of ion valency than with increase of crystal lattice energy. Therefore, the solubility is greatly reduced with an increase in valency. Because of this reason, the sulphides or oxides of bivalent metals are much less soluble in water than the corresponding monovalent metal compounds. With the same cation, the solubility of a metal compound is proportional to the radius of anion.

11.4.7 REACTION BETWEEN SULPHIDE MINERALS AND OXYGEN AND CARBON DIOXIDE

Chemical reaction between sulphide minerals and dissolved oxygen in water is extremely important in flotation process. Oxygen may cause fundamental changes in the composition of mineral surface by chemical action. Under normal pressures, 1 litre of water may contain dissolved oxygen up to about 10 mg. Rain water regarded as air saturated, contains 25–30 ml of gas containing 30 per cent O_2, 10 per cent CO_2 and 60 per cent N_2. In the presence of water, the action of oxygen on sulphide minerals causes their oxidation into sulphates through intermediate stages. For extample lead, copper and zinc sulphates are formed respectively when galena, chalcopyrite and sphalerite are oxidised. Thus the surfaces formed are much more soluble than their corresponding sulphides. The smaller particles of mineral and product of decomposition passed into solution have a strong tendency for oxidation with oxygen. Oxidation of sulphide minerals produce two important results, i.e. (i) fundamental changes produced in composition of surface layers and (ii) the surfaces formed are highly soluble than their corresponding sulphides. Thus, the reaction of oxygen on sulphides causes the pulp to become richer in sulphate and heavy metal ions, which react with the mineral surface and the flotation reagents.

The presence of CO_2 dissolved in water also affects composition of surfaces of individual minerals and the ionic composition of the pulp. The corresponding carbonates may be formed from sulphates. The formation of new compounds from sulphides results in lowering the pH value of pulp.

11.4.8 REACTIONS BETWEEN MINERAL SURFACE AND DISSOLVED COMPONENTS

These reactions may be considered as the adsorption of electrolytes dissolved in water on the solid-liquid interface. These electrolytes may be either added to the pulp or formed in it due to the passage of certain amount of mineral into solution. These reactions may be classified as the following:

a) Molecular adsorption b) Ionic adsorption c) Specific adsorption.

a) *Molecular Adsorption*

This is also known as non-polar or equivalent adsoption in which adsorbent takes up an equal amount of cations and anions from solution giving an effect equivalent to the adsorption of complete electrolyte molecules by the solid phase. Therefore, the state of electrical neutrality on the phase boundary is not changed. If a weak electrolyte is adsorbed on the solid-phase surface, it will be adsorbed directly in molecular form due to its poor dissociation. When strong electrolytes present in dissociated form in solution are adsorbed, the varying susceptibility to adsorption of any electrolyte's anions and cations is of great importance. The amount of electrolyte adsorbed is dependent on its solubility-adsorption (i.e. inversely proportional to solubility). The susceptibility of any ions to adsorption depends on their valency, hydration, and the solubility of complexes formed on adsorption.

In general, the mineral acids are adsorbed to a greater extent than their alkaline salts, indicating that H-ions are adsorbed more easily than alkali metal ions. Similarly, the OH-ions are the most susceptible to adsorption of the monovalent inorganic anions.

b) *Ionic Adsorption*

This is also known as polar adsorption in which ions of the same sign are transferred between the solution and the solid-phase surface, resulting in the surface charge remaining unchanged. Ionic adsorption may be anionic or cationic depending on the type of ion getting exchanged. This may be expressed as $XM_1 + M_2 = XM_2 + M_1$

where XM_1 and XM_2 are the solid surfaces, and M_1 and M_2 are the ions present in solution. If the surface concentration of the compounds XM_1 and XM_2 are represented by a_1 and a_2 respectively, and the bulk concentrations of M_1 and M_2 are represented by C_1 and C_2, the exchange adsorption equation may be represented by the following expression:

$$a_1 C_2 = K a_2 C_1 \quad \text{or} \quad \frac{a_1}{a_2} = K \frac{C_1}{C_2}$$

c) *Specific Adsorption*

It is characterised by the selective uptake of anions or cations from solution by the solid phase, resulting into formation of difference in potential on the phase interface, leading to the establishment of double electric layer. The ions or groups of ions already existing in the mineral crystal lattice or capable of replacing these ions will most readily be adsorbed in the inner skin of double layer. The most important condition for this exchange is that, the ions replacing each other should possess the same or nearly the same dimensions. The ions fulfilling these requirements can be adsorbed by the mineral surface forming a double electric layer and are known as potential determining ions. These ions include both, the ions which attach themselves to the mineral surface and form the double layer and their antiions.

The double electric layer has a marked effect on wettability, as its outer layer is formed in the zone of hydrated layers which are formed by surface adsorption of water. Electrolytes not containing potential determining ions, show comparatively less effect on thermodynamic potential, but may cause drastic changes in the zeta potential.

It corresponds to the potential that exists at the so-called shear plane, which is created when a charged particle moves under the influence of an external electric field.

Zeta potential can be measured by electrophoresis, sedimentation potential, streaming potential, and electroosmosis. In the first two of these methods the shear plane is created by the motion of the particle; in the other two, the particle (more precisely, a porous bed consisting of many charged particles) is fixed.Following equation correlates the zeta potential with the electrophoretic mobility of the particles under the influence of an external electric field :

$$\zeta = \frac{4 \pi \eta}{\varepsilon} \frac{V_e}{E}$$

where ζ = zeta potential

η = viscosity of the aqueous medium

ε = dielectric constant

V_e = electrophoretic velocity of particle

E = electric field between electrodes applying the field

At 25°C and $\eta = 0.895$ mPa.s, the equation reduces to $\zeta = 12.8$

$V_e = 12.8\ u$

where u = electrophoretic mobility.

The significance of the zeta potential becomes apparent where, electrostatic interactions play a major role in flotation reagent adsorption. The condition at which $\zeta = 0$ is known as the isoelectric point (iep), sometimes, less accurately, the zero point of charge (zpc). In principle, although an iep can be defined for each solid for numerous conditions, the variable most frequently used to describe the ionic conditions at which iep occurs is pH.

Adsorption mechanisms correlate with zeta potential; that is, physical adsorption affects the magnitude of the zeta potential. A change of sign in the presence of an appropriate concentration of absorbate is taken as an indication of a chemisorption-type reagent uptake. The iep does not necessarily indicate zero potential at the surface of the particle, only that the potential at the shear plane is equal to zero.

Depending on the mechanism of reagent-solid interaction, many systems exhibit correlations between system variables and zeta potential.

The following isoelectric points ($\zeta = 0$) are observed in the absence of other substances.

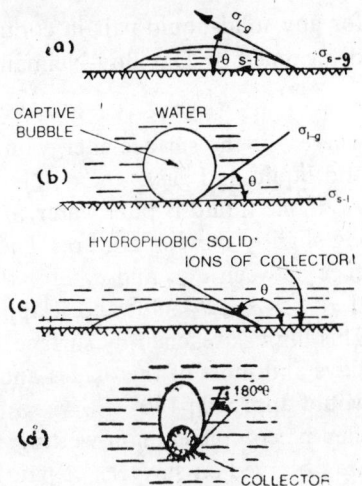

AgCl	$pAg^+ = 4$
AgBr	$pAg^+ = 5.4$
Ag_2S	$pAg^- = 10.2$
$CuSiO_3 \cdot 2H_2O$	$pCu^{2-} = 4$, at pH = 7
CaF_2	$pCa^{2+} = 3$
Fe_2O_3	pH = 6.5 and 8.2
FeS_2	pH = 7
FeS_2	pH = 5 (30 mg NaCN/L)
PbS	pH = 3.5
SiO_2	pH = 2–3.7

The addition of electrolyte, reduces the zeta potential and also reduces the diffused layer thickness. In these circumstances, the extent of mineral surface hydration must also be reduced.

Fig. 11.2. Relationship between contact angle and surface tension with effect of collector..

In specific adsorption, the adsorbed ions may pass directly into the mineral crystal lattice, altering the structure of mineral surface layer. The reactions between ions attached in a double electric layer inner skin and a mineral surface, do not differ considerably in nature from the reactions occurring in the mineral crystal lattice itself.

The adsorption of potential determining ions common to the mineral crystal lattice is of specific interest. A crystal immersed in a solution containing ions similar to its own, will actively adsorb one or another of these ions, resulting in an increase of its electrokinetic potential. For example, AgI crystal will selectively adsorb Ag ions from $AgNO_3$ solution and iodide ions from KI solution.

11.4.9 WETTABILITY AND CONTACT ANGLE

The actual surface property of interest is its chemical and crystallographic capability of interacting with particular organic ions, resulting in a very thin layer (probably a monomolecular layer) of these ions attached to the surface. This condition results in a high effective surface tension and is recognised as *wettability*, which is measured quantitatively in terms of contact angle, θ as shown in Fig. 11.2. If a drop of liquid is placed on a solid surface, the angle towards the liquid phase is called the contact angle. Theoretically, the contact angle may vary from 0° to 180°, where zero corresponds to complete water-avidity (fully wettable) in the mineral and 180° corresponds to the complete water-repellence (completely unwettable). However, in practice such extreme cases do not occur since there is always certain amount of adhesion.

The contact angle cannot be measured easily as it is sensitive to contaminants at the interface, surface irregularities, and the direction of water flowing across the surface. However, there is an equilibrium contact angle for any solid-liquid pair in contact with air at any temperature, and can be determined by Davidov-Neiman equation

$$\sigma_{s-g} - \sigma_{s-l} = \sigma_{l-g} \cos \theta \tag{11.1}$$

where σ is the surface energy on the interface (solid and gas, solid and liquid, and liquid and gas).

If the liquid is pure water, σ_{l-g} has a value of about $0.072\ NM^{-1}$, depending slightly on temperature. The contact angle then depends on the difference between σ_{s-g} and σ_{s-l} for the solid surface. If these are equal, $\theta = 90°$ If $\sigma_{s-g} > \sigma_{s-l}$ $\theta < 90°$ and it approaches zero as $\sigma_{s-g} - \sigma_{s-l}$ approaches σ_{l-g}. This limit is reached by surfaces which are very easily wetted by water, e.g., clays and most of the oxides and hydrates. When $\sigma_{s-g} < \sigma_{s-l}$, $\theta > 90°$ and would approach $180°$ if $\sigma_{s-l} - \sigma_{s-g}$ becomes equal to σ_{l-g}. The surface would then be completely non-wetting. In practice the largest angles of about $110°$ are observed on surfaces of paraffin wax. For wetting of a mineral, the work of adhesion of liquid to solid should exceed the work of *cohesion* of the liquid, which is equal to double the surface tension of the liquid, i.e.

$$W_{ad} = \sigma_{s-g} + \sigma_{l-g} - \sigma_{s-l} > 2\ \sigma_{s-l} = W_{co} = 0.144\ NM^{-1} \text{ for water} \tag{11.2}$$

By combining Eqs. (11.1) and (11.2), $W_{ad} = \sigma_{l-g}\ (1 + \cos \theta)$ (11.3)

For wetting $W_{ad} > \sigma_{l-g}\ (1 + \cos \theta)$ and for non-wetting $W_{ad} < \sigma_{l-g}$ $(1 + \cos \theta)$.

The right hand components of Eq. (11.3) are measurable and can therefore be used to calculate the force of liquid adhesion to a solid surface. In practice, the values of θ usually lie in the range of $25°$ to $80°$ corresponding to the 0.14 to 0.85 JM^{-2} range of W_{ad}. Figure 11.2 represents the different values of contact angles for attachment of air bubbles. If θ is small, the air bubble tends to become detached [Fig. 11.2 (b)], whereas, if θ is large, the bubble becomes firmly attached to the surface [Fig. 11.2 (c)]. If the surface is part of a small particle, the net density of the particle and the bubble together may become less than that of the water and thus the air attached particle will rise to the surface [Fig. 11.2 (d)]. Contact angles of solids in water and various aqueous media are shown in Table 11.2.

11.5 Floatability Test of Minerals

In the flotation system of air-water-mineral, the surface energy of the phases in contact determines the degree of floatability. For a liquid, the surface tension can be directly measured, whereas for solids its value can only be assessed indirectly. The various techniques used for the measurement of surface tension are capillary rise, bubble pressure, dropping weight, and adherence measurement of a ring, or of a mica plate. The latter two techniques depend on the measurement of force required to detach a completely wetted wire loop or vertically suspended thin plate. Unfortunately, at present no direct means are available for measuring the surface tension at a solid surface.

In the laboratory, studies of surface reactions are carried out by contact angle measurement. For determination, a clean smooth surface of mineral

Table 11.2. Contact angles of solids in water and various aqueous media

Solid	Contact angle (degrees)	Solution conditions
Colemanite	43	5×10^{-3} M oleate* solution
Copper metal	93	in 50 mg/L oleate* solution
Fluorite	91	in 10^{-5} M oleate*, pH = 8.1
Galena	60	in 10^{-3} M ethyl xanthate solution
Graphite	96	water
Ilmenite	80	treated with oleate*, pH = 8, T = 75°C
Colorado oil shale with 28% organic carbon	59.5	water
Paraffin wax	108	water
Silica	81	2.5 mg dodecylammonium chloride per L solution, pH = 10
Teflon	160	water
Teflon	0	methanol-water solution with surface tension <20 mN/m

* Sodium oleate.

is placed in distilled water and a bubble is pressed down upon it. If after half an hour (induction time) no adhesion is observed, the surface is considered to be clean (i.e. completely wet). An appropriate reagent is then added to the water at a given pH, and after a short reacting time the air bubble is again pressed to the surface. Adherence of air bubble indicates the surface to be aerophilic.

For the given surface condition of the mineral particles, contact angle (θ) is the index of surface energy. If $\theta = 0°$ ($W_{ad} \simeq 0.144$ JM^{-2}), the mineral will be completely wetted, whereas if $\theta = 90°$ ($W_{ad} = 0.072$ JM^{-2}), the mineral will not be wetted and float completely. Even a contact angle of few degrees indicates some floatability. However, in a conditioned system, θ is not a characteristic of the mineral phase, but is a characteristic of an aerophilic organic molecule (collector) adsorbed on particle surfaces. In such a case, floatability is determined by the surface coating and not by substrate. For example, contact angle of all minerals (Pb, Zn, Cu, etc.) conditioned by xanthate is about 60°.

The contact angle method for establishing the conditions of floatability suffers with certain experimental problems, some of which are as follows:

a) Problems of obtaining a true representative crystal of the required mineral of a sufficiently large size (over 0.5 cm² in plane area).

b) Characteristic of mineral may change after its intensive polishing with abrasives under water.

c) The test is carried out under static condition, which differs greatly from the dynamic one in which the mineral particle must adhere to a rising bubble towards the surface of the test fluid.

d) Maintenance of complete cleanliness becomes a problem when new surface is prepared many times for studies of variables such as pH and reagent concentrations.

Difficulties in contact-angle measurements on powdered solids, as used in flotation practice, can be overcome by a technique where flotation recoveries are plotted against solution surface tension. The $\sigma_{l\text{-}g}$ at which flotation recovery is equal to zero, is taken as the critical surface tension of wetting, σ_c of the solid under investigation.

Many freshly fractured surfaces show a natural contact angle of a few degrees. For example sulphur, ozokenite, graphite, coal, etc., exhibit contact angle high enough to cause their flotation without the use of collector agents. The greater difference between the natural contact angles of the various minerals and gangue facilitates better separation by flotation.

Two techniques, i.e. bubble pick-up and phase inversion method have been developed, in which most of the experimental problems are overcome.

11.5.1 BUBBLE PICK-UP METHOD

The method has been developed by Cooke and Digre. In this method, a small amount of the test mineral is crushed, sized to the required mesh range, and cleaned by shaking in distilled water. An air bubble is pressed against particles and then lifted. The quantity of grains adhering is recorded. A good correlation with contact angle tests on similar material has been found. The bubble pick-up method is more sensitive than the measurement of contact angle to changes in reagent concentrations. An improvement of this method has been reported by Sun and Troxell in which 0.5 g of sized and cleaned mineral is placed in 200 ml of distilled water. After adjustment of test chemicals and allowing the required conditioning period for reaction to occur, the particles are swirled to the centre of the beaker, pressed on by an air bubble held in the concave end of a glass rod. The glass rod is then raised gently and moved to a clear part of the beaker and tapped. The dropped particles are then examined. The strongly aerophilic particles remain attached and are usually visible. The bubble pick-up technique makes it possible to relate changes of reagent and pH at all stages between non-attachment and strong attachment of particle to air-water interphase. A series of tests can be made quite rapidly on a small number of particles.

For bubble pick-up method, a modified Hallimond tube (Fig. 11.3) can also be employed. In this the particles are held on a porous surface of sintered glass at the bottom of the tube containing distilled water and testing reagents. A stream of air bubbles is blown upward through the sinter glass and particles adhering to each bubble explode and the associated particles slide down to the receiving pocket, from which they are removed as desired. In this case floating of particles is fully dynamic and the conditions are well comparable with conditions of commercial flotation cell. A small sample is taken and separated into floating and non-floating fractions which are separately weighed and examined. However, Hallimond tube does not take care

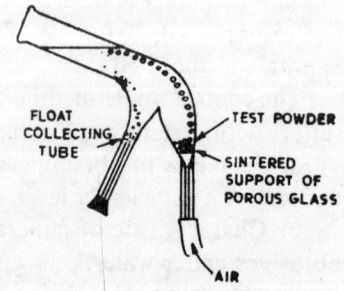

Fig. 11.3. A modified Hallimond tube for floatability test.

of only one factor, i.e. the effect of the frothing agent and its associated froth column which can be left out for many purposes.

11.5.2 PHASE INVERSION METHOD

This is based on the mechanical stabilisation of an emulsion (water in oil or oil in water) by an insoluble interphase. This is a simple method and requires only the screening of a small sample of mineral through 50 micron sieve, shaking of the phases (e.g. 2 ml each of benzene and water) and then observing the effects of reagents or pH.

11.6 Mineralisation of Air Bubbles in Flotation

Froth flotation process mainly depends on mineral particle attachment to air bubbles. The most favourable conditions for selective, strong, and rapid attachment of certain mineral particle to air bubbles are obtained by pretreatment of ore-pulp. In flotation process the bubble attachment may be achieved basically by two methods. i.e., (a) due to particle coming into contact with bubbles in the pulp, i.e. collision between particle and bubble, and (b) due to the formation of gas bubbles on the particle surface from solution. Thermodynamic and kinetic methods may be used to study the attachment of mineral to air bubble. By the former method, possibility of bubble mineralisation can be studied, whereas, the mechanics of the process can be studied by the latter method.

In the flotation cell, coursing of bubbles may be due to diffusion of air blown, in this instance it starts in a state of rapid vibration with changing the shape and area as many as 100 times per second. It may be sheared into the pulp near the hub of the impeller of a mechanical cell and during milling with the solids in turbulent zone, swept by the tips of the impeller. Precipitated bubbles can be pressed against the particles settling below the impeller and drawn back towards the hub for recirculation.

In froth flotation, the energy barrier between the hydrated surface of mineral and the bubble must be overcome by certain force, which may be caused either due to the adsorptive attraction of a collector or to collision resulting from mechanised energy. Acceptance of mineral particle by an air bubble may be complete in less than 0.005 sec.

Under correctly selected flotation conditions, desired mineral particles will adhere to air bubbles in the pulp. During flotation, a relatively unstable system (separate bubbles and particles prepared for flotation) undergoes a transition to a relatively stable system (mineral-bubble groups). The probability of mineral-bubble formation can be assessed by making use of second law of thermodynamics, according to which any reaction can occur spontaneously and at a definite speed in the direction corresponding to a reduction in free energy of the given system. The system attains its maximum stability having its free energy (surface energy) at minimum.

11.6.1 BUBBLE ATTACHMENTS TO MINERAL PARTICLE DUE TO COLLISION

Before bubble attachment, the free energy of the system may be given as follows:

$$W_1 = S_{l-g}\, \sigma_{l-g} + S_{s-l}\, \sigma_{s-l} \qquad\qquad (11.4)$$

where S_{l-g} and S_{s-l} are the liquid-gas and solid-liquid interface areas, and σ_{l-g} and σ_{s-l} correspond to the surface energies on these interfaces.

After bubble attachment, the free energy of the system to form an attachment area of 1 cm² may be written as follows:

$$W_2 = (S_{l-g}-1)\, \sigma_{l-g} + (S_{s-l}-1)\, \sigma_{s-l} + \sigma_{s-g} \tag{11.5}$$

Then reduction in free energy of the system

$$\Delta W = W_1-W_2 = \sigma_{l-g} + \sigma_{s-l}-\sigma_{s-g} > 0. \tag{11.6}$$

or $\qquad \sigma_{l-g} + \sigma_{s-l} > \sigma_{s-g}$

Putting the value of σ_{s-l} and σ_{s-g} in terms of measurable quantities σ_{l-g} and contact angle θ from Eq. (11.1).

$$\Delta W = \sigma_{l-g}\,(1-\cos\,\theta) \tag{11.7}$$

Equation (11.7) is valid for a contact area of 1 cm². It indicates that the decrease in free energy in the system is proportional to the contact angle (a measure of bubble attachment) and σ_{l-g} (flotation activity of a surface falls when σ_{l-g} decreases). However, in practice, frothers are found to increase flotation in spite of their somewhat reducing effect on σ_{l-g}, which is mainly due to the size reduction and stability of air bubbles caused by reagents.

The kinetics of mineral attachment to bubbles due to collision are determined by the time taken in disintegration of the water layer, separating the bubble and the mineral particle.

In a flotation machine, innumerable collisions take place in a second, and thus the process of bubble attachment is subjected to statistical rules. A definite proportion of these collisions result into bubble attachment and in flotation the conditions ensuring the maximum probability of attachment are produced by controlling the various physicochemical and physical factors. In practice it is impossible to obtain 100 per cent separation of minerals into concentrate and gangue, which are absolutely free from each other. Therefore, the main problem in flotation is to obtain conditions ensuring the maximum possible mineral separation.

11.6.2 FORMATION OF GAS BUBBLES ON A MINERAL PARTICLE FROM SOLUTION

The amount of dissolved gases is proportional to pressure and therefore, tiny gas bubbles will be formed on reducing the pressure. The bubbles so formed will possess free surface energy and therefore, work W_1 must be expended in forming them. In the general case

$$W_1 = W_a + W_b + W_c \tag{11.8}$$

where W_a, W_b, and W_c are the work needed to form a new phase interface, to break the continuity of water (to form the cavity), and to fill the cavity with water vapour, respectively. Substituting the values of W_a, W_b, and W_c

$$W_1 = 4\,\pi\,R^2\,\sigma_{l-g} + 4/3\,\pi\,R^3\,P-4/3\,R^3/\rho_v \tag{11.9}$$

or $\qquad W_1 = 4\,\pi\,R^2\left[\sigma_{l-g}-\dfrac{R}{3}\,(\rho_v-P)\right] \tag{11.10}$

where R = bubble radius, P = pressure in bubble, and P_v = elasticity of the liquid vapour.

If a bubble appears on a solid surface instead of the body of the water, the work W_2 must be done, where

$$W_2 = S_2\, \sigma_{l-g} + S_1\, \sigma_{s-g} - S_1\, \sigma_{s-l} + W_b + W_c \qquad (11.11)$$

where, S_1 = attachment surface area, and

S_2 = bubble surface area (shaped like a sphere segment).

By neglecting the values of W_b and W_c as W_1 is basically determined by the value of W_a and carrying out a series of conversions, final equation may be obtained as the following:

$$\frac{W_1}{W_2} = \frac{1 + \tan^2 \dfrac{\theta}{2}}{\sqrt[3]{3\,\tan^2 \dfrac{\theta}{2} + 1}} \qquad (11.12)$$

By substituting the values of W_1/W_2 for different values of θ, it will be found that $W_1/W_2 > 1$ indicating the requirement of less energy for the formation of bubble on a solid surface than necessary to produce a bubble in the body of the water. From Eq. (11.12), it may be followed that greater the value of θ, easier is the gas bubble formation on a given surface.

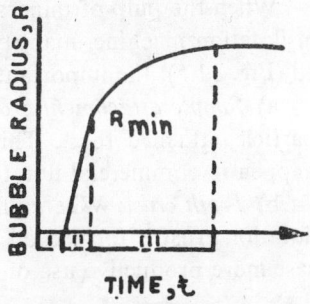

Fig. 11.4. Course ot bubble formation from solution.

The course of bubble formation from solution is shown in Fig. 11.4. During the period-I (extremely short time), the pulp water becomes oversaturated and no bubbles are formed. During this period, the gas molcules are displaced and moved to areas where they can combine more easily and break the bonds existing between the water molecules. When an adequate number of gas molcules accumulate in certain areas during the period-II, the molecules unite due to the action of Van'der Waals (molecular) forces and form bubble molecules. This process is very rapid and the period is about 10^{-12} sec only. The bubble nucleus then grows due to diffusion of dissolved gases into it during the period-III. The bubbles develop more easily on less hydrated mineral surface.

Bubble formation from solution onto particle surface is also governed by statistical rules. Probability of appearance of bubbles on given mineral particles determines the results of flotation. This probability increases with the increase of water-repellent properties of the particle surface and over-saturation of solution by the gases. Though the bubbles produced from solution are usually very small, and incapable of carrying the mineral particles into the froth, these bubbles help considerably the flotation process.

11.6.3 COMBINED BUBBLE ATTACHMENT TO A MINERAL SURFACE

This includes the combination of the above two methods of bubble formation. In this case, initially a small bubble is formed from solution on the mineral particle. A larger bubble having enough lifting power to carry the particles into the froth attaches itself to the earlier bubble. The large and small bubbles coalesce and form a larger bubble, which sticks to the mineral particle. The small bubble formed from the solution activates the mineral surface and improves its attachment to other bubbles.

11.6.4 Bubble – Particle Interactions

The ultimate objective of a flotation process is the selective removal of solid particles from the aqueous medium, which is accomplished by the adhesion of air bubbles to the hydrophobic particles. Particle flotability can be treated as a probability : $P_f = P_c \cdot P_a \cdot P_s$ where P_f = probability of flotation

P_c = probability of particle–bubble collision

P_a = probability of particle–bubble adhesion

P_s = probability of formation of a stable particle–bubble aggregate

In some methods, such as vacuum flotation, where dissolved gases become the bubble source, or in situ bubble-generation processes, where acids generate bubbles of carbon dioxide from carbonate-containing pulp, the probability treatment needs to be modified. Above equation, however, is widely applied, because most flotation systems rely on extraneously introduced air bubbles.

When the pulp of mineral particles and bubbles is treated with reagents in flotation machine, many types of mineral particle-bubble groups are formed (Fig. 11.5), the important ones of which are described below:

a) *Simple attachment group*: This consists of one bubble with a single particle attached to it. This is the simplest possible group and does not happen in commercial flotation processes.

b) *Froth crust*: When a number of particles in the pulp are increased, flotation crust is formed on lower surface of the bubble (Fig. 5a). In this case more productive use of bubble is made. Usually 1–2 per cent of bubble

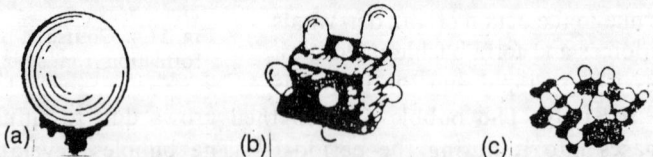

(a)   (b)   (c)

Fig. 11.5. Different froms of particle-bubble aggregates (a) Particle attached to bubble. (b) Multiple bubble attached to large particle (common in vacuum flotation). (c) Particle-bubble flocculation in pulps containing fine particles.

surface is occupied by mineral particles in basic flotation of poor ores to 20–30 per cent in case of coal flotation or purification process depending on the conditions of flotation.

c) Single large particle floated by a group of bubbles (Fig. 5b).

d) *Air flocs*: In this case a group of particles and bubbles coalesce to form *air flocs* (Fig. 11.5c). Formation of these mineral particle–bubble groups requires special conditions, such as high degree of mineral surface water repellence, a high degree of pulp aeration, relatively little pulp mixing, and a large number of float particles per unit of pulp volume.

11.6.5 FACTORS AFFECTING THE BUBBLE ATTACHMENT

Various factors affecting the mineral particle–bubble attachment may be considered as the following:

a) *Distance of the particle from the vertical axis of rising bubbles*: The attachment of bubbles to particle decreases with increase of distance.

b) *Size of bubbles*: A small bubble attaches more rapidly to a polished surface than a large one and after getting attached, the condition becomes favourable for the attachment of larger bubbles.

c) *Shape of particles*: A sharp projection on the particle favours the attachment, whereas large flat particle adversely affects the attachment.

d) *Surface condition of particle*: A rough surface of a hydrophillic particle favours wetting by trapping water, whereas a surface shielding a pocket of air helps bubble attachment.

e) *Oxygenation sulphide surface*: Freshly cleaved sulphide minerals hydrate very strongly, whereas, surfaces exposed to air or oxygen for some time show some degree of hydrophobicity.

f) *Frictional, gravitational and inertial forces*: In flotation machines these forces are set up and tend to detach the mineral particles from the bubbles.

11.7 Aeration and Froth Formation

Aeration may be defined as the introduction of air bubbles in the pulp in a manner to result in the stable froth formation. The air bubbles are introduced in the pulp in the form of *neo*-bubbles (N-bubble) which form free-bubbles (F-bubble) on emerging.

11.7.1 CONCEPT OF NEO-BUBBLE (N-BUBBLE) AND FREE-BUBBLE (F-BUBBLE)

N-bubble refers to an immersed bubble of air having a characteristic such that when it arrives in water, the surface tension at the air–water interface is at its maximum value for the system ($\gamma = 72.3$ dyne/cm for pure water and air). The surface tension is then continuously reduced until the bubble emerges out from the water as an independent bubble. This change in surface tension is important in several ways in flotation of minerals from the pulp. The N-bubble has only half the surface area of a free-bubble (F-bubble). Free-bubble is that which has risen clear of the aqueous phase and has achieved independent existence as a hollow liquid spheroid with two air–water faces. For the existence of F-bubble, a surface tension considerably below that of water is essential. This condition is generally achieved in flotation by sorption of molecules containing hydrocarbon groups to the interface.

A dynamic difference also exists between N- and F-bubbles. The N-bubble is an air pocket which is pressed by surrounding pulp. The N-bubble moves through the pulp according to its buoyancy. N-bubble vibrates violently when it is at considerable depth in the pulp and gets distorted during its upward drift. The area of N–bubble is variable without having much change in volume (only slight expansion occurs due to decrease in hydrostatic pressure). On the other hand, F-bubble is free from hydrostatic pressure and turbulent movements. The fundamental difference between the N-bubble and F-bubble is that the former is formed before emergence, whereas the latter is formed after emergence and is of completely different nature.

The air introduced must search the entire pulp and each N-bubble should be of optimum size. Too small N-bubble (less than 0.5 mm in diameter) tends to promote too stable a froth, whereas, too large an N-bubble rushes to the surface (due to high buoyancy) and reduces the time available for fixing particles in the N-bubble wall. Further, large N-bubble sets up turbulence in the frothing zone, which is highly undesirable. The normal size of N-bubble is 1 to 4 mm. The size of N-bubble depends on (a) size of the aperture from which it emerges, (b) hydrostatic head acting against it, (c) surface tension of the interface with pulp on emergence, (d) speed, volume, and pressure of air, and (e) turbulence of the surrounding pulp.

Despite distortion, vibrations, and collisions, bubbles do not appreciably coalesce while rising. Each bubble has a hydrated enveloping barrier surrounded by a slip stream, and an air–water interface toughened by surfactant molecules. These conditions reduce chances of collisions and thus size growth of bubble in the body of the pulp is not favoured, but the situation changes suddenly on emergence of bubble. The bubbles between 2.5 mm and 6 mm rise freely through water and the bubbles below this size change their shape rapidly during ascent (periodicity may be as low as 1/1,000th second). Bubbles larger than 6 mm have quite irregular shapes. The size of bubbles is largely dependent on the mechanism of introducing air.

11.7.2 PHASE SYSTEM IN FLOTATION

The flotation system consists of three phases, i.e. water, air, and mineral particle. Before introducing air, the system consists of water, frother and aerophilic particle. When the air is just introduced, the newly born N-bubble is momentarily in rapid vibration with maximum surface tension for the system. Frother molecules and mineral particles compete for acquiring the positions in the interphase. Completely wetted particles do not rise by air transport but may be carried over into froth by agitation. For each concentrate loaded bubble to appear in froth, hundreds of mineral bearing N-bubbles emerge from the pulp, since a great majority of N-bubbles fail to become bubbles and burst at the surface of the froth. Therefore, a great excess of N-bubbles over that actually needed to transport the concentrate out of the system is necessary in commercial flotation. Oversized particles fail to overcome the surface tension force holding them in the interphase and thus may drop out during the period of bursting. Lightly held particles may be shaken or torn away by the collisions and turbulence in the agitated pulp. Another important feature of the change from N- to F-bubble state is loosing the aid of the pulp density, as they move from water–solid system of pulp to the froth column which is mainly gaseous.

There is a sudden change in ratio of air to liquid, from the predominant liquid pulp to the predominant gaseous froth. Figure 11.6 represents the sluicing effect of the water draining down, resulting into further concentration of valuable minerals.

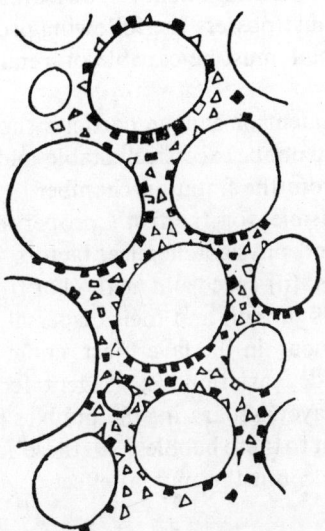

Fig. 11.6. Sluicing effect of water in bubble column to effect
further concentration.

The degree of submerging the particle in the wall of a F-bubble depends mainly on contact angle. Strongly hydrophobic (such as oxide slimes) may project in the air and form a *dry froth*.

The bubble increases in size upward through the froth, and the mechanism of this increasing in size is mainly entropic. When several small bubbles merge, the total air volume will be the same, but the total surface area will reduce considerably. Mathematically this can be stated as follows:

$$\frac{A_1}{A} = \sqrt[3]{n} \qquad (11.13)$$

where n = number of bubbles, each having the same radius,

A_1 = combined area of n bubbles, and

A = area of bubble formed on merging n bubbles.

For example 100 bubbles of 1 mm diameter would shrink to form one bubble of 4.6 mm diameter and the total surface area would drop from 6.28 to 1.35 cm². Thus this decrease of total surface will result in decrease of free energy.

The particles carried on the top find a decreasing available area of interphase, in which the particles cling. In such a situation the particles having highest tenacity (conditioned by collector) in terms of contact angle will only persist and the other gangue particles will drop out from the burst bubbles. Persisted particles are withdrawn as concentrate. The entropic change in the froth layer may be aided in some flotation cells by a gentle rotating action of the pulp.

11.7.3 FROTH PROPERTIES AND THEIR IMPORTANCE IN FLOTATION

A flotation froth must possess the following properties:

a) The froth formed must be capable of retaining the mineralised air bubbles.

b) Selective detachment of gangue particles should occur in the froth.

c) The froth should not be excessively stable and must break down easily after being removed from the flotation chamber.

The factors responsible for the froth properties may be fundamentally classified as physical and physico-chemical factors. The physical factors are: (i) froth layer thickness, (ii) speed and method of froth removal, (iii) amount of bubbles and minerals present and their dispersal in the froth, and (iv) intensity of pulp movement in the layer just under the froth. The physico-chemical factors include variations in reagent feeds, which influence the stability of the liquid layers separating the bubbles and the strength of mineral particles attachment to these bubbles. All these factors should be selected carefully, taking into account the mutual effect.

11.7.4 STRUCTURE OF FROTHS

Flotation froths consist of mineralised air bubbles separated by layers. The froths formed may possess the structure of following three types, i.e. film structural, aggregate, and film froth (Fig. 11.7).

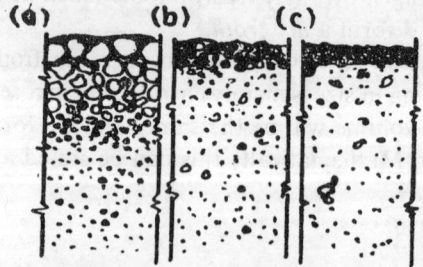

Fig. 11.7. Structure of froths:
(a) film structural froth, (b) aggregate froth, and (c) film froth.

a) *Film Structural Froths*

This type of froths is most common in flotation of normal size particles, and are characterised by the following structural features:

i) The air bubbles in the lower layer of froth are smaller than those in the upper layers.

ii) The water layers separating the air bubbles in the froth diminish in thickness towards the froth surface.

iii) The froth layer is considerably thick, i.e. 5–20 cm.

iv) The larger bubbles of froth are generally deformed.

v) Froths contain relatively more water than the other types, particularly in the lower layers.

vi) These froths are extremely mobile and their stability varies over a wide range.

b) *Aggregate Froths*

These comprise of relatively large particles firmly attached to each other by numerous air bubbles which are smaller than the bubbles in film structural froths. The froths contain relatively less water and are quite stable, but readily break down with a characteristic sound during their fall into the trough.

c) *Film Froths*

These are fundamentally quite similar to aggregate froths but are thinner (of the order of several particles). The mineral particles making up the film froth are very large, having a low specific gravity and are normally strong water-repellent, e.g. coal, graphite, etc.

The extent and efficiency of flotation can be judged from the appearance of the froth, i.e. its structure, mineral content, colour, etc.

11.7.5 FROTH STABILITY

The stability of froths determines their basic properties. On standing, the froth breaks down to some extent, losing a considerable amount of the attached mineral particles. The gangue particles are the first to drop from the froth. Breaking of froth takes place due to coalescence of mineralised bubbles which form the froth. The coalescence takes place only when water layers dividing bubbles become sufficiently thin. Thinning of water may be caused due to the following reasons:

a) The water present in the layers dividing the bubbles tends to run in a downward direction due to gravity. This process is further aided by the pressure in the froth produced by the continuous impacts of the bubble masses from below and by the pressure of the upper froth layers.

b) The water in the layers evaporates, particularly, from the froth surface into the surrounding atmosphere.

Following factors increase the froth stability and are of equal importance:

a) The frother molecules adsorbed on the surfaces.

b) In a three-phase froth, float particles are attached to the bubble surfaces, which prevent the bubbles forming too close together and control excessive decrease in the thickness of the water layers separating them.

c) The stabilising effect of floated particles is proportional to their water repellent action. Thus froth stability increases to a considerable extent by the presence of small and flat particles due to more covering of bubble surfaces.

d) Froth stability is also affected by addition of some reagents. The effect is caused due to alteration in (i) structure and composition of the adsorption layers on the bubble surfaces, and (ii) the nature of the mineral coating on these surfaces. Froth stability agents may be divided into two categories, i.e. (i) water soluble reagents (lower alcohols, turpentines, etc.), and (ii) non-polar substances (kerosene).

11.8 Flotation Kinetics and Speed of Flotation

This covers the studies on the effect of flotation time on the amount of froth produced and the quantitative identification of all rate controlling factors. In flotation there are hundreds of operating variables, in addition to the varying pulp constituents and the interaction of these variables leads to permutations into almost infinite figures. Therefore, the approach to these studies can only be empirical and confined to the control of only key variables controlling the process to a major extent.

Under precise laboratory control of size, purity of mineral, and aeration of the pulp, the flotation rate for a single mineral of given size is first order. The overall rate for a mixture of different sizes will depend on the weight proportions and the flotation rates of individual sizes. For readily responsive minerals having inhibited conditions, the flotation rates of fine size are proportional to the square of the particle radius, whereas for intermediate sizes flotation always depends less on particle size. In floating mixed minerals of wide range, the concentrate grade increases with increase of particle size.

Conditioning of pulp may be accompanied by several reactions simultaneously, such as ionisation of reagents, diffusion, chemical surface reaction, desorption or diffusion of reaction products, sliming, flocculation, etc. It has been reported that recovery of floatable mineral from a pulp follows first-order law defined by the following equation:

$$-\frac{dN_m}{dt} = k_1 N_m \qquad (11.14)$$

where N_m is the number of particles in the cell at any time of flotation t, and k_1 is the rate constant depending on floatability.

Flotation speed determines the efficiency of the process and can be defined as the mean speed V_m, according to the following equation:

$$V_m = \frac{p}{t} \qquad (11.15)$$

where p is the percentage of float mineral extracted into the concentrate in the time t (in minutes).

Though the calculation is simple, this mean flotation speed does not give a correct picture of the process. It does not take into account the changes taking place in flotation speed within the time interval t.

In another method, flotation speed can be based on results of fractional flotation, in which the froth produced is removed after equal-intervals of time and transferred to separate vessels. The different portions of concentrate are then weighed, and their float mineral content is determined.

A mathematical treatment of flotation speed by fractional flotation may be considered as follows.

Lat n be the number of particles subjected to flotation in the initial pulp, and x is the number of particles which have passed into the froth product by the time t. Then $n-x$ will be the number of particles left in pulp after time t.

If N is the number of bubbles passing through the pulp in a unit time, then Ndt will be the number of bubbles passing through the pulp during the time dt, and dx will be the number of particles floated.

The number of particle–bubble collisions during the time dt will be proportional to $N(n-x)dt$. The number of particles dx floated in the time dt may be determined as follows:

$$dx = k N (n-x) \phi_{att} dt \qquad (11.16)$$

where ϕ_{att} is the probability of stable attachment taking into account all the circumstances influencing the flotation, and k is a coefficient.

By transposing the value $(n-x)$ to the left and integrating,

$$\int_0^x \frac{dx}{n-x} = k \int_0^t N\phi_{att}\, dt \qquad (11.17)$$

or
$$\log \frac{x}{n-x} = k \int_0^t N\phi_{att}\, dt \qquad (11.18)$$

or
$$\log \frac{x/n}{(n-x)/n} = k \int_0^t N\phi_{att}\, dt \qquad (11.19)$$

Substituting e for $\dfrac{x}{n}$ the extraction in fraction of a unit

$$\log \frac{e}{1-e} = k \int_0^t N\phi_{att}\, dt \qquad (11.20)$$

The value $\dfrac{e}{1-e}$ may be referred as the specific flotation speed coefficient and its value is independent of the number of particles floated in the time preceding that under test.

Figure 11.8 represents the curves showing changes in the specific flotation speed coefficient as $\log \dfrac{e}{1-e}$ with time (t) on the basis of fraction flotation results. These curves may have three shapes, differing fundamentally from each other. The curve-1 represents a rectilinear relationship which indicates the flotation with continuously constant speed. The curve-2 represents a convex form indicating a reduction in flotation speed towards the end of the process. This may be due to the removal of the most readily floatable particles during the first minutes, reduced reagent concentration in the pulp and some other factors. The curve-3 represents the concave form showing an increase in flotation speed by the end of the process. This may be due to the removal of fine slimes from the pulp during the first few minutes of flotation, increased separation affected by longer mineral contact with reagents, and better pulp aeration in the last chambers of the flotation machine.

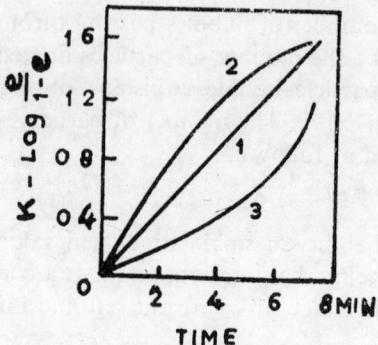

Fig. 11.8. Changes in the specific flotation speed coefficient
$$\left(\log \frac{e}{1-e} \right) \text{ with time } (t).$$

11.9 Factors Affecting Flotation

The efficiency of flotation process and the results obtained are influenced by a large number of factors such as characteristics of the ores (structure and origin), surface conditions, particle size, specific gravity of mineral, shape of particles, density of pulp, temperature of pulp, composition of process water, reagent feeds, flotation machine design, etc. In order to get best possible results, the individual factor should be maintained to its optimum value, since even one of these factors may completely upset the entire process while maintaining the remaining conditions to an optimum level.

11.9.1 Effect of Quality of Minerals to be Concentrated

Characteristics of minerals such as mineralogical composition, structure, presence of impurities, etc., predetermine the required conditions of flotation. The separation of mineral depends on the associated minerals. For example, separation of fluorite and phosphate from quartz is much easier than their separation from calcite or dolomite. Similarly, the separation of sulphides from non-sulphide minerals is usually quite simple, but separation of sulphide minerals or partially oxidised sulphide minerals is much more difficult.

11.9.2 Effect of Size and Shape of Mineral Particles

The optimum particle size depends on the surface properties, specific gravity and shape of the particles. The minerals possessing stronger water-repellent properties and lower specific gravity, can be floated in much coarser sizes, for example coal particles can be floated in size range of 1.5 to 2 mm. On the other hand, flaky minerals of even high specific gravity (such as gold leaf) can also be easily floated. However, the most common size range of minerals for flotation is usually 0.1 to 0.02 mm.

The presence of too fine particles (slimes) in the pulp usually causes adverse effects in flotation, reduces speed of flotation, increases the consumption of reagents, and decreases the selectivity. On the other hand, too large particles cannot usually be floated efficiently.

In case of sulphide minerals the effect of particle size is much more important, since the small particles are oxidised more quickly and to a greater extent than large ones which influences the reaction with collector. The passage of minerals into solution increases with decrease in particle size. This situation leads to the accumulation of *inevitable* ions in the pulp resulting in reduced selectivity, low quality of concentrate, and unproductive consumption of reagents.

11.9.3 EFFECT OF MINERAL SURFACES AND ISOMORPHISM IN CRYSTAL LATTICES

The various faces of a true crystal possess different free surface energy levels and thus can absorb water or reagents to different extents. Further, the adsorption activity of mineral surface atoms depends on their position on the surface. This adsorption is further affected by the presence of numerous micro and macroscopic cracks, crystal lattice defects, and electrochemical heterogenity. The behaviour of a mineral particle in flotation must therefore be determined by the overall result of the reaction of reagents on the mineral.

Isomorphism (mutual substitution of atoms or ions in the crystal lattice forming solid solution) is a common process in mineral formation. The amount of isomorphism impurities in minerals may vary widely depending on the concentration of individual elements in the initial magma or solution and on the extent of isomorphic solubility. This variation often produces fundamental changes in composition and structure of the same mineral depending on the circumstances of mineral formation and subsequent history. This situation greatly affects the reaction of reagents and thus flotation process.

11.9.4 EFFECT OF PULP DENSITY

Pulp density may be considered as percentage of solid contents of the pulp expressed on weight basis in unit volume. The pulp density exerts a great effect on flotation. With increased pulp density, both retention time of pulp in flotation machine and the volumetric reagent concentration increase, and thus it would appear that best results may be obtained by employing pulp of maximum density, since this will result in increased output of flotation machine. However, excessive pulp density will result in adverse effects, such as deterioration in pulp aeration and floatability of large particles, increased floatability of fine gangue particles, and reduced quality of concentrate. The usual pulp density employed varies from 15 to 40 per cent solid. Higher pulp density is employed in roughing operations and concentration of rich ores, whereas low pulp density is employed in concentrating poor ores and repurifying operations.

11.9.5 EFFECT OF PULP TEMPERATURE

The temperature of pulp influences the speed of reactions to some extent, between reagents and phase interfaces. In general, a rise in pulp temperature improves flotation, but reduces the selectivity. The reagent consumption is particularly reduced when employing less soluble or slow reacting collector, e.g. fatty acids and their soaps. The effect of temperature is much less when using xanthates compared to the fatty acids.

11.9.6 EFFECT OF COMPOSITION OF PROCESS WATER

The chief constituent of the pulp is the mill water, composition of which has a fundamental effect on the flotation process. The mill or process water contains various ions (drawn from river or lakes) such as Cl^-, SO_4^{--}, HCO_3^-, Na^+, K^+, Ca^+, H^+, etc. (these ions affect water pH value), dissolved gases, and various colloidal organic impurites. Further, water gets saturated in mineral ions as a result of contact with the minerals during the various stages of concentration. Many of the minerals when slightly soluble, change the pulp pH. Water may also get contaminated with many impurities such as oil, grease, etc., with their careless use. The amount of reagent added for flotation is usually 0.05 kg/tonne of pulp and thus a little amount of undesirable constituents in water will upset the whole process. In general, the ions and impurities in mill water may depress or activate the pulp constituents. The effect of composition of process water is more pronounced in flotation of non-sulphide minerals (such as oxides, silicates, aluminosilicate, alkali minerals, etc.) with fatty acids and their soaps, where even an extremely small amount of inevitable ions may alter the whole course of the flotation process. In case of xanthate flotation of zinc blende, a small amount of copper ions dissolved in the pulp gives powerful activating effect.

The process water ions may affect the flotation in following two ways:

i) The ions present attach themselves to mineral surfaces resulting in the alteration of flotation properties.

ii) The ions may react chemically with flotation reagents (collectors, modifiers, etc.).

The adverse effect of inevitable ions on flotation may be prevented or controlled by the following methods:

i) The recirculating water must be checked and corrected by suitable treatment.

ii) Mill floor washings should be prevented from entering the circuit as oil and grease are likely to be carried.

iii) Frothing agents not appreciably adsorbed by concentrates are carried by water and thus care should be taken for the same.

iv) Earth salts (particularly dissolved Ca-ions) should not be allowed to be build up beyond a certain limit, as they may precipitate scale in pipes, metering devices, filter cloth, etc.

11.9.7 EFFECT OF REAGENT FEEDS

The term *reagent feeds* includes variety, amount, and the sequence of reagents employed in flotation, and the time of contact between pulp and reagent. Usually the reagents are added in the following sequence:

i) regulators to alter the pH of the medium,
ii) depressants to depress certain constituent,
iii) collector, and lastly
iv) frother.

However, this general sequence may have many variations depending on the mineral to be floated and other factors. For example, in flotation of non-sulphide ores employing fatty acids, soda may be added to the mill to transform the iron into relatively insoluble hydroxides. Sometimes, water glass addition to the mill discharge (containing iron ions) may be useful to peptize the pulp in the classifiers, etc. In order to maintain a specific reagent consumption, it is necessary to have the periodic checks of feed and the pulp properties.

11.9.8 RATE OF PULP ENTERING THE FLOTATION MACHINE

For a given system, the amount and density of pulp entering the flotation machine must be maintained at a rate to ensure the optimum flotation time. In case of excessive feed rate, flotation time will be less than that required, which will lead to losses in tailings and to a low level of mineral extraction. On the other hand, with insufficient feed, flotation time will be too long, and thus may lead to the flotation of gangue particles and production of low-grade concenrtate. Further, longer retention of pulp in machine will result in requirement of additional power and pulverisation of material to form slimes.

Decreased rate of pulp entering the flotation machines results in the drop of pulp level, whereas increased rate of the pulp feed causes the pulp level to rise. In both the cases efficiency of flotation suffers.

11.9.9 EFFECT OF PARTICLE AGGLOMERATION IN THE PULP

Flotability of mineral particles depends considerably on agglomeration (aggregation) occurring in the pulp. Agglomeration (quite-large clumps) may occur in the following cases:

i) Highly water-repellent particles, e.g. finely crushed native sulphur forms small solid clumps in water.

ii) In the presence of sufficient concentrations of certain electrolytes (such as $CuSO_4$).

iii) When certain collectors are added to react with the mineral surfaces.

Mineral particle agglomeration is caused by molecular forces which are not compensated for in the surface layers. The dejoining surface pressure exerted by the thin layer of water between the particles resist the agglomeration process and therefore, the dejoining pressure exerted by the water layers should be minimum for agglomeration. All processes of agglomeration may be classified under *coagulation*, i.e. result of reaction between mineral surfaces and inorganic electrolytes, or *flocculation*, i.e. result of linkage of non-polar groups from reagents on the particle surfaces. Coagulation of particles in no way indicates the susceptibility of these particles to flotation as inorganic electrolytes do not generally reduce the hydration of the surface. Flocculation becomes more intensive when the reagent renders the surface highly water-repellent, and thus the presence of flocculation indicates the susceptibility of mineral surfaces to flotation. However, excessive flocculation results in the formation of very large and heavy aggregates which affect the flotation adversely.

Important effects of agglomeration of particles on flotation may be summarised as the following:

i) Agglomeration of non-selective particles greatly reduces the selectivity of flotation resulting into a poor grade of concentrate.

ii) Adhesion of small slime particles to larger particles generally affect the flotation adversely.

iii) The formation of heavy large clumps may lead to increased losses in tailings.

iv) In some instances, the agglomeration of fine slimes into bigger clumps may lead to their removal from the process, and thus reducing their harmful effect on flotation.

In general, particle agglomeration has an adverse effect on flotation, and thus it should be prevented as far as possible. Agglomeration may be prevented either by physico-chemical or physical methods. The former methods involve the addition to the pulp of peptizing reagents such as silicates, carbonates, sulphides, hydroxides, cyanides of alkali metals, starch, gelatin, sulphite liquor, etc. Water glass and sodium pyrophosphate are most common. The action of these reagents mainly depends on the pH of the pulp. Depending on pH, the same reagent may have either peptizing or coagulating effect.

The physical methods involve a reduction in probability of particle collision by employing somewhat dilute pulps. The methods employed include (i) removal of fine slimes before flotation, and (ii) variation of pulp mixing speed, i.e., the mixing rate is decreased or increased to break the particle clumps.

11.9.10 EFFECT OF FINE SLIMES ON FLOTATION

The slime particles are usually of 3–10 microns size and have a varied effect on flotation process. The presence of fine slime in the pulp usually gives rise to (i) contamination of froth product by small fraction of gangue, (ii) increased reagent consumption, and (iii) decreased flotation speed. These effects are caused due to the following facts:

    i) Fine slime coatings are formed on the particle surfaces resulting into non-adherence of air bubbles.

    ii) Fine slime particles cover air bubble surfaces, to which particles of normal flotation size do not readily adhere.

    iii) Fine slime particles possess a high specific surface area and thus absorb more amounts of reagents and enormous number of bubbles from the pulp than do large particles.

    iv) In case of sulphide minerals, the surfaces of fine particles are more heavily oxidised than those of large particles and this in turn will lead to low recovery.

    v) Fine particles possess increased flotation activity, which will impede the separation of individual minerals.

The adverse effect of fine slimes on flotation can be prevented by the following methods:

    i) By the addition of peptizing agents.

    ii) By making use of more dilute pulps.

    iii) By fractional (stage by stage) feeding of collectors throughout the process.

    iv) By preliminary separation of slime and granular fractions and treating them separately.

CHAPTER 12

Flotation Reagents and Their Action

In order to produce optimum conditions for highly selective and efficient flotation process, various flotation reagents are added. In practice, flotation is almost impossible without the use of flotation reagents. Flotation reagents alter the surface properties of minerals over a wide range and make the mineral particles either water-avid or water-repellent. A great variety of flotation reagents is widely used, which differ in composition. Common reagents employed are organic and inorganic compounds, acids and alkalis, various salts, water soluble substances, and materials practically insoluble in water. The reagents used may be broadly classified as collectors, frothers, and modifiers (regulators) according to their function in flotation.

12.1 Mechanism of Reagent's Function in Flotation

The reagents in flotation process are effective due to their attachment to the surfaces of either air bubbles or mineral particles and the various aspects involved in the process are described as follows.

12.1.1 ADSORPTION OF REAGENTS

In most instances, reagents get adsorbed at the appropriate phase boundary surfaces of bubble and mineral grain. Adsorption results from the interaction of electrical forces between the adsorbent and the reagent adsorbed. The adsoption may be physical and chemical in nature. In both the cases, the adsorption is spontaneous and is accompanied by a decrease in the free energy of the system and the release of certain quantity of heat.

Chemical and physical adsorption differ in the following respects:

a) In physical adsorption, adsorbed substance and the crystal lattice of the mineral are two independent systems, whereas in chemical adsorption, adsorbed substance and the crystal lattice are regarded as a single system with respect to their energy relationship. Thus chemical adsorption is due to transition of electrons from the adsorbed atom to the lattice or vice versa, whereas, in physical adsorption, the bond with the crystal lattice results due to forces of inter-molecular attraction.

b) In physical adsorption, the heat effect, and consequently the strength of the bond is much less, whereas in chemical adsorption, bond strength (heat effect) is considerable.

c) Physical adsorption is a simple process and the reagent can be caused to pass from the mineral surface into solution. In chemical adsorption, the adsorption layer cannot even be removed by repeated washing with water.

d) Chemical adsorption is highly selective in action of the reagent on the mineral, which is of fundamental importance in flotation.

e) Physical adsorption is remarkably fast, whereas speed of chemical adsorption varies within wide limits.

f) Physical adsorption gives more-even distribution on the adsorbent surface. In chemical adsorption the reagent gets attached first to most active positions of mineral and the adsorption layer may be formed on the remaining surface.

12.1.2 ATTACHMENT OF REAGENT TO AIR BUBBLE SURFACES

Free surface energy at an air-water interface is due to the monomolecular layer of water. The boundary layer molecules are extremely mobile and change their place continuously. The time spent by each boundary layer molecule on the surface of the liquid may be less than a millionth of a second. The surface-active substances introduced, concentrate at the air–water interface by the process of physical adsorption and gives rise to a heteropolar molecular structure which is of the greatest importance in flotation. These compounds (high in molecular weight) have non-polar hydrocarbon group and polar group (hydroxyl or carboxyl). These compounds are responsible for high dipole moments and try to concentrate at the air–water interface, resulting into decrease of surface tension of the water.

When the concentration of a surface-active substance in the pulp is considerably high, its molecules form a saturated adsorption layer of oriented molecules at the interface (Fig. 12.1). The water dipoles work actively with

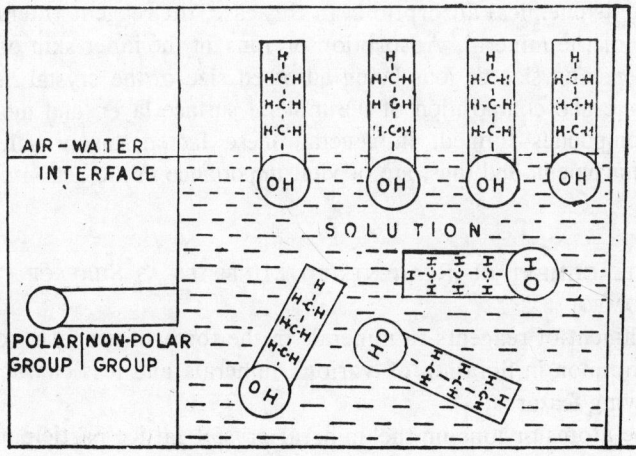

Fig. 12.1. Saturated absorption layer of oriented molecules at the interface.

the polar groups in the surface-active molecules and do not react with the non-polar hydrocarbon group, and thus the latter is ejected into the air.

Concentration of surface-active reagents adsorbed at the air–water interface is extremely low, and therefore, the molecules of these reagents must be separated from each other in the surface layer by distances much greater than the size of the molecule. In most cases the molecules of reagents lie in the plane of the surface layer itself or at certain angles to it (not perpendicular).

Activity between the polar groups of surface-active molecules and the water dipoles results into hydration of these groups by the water dipoles and in some cases a peculiar casing is produced, due to which the surface adsorption layer becomes considerably rigid and stable. However, the stability of adsorption layer of surface-active reagents may be reduced due to heat movement and high molecular mobility of water molecules.

12.1.3 ATTACHMENT OF REAGENT IN A DOUBLE ELECTRIC LAYER

The surface charge of minerals also has some effect on attachment of mineral particles to a gas bubble. In a double electric field, concentration of different reagents change the particle charge, which affects the floatability. Ion adsorption in the outer skin is not selective with respect to mineral, and adsorption of ions is proportional to their respective concentrations. Attachment of flotation reagent in the outer skin of a double electric layer is referred to *physical adsorption*, characterised by instability and reversibility. Thus the flotation reagent ions adsorbed in the outer skin will begin to pass from the outer skin into solution on diluting the concentration of ions in solution and can be easily detached by washing operation.

Attachment of reagent (of active organic compound) in the inner skin of a double layer (in contrast to the outer skin) is quite selective in nature and is referred as chemical adsorption. In this case, the reagent enters the crystal lattice of the mineral. Adsorption of ions in the inner skin of a double layer depends on size of ions being adsorbed size of the crystal lattice ions, the structure and composition of the mineral surface layer, and the solubility of the compounds formed. In general, these factors favour selectivity of reagent attachment, and thus, are of vital importance in the action of flotation reagents.

12.1.4 ATTACHMENT OF REAGENTS TO MINERALS AS SURFACE COMPOUNDS

Attachment of reagents to minerals in the form of surface compounds is most common in flotation of various minerals and it is characterised by the following features:

a) The atoms or ions in the mineral crystal lattice participating in the formation of the surface compound are the links between the main body of ions or atoms forming the lattice and the ion or atom attached to this lattice.

b) In an ionic crystal lattice, the surface compound is formed by the attachment of cation from the solution to an anion in the lattice or vice versa, whereas in case of an atomic lattice, this restriction does not operate.

c) There is no stoichiometric relationship in ions or atoms forming the compounds.

d) Flotation reagent may be attached to mineral surfaces even when the solubility of the reagent-mineral reaction compound is high.

Figure 12.2 represents surface compound formation with an ionic crystal and an atomic lattice (attachment of oxygen to carbon atoms in the coal crystal lattice). In formation of surface compounds, mineral crystal lattice is not damaged, since the energy required to form the surface compound is smaller than the lattice energy.

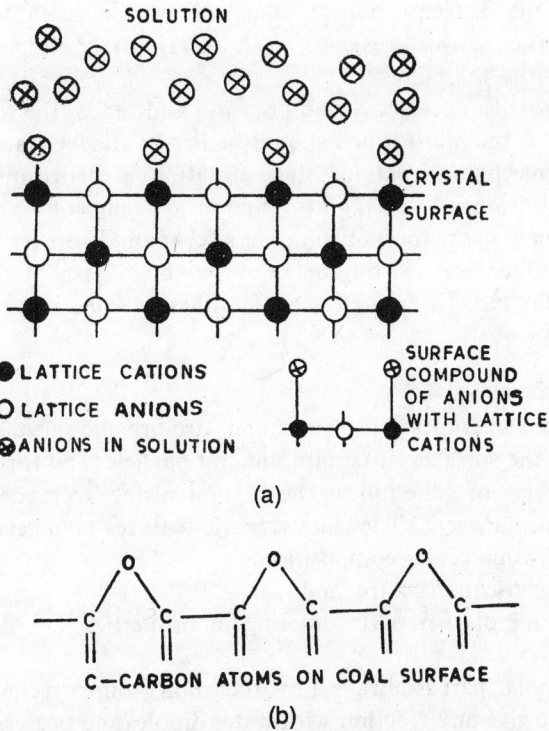

Fig. 12.2. Surface compound formation:
(a) with an ionic crystal, and (b) with an atomic lattice.

The process of surface compound formation is a process of chemical adsorption and is generally highly selective and specific. The suitable atoms or ions are firmly attached to the mineral surface and the process is not reversible. Under certain conditions, several complex modified layers of adsorption may be formed on certain areas. Surface compound formation in the form of ion exchange adsorption is most common in flotation.

12.1.5 ATTACHMENT OF REAGENT TO MINERAL SURFACES AS FILMS

In many cases, the reagent–mineral action forms the films of various compositions on mineral-particle surfaces. Sometimes these films may be quite thick and can be seen on the mineral surface. For example, the original colouring is drastically changed and darkened when cerussite or malachite is acted upon by Na_2S (PbS or CuS film is formed). These films can be removed mechanically from the main body of the mineral and thus exposing the fresh surface of mineral. These films are important in flotation as their appearance on a mineral surface changes the composition and flotation properties of that mineral. The strength of attachment depends on the relationship between the parameters of the film-crystal lattice and the mineral surface.

The formation rate of film depends on (a) reagent concentration in the solution, (b) the concentration of mineral matter in the diffusion layer, (c) speed of reaction between reagents and mineral, (d) the temperature of pulp, and (e) mixing of the pulp.

In order to ensure selectivity in flotation and render the mineral surface water-repellent, the film formed should be firmly attached and should have resistance to mechanical action. Since flotation is carried out with low reagent concentrations (20–50 mg/litre) and at low temperature (15–20°C), reagent attachment in the form of film is less common. However, in spite of the considerable differences existing between the above types of reagent attachment, they may pass from one type to another in some cases and may also occur simultaneously.

12.2 Collectors

Collectors, sometimes called promoters are organic substances which act selectively on the surfaces of certain mineral particles and form a thin coating by adsorption or adhesion on the mineral surface to render them water-repellent or air adherent. The characteristic features of collectors are:

   a) Complex molecular composition,

   b) Asymmetrical structure, and

   c) Consisting of two parts differing in properties, i.e., polar and non-polar groups.

The non-polar part is always a hydrocarbon group or a chain and practically does not give any reaction with water dipoles and possesses pronounced water-repellent properties. The polar group possesses the property to react with water. During flotation, the polar part is adsorbed on the mineral surface while the non-polar is oriented outwards, and this condition makes the mineral surface water-repellent. Polar and non-polar parts of a typical collector are shown below:

$$R-O-C\begin{smallmatrix} SX \\ \\ S \end{smallmatrix}$$

Non polar Polar

Non-polar hydrocarbon liquids without a heteropolar structure and not dissociating into ions may also be used in flotation practice.

Collectors may be classified according to their ability to dissociate into ions in aqueous solutions and depending on the ion (anion or cation) responsible for water-repelling effect. Figure 12.3 represents the classification of collectors depending on dissociation properties, anion and cation activity with mineral surface and solidophil group structure.

12.2.1 IONISING COLLECTORS

The compounds which dissociate into ions in water are known as ionising compounds. These collectors should possess the following characteristics:

i) Heteropolar structure of collector molecules.

ii) The solidophil group must be attached strongly and selectively to the mineral surface.

iii) Non-polar group of the collector should be long enough to give sufficient collector effect.

iv) The collector should be non-toxic, reasonably soluble in water, stable in composition and commercially available.

Ionising collectors may be anion or cation collectors, depending on the ion of collector taking part into reaction.

A) *Anion Collectors*

Anion of these compounds renders the mineral water-repellent, whereas cations do not take any significant part in the reagent–mineral reactions. Presently, anion collectors are most widely used. Most of the anion collectors are remarkable for their selectivity and strong attachment to the mineral surface. Anion collectors may be further classified into two basic forms depending on composition and structure of solidophil group.

a) *Anion collectors in which solidophil group consists of organic and sulpho-acid anions*: In this group the solidophil group consists of carboxyl

$$C \overset{= \ 0}{\underset{\diagdown \ 0\text{-}}{}}$$ and the anion has the form $\left[R - C \overset{\nearrow 0}{\underset{\diagdown 0}{}} \right]^{-}$. Most common collectors

of this group are organic acids or soaps (salts of alkali metals and organic acids). Collectors having the sulphurous and sulphonic acid anions as the solidophil group are rarely used.

b) *Anion collectors in which bivalent S is in solidophil group*: This group

mainly consists of xanthogenates having anion $\left[R - 0 - C \overset{\nearrow S}{\underset{\diagdown S}{}} \right]^{-}$ known as

Xanthates and dithiophosphates having anion $\left[\overset{RO}{\underset{RO}{}} {>} P \overset{\nearrow S}{\underset{\diagdown S}{}} \right]^{-}$ known as

aerofloats. These collectors differ in the composition of solidophil group, the

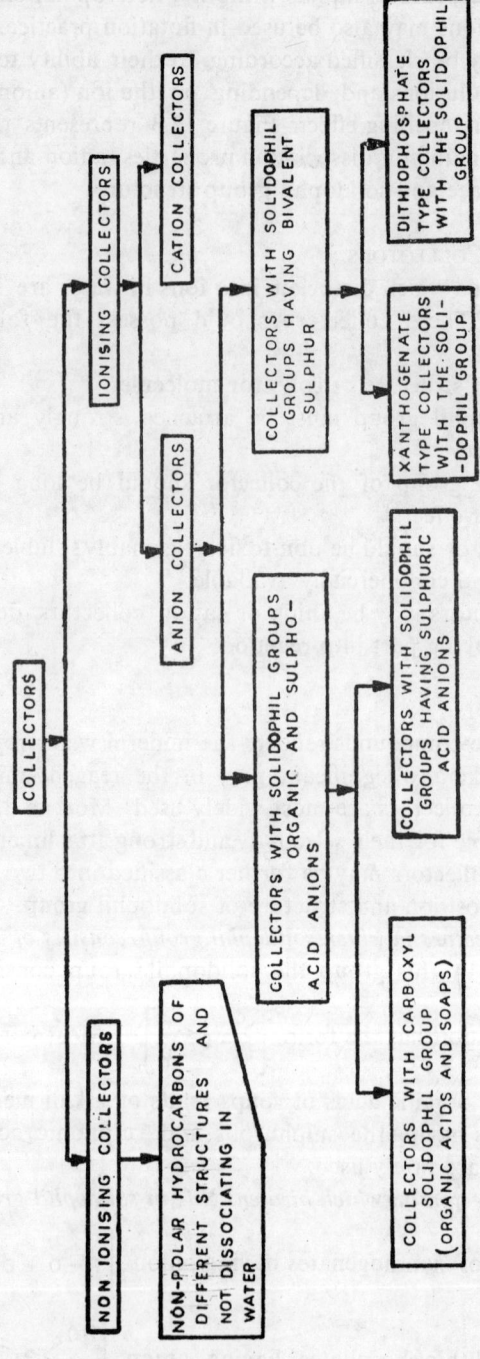

Fig. 12.3. Classification of various collectors.

latter has pentavalent P in the solidophil group composition, instead of tetravalent carbon in the former. These collectors usually have H, Na or K as their cations. These collectors attach themselves very strongly to sulphide mineral surface. Xanthogenates are most selective and effective collectors in sulphide mineral flotation.

B) Cation Collectors

In cation collectors, cation (positively charged ion) is the water-repellent ion. These consist of a hydrocarbon radical to which solidophil group is bound chemically. The anions of cation collectors are usually halides and sometime a hydroxyl which do not take any active part in the reaction with mineral. In contrast to anion type collectors (characterised by strong attachment to the mineral surface and irreversibility), cation collectors are weakly attached and can readily be desorbed by reducing their concentration in the pulp (i.e. by pulp dilution).

Most cation collectors are derivatives of amines and ammonium salts. In an alkaline medium, amines form compounds having a hydroxyl group, whereas in an acid medium, the corresponding amine salts are formed. These salts can dissociate into ions and thus are sensitive to pH of the medium.

Cation collectors can be further subdivided into the following groups depending on their chemical composition and structure.

a) *Primary, secondary and tertiary amine derivatives*: In this case amino group is connected to the water-repellent radical, either directly or through any other group (ether, amide, etc.). Laurylamine hydrochloride ($C_{12} H_{25} NH_3 Cl$) and cetyl trimethyl ammonium bromide ($C_{16}H_{33} \cdot CH_3 \cdot CH_3 N Br$) are examples of amine derivatives.

b) *Ammonium, sulphonium, and phosphonium bases*: In this case multivalent N (pentavalent), S (tetravalent) or P (pentavalent) joined to hydrocarbon radicals together form the reagent cation. The anion may be a halide (Cl or Br) or sulphate. A special case in this group based on N is that of pyridine in which N forms a ring in the molecule as the following, where A is an anion.

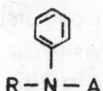

$$R-N-A$$

c) *Cation collectors of various types of structure and not covered by the groups (a) and (b)*: These are characterised by the oximide group ($=N-OH$). The example is aldoximes ($R-CH=N-OH$).

In general cation collector behaviour in flotation is mainly due to the chemical properties of amines and N. Thus, primary amines are the most active collectors.

When cation collector is dissolved in solution, a complex cation with pentavalent N at the nucleus will be formed, in which cation is connected to the hydrocarbon radicals for four covalent bonds and has one ionic bond and thus behaves as a monovalent cation. Since amines readily pass into an ionic form in solution, the efficiency of cation collectors depends largely on the variation of pH. Thus in flotation with cation collectors, a narrow range of pH has to be maintained. By employing cation collectors, flotation cannot be affected in strongly alkaline and strongly acid media. Many cation collectors also possess frother properties, but in practice other frothers are used to give better results. A very useful characteristic of cation collectors is their insensitivity to hard water.

Cation collectors can be employed for flotation of sulphide as well as non-sulphide minerals. However, cation collectors do not show any advantage over xanthogenates and dithiophosphates used for flotation of sulphide minerals. Silicate minerals and quartz can be effectively floated with cation collectors and thus these are very useful in separation of quartz from various minerals such as sulphides, phosphite, garnet, tourmeline, coal, graphite etc.

12.2.2 NON-POLAR AND OTHER COLLECTORS

These reagents are practically insoluble in water and do not dissociate into ions. These collectors render the mineral water-repellent by covering its surface with a thin film. These collectors are mainly hydrocarbon liquids obtained from petroleum and coal as by-products of coking or the coal-tar industry (kerosene, hydrocarbon oils, cresols, carboxylic acid, diesel oil, etc.). Non-polar reagents have the following characteristic features:

a) These are hydrocarbons having no polar group.

b) These have covalent bonds between atoms of molecules.

c) These do not react with water dipoles.

d) These react very weakly with ionic lattice of minerals.

e) These do not have any solidophil group.

f) Their attachment to minerals is due to adhesion rather than adsorption.

g) Attachment of collectors is easier to naturally water-repellent, i.e. weakly hydrated minerals, such as coal, graphite, sulphur molybdenite, diamond, etc.

h) These collectors form emulsions in the pulp and this property helps in attaining the required degree of dispersion.

Flotation with anion collector is greatly improved when non-polar collectors are added along with that. In this case, most probably an increase in water-repellence is due to the concentration of the non-polar collector in the adsorption layer zone of anion collector. For example, when hydrocarbon oil is added in xanthogenate flotation of copper and lead minerals, flotation rate and separation are found to be increased greatly. This joint effect may be attributed to the reaction between hydrocarbon chains of anion collector oriented towards water and the non-polar reagents along three phase line.

12.2.3 MECHANISM OF REACTION OF COLLECTOR WITH MINERALS

Reaction between minerals and collectors in an aqueous medium is basically of adsorption in nature. The collectors dissociating into ions may result into a monomolecular adsorption layer due to ion-exchange adsorption. The decrease in ion concentration of collector in the solution corresponds stoichiometrically to the increase in the concentration of mineral ions passing into the solution from the mineral surface. Figure 12.4 represents oriented adsorption layer of collector on the mineral surface. The non-polar hydrocarbon groups are turned towards the liquid phase, whereas the polar groups are attached directly to the surface of mineral. This orientation of collector molecules renders the mineral surface water-repellent. In this case the adhesion forces between the water and mineral are low and the free surface energy on the mineral–water interface is adequate.

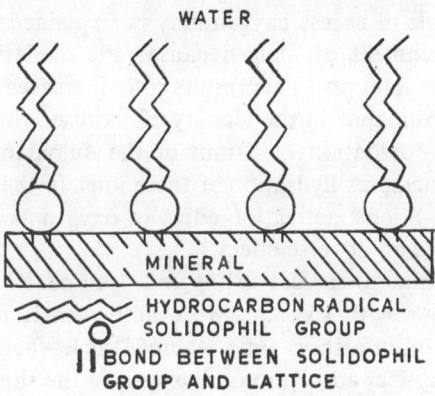

WATER

MINERAL

〜〜〜〜 HYDROCARBON RADICAL
○ SOLIDOPHIL GROUP
‖ BOND BETWEEN SOLIDOPHIL
GROUP AND LATTICE

Fig. 12.4. Oriented absorption layer of collector on the
mineral surface.

In case of an anion collector, its anion is chemically linked through its solidophil group to the mineral crystal-lattice cation. The nature of the chemical bond between collector anion and mineral cation and the energy of this bond depends on the composition and structure of the solidophil group. The energy of the bond is considerable in case of a xanthogenate solidophil

group $-O-C\underset{\diagdown S}{\overset{\diagup S}{\lessgtr}}$ and heavy metal cation (Pb, Ni, Cu, Fe, etc.) and thus

attachment of xanthogenate anions to surface of heavy metal minerals is strong. Insoluble heavy metal xanthogenates show poor collector properties than the corresponding water soluble xanthogenates.

The mechanism of reaction between minerals and carboxyl type anion collectors is more complex.

In flotation technology, the reaction between collectors and sulphide minerals is most important. In these cases the action of oxygen is considered to be of fundamental importance, since it has been established that sulphide is not linked to the collector anions without the previous action of oxygen. However, an excess of oxygen is unfavourable.

The favourable role of oxygen in sulphide–xanthogenate reaction may be expressed as the following:

a) By the action of oxygen, the thickness of hydrated layer gets reduced, which facilitates penetration of the xanthogenate anion to the mineral surface layer.

b) Adsorbed O_2 on the sulphide surface during the ionisation process takes up electrons from the free zone $(1/2\ O_2 + 2e = O^{-2}$ and $O^{-2} + H_2O = 2\ OH^-)$ and thus the electrochemical potential level is gradually reduced as the concentration of dissolved oxygen in the pulp increases. Under these conditions, the probability of overcoming this barrier by xanthogenate anions increases gradually and the attachment of xanthogenate to the sulphide surface commences.

Unfavourable role of excess oxygen may be explained as follows:

The excess attachment of oxygen causes the oxidation of S to SO_4^{2-} ions which pass into solution. This results into increased chemical adsorption activity to metallic ions in the vicinity of oxidised S-atoms passed into solution. Therefore, oxidation of S-ions in the sulphide crystal lattice up to the SO_4^{2-} stage increases hydration of these ions. In the presence of large number of oxidised S-ions, activation effect of oxygen may be blocked and thus oxygen will give an adverse effect.

In practice the mineral surface is not uniform and its different sections give varying reactions with oxygen. More active areas may pass through all stages of oxidation in a short time, whereas the less active areas may only pass the stage of physical adsorption of oxygen in the same time. However, the final effect will be determined by the overall effect of oxygen, water, and collector.

For flotation of sulphide minerals the following points may thus be drawn:

a) A slight oxidation resulting in the formation of sulphide-sulphate is essential.

b) Pure sulphide (free from oxidation) does not respond to a collector and thus cannot be flotated.

c) A higher degree of oxidation accompanied by formation of complete sulphates prevents flotation, since sulphate readily reacts with xanthogenates.

d) Flotation of sulphide minerals is drastically reduced in a strongly acid medium (pH 3–4), since a major part of the xanthogenate gets converted into xanthogenic acid (in molecular form) having weak collector properties.

12.2.4 Factors Affecting the Reaction of Collectors with Minerals and Their Water-Repellent Effect

a) *Molecule Length*

This varies from a few angstroms (5–7 Å) for low molecular weight xanthogenates to several dozen angstrom unit for carboxilic acids and their soaps. Most of the molecule length is made up of hydrocarbon chain, whereas the solidophil group ($-SH$) is comparatively short, but broader than the hydrocarbon chain. With increasing hydrocarbon chain, water solubility of compounds decreases (one CH_2 chain reduces by 4.25 times) and the water repellent effect increases. The contact angle formed with potassium ethyl xanthate is about 60° whereas with amyl xanthate it is 85–90°. The practical working limit is reached with hexyl xanthate. Decreased solubility of collectors having larger hydrocarbon results in (i) difficulty in even distribution of collector in water, and (ii) less attachment of collector to mineral resulting into higher consumption of reagent. Branching of a collector hydrocarbon chain (transition from normal to isostructure) has an adverse effect on collector action due to reduction in dispersion reaction and increased solubility of reagent. More the number of hydrocarbon chain branches more is the adverse effect on collector action.

b) *Structure of Collector*

Open or cyclic (aliphatic or aromatic) chain influences the collector's action. Collectors having cyclic non-polar group possess no specific advantage over those having open structure hydrocarbon chain and thus these do not find much use in industry. Xanthogenates obtained from secondary alcohols are less effective than collectors synthesised from primary alcohols. However, isopropyl and isobutyl xanthogenates are exceptions, which are more active than their corresponding xanthogenates obtained from primary alcohols.

c) *Composition*

The introduction of a substitute into a chemically active solidophil group greatly influences the collector action. Monothiocarbonate $\left(R.O.C\diagup_{\diagdown S-}^{\diagup O}\right)$ has a great affinity for water and forms relatively soluble compounds with heavy metal ions and thus are weak collectors. The introduction of S into the group $-\overset{O}{\underset{\|}{C}}-$ in place of oxygen results in a great increase in the collector capacity by reducing their affinity for water, and xanthogenates, i.e. dithio-carbonates $\left(R.O.C\diagup_{\diagdown S-}^{\diagup S}\right)$ are extremely powerful collectors. Further transformation of dithiocarbonate into the trithiocarbonate $\left(R.S.C\diagup_{\diagdown S-}^{\diagup S}\right)$ by the introduction of a third S instead of the oxygen also increases the collector effect, but to a much lesser extent than the former case. The central nucleus of solidophil group also has an important effect on collector properties. This may be illustrated by comparing xanthogenates (having C-nucleus) with dithiophosphates (having pentavalent P nucleus).

d) *Concentration of Collector*

Up to a certain limit, collector concentration affects the speed of reaction between mineral and collector and the rate of formation of adsorption layer. Afterwards higher concentrations are ineffective.

e) *Duration of Contact between Mineral and Collector*

When ionising collectors are employed, a short duration (2–10 min) for pulp-collector contact is sufficient, whereas collectors not readily soluble or forming emulsions need prolonged contact with the pulp.

f) *Entry Point of the Collector*

In the flotation process, feeding of collector at the suitable point is an important factor for collector efficiency. The reagent should preferably be fed in the grinding circuit (i.e. acting on freshly exposed surfaces).

g) *Use of Mixture of Collectors*

In practice it has been observed that use of a mixture of two or more collectors may in some cases give better performance (increased flotation speed, increased mineral extraction, and reduced reagent consumption). For example, when galena is floated with ethyl xanthate, amyl xanthate and their mixture in the ratio of 1 : 2 to 1 : 1 (the amount of each collector and of mixture employed being the same, i.e. 50 g/tonne), the extraction has been found to be 70, 73, and 83 per cent, respectively. The extraction is greatly improved when combinations of collectors differing greatly in composition and properties are employed. For example, when sodium oleate, butyl xanthogenate and their mixture (30 per cent butyl xanthogenate and 70 per cent sodium oleate) are employed (in each case total collector being 75 g/tonne), the extraction has been found to be 50, 60 and 90 per cent, respectively. When a mixture of a weak and a strong collectors is employed, their anions are attached selectively to those parts of mineral which provide best condition to the respective collector. By employing a mixture, adsorption layer density and the amount of reagent adsorbed are usually greater and the mineral is rendered water-repellent more quickly.

h) *Collector Layer Density on the Mineral Surface*

Collector layer density may be referred as the amount of adsorbed reagent divided by the area of mineral surface exposed to its action. The layer formed is usually monomolecular, and thus/when the whole mineral surface is covered by a continuous monomolecular collector layer, the adsorption layer density is referred to as 100 per cent. In chemical adsorption, the layer density is generally 20–30 per cent, which is due to lack of uniformity in the mineral surface. In practice, a monomolecular coating covering less than the whole mineral surface is sufficient for an efficient flotation. In general, increased collector coating leads to improved floatability, but after a certain limit (optimum density) a further increase in adsorption layer density does not improve flotation efficiency.

12.2.5 VARIOUS COLLECTORS AND THEIR APPLICATIONS

The choice of collector for a specific process is generally determined by experience and experiments. It depends on many factors, particularly the nature of minerals involved, their degree of oxidation and presence of other heavy metals, even in traces, in the pulp. The important collectors are given in Table 12.1, with necessary details.

12.3 Frothing Agents (Frothers)

Frothers are the reagents which are added to the pulp to permit the production of a sufficiently stable froth to hold the floating mineral particles until the froth is removed from the flotation cell. These are the surfactants of low solubility and surface tension, and thus increase the life of the bubbles produced. The solid particles of mineral also form a network around the pulp and thus a more stable froth is produced.

12.3.1 MECHANISM OF FROTHING ACTION

Frothing agents are heteropolar surface-active substances which are adsorbed at the air–water interface. After mineral particle surfaces are rendered water repellent, the presence of frothers keep the air-bubble dispersed and increase the stability of flotation froth. Adsorption of frothing agent at air–water interface is due to its surface activity and ability to reduce water surface tension.

In the pulp, polar groups of frother molecules combine (become covered) with water dipoles' whereas their non-polar ends become oriented toward the air phase. The frother adsorption layer (surface layer with frother molecules) thus formed gives the frothing action as follows:

a) It increases bubble resistance to various external forces, i.e. increased strength of bubble is achieved.

b) It reduces the speed of bubble movement in the pulp and thus bubble contact with mineral particles is more, which results in better conditions for attachment.

c) The reduction in bubble speed also reduces the force of collisions between the bubbles and thus a favourable effect on froth stability is achieved.

12.3.2 FACTORS AFFECTING THE ACTION OF FROTHERS

The various factors which influence the action of frothing agents are the following:

a) *Concentration of Frother*

The number of water dipoles combining with polar group of frother

Table 12.1. Details of important collectors[8, 22-27] (preparation, applications, amounts used, etc.)

Type of collector	Collector in general	Chemicals employed	Preparation/Source	Applications	Amount of reagent used in g/tonne	Remarks
1	2	3	4	5	6	7
A. Ionising collectors						
i) Anionic collectors	(a) Fatty organic acids and soaps $CH_3 (CH_2)_n X$, X = Carboxylic, sulphate or sulphomate, n is usually more than 9	Oleic acid $C_{17}H_{33}COOH$, Stearic acid $CH_3(CH_2)_{16} COOH$, Lauric acid $C_{11}H_{23} COOH$, Palmitic acid $C_{15}H_{33}-COOH$, Na soaps of above acids,	By-product from wood pulp industry and synthetically prepared. For making the salts, respective acid is reacted with NaOH	Oxidised rare, ferrous and non-ferrous minerals (carbonates, oxides, and sulphates) such as of W, Mo, Be, Li, Sn, silicates, fluorspar, phospharite, apatite, barite and other oxide minerals. Flotation of water soluble salts of alkali and alkali earth metals	200–500	Collector properties of these chemicals are much weaker than those of xanthogenates and aerofloats. These are sensitive to temp (higher temp. is preferred). These are used by mixing them with non-polar compounds such as kerosene to give better dispersion. These collectors cannot be used in hard water. These also act as frothers. Alkaline medium is used in flotation.
		Nephthenic acid $C_5H_9 COOH$ and its soap	Found in low paraffin crude oil extracted from petroleum products	Flotation of whole series of non-sulphide minerals	300–800	Readily dissolves in water. These possess both collector and frother properties. These are less sensitive to low pulp temp. and hard water. Cost is also low.

b) Reagents with a H₂SO₄ radical

Alkyl sulphates $R(SO_2)OH$, Alkyl (aryl) sulphonates $(R = 12\text{-}16\,C\text{'s})$ $R(SO_2) ONa$, H_2SO_4 esters $RO (SO_2) ONa$ $(R = 10\text{-}12\,C\text{'s})$, Alkyl—aryl sulphonates

R——◯——SO_3Na

$(R = up$ to $12\,C\text{'s})$,

H_2SO_4 esters of fatty acids

CH_2COOCR

CH_2CSO_3Na

$(R < 10\,C\text{'s})$,
Dialkyl sulphonates

$R\ COOCH_2$

$R\ COOCHSO_3Na$

$(R = 4\text{-}8\,C\text{'s})$
Sulphonated esters of amines

$R\text{-}CONC_3H_4OSO_3Na$

Alkyl sulphonates with amide interlink $(R = 15\text{-}18\ CH)$

CH_3

$R\text{-}CONC_2H_4SO\cdot Na$

By the action of H_2SO_4 on the appropriate hydrocarbons, alcohols, carboxylic acids or other organic compounds

Selective flotation of oxidised minerals with alkali earth cations. Flotation of barite from scheelite–barite concentration, flotation and separation of fluorite from cassiterite at pH 6.8, selective flotation of zircon and pyrochlore in acid medium (pH 1.5–3)

500–900

These reagents are highly soluble in water and dissociate into ions being relatively strong acids. These have collector action as well as frother action and are not sensitive to hard water.

(Contd.)

Table 12.1 (*Contd.*)

1	2	3	4	5	6	7
c) Xanthogenates	$R-O-C{\diagup}^{SX}_{\diagdown S}$ X = Na or K	Potassium ethyl xanthate C_2H_5OCSSK, Potassium isopropyl xanthate C_3H_7OCSSK, Potassium butyl xanthate C_4H_9OCSSK, Potassium amyl xanthate $C_5H_{11}OCSSK$, Potassium hexyl xanthate $C_6H_{13}OCSSK$	By the reaction between CS_2, KOH and respective alcohol $ROH+KOH+CS_2=ROCSSK+H_2O$	Selective flotation of sulphides and precious metals		Xanthates decompose in storage due to moisture. Higher homologues are more powerful collectors but their cost limits their application to specially difficult conditions.
d) Dithiophosphates	$RO{\diagdown}_{RO}{\diagup}P{\diagup}^{S}_{\diagdown SX}$ X = Na, K or H	Cresyl dithiophosphate $CH_3C_6H_4O{\diagdown}P{\diagup}^{S}_{\diagdown SH}$ $CH_3C_6H_4O{\diagup}$ Cresyl dithiophosphate with 6% thiocarbamilide Xylenol dithiophosphate $(CH_3)_2C_6H_3O{\diagdown}P{\diagup}^{S}_{\diagdown SH}$ $(CH_3)_2C_6H_3O{\diagup}$ Secondary butyl dithiophosphate sodium aerofloats	By the reaction between alcohol/ phenol, and P_2S_5 $4ROH+P_2S_5=$ $RO{\diagdown}P{\diagup}^{S}_{\diagdown SH}{\diagup}RO{\diagup}+H_2S$ For Na or K salts neutralisation is carried out with Na/ K carbonate.	Acts as collector as well as frother for non-ferrous sulphides free from iron. Used as frother in gold bearing ores. Used for extraction of Pb-glance and Ag-bearing minerals. Flotation of Pb-Zn-Cu ores and precious metals. Cu-minerals and Zn blende where minimum pyrite is required.	20-200 25-125	These are strong frothers but weak collectors giving a low selectivity. These may be used before or after the use of xanthates. Xylenol dithiophosphates are designated by number (e.g. D-25, D-15, etc.) based on percentage of P_2S_5. Xylenol dithiophosphates

(Contd.)

(ii) **Cationic collectors**	(a) Amine and amine derivatives	Primary $\overset{H}{\underset{H}{R-N-R}}$ Secondary $\overset{H}{\underset{R}{R-N-R}}$ Tertiary $\overset{R}{\underset{R}{R-N-R}}$ **Quaternary** $\overset{R}{\underset{R}{R-N-OH}}$ Dodecylamine CH₃(CH₂)₁₁NH₂		

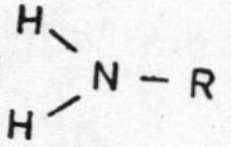

Quaternary ammonium salts are reaction products of tertiary amine with alkyl halide

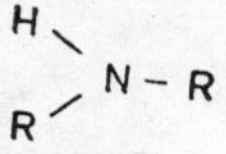

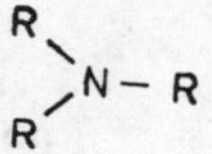

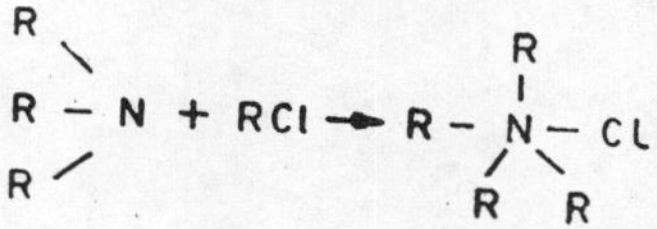

$$R-\overset{R}{\underset{R}{N}} + RCl \longrightarrow R-\overset{R}{\underset{R}{N}}-Cl$$

Cleaning of silica from sulphides, phosphites. Concentration and removal of micaceous contaminants from china clay. Flotation of silicates, alumino-silicates, certain oxide, etc.

Separation of feldspar from talc, mica, kaolin, silicate, and can-

50–500

contains S > 8.5% and P > 5.5% and replaces xanthates and pine oil. Sodium aerofloats do not possess frother properties.

This reagent is not affected by hard water and can be used even with sea water.

Table 12.1 (*Contd.*)

1	2	3	4	5	6	7
		Guanidine $HN-C(NH_2)_2$		tonates. Separation of chromite from olivine. Flotation of oxide, carbonate of Pb and Zn minerals		
b) Non-polar collectors	By-products of coal tar and petroleum industry	Kerosene Hydrocarbon oils Diesel oil Cresols		Flotation of graphite, coal, talc, molybdenite, sulphur. Used as solvent for soluble anion collectors.		
	Dimethylglyoxime type organic compounds	Dimethylglyoxime		Flotation of Cu–Ni minerals from partly oxidised pyrite–pyrrhotite, Ni–Cu ores not readily concentrated	200-600	
c) Other collectors of industrial importance	Oxidised kerosene	Oxidised kerosene (complex composition)	Obtained by oxidation of kerosene by atmospheric O_2 at elevated temp. in the presence of a catalyst	Flotation of coal		To increase solubility in water, oxidised kerosene is saphonified with alkali Fluctuation in composition is an unfavourable feature. It
	Peat tar, shale oil and its products	Complex composition consisting of neutral compounds, organic acids, phenols, and	Product of peat coking	Flotation of apatite, and other non-metallic minerals, oxi-		

	organic sulphur compounds		dised lead ores, and certain polymetallic ores	400–800	is both frother and collector.
Sulphonate kerosene	Sulphonated kerosene	By reaction of strong H_2SO_4 on kerosene at 70–80°C	Flotation of coal	300–700	—
Minerec	Minerec	By oxidation of xanthates	Reducing sulphide minerals	—	—
Green acids and mahogany soaps	Green acids and mahogany soaps	Mixed sulphonation products from the cracking process	Flotation of oxide and silicate ores notably iron ores	—	—
Reagents having S in direct combination with C	Diphenyl thio urea $$\overset{S}{\underset{\|}{}}$$ $C_6H_5\text{-HN-C-NHC}_6H_5$		Selective flotation of galena from polymetallic ores and copper sulphide	20–100	Used in the form of solution in org. solvent such as ortho-toluidine (T-T mixture)
Thioalcohols and their Na/K salts	Thiophenols RSH or RSK mercaptans		Selective flotation of Cu and Zn sulphide minerals in the presence of pyrite. Flotation of oxidised Cu minerals. Flotation of sulphides	—	Mercaptans give strong unpleasant odour.
Dithiocarbamates $$\overset{S}{\underset{}{R-N-C-SK}} \atop R$$	Potassium diethyl dithiocarbamate $$C_2H_5-N-C\overset{S}{=} \atop C_2H_5 \quad SK$$				Are similar to xanthogenate in structure and composition but more expensive

(Contd.)

Table 12.1 (Contd.)

1	2	3	4	5	6	7
	Flotegen (proprietary name)	Sodium ethyl monothio-carbamate Sodium ethyl dithiocarbamate Sodium ethyl trithiocarbamate Mercaptobenzothiozale 	Derived from carbamate	Flotation of oxidised lead ores in alkaline circuit. Flotation of Cu and Zn sulphide minerals.	100–300	Used only in alkaline media. When used with dibutyl dithiophosphate and butyl xanthogenate, it is used for flotation, of cerrusite without sulphidizers, and oxidised copper ores with Na_2S.

Organic disulphides	Dixanthogenide	Oxidation of xanthogenates	Used as an additive to sulphydryl collectors to reinforce their collector effect in flotation of oxidised heavy metal sulphide minerals. Flotation of cement copper.	Can be used in weakly acid or neutral medium. It is very slightly soluble in water and thus fed into the mill or contact vats.
	$$R-O-\overset{\displaystyle S}{\underset{\displaystyle \|}{C}}-SS-\overset{\displaystyle S}{\underset{\displaystyle \|}{C}}-O-R$$		300–500	

depends on the concentration of frother. With low frother concentrations, effect of bubble dispersion is more effective, whereas increased concentrations assist coalescence and are thus undesirable.

b) *Size of Bubbles*

Frother will reduce coalescence of bubbles which do not vary too greatly in size among themselves.

c) *Number of Polar Groups*

Usually the presence of 1–2 polar groups in a molecule is adequate. An increase in the number of polar groups with the same hydrocarbon radical does not improve frother efficiency and sometimes even an adverse effect may be obtained.

d) *Structure and Molecular Length of the Non-polar Group*

Frothing action increases considerably with increase in the length of non-polar group (hydrocarbon), being zero for CH_3OH and then increasing with C_2H_5OH to propyl, butyl and amyl alcohol. For a frother to be effective, at least six C-atoms should be present in the hydrocarbon group. However, with too long a chain, solubility and frother effect are greatly reduced.

e) *Composition of Polar Group*

Combination of the same non-polar group with polar groups of different composition may give the reagent good frother properties. For example, cetyl alcohol ($C_{16}H_{33}OH$) does not possess frother properties due to its extremely low solubility, whereas cetyl alcohol bisulphate ($C_{16}H_{33}SO_4H$) gives good results.

f) *Presence of Certain Minerals and Colloidal Suspensions of Metals*

Froth stability may be considerably enhanced in the presence of small quantities of these suspensions. For example, suspensions of chalcopyrite, galena, and molybdenite improve froth stability while employing propyl alcohol, phenol, and pyridine as frothers.

g) *Collector Properties in Frothers*

Reagents having collector as well as frother properties may make flotation control difficult (particularly selective flotation). Therefore, frothers with a hydroxyl group (alcohols) having practically no collector properties are usually preferred.

h) *Proportion of Collector and Frother*

Frother efficiency also depends to some extent on correct proportions of collector and frother to be used.

i) *Temperature*

Temperature of the pulp affects the number of water dipoles combining the polar group of frother efficiency.

12.3.3 TYPES OF FROTHERS

In order to have the frother's surface active properties fully effective, they must be soluble in water to some extent (to have an even distribution). Frothers are usually soluble in the range 0.001–4 per cent. In the aliphatic series, the acids, amines and alcohols are the most soluble, whereas in aromatic series the alcohols, amines, and acids have maximum solubility. A great variety of substances having varied composition of polar group are used as frothers. The most effective frothers consist of hydroxyl ($-OH$),

carboxyl $\left(-C\underset{\displaystyle \diagdown OH}{\overset{\displaystyle \diagup O}{}}\right)$, carbonyl ($=C=O$), nitrogen (N), an amino group

($-NH_2$), or a sulpho group ($-SO_2OH$, or $-OSO_2OH$) as their hydrophilic groups (non-collectors). Among these, hydroxyls are the most common, e.g. terpinols ($C_{10}H_{17}OH$), cresols ($CH_3 \cdot C_6H_4OH$) and alcohols ($C_5H_{11}OH$), etc.

12.3.4 APPLICATIONS OF FROTHERS

Important frothers of commercial importance are given in Table 12.2 along with their applications, properties, sources, etc.

12.4 Modifying Agents and Their Action in Flotation

These reagents, also known as regulators or conditioners, may be defined as reagents used in flotation to intensify the specific action of a collector on the mineral surfaces, i.e. to decrease or increase water-repellent effect on specific minerals. In the presence of regulators, the collector activates selectively only specific minerals required to pass in the froth. The function of regulators (modifying agents) involves the reactions (physical and/or chemical) with minerals, collectors, and ions present in the pulp. The principal requirement of using regulators is to achieve most selective separation between the mineral species floated and that left in the pulp.

12.4.1 MECHANISM OF ACTION OF MODIFYING AGENTS

The net effect of regulators may be considered in terms of wetting and drying out forces. Reactions resulting in an increase in hydrated areas lead to surface hydration or wetting which favours the retention of mineral particles in the pulp, whereas reactions resulting in drying out or air-avidity favour the attachment of the particles to air-bubble. Thermodynamically, the replacement of gas by liquid (or vice versa) on a solid surface is due to a net decrease in the total interfacial free energy of the system. The regulating chemicals used in flotation maintain the desired surface activity of specific minerals in the pulp by reinforcing the adsorption of collectors in

Table 12.2. Details of important frothers[8,22-27] (preparation, applications, amounts used, etc.)

Type of frother	Name of frother and its chemical composition	Source/preparation	Properties	Application	Amount used in g/tonne	Remarks
1	2	3	4	5	6	7
Hydroxyls Aromatic alcohol	(a) Pine oil containing > 44 per cent terpineol ($C_{10}H_{17}OH$) and other compounds (limonene $C_{10}H_{16}$, dipententene $C_{10}H_{16}$, borneol $C_{10}H_{17}OH$, phenol, organic acid, etc.).	From distillation of pine wood and then fractional distillation of the crude oil so obtained.	Transparent and yellow in colour, odour of terpentine, Sp. gr. 0.915–0.935, b.p. ≤ 170°C, solubility 2.5 g/litre (at 25°C)	Sulphide minerals	5–120	The frother action is due to its alcohol (terpineol) content. It is difficult to get the pine oil of constant composition. Presence of small amount of organic acids gives a slight collector effect.
	(b) Eucalyptus oil	From eucalyptus wood	—	Oxide minerals	10–100	—
	(c) Wood tar. Heavy wood tar contains minimum 40 per cent, phenols and light wood tar contains minimum 15 per cent phenols.	Prepared by extraction from products of destructive distillation of wood	It is dark coloured mobile oily liquid	Heavy mineral grains	5–100	—
	(d) Cresols ($CH_3C_6H_4OH$) containing considerable amount of neutral hydrocarbons, phenols, xylol derivatives and higher aromatics.	By distillation of raw coal tar and from light cresote oil fraction.	Solubility in water is 2 g/litre at 25°C.	Heavy mineral grains or large mineral particles	5–160	Raw cresol used for flotation usually contains 3-isomers (*para* cresol, *ortho* cresol, and *meta* cresol). When it contains a certain amount of toluol and xylol, it is a very strong frother. Cresols are toxic and cause burns on skin.

Synthetic frothers having hydroxyls	(a) Polypropylene glycol methyl esters $CH_3-(O-C_3H_6)_x-OH$	Prepared from propylene oxide and CH_3OH	Water soluble	Flotation of sulphides	30-60	These are also known as Dowfroths (nos. 200 and 250). The number refers to the molecular weight.
	(b) Methyl amyl alcohol	Prepared synthetically.	Water solubility 17 g/litre at 20°C		25-60	Advantage of synthetic frothers over industrial products is stability of composition which makes the flotation control easier.
	(c) Capryl alcohol	Prepared synthetically.	Water solubility 1.25 g/litre at 25°C.		20-80	
	(d) Methyl isobutyl carbinol	Prepared synthetically.	—		25-120	
	(e) Aerofroth 65 based on higher alcohols	Product of American Cyanamid Company	Water soluble		30-60	
Reagents with sulpho-groups. Sulphates $(ROSO_2OH)$ and Sulphonates	(a) Cetyl alcohol bisulphate ester $C_{10}H_{12}SO_3$	Prepared synthetically			—	The radical should contain 8-12 C-atoms for frother properties.
	(b) Stearyl glycerol sulphate $C_{17}H_{35}$ COO (CH_2) CH(OH) CH_2SO_4·Na				—	If C > 12, the reagents are collectors
	(c) Reagent 'DS' a mixture of various chemical	Obtained from petroleum	This is detergent in nature and readily	Flotation of sulphide ores		Its frothy properties are due to the presence of

(Contd.)

Table 12.2. (*Contd.*)

1	2	3	4	5	6	7
(RSO$_2$OH)	compounds, basically being alkylaryl sulphonate	products by sulphonation of the kerosene-gas oil fraction obtained from petroleum distillation	dissolves in water			sodium alkyl-aryl sulphonate. This is cheap and non-toxic
Contain-ing-N	Piridine C$_5$H$_5$N and Quinoline C$_9$H$_7$N	Product of coking	—			Heavy pyridine bases also include other compounds.

the case of floating species or hydration in case of depressants.

Modifying agents may perform their function in one or more of the following ways:

a) They may act directly on the surface of a specific mineral and change its chemical composition, resulting into better interaction of the mineral and the collector, and thus results of flotation are improved (activation effect). For example, the action of $CuSO_4$ on sphalerite (ZnS) leads to the formation of CuS on the surface and thus attachment of xanthate on ZnS increases.

b) In some cases they may produce an adverse condition for mineral activation by the collectors (detach a collector from mineral surface) which leads to relatively poor results in flotation (depression effect). For example, addition of Na_2S in pulp leads to the detachment of xanthate from the surfaces of galena and other non-ferrous sulphides.

c) They may give a direct effect on the stability of the hydrated layer near the mineral surface, which may alter its floatability irrespective of collector action. For example, $K_2Cr_2O_7$ gets adsorbed on galena surfaces which are free from collector and thus hydrate these areas. This leads to decreased floatability without detachment of xanthate.

d) In some cases the regulators maintain a suitable ionic structure of pulp by altering the concentration of H-ions or soluble salts (pH regulation). As a result of this, an environment is produced which may be suitable for the flotation of one mineral and unsuitable for another. For example, pyrite cannot be floated in an alkaline medium, but can be floated in neutral or weakly acid medium.

e) Certain reagents keep the gangue particles dispersed to improve flotation condition (dispersion effect). For example, Na_2SiO_3 and polyphosphates disperse the masking slime from mineral surfaces.

12.4.2 TYPES OF MODIFYING AGENTS

The modifying agents employed in general are of great variety of inorganic and organic chemicals, such as salts, acids, alkalis, electrolytes, and non-dissociating compounds differing greatly in composition, structure, and properties. Various modifying agents may be classified as depressants, activators, wetting agents, flocculants, dispersants, pH regulators, surfactants, etc., depending on their function in the flotation. These are discussed below.

a) *Activators and Their Action*

These chemicals are generally inorganic compounds, addition of which leads to changes in the chemical composition of mineral surface layer as a result of forming a surface compound or passage of activator ions (usually cations) into the mineral crystal lattice. This usually happens as a result of exchange adsorption between activator and mineral ions. These changes reduce surface hydration, increase the amount of collector attached to the surface and improve flotation. The active ion of the activators has the opposite charge to that of the collector ion. Various activators in common use are—water soluble salts of heavy non-ferrous metals such as $CuSO_4$ (to activate sphalerite, pyrite, quartz, etc.) and water soluble salts of alkali metals (to activate quartz and other non-sulphide minerals), Na_2S and other water-soluble sulphides used for activation of oxidised non-ferrous metal minerals (cerrussite, malachite, etc.), and oxygen (usually in the form of atmospheric oxygen) to activate many sulphides. Some important examples of activation are given below:

i) Quartz can be floated with anion collector only after the treatment with solutions of heavy metal salts (ferrous and copper sulphates). In this case a surface compound containing Fe or Cu cations is formed, which acts as a connecting link between the mineral and the anion collector. However, no activator is required when employing cation collector.

ii) Sphalerite can be floated with xanthate collectors, but for satisfactory results, xanthates having comparatively long non-polar groups are to be used in large quantities. If water-soluble salts of heavy metals (Cu, Ag, Pb, etc.) are used as activators, the metal ions of these salts will replace the zinc ions in the mineral surface as follows:

$$Cu^{2+} + ZnS \rightarrow CuS + Zn^{2+} \tag{12.1}$$

Since xanthate reacts more readily with Cu-cations, the xanthate–copper compound so formed is more resistant to the dissolving action of water than the compound formed with zinc. Thus, use of xanthate in aqueous medium with $CuSO_4$ as an activator will give better flotation results on sphalerite. In this case, copper enters the sphalerite lattice easily due to the facts that, (i) ionic radii of Cu and Zn are similar, and (ii) sulphur has greater affinity for Cu than Zn.

iii) Flotation of oxidised non-ferrous metal minerals (sulphates, carbonates, etc.) and heavily oxidised sulphide minerals with xanthates is unsatisfactory and requires a large quantity of the collector. Addition of Na_2S results in sulphidisation of mineral due to passage of sulphur ions into the crystal lattice of the oxidised minerals, replacing sulphate or carbonate ions and transforming the mineral into sulphide. Thus, flotation can be carried out with xanthates. Na_2S reacts with cerussite according to the following reaction:

$$Na_2S + PbCO_3 = Na_2CO_3 + PbS \tag{12.2}$$

When sodium concentration is sufficient, a thick layer of Na_2S is formed, which gives depressing effect (since xanthates do not react with pure sulphides). An optimum pH value for cerussite is 9–9.5, at which maximum adsorption of S-ions takes place.

iv) Silica can be floated with oleic acid when a soluble salt of barium is used as an activator. In this case barium-ions are adsorbed on silica. Whereas, barium salt would prevent collection of silica by docecylamine.

v) Water glass when used in comparatively small amounts, acts as an activator for apatite, fluorite, and even for malachite and cerussite when the latter are floated with iso-amyl xanthate.

b) *Depressors and Their Action*

These chemicals (organic or inorganic) also known as depressants and supressors, are the chemicals having opposite effect on a mineral to that of activators. These chemicals render a mineral surface inactive to a collector and thus mineral becomes unfloatable even in the presence of suitable collector by forming a coating on them. The working ions of these reagents have the same charge as that of the collector ions. For example, cyanide is a depressant with an anionic collector, whereas sodium is a depressant with cationic collector.

The reaction of depressants with the mineral is identical to that taking place in activation. Depressants impede or check collector attachment (since mineral–collector compound becomes unstable), increase mineral hydration and detach collector ions from the mineral surface.

Various depressants used in practice are Na_2S and water soluble sulphides for depression of sulphide minerals, cyanides for selective depression of sulphide minerals, sulphites, hyposulphites and certain sulphates for selective flotation of sulphide ores, K_2CrO_4 and $K_2Cr_2O_7$ for depression of PbS, water glass ($Na_2O \cdot mSiO_2$) for depression of quartz and other gangue minerals, organic non-ionising reagents (starch, dextrin, tannic acid, etc.) for depression of non-sulphide minerals, and lime for specific depression of iron bearing sulphides. Some important examples are given below:

i) *Na_2S and water soluble sulphides*: Na_2S acts as a depressant for most sulphide minerals as well as to oxidised minerals after sulphidisation. This depressing action takes place due to the presence of S and HS ions under these conditions. Similarly, quartz activated by Fe or Cu ions can be effectively depressed with Na_2S (since colloidal Fe or Cu sulphides are formed on the mineral surface which are incapable of reacting with collectors). S and HS ions are freely adsorbed on sulphide surfaces (a high negative charge is attained) even in very low concentrations, as a result of which adsorption of collector ions as well as other ions will be prevented from the surface. This condition also prevents the activation of sulphides by oxygen.

In practice Na_2S is used in flotation of molybdenite when associated with considerable amount of other sulphides to depress all sulphides except molybdenite under suitable conditions.

ii) *Cyanides*: Cyanides have been very widely used in the selective depression of certain sulphide minerals such as sphalerite, copper minerals, pyrite, nickel sulphide, silver minerals, etc. Addition of a reagent in the pulp leading to breakdown of cyanide complexes and regeneration of free cyanide ions increases the depressive action of cyanide, reduces its consumption, and improves the selectivity of Cu and Zn minerals. Na_2S is used for this purpose, since it has a dual role, i.e. (i) it forms an insoluble compound with the activator ions (copper) and (2) frees the cyanide and thus giving increased concentration of cyanide in the pulp. Use of $ZnSO_4$ with cyanide is highly effective due to the formation of zinc cyanide, $Zn(CN)_2$ which is relatively insoluble (precipitates on the mineral surfaces to be depressed). This provides sufficient depression of ZnS for a low cyanide consumption.

The depression effect of cyanide depends on its concentration, composition of the minerals to be separated, the ionic composition of the pulp, and the length of xanthate hydrocarbon chain. The longer heavy metal-xanthate hydrocarbon chain has greater stability in relation to cyanide and thus a relatively high concentration of cyanide is needed to depress a mineral in the presence of a xanthate with a long hydrocarbon chain. Thus xanthates with short hydrocarbon chain should be employed for sulphide flotation while using cyanide as depressing agent. Depressive action of cyanide may be given as the following:

The depressive action of cyanide is mainly due to its ability to react and form stable compounds with heavy metal cations. In the reaction between metal cation and cyanide, first precipitates of comparatively insoluble cyanide salts are formed according to the following reaction:

$$MS + 2NaCN \rightarrow Na_2S + M\,(CN)_2 \qquad (12.3)$$

and then salts are dissolved by the action of excess cyanide and complex cyanogen ions are formed according to the following reaction:

$$M\,(CN)_2 + 2\,(CN)^{2-} \rightarrow M\,(CN)_4^{2-} \qquad (12.4)$$

These cyanogen ions are responsible for depressive action of cyanide. In the presence of sufficiently high concentration of cyanogen ions, xanthates do not attach themselves to the mineral surfaces susceptible of cyanide depression. Most common cases are the depression of sphalerite and copper sulphides by NaCN.

Though there are certain problems associated with the use of cyanide, particularly toxicity, relatively high cost, and their ability to depress and dissolve gold, even then cyanides are widely used due to their high selectivity in flotation.

iii) $K_2Cr_2O_7$ *and* K_2CrO_4: The depressive effect of chromate ions is due to their chemical attachment to mineral surfaces (due to adsorption of CrO_4^{2-} anions) which is independent of xanthate ion attachment and apparently takes place in different areas. The areas where chromate ions get attached become hydrated and the surface hydration is so much that the mineral becomes unfloatable, in spite of the presence of xanthogenate adsorption layer on mineral surface to the extent of 30–33 per cent of a complete monolayer.

iv) *Water-glass*: Sodium and potassium silicates ($Na_2O \cdot m\ SiO_2$ and $K_2O \cdot m\ SiO_2$, where $m = 2$-4.5) are referred as water-glasses. Sodium silicates are preferred owing to their lower cost. Aqueous solutions of water-glass give an alkaline reaction due to the hydrolytic decomposition of sodium silicates (as these are the salts of a strong base and a weak acid) according to the following reaction:

$$Na_2SiO_3 + 2 H_2O \rightleftharpoons H_2SiO_3 + 2 Na^+ + 2 OH^- \qquad (12.5)$$

Water-glass is commonly used as a depressant for separation of quartz and silicates in soap-flotation, as well as separation of non-sulphide minerals having similar flotation properties (e.g. calcite and fluorite, calcite and scheelite, etc.

The water-glass employed in flotation generally has the value of m equal to 2–3. Water-glass of low m value gives higher alkalinity in aqueous solution and is weak as depressing agent for quartz. However, water-glasses with $m > 3$ do not readily dissolve and give rise to formation of a large number of coarse silica-gel particles. For consistent results of flotation, solution of water-glass should be prepared fresh every time.

Action of water-glass is highly selective and produces changes in the mineral surfaces rendering them water-avid. However, the selective effect goes on reducing with increasing concentrations of water-glass (becomes non-selective when used in more than usual amounts, i.e. 1.1–1.2 kg/tonne). This selective depression is due to the facts that (1) non-sulphide minerals adsorb different amount of water-glass, (2) water-glass attachment to minerals varies in strength, and (3) depressive effect is not uniform for all non-sulphides.

The OH^- ion concentration is inversely proportional to m. OH^- ions are readily adsorbed by many minerals and can detach collector ions attached to the mineral surfaces. For example, in flotation of calcite and fluorite, amount of sodium oleate is reduced.

Use of Na_2CO_3 and polyvalent metal salts with water-glass greatly increases the selectivity of depression.

Attachment of heavily hydrated micells (colloidal particles) of silicic acid plays an important role in depressive action due to their hydration action on mineral surfaces. The action of water-glass may be considered purely physical in nature, since it can depress a large number of minerals under suitable conditions and can be readily desorbed on dilution of the solution or washing of minerals with water.

v) *Organic reagents which do not dissociate*: This group includes reagents of varied chemical structure. Tannin (tannic acid), and quebracho (tannic extract), dextrin and starch, albumin, gelatin, gum arabic, etc., are the common reagents of this group. The depressive effect is mainly due to the formation of colloidal particles (micells) on the mineral surfaces in the pulp rendering their hydration. By using large amounts of these reagents, all minerals can be completely depressed but selectivity is achieved by feeding them in small amounts.

vi) *Electrolytes having ions similar to the mineral ions*: The addition of these electrolytes lead to an extensive attachment of these ions to the mineral crystal lattice to the least saturated cations. This condition results in the depression of minerals, since the sites already occupied by electrolyte anions do not respond to collector reaction. The attachment of the similar ions (particularly anions) will increase with decreasing solubility of minerals and thus a stronger depressive effect of these anions on the mineral will be achieved. It has been established that Na_2SO_4, K_2SO_4, and H_2SO_4 have a depressive effect on the flotation of minerals containing sulphate ions (barite, celestite, anglesite, etc.) whereas electrolytes not containing sulphate ions do not depress these minerals. Similarly, $ZnSO_4$ can be used for the depression of ZnS where similar cations (Zn) are involved. Other example is Na_2S (discussed earlier in detail) to suppress sulphide minerals.

vii) *Differential depression*: For the recovery of minerals separately from bulk flotation product, it is reconditioned and refloated. It needs selective desorption or destruction of original collector, and specific reactivation. Use of permanganate, dichromate, or ferricyanide destroys the xanthate film by oxidation of xanthate to di-xanthogen. Differential depression can also be obtained by controlled heating prior to refloating to volatilise alcohol group of the collector (e.g. separation of molydenite float from a copper sulphide sunk fraction). Other examples are (1) use of NaOH to dissolve anionic fatty acid collector for removal of silica from a phosphate silica float, where silica is then floated with a cationic collector, and (2) use of NaCN to dissolve copper used for activation of sphalerite.

viii) *Lime*: It forms a mixed surface film with ferrous and ferric hydroxides and insoluble lime salts ($CaSO_4$) due to the reaction between lime and pyrite (also oxidised products of pyrite such as thiocompounds) which help to depress the pyrite by reducing the adsorption of xanthate. Lime is a more powerful depressant than NaOH for pyrite, since both OH^- and Ca^{++} ions participate in the depressive effect. It has been observed that Na^+ ions do not attach themselves strongly to pyrite and are readily detached by simple washing, whereas Ca^{++} ions attach themselves very firmly to the pyrite.

c) *pH Regulators*

The purpose of adding pH regulators (organic as well as inorganic) is to control the pH of pulp to the required optimum value, since pH of pulp has a critical controlling effect on the action of flotation regents as discussed earlier. Careful control of pH helps in selective flotation of various minerals, such as lead, zinc and iron sulphides. Common pH regulators for creating alkaline medium are lime, Na_2CO_3, Na_2SiO_3, NaOH, NH_4OH, etc., and for acid medium are H_2SO_4, HCl, oxalic acid, etc. Certain collectors such as Na_2S, NaCN, and Na_2SiO_3 also give alkaline medium on hydrolysis. Value of pH of the medium depends both on the composition and amount of reagents used and on the nature of minerals. For example, calcite, magnesite, and dolomite give an alkaline medium by reacting with water, whereas barite and anglesite produce an acid medium. Use of pH regulators are made to increase the attraction of the collector towards the surface to be floated.

H^+ and OH^- ions alter the mineral surface hydration due to their adsorption on the mineral. These ions may be adsorbed even in the absence of collector, as the collector and these ions are attached to different parts of the mineral surface. H^+ and OH^- ions have a potential determining effect on some oxide minerals such as quartz, silicates, and the hydroxides, where they are readily adsorbed. Thus, these minerals are highly sensitive to changes in pH of pulp which is an important factor in control of flotation.

Lime in the form of lime suspension (1.25 g lime per litre) is used to depress pyrite and prevent the adverse effect of soluble salts on flotation. NaOH has limited use mainly in the flotation of gold. Soda ash has the advantage of precipitating Ca^{++} ions from hard mill water and thus reducing their plating effect (this may lead to random activation of unwanted minerals). Na_2CO_3 is employed as pH modifier, while using fatty acid soaps as collector. Acid is used to remove the residual cyanide ion from auriferous pyrite for further conditioning.

Generally, there is a critical pH above which a mineral will not float, e.g. for floating galena with potassium ethyle xanthate this value of pH is 9.5. Different minerals have different critical pH value for the same collector and thus some degree of differentiation between minerals is possible by control of pH alone. The choice of pH is very important and search for the optimum value may lead to a change in the choice of collector, if results are not satisfactory with respect to the cost of collector.

In general, acid media (pH < 7) is used for floating oxide minerals whereas alkaline media (pH > 7) is used to float most sulphide minerals.

d) *Dispersing Agents*

These reagents are used to disperse the masking slimes (extremely fine) from mineral surfaces. These slimes would otherwise adsorb on the surfaces of minerals to be floated and in such cases flotation reagents will be ineffective and difficulties will be faced in flotation. The reagents commonly employed are sodium silicate, Na_2O m SiO_2 with $m = 2.3$–3, starch, and polyphosphate.

e) *Wetting Agents*

These are added to the pulp to increase the wettability of the surfaces not to be floated. Their addition ensures proper wetting of gangue minerals. Common reagents used for this purpose are di-ethyl-hexyl sodium sulphosuccinate (aerosol), sodium silicate, and organic colloids. Wetting effect is obtained due to surface modification, in which hydroxyls or other wetting ions increase their hydration of counter-ions bound in the discontinuity lattice.

f) *Resurfacing Agents*

These reagents resurface or plate selected minerals. The reagents commonly employed are salts of Ba, Ca, Cu, Pb, Zn, and Ag. In the electromotive series (K, Na, Li, Ba, Sr, Ca, Mg, Al, Mn, Zn, Cr, Cd, Fe, Co, Ni, Sn, Pb, H, Cu, As, Bi, Sb, Hg, Ag, Pd, Pt, Au) a metal can displace the lower one from solution. Thus copper salts being lower in series displace zinc from

Table 12.3. Important modifying agents, and their application[8,22-27]

Modifying agent	Chemical formula	Active agent	Applications	Collector used	Amount kg/tonne of ore
1	2	3	4	5	6
1. Activator	$CuSO_4/FeSO_4$	Cu^{++}/Fe^{++}	Quartz	Anion collector	0.045–0.9
	$CuSO_4$	Cu^{++}	Zn, Fe, Co, Ni sulphides	Xanthate	0.045–0.9
	$Pb(CH_3COO)_2$	Pb^{++}	Stibnite	Xanthate	0.045–0.14
	$Pb(CH_3COO)_2$	Pb^{++}	Halite	Fatty acid	0.25–4.5
	$Ca(OH)_2$	Ca^{++}	Silica	Fatty acid	0.22–0.9
	Na_2S	S^{--}	Pb & Cu oxides	Xanthate	0.22–2.2
	Na_2CO_3	CO_3^{--}	Pb, Fe sulphides	Xanthate	0.045–0.2
	Na_2SiO_3	SiO_3	Apatite, fluorite, malachite, cerussite, and silicates.	Isoamyl xanthate	
2. Depressor	Na_2S	S^{--}	Sulphides when used in excessive amount	Xanthate	2.0–8.0
	$NaCN$	CN^-	Cu, Zn, Fe sulphides	Xanthate	0.23–0.45
	Na_2SiO_3	SiO_2	Calcite, quartz, gangue slime	Cationic collector	0.23–0.9
	Na_2CO_3	CO_3^{--}	Gangue	Fatty acid	0.45–4.5
	$K_2Cr_2O_7/K_2CrO_4$	CrO_4^{--}	Lead glance	Xanthate	—
	$Ca(OH)_2$	Ca^{++}	Pyrite	Xanthate	1.5–2.5
	$ZnSO_4$	Zn^{++}	Sphalerite	Xanthate	0.1–0.9
	Air	O^{--}	Pyrrhotite	Xanthate	—
	Na_2SO_3	SO_3^{--}	Sphalerite	Xanthate	0.25–0.9
	$KMnO_4/K_2CrO_4$	MnO_4^-/CrO_4^{--}	Differential depression	Xanthate	

	Reagent / Examples	Ions	Action	Used with collector	Dosage
	Organic colloids (dextrin, starch, lignin, sulphonate)	—	Gangue slimes, especially carbonaceous slimes in gold flotation	Xanthate/Fatty acid	0.05–2.2
	Tannic acid	—	Ca salt in fluorite	Fatty acid	0.05–1.0
3. pH Regulators	Que bracho H_2SO_4, HCl, oxalic acid, etc.	—	To decrease the pH, i.e. for acidic medium	—	—
	NaOH, $Ca(OH)_2$, NH_4OH, Na_2CO_3 Na_2SiO_3	—	To increase the pH, i.e. for alkaline medium	—	—
4. Dispersing agents	Na_2SiO_3, starch, polyphosphate	—	To disperse masking slime from minerals	—	0.1–1.0
5. Wetting agents	Na_2CO_3, organic colloids & diethyl hexyl sodium sulphosuccinate	—	Wetting of gangue minerals	Xanthate	0.2–0.5
6. Resurfacing agent	$CuSO_4$	—	Zn sulphide	—	0.04–0.8
7. Precipitating agents	Compounds containing Ca, Ba, CN, CO_3, PO_4, SO_3 ions	Ca^{++}, Ba^{++}, CN^-, CO_3^{--}, PO_4^{3-}, SO_3^{--}	Removal of interfering ions from the pulp water	Xanthate	0.02–0.4

the sulphide surface. Resurfacing agents promote the adsorption of hydrophobic hydrocarbon groups to preferred lattice points and thus the aerophilic hydrophobic ratio of the surface area is increased resulting into increased floatability.

g) *Precipitating Agents*

These chemicals remove interfering ions from the pulp water and form insoluble or non-ionised compounds. The usual chemicals are those containing Ca^{2+}, Ba^{2+}, CN^-, CO_3^{2-}, PO_4^{3-}, SO_3^{2-} ions.

12.4.3 APPLICATIONS OF MODIFYING AGENTS

One chemical may give more than one effect depending upon the flotation conditions. Table 12.3 summarises the various modifying agents with reference to their effect, applications, source, and amounts used.

12.5 Storage and Handling of Reagents

Flotation reagents should be arranged in such a way that the oldest one is used first with minimum handling. Necessary precautions should be taken to protect the reagents from wetting, heat, cold, smelter fumes, corrosion of containers, and dangerous leakage of poisons and acids. Reagents may be packed in sacks, multiwalled paper bags, drums, barrels, etc., which are not affected by rains. Special security measures are needed for deadly poisonous cyanide. Precautions must also be taken to ensure that water discarded as mill effluent does not carry poisonous and harmful chemicals. Periodic personal check of stored chemicals is needed, particularly of less stable chemicals.

12.6 Factors Affecting the Addition of Reagents

There is no fixed rule to the type and qualities of the reagents used even for a particular mineral. Each plant works out its own most suitable combination of reagents, sequence of addition, concentration and staging time. Quantity of reagents added mainly depends on the assay value of the reacting mineral and the fineness of the ground ore. When the feed consists of excessive fines, (particularly of earths and oxidised ores), a dry froth may be produced. In such cases frothing agent should be reduced, otherwise there is a risk of high tailing loss and poor grade of concentrate. Type of flotation cell also affects the consumption of reagents. In flotation of a given ore, a specific amount of each reagent is used for maximum efficiency. This specific amount may vary with the method of air introduction, and the size and rate of bubble formation. Any improvement towards the air-bubbles will reduce the consumption of reagents.

CHAPTER 13

Flotation Practice and Machines

13.1 Introduction

In plant practice. a clear-cut procedure is laid down for flotation of particular ore, keeping in view the overall economy of processing. Various factors affecting the steady control of flotation process are as follows:

a) Quality of ore (its texture, interlocking of minerals, variation in composition).

b) Process water (seasonal variation, progressive fouling on recirculation, etc.).

c) Reagents used (nature, reaction rate, appropriate amounts, etc.).

d) Type of flotation machines (possible adjustment for capacity, mechanical, pneumatic, etc.).

e) Quality of pulp (solid–liquid ratio, size range of particles, slime content, pH stability, etc.).

f) Aeration rate.

g) Grades of various products.

h) Amount of recirculating middlings.

i) Sampling accuracy, assay speed, monitoring, automatic control, etc., for required adjustments.

These factors may also show some interdependence. In order to carry out the process efficiently, periodic check of above factors is necessary.

The first step in flotation is the grinding of ore to the optimum size giving maximum liberation, making it easily flotable with minimum incurred cost. The optimum mesh of grind is mainly dependent on the crystal structure, interlocking of minerals, and properties of associated minerals. Primary metal sulphides are usually stable and float readily, whereas secondary metal sulphides occurring as coatings on primaries or penetration along cleavages are less reliable (e.g. when covellite coats pyrite crystal). Since flotation is based on the surface properties, completely or partially coated particle of foreign mineral may float, acting as true concentrate. Oxidised ores are usually soft resulting into overgrinding, increased reagent consumption, and production of slime, and these all affect the flotation process adversely. The overall flotation system is a complex combination of variables outline in Figure 13.1.

13.2 Flotation Process

The basic steps involved in flotation process are (a) pulp preparation, (b) conditioning, and (c) aeration to obtain the froth.

13.2.1 PULP PREPARATION

The most important factor in preparation of flotation of pulp is correct wet grinding, i.e. grinding to suitably flotable and liberated 'size. Mineral particles as coarse as 28 mesh (600 microns) can be floated (light and easily floatable minerals such as coal), but for coarse particles, thick pulp is needed to produce the stable froth, whereas for fine particles, a thin pulp is preferred. Therefore, the milling operations should be controlled precisely with the help of analysis and micro-observation of each screened fraction in a representative tailing and concentrate samples.

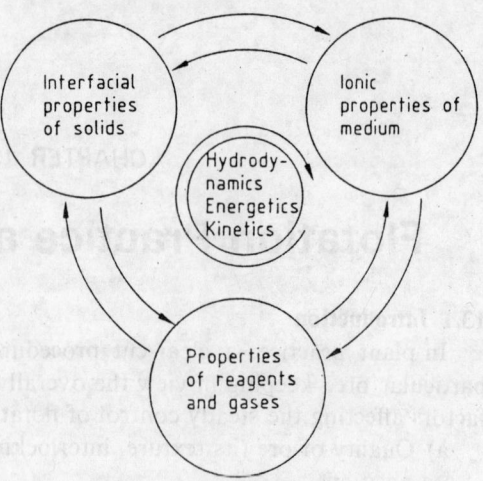

Fig. 13.1. System variables that affect the flotation process.

If ore is crushed too coarse, adequate liberation may not result and if ground too fine, losses may occur due to overgrinding. The maximum floatable size of a fully liberated mineral is dependent mainly on its density, aerophilic attraction, and shape. In general, minerals are ground below about 300 microns, since particles coarser than that are difficult to be carried out of the pulp by the air-bubbles. Particles of less than 30 microns size should be held at a minimum.

Coal being light and easily floatable, can be floated easily with as coarse as 10 mesh (1.2 mm) size. In case of metal sulphides, the upper limit is 48–65 mesh (300–200 microns) and to metallic gold 100–150 mesh (150–100 microns). With increasing density and decreasing collection quality, the size needed to be floated decreases.

The gangue particle size in flotation is limited only by liberation size and smooth transport through the plant after treatment, since larger particles having greater tendency to settle cause problems. Difficulties in flotation of very fine particles (slime) are mainly due to the disproportion between their mass and the surface area (fine particles behave as their surface tension is very high). Consequently, fine particles react prematurely becoming oxidised, flocculated (heavy to float) or slimed before they reach the conditioning section. These conditions prevent the collector from acting on the particles satisfactorily. Further, the small particles (4–5 microns) tend to overflow from the froth column with concentrate irrespective of their composition. Therefore, sometimes the pulp may be classified to remove at least the coarser gangue particles. This will reduce the volume of the pulp and thereby the amount of reagents needed.

The ore can be processed economically by coarse grinding with the recovery of valuables followed by fine grinding to concentrate to float it further as following:

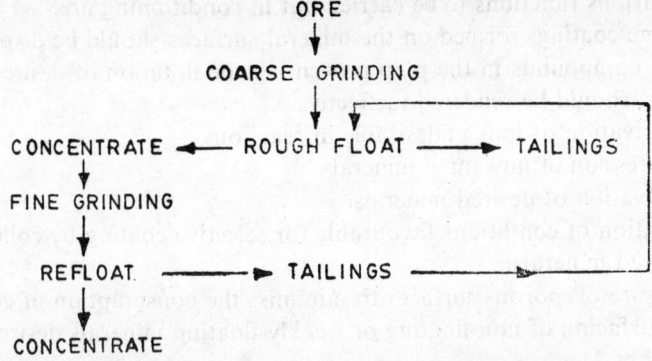

The operations of grinding and reconcentration can be repeated as many times as required for maximum recovery economically.

The rate of travel of correctly liberated particles through the circuit of the flotation must be adjusted such that a correct dwelling time is obtained in each conditioning stage as well as in the flotation lines.

Grinding to suitable mesh size can be obtained by a suitable combination of crushing and grinding mills as discussed in Ch. 5 and 6.

Density of pulp, i.e. solid–liquid ratio, plays an important role in flotation and this should be fairly high. The pulp densities employed in flotation greatly depend on mineral density and particle size. Up to a certain point increasing pulp density (increasing percentage of solids) will require less reagent, less stirring power to keep the particles suspended, and shorter conditioning time. A thick pulp is preferred to get a stabilised froth. Therefore, the working principle in flotation is to carry out the roughing flotation in thick pulp and cleaning operation in dilute pulp, in order to obtain a high-grade concentrate. The pulp density of roughing stage is also affected by its slime content, i.e. more the slime content, more dilute must be the pulp at its roughing stage. Further, the density of pulp also depends on type of flotation machine, e.g. pneumatic machines require dilute pulps.

Heavy sulphide minerals such as chalcopyrite, galena, etc., are roughed at 30–50 per cent solids and finely disseminated gold pyrite would need pulp density down to 15 per cent solids. Whereas, cleaning treatment is carried out at 30 to 8 per cent solids in appropriate stages.

13.2.2 CONDITIONING

This is an operation in which pulp is continuously contacted for an optimum time with appropriate reagents at different stages to provide the conditioned pulp giving the best results in flotation. The conditioning efficiency depends on the following factors:

a) Complete mixing and dispersion of all reagents throughout the pulp uniformly.

b) Time of contact needed to produce the desired reactions.

c) Number of contacts made between molecules of reagent and concerned mineral particles at each conditioning stage.

The various functions to be carried out in conditioning are:

i) Slime coatings formed on the mineral surfaces should be dispersed.

ii) The compounds in the pulp preventing the flotation of desired minerals, should be rendered ineffective.

iii) Passivation of ions undesirable in reactions.

iv) Depression of unwanted minerals.

v) Activation of desired minerals.

vi) Creation of conditions favourable for selective coating by collector on desired minerals.

vii) Closure of porous surfaces to minimise the consumption of collector.

viii) Resurfacing of non-floating or weakly-floating minerals desired in the froth.

ix) Finally the stabilisation of correctly mineral loaded bubbles at the top of the froth column by the addition of frother.

The conditioning steps, concentration and amounts of reagents, and time of reaction are worked out in laboratory and applied in the plant practice.

The first step in conditioning is to depress the slimes since they sorb expensive reagents, contaminate froths and stabilise them, and give problem in settling of pulps after treatment. To handle the problem of slimes, the preventive method is preferred to chemical method. If it is possible to remove slimes before fine grinding without undue loss of values, the problem can be overcome by keeping the grinding to a minimum. Further, if collector is added in the grinding circuit, it will be sorbed before particles become slime coated.

The next important step in conditioning is the addition of required amounts of reagents in suitable concentrations. The methods employed for addtion of reagents should ensure their thorough and uniform mixing in the pulp. Readily soluble reagents such as soda ash, disperse quickly into the pulp and thus can be added with the feed as a solid or dispersed in aqueous solution.

In treating sulphide ores, lime can be added as a slurry to the grinding circuit to produce the required pH in the pulp. In case of ores tending to oxidise in their way from the mine, addition of some dry lime to ore before its entry to the bins may be beneficial (this can also be used for the first stage of pH control, where final pH control is made by adding slurried lime).

Relatively insoluble oils should be thoroughly mixed, otherwise a small amount of mineral will attract an undue proportion. Usually it is found feasible to emulsify the oil by mixing them with hot solutions of sodium carbonate, and then emulsions can be easily mixed uniformly.

When there is a chance of reaction of reagents with each other, instead of with minerals, it becomes essential to use the reagent in successive steps. The collector is used at the end. For example, activation of sphalerite by $CuSO_4$ should be completed before the pulp is transferred to the xanthate conditioning tank, otherwise stoichiometric reaction will take place between copper and xanthate, resulting into killing of reagents. As a result, sphalerite will remain unconditioned. Copper sulphate should not be allowed to plate out on the steel structure of machines and piping. Similarly, sodium silicate should not be used in a pulp made strongly alkaline with lime, as calcium silicate will be precipitated in such cases.

Addition of collector in stages, i.e. addition to starvation is usually helpful in making a high-grade concentrate, due to improved selectivity of collector. When the valuable mineral is present both as its sulphide and oxide, a strict control is needed in using soluble sulphide for conditioning, since slight excess of this reagent will depress both minerals. Therefore, in such cases the sulphide mineral should be floated first before conditioning with soluble sulphide, which is followed by flotation of oxide mineral after conditioning with soluble sulphide.

When fatty acids (oleic acid) are employed, freezing of reagent may occur in cool pulps. This problem can be handled either by warming the pulp or emulsifying the reagent. Some modifications of oleic acid structure have been developed by inducing the partial sulphonation, which can be used in a fairly stable dispersed state.

The action of collector mainly depends on the pH of the pulp, but further specific conditions may be required before the addition of collector, in order to increase the difference between aerophilic and hydrophobic mineral surfaces. This is particularly of great significance in differential flotation, where the desired products are floated successively. For flotation of each series of valuables, a further conditioning is required.

The reactions involved in conditioning require only small amounts of reagents, since surface areas to be rendered water repellent are only a small fraction of the particle area. Since the quantities of reagents added are as low as 0.05 kg/tonne of pulp, the addition should be made with high precision. The conditioning period provided should be sufficient for the reaction to be almost complete. Sometimes warming of pulp may be needed for reaction, but usually the mill temperature is adequate. After making the addition of necessary reagents, the pulp should be agitated (conditioned) in a special tank, known as contact vats or conditioning tank to distribute them thoroughly and uniformly. Except frother, all the reagents may be added during conditioning. Frother is usually added after the conditioner and just before the entry of pulp to the first cell. Sometimes reagents may be added in the ball mill just preceding the flotation cells, particularly when freshly fractured surfaces give better collector reaction.

After conditioning is over, i.e. the pulp is first activated and dosed with a collector, specific to activated mineral to be floated. The pulp is ready for aeration after making the addition of frother.

13.2.3. AERATION

Finally the conditioned pulp is agitated in flotation machines of suitable design and size. The air admitted forms many bubbles, which rise through the cell and collect the mineral particles. The mineralised bubbles collect as a froth on the top. The froth can be discharged into a launder with the help of scraper and water spray. In plant practice, a series of flotation cells (usually 10–12 in number) is employed, where the pulp is continuously pumped from one into the next cell for repeated treatments. The floats in successive treatment become poorer and are returned as a middling for further treatment. The pulp discharging from the last cell is barren and can be discarded.

In flotation, the extent of aeration and size of air-bubbles formed are of extreme importance. The extent of aeration depends on size, number, and even distribution of the air-bubbles in the pulp. Pulp aeration determines the flotation speed (flotation machine output), grades of products, and the reagent consumption. Best flotation conditions are obtained with an optimum degree of pulp aeration. In mechanical type machine, the usual bubble size is 0.8–1.0 mm in diameter, for which an amount of optimum frother is needed. Whereas, in pneumatic type cells, the bubble size varies from 2.5 to 4 mm in diameter.

The amount of air-bubbles in the pulp is directly proportional to volume of air entering the flotation cell and inversely proportional to the speed of air passing through the pulp, i.e.

$$V = \frac{100\, t\, Q_a}{Q_b} \text{ per cent} \qquad (13.1)$$

where V = total volume of bubbles in a unit of pulp volume as percentage of the total effective machine capacity,

t = average time in seconds spent by bubbles in the pulp,

Q_a = air flow in m³/sec and

Q_b = effective capacity of the chamber in m³.

Therefore, it is essential to introduce a large volume of air into the pulp for high output of flotation machine. This air must remain in the pulp for considerably long time, which can be achieved by effective air dispersion, sufficient pulp layer thickness, and turbulent motion. The formation of small bubbles is intensified by the addition of frother.

13.3 Flotation Machines

Flotation machines or cells are the devices used to carry out the flotation process. The important features of a flotation machine are as the following:

a) Aeration of the pulp should be as complete as possible, without letting too large bubbles or bursting of air create.

b) Solids should be distributed evenly throughout the pulp without settling.

c) The machine should work continuously.

d) Removal of mineralised froth (float) and tailings (sink) by separate channels.

e) Avoidance of pulp entry to discharge channel without being processed.

f) Machine should have a zone where quiet blanket of mineralised froth is formed and gangue particles should drop back from it into the pulp.

g) Pulp level and height of froth column in the flotation machine should be controllable.

h) Easy re-start of machine after mechanical failure should be possible without causing the standing-up of mechanical parts.

i) Power and mill space should be economically utilised.

j) Machine should not have odd corners to prevent the accumulation of undesired materials such as wood debris, lime scale, etc.

k) Provision for periodic discharge of coarse sand accumulated at the bottom.

Since the flotation process has come into practice, a variety of flotation machines have been designed. Three types of machines (mechanical, pneumatic, and reduced pressure) were developed almost simultaneously, but the extent of their development varied greatly. Mechanical type machines received maximum attention for developmental work, pneumatic type received comparatively much less attention, whereas reduced pressure type hardly had any development in its original form.

In case of mechanical type of machines, maximum efforts have been made to develop most suitable impeller designs, larger size of chambers and to have controlled flow of pulp. Various developments in flotation machines can be summarised as follows:

a) In the beginning, simple mixers were used as impeller. It was followed by fitting the discs above the impeller. Afterwards, radial impellers, mechanical aerators and other devices have been developed. The most successful design is considered to be the squirrel cage rotor and stator with axial type impeller.

b) In mechanical machines, in the beginning separate and distinct mixers and flotation chambers were employed. It was followed by connecting the chambers but separated by grids.

c) Another important development has been towards the use of large flotation chambers. Several cubic metres capacity of chambers is the characteristic of modern flotation machines, which is very useful in high output concentration plants.

d) Recent trends are to combine the chamber and the trough—flow principle in the same unit with automatic/semi-automatic pulp level control.

e) In case of pneumatic machines, increased depth of chamber has been the main development.

13.3.1 GENERAL ARRANGEMENT OF FLOTATION CELLS

Flotation cells are arranged in series in such a way that each cell in a *bank* (set of cells) receives the tailings from the one preceding it. The height of the froth column is fixed for individual cell by adjusting the height of the tailings overflow weir, since the overflow lip for the froth is fixed (difference between the overflow lip and the weir level determine the height of the frothing column). The whole system of cells consists of three sections, i.e. rougher, scavenger and cleaner. The fresh feed enters the rougher section of a typical tank. The froth column is maintained highest at the head of the section and then lessened from cell to cell by progressive raising of weir hight. The scavenger cells are used to treat the product containing little mineral, with which froth is to be stabilised and therefore, their weirs are raised till the pulp becomes almost spilling at the end of the scavenger section. In the cleaner section the pulp is thinned and the weirs are kept low in order to maintain a thick layer of froth and to obtain the maximum possible cleaning action.

As an alternative to these individual tanks, a series of long troughs can be used, each of which contains an appropriate number of aerating and agitating devices. Each trough is controlled by a single weir. Recent practice favours the latter system.

13.3.2 CLASSIFICATION OF FLOTATION MACHINES

Flotation machines or cells can be classified mainly on two basis, i.e. (a) based on the way of pulp flow, and (b) based on the method of pulp aeration. The latter is simplest and most important, since aeration is the basic feature and essential requirement in flotation machines. General classification of flotation machines is given in Table 13.1.

a) *Flotation Machines Based on Pulp Flow*

i) *Trough type flotation machines*: These are elongated chambers (as shown in Fig. 13.2 (a)), in which flotation pulp enters the lower part of the charging end and the tailings leave the chamber at the opposite end. The froth flows over the trough sides into channels along its entire length. The pulp level remains the same throughout the cell. In these cells pulp is aerated by blowing air into it.

ii) *Through-flow (common level) machines*: These are divided into separate compartments by a series of transverse boards [Fig. 13.2 (b)] which do not extend to the bottom of the trough. In general, the upper parts of boards remain slightly below the pulp level. Therefore, the compartments of these machines are open to each other at the top as well as at the bottom. The pulp level remains the same throughout the machine. Each compartment is provided with the aeration provision which may be carried out either by impeller or blowing the air into the pulp.

iii) *Chamber type machines*: These machines consist of separate chambers [Fig. 13.2 (c)]. However, in this case also the pulp passes from one chamber to another. A special device is installed between the chambers to regulate the flow of pulp, and thus it is possible to maintain different pulp levels in different chambers having different flotation conditions. Aeration may be carried out by impellers, or mixing the pulp with air.

b) *Flotation Machines Based on the Method of Pulp Aeration*

i) *Mechanical type (self-aerating and mechanically agitated)*: This type

Table 13.1. Classification of flotation machines[26]

Basis of classification	Type	Method of pulp aeration	Design features
A. Based on the way of pulp flow	I. Trough (box) type	Same as in pneumatic cells	Single elongated chamber
	II. Through-flow (common level) type	Same as in mechanical type	Divided into separate compartments
	III. Chamber type	Same as in mechanical type	Consists of separate chambers
B. Based on method of aeration	I. Mechanical cells	By mixing the pulp with air, using impellers of different designs	1. With impeller having radial blades 2. With special impellers 3. With squirrel cage
	II. Pneumatic cells	By blowing of air into the pulp	1. With porous mats (fixed or movable) 2. Air-lift
	III. Pressure variation cells	By pressure difference to make gases evolve from pulp	1. Vacuum (pressure above the pulp is less than atmospheric) 2. Compressor (pulp is discharged under pressure)
	IV. Cascade (air entrained) cells	Air is entrained during the fall of splashing pulp into the cell	Simple home made devices

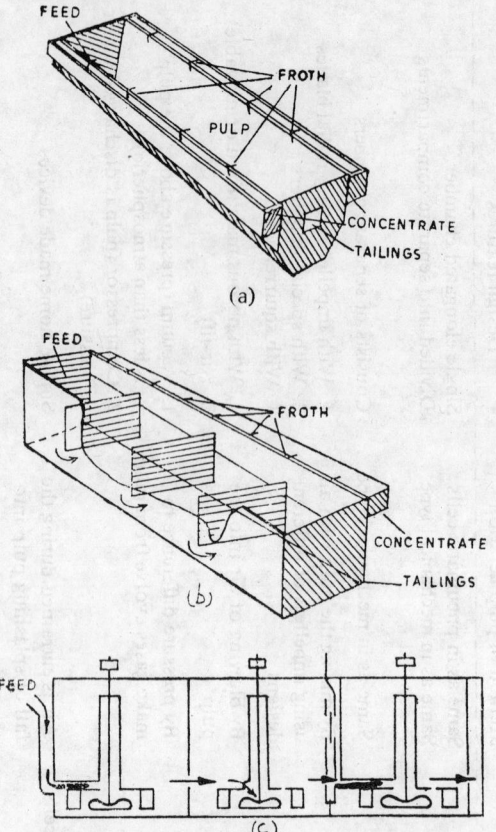

Fig. 13.2 . Flotation machines based on pulp flow:
(a) trough type, (b) through flow (common level), and
(c) chamber type.

of machine mainly consists of a chamber (square, rectangular or round) and a radial impeller, mounted with a vetrical drive shaft. The impeller in motion maintains the solids in suspension and also draws in the air. The aerated pulp is pushed out by the centrifugal force of the impeller and the mineralised froth floats at the top, from where it is removed. This type of flotation machine possesses the following features:

1) It consists of a mixing chamber where the pulp is fed.
2) An impeller mounted in the mixing chamber to give basic aeration of the pulp.
3) A flotation tank isolated from the mixing chamber in which bubble-pulp mixture is injected from mixing chamber.
4) The impeller provided acts only as a simple mixer without affecting the flow of pulp streams.
5) The volume of aerated pulp is small compared to the total pulp volume.

6) Production of air-bubbles occurs due to mechanical action on the air, evolution of gases from solution and reduction in pressure in the flotation tank.

7) Froth product is discharged from the top lip of the machine, whereas chamber product (tailings) is discharged through an opening at the bottom.

8) Capacity of the flotation tank is large enough to check the turbulent motion of the pulp having a large surface area.

The category includes various proprietory machines such as Denver, Fagergren, Humboldt, Massco, Knapp and Bates, which differ from each other in some specific variations having the same basic principles. The first two types are more common in industries, and thus these are discussed below in detail.

Denver flotation cell: A standard Denver cell is shown in its cross-section in Fig. 13.3. It consists mainly of a square cross-sectioned tank and an impeller. The impeller agitates the pulp violently and effects some conditioning. The rotation of impeller causes the suction of air (pressure air or atmospheric air) downward through a stand pipe surrounding the impeller shaft.

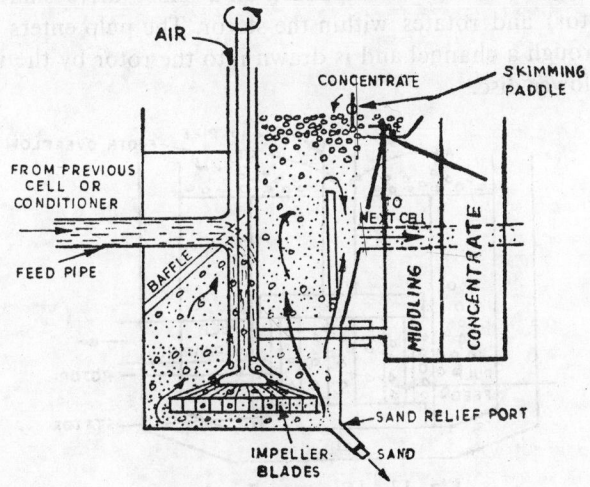

Fig. 13.3. Standard Denver cell.

The air is drawn in to the vortex near the hub of the impeller where a partial vacuum allows air to be drawn down to the impeller blades, where the air gets mixed with the pulp. In due course of processing, the tailings from the cell will pass under a baffle through a tailing pipe or over a weir to the next cell in line. Sand, which is too coarse to rise to this tailing discharge weir, can be removed from the cell by means of the sand-relief ports provided at the bottom of the cell.

A pipe is provided at the side of the cell which may be connected to the middlings return system. Adjustable baffles are provided to push the rising bubbles forward towards the skimming paddle which removes a froth layer at desired rate from the cell. The horizontal portion of the impeller near the bottom of the cell acts as a centrifugal pump giving a pushing force to the pulp. The impeller is driven by a vertical shaft from an overhead motor at a peripheral speed of 400–600 m/min. The feed (from conditioning section, middlings or from previous cell) is introduced near the centre of the impeller. Stationary baffles are provided to check the swirl of the outflowing pulp.

Fagergren flotation cell (*with squirrel cage rotor and stator*): In most instances it is used in the form of a chamber machine, but in some cases it is also used as a through flow machine (long rectangular trough divided into sections). The chamber may be round or square in shape made of steel. A simplified sketch of this machine is shown in Fig. 13.4. For aeration and agitation this cell has a rotor and a stator assembly in place of impeller and hood plate (as in Denver cell). The stator consists of a series of cylindrical spacers (vertical rods) sleeved with rubber, which are mounted between two rings rigidly fastened to the tank. The construction of rotor is similar to that of stator except that its upper and lower bladed impellers are mounted within the rings. The rotor is suspended on a short drive shaft (driven by electric motor) and rotates within the stator. The pulp enters directly into the tank through a channel and is drawn into the rotor by the impeller blades on the lower disc.

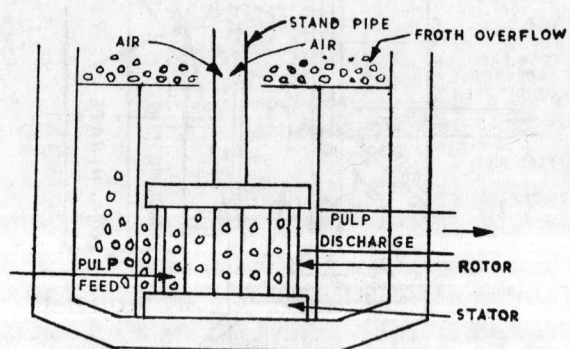

Fig. 13.4. Fagergen flotation cell.

The rapid pulp circulation (displacement) within the chamber creates a partial vacuum which causes the atmospheric air to be drawn into the rotor through the stand pipe. The rotatory motion of the rotor cage ejects the pulp–air mixture into the chamber. The bubbles dispersed in the pulp ascend along with attached mineral particles, forming a flotation froth which overflows into the annular trough surrounding the chamber or removed by a rotating skimmer. The chamber product (tailings) is withdrawn into the tailing tank through a slit at the bottom of the chamber. When partitioned chamber is used, the feed enters below the first partition. Tailings pass over partition from one compartment to the next. The froth level is adjusted at the last tailing weir, which has a slit (sand gate) at the bottom for discharge of accumulated coarse material.

The chambers are made usually in sizes of 60–150 cm diameter. Pulp level and froth weir height in the chamber can be controlled independently. Therefore, the thickness of froth layer and the speed of its removal can be altered independently of the pulp level. The special features of Fagergren cell may be summarised as follows:

a) Though power required to grind air into the cell is much higher than for Denver cell of equal capacity, it is less than that for many other types. Usual power consumption is about 2 KWH/tonne of ore processed.

b) Dispersion of air is highly effective for bubble formation, which helps in floating reluctant and slightly aerophilic particles.

c) Good froth removal system which makes the flotation of relatively large particles.

d) There is no self-positive pull from other cells to the impeller, and thus for moving the pulp from one point to another, independent pump system is needed in the absence of required gravity flow.

Mechanical machines fed with blower and agitation mechanism: The next development in the mechanical type machines was to introduce air at the compartment bottom through a hollow pipe from a blower. These machines are also called *subaeration* machines. These machines in construction are identical to the mechanical type having self-aeration as described above. These machines essentially consist of a radial impeller working only as an agitating device and a chamber for mixing and flotation. In this case the pulp (either freshly fed in or from the preceding chamber) enters below the bulb of the impeller to the mixer and the air is blown in at low pressure (0.05–0.2 kg/cm²) from a blower or similar device, and the air is ground with the pulp. Froth is skimmed off to the trough by the rakes/paddles, whereas tailings are allowed to flow to the next cell through a weir. The chamber product (sand) is removed through a pipe.

The working machine consists of a long trough divided into several compartments and the pulp moves from feed end to the discharge. In each compartment an impeller is provided which rotates inside a baffle system. Impeller can be rotated with controlled peripheral speed varying from 250 to 500 m/min. Control of speed and volume of air permit the operational control of the froth column to provide its flow counter-current to the pulp. This type of machine has the following special features:

a) These cells produce copious froths with reasonable power consumption. Thus the ores of poor floatability requiring a large volume of froth (to help the mildly aerophilic particles to overflow) can be treated efficiently by this type of machine.

b) Bulk of the air is supplied under pressure.

i) *Cascade type machines (air entrained)*: In this type of machines air gets entrained during the fall of splashing pulp into the cell. This entrained air effects the flotation. The use of cascade type machine is mainly confined to simple home-made devices used at the tailings end of a process to cascade the finished pulp running down to waste.

ii) Pneumatic type flotation machines (*air is blown to aerate as well as agitate the pulp*): This is the original form of flotation cell. In general, a pneumatic cell consists of a long trough and provision for admitting the compressed air from below the cell. The feed enters at one end and the tailings are removed from the other. In pneumatic cells the small bubbles of air passing upward through the pulp induce a mild agitation and become adsorbed or attached to the mineral particles to be floated. The froth maintained to a depth of 20–30 cm in the trough overflows into the concentrate launder surrounding the cell. Some varieties of pneumatic cells are described below:

Pneumatic cells with fixed porous bottom: This is the oldest type of flotation machine designed in 1914 and widely used earlier due to its simplicity. The cell consists of a long rectangular tank with inclined bottom. The bottom of the cell is covered by a porous mat divided into several sections through which the compressed air is admitted from a blower. The froth flows over the longitudinal sides of the tank into troughs. The chamber product (coarse tailings) is discharged from the bottom of the tank. It has become obsolete now due to two specific disadvantages, i.e. (a) pores soon become clogged with fine slime particles, and (b) there is no pulp mixing, restricting the flotation of coarse material due to settling action of coarse particles.

Pneumatic cell with a porous rotor: This was a development over the previous one. Up to about 1925, this was common in use. It consists of an elongated tank having a narrow bottom section, and rotating hollow shaft fitted with a mesh cylinder covered with porous fabric or perforated rubber sheeting. Compressed air is admitted into this cylinder from a blower. The specific features of this type of machine are:

a) The aerator rotates slowly and is washed continuously from inside to prevent blockage of the pores and to give uniform and good pulp aeration.

b) Larger material can be floated in this machine due to very weak pulp mixing action in the lower part of the cell.

However, this was associated with certain disadvantages such as frequent replacement required of the rotor's porous cover and extremely low level of pulp mixing.

Shallow pneumatic type air-lift cell: It was developed in 1921, and since then many variations have been worked out differing mainly in design aspects. A typical shallow air-lift pneumatic cell (Fig. 13.5) consists of a shallow, long, pyramid-shaped tank. The baffles run the whole length of the tank but do not extend to the bottom. The space between these baffles is called the aeration space. Air is blown down to vertical pipes from a distributing header (receiver) which is above the aeration space. This air agitates the pulp.

In this type of cell, the pulp enters the lower part of the tank. Air rising at a high speed through the pulp breaks down into large bubbles and draws the pulp upwards. A voluminous mobile froth formed by mineralised air bubbles overflows to the side troughs of the machine. The pulp which has

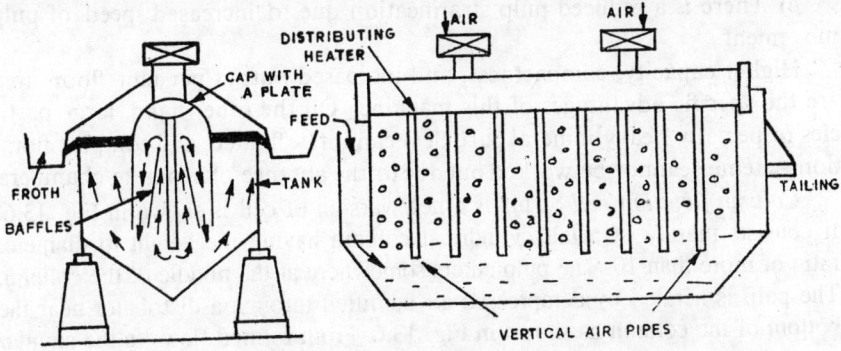

Fig. 13.5 Typical shallow pneumatic air-lift machine.

risen to a top baffle under the pumping action of air and is practically bubble-free, drops back and is again lifted a little lower down the trough. This causes the continuous circulation of pulp in the vertical plane, and slow movement towards the discharge end. Thus, pulp moves slowly in spirals. Tailings are removed by a similar device as used in mechanical flotation cell.

Some specific features of shallow air-lift pneumatic cell are as follows:

a) The cell is simple in design.

b) Air is blown under pressure which aerates and mixes the pulp.

c) Air-lift principle gives high rate of pulp circulation.

d) The bubbles are produced by a double change in its direction of movement.

e) Low rate of pulp mixing which reduces the extent of air breaking.

However, the conditions suitable for the flotation of small particles and large particles are not obtained in this cell. This type of machine is mainly used for flotation of readily floatable and cheap minerals, such as coal, native sulphur, talc, etc.

Deep pneumatic air-lift cell: In this type of machines the same working principle is used as in case of shallow pneumatic air-lift cells. This differs mainly in depth which is three to four times (2-4 m) of the shallow type. In this case, the pulp moves much more quickly in the aeration section compared to shallow air-lift type due to the relatively great depth and the difference in hydrostatic pressure between the columns of aerated and non-aerated pulp. As a result, there is a greater degree of pulp turbulence obtained in the bubble formation zones. The bubbles formed in deep machines are much smaller compared to the shallow type (size of bubbles is almost equal to that produced by mechanical machines). These machines vary in depth from 2.0 to 4 m. The total air pressure is somewhat higher (0.25–0.3 kg/cm^2) than in the shallow type. The principal feutures of deep air-lift type are as follows:

a) Increased depth (several times deep as the ordinary shallow type) gives a great improvement in pulp aeration.

b) Special valves are employed to prevent clogging of the cell caused by entering the pulp to the air nozzles.

c) There is a reduced pulp stratification due to increased speed of pulp movement.

Higher capacity, compactness, and increased output per unit floor area are the specific advantages of this machine. On the other hand, large particles (especially heavy mineral particles) cannot be floated, and complex flotation schemes cannot be worked out due to the absence of separate chambers.

Column flotation cell : This is a new version of cell is shown in Fig. 13.6. It consists mainly of a tall cylindrical column having the height to diameter ratio of more than 10. The pulp enters somewhere at the middle of the column. The pulp is aerated by compressed air admitted through a distributor near the bottom of the column as shown in Fig. 13.6. Froth formed flows to the annular trough surrounding the column at the top, from where, it is removed through a launder under gravity. The tailings are removed from the bottom through a gate valve. This is the simplest form of pneumatic type flotation machine. This needs minimum floor space and requires minimum maintenance. However, this machine can be used for easily floatable minerals, where low pulp mixing is required. Column flotation has been successfully used for concentration of graphite and sulphides. It can be used to treat.

Column flotation offers the following advantages :

1. reduction of gangue entrainment in froth,

2. increase of bubble residence times in the pulp,

3. improved selectivity,

4. energy economy per unit weight of values recovered, and

5. improved control and instrumentation

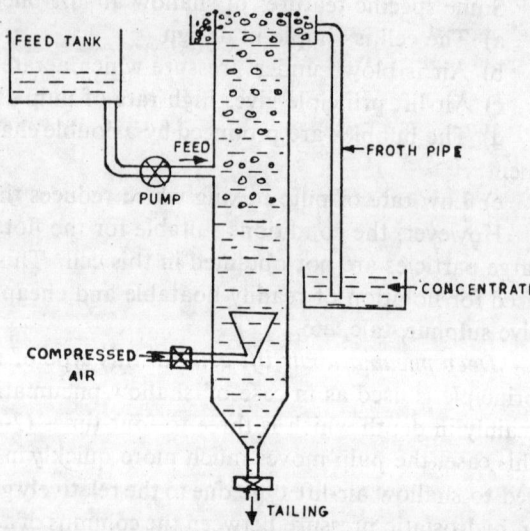

Fig. 13.6. A simplified sketch of column flotation cell.

iii) Flotation machines using reduced pressure: Flotation machines using reduced pressure (gas evolved from solution) above the pulp were developed quite early (in 1906) and were in use in the past for coal concentration at a number of places. The machine consists of a chamber in the shape of two truncated cones having a cylindrical section on the top, which is encircled by a trough. The cylinder and trough are hermetically sealed by a cap from which air is extracted by a vacuum pump (500–650 mm Hg). The pulp is first mixed with air (to increase the concentration of dissolved gases) and fed into the machine chamber from the sump through a pipe. On application of vacuum, the dissolved gases evolve from the pulp forming bubbles on the mineral particles. The mineral bearing froth fills the cylindrical vessel and then overflows into the annular trough. The chamber product is removed through a sand pipe having a gate valve at its bottom.

13.3.3 Selection of Flotation Machines

Selection and assessment of flotation machines are made by comparing their metallurgical performance, plant output, power consumption, reagents consumption, and ease of operation, for the treatment of same pulp by different machines. The metallurgical factors are linked with the properties and composition of the ore to be concentrated and reagents used. Further, the choice of flotation machine depends on the following factors:

a) The Size Classification of the Mineral to be Floated

Coarse particles require a shallow machine, which provides high rate of pulp mixing in the lower part of the machine, turbulence free condition in the upper part of the machine and good aeration. Whereas, fine slimes require deep flotation machine which provides lower rate of pulp mixing, and pulp aeration with extremely small bubbles.

b) Density of Ore

Coarse ground minerals of high density have faster rate of settling and thus can be treated more effectively with shallower machines where a higher rate of pulp mixing is achieved.

c) Water Repellent Properties of Minerals

Highly water-repellent minerals can be floated easily in pneumatic cells having lower rate of pulp mixing.

d) Volume of Froth Formed

Flotation of minerals providing voluminous froth requires machines with large froth formation area compared to pulp capacity along with special methods of quick froth removal.

e) Tendency of Slime Formation During Mixing

If the minerals readily disintegrate to fine slimes during mixing, flotation will be adversely affected. In such cases, the machines with minimum rate of pulp mixing (pneumatic type) can be employed.

f) Flotation Scheme

When complex flotation schemes are used with multiple recirculation of intermediate products and fractional reagent feeds, chamber type machines are preferred. In this the products are sucked in by the impeller.

g) *Dissolved Gases*

If the flotation results are better with dissolved gases, the pulp needs intensive aeration with very small bubbles and slow rate of flotation should be adopted where oxidation of minerals gives a harmful effect.

h) *Effect of Reagents on Machine Parts*

When lime is used, clogging of porous aerator may be caused. Rubber aerators may be damaged by the use of kerosene, and similar products. The use of relatively insoluble reagents (non-polar oils, etc.) forming emulsions in the pulp requires the machines giving higher rate of pulp mixing.

i) *Flotation Machine Output and Power Consumption*

These also have an important bearing on selection of flotation machines, particulary in concentration of relatively cheap minerals. Deep air-lift machines are most economical with respect to power consumption.

In addition to the above factors, following conditions also determine the efficiency of flotation and thus indirectly affect the choice of flotation machine:

    i) Availability of a quiet place for forming and cleaning the froth column.

    ii) Easy and cheap maintenance.

    iii) Low reagent consumption.

    iv) Flotation machine should be easily fitted into the circuit.

    v) Positive pull should be created to transport the pulp from the previous cell.

    vi) Continuity of action.

    vii) Effect of variation in quality and output on operation.

    viii) Long working life of individual components of machine.

    ix) Less floor space and building required.

    x) Possibilities for automatic control.

13.3.4 FACTORS AFFECTING THE SIZE OF FLOTATION MACHINE

The important factors affecting the size of flotation machine to be used, may be summarised as follows:

    a) Total flotation time required for a particular ore.

    b) Space and handling arrangements required.

    c) Capacity of the plant.

In terms of capacity, a cell has four dimensions, i.e., length, breadth, height and dwelling time (rate of throughput). Dwelling time can be partly distributed to the conditioning section, and a part can be varied by varying the number of cells (depends on aeration) used in line. The length and breadth of an individual cell depends on the agitating and searching power generated of the impeller, which is a function of the square of its diameter at a given rpm, whereas power consumption in unaerated pulp is a linear function of this diameter. In practice, diameter of impeller has to be limited, due to high rate of wear caused when peripheral speed exceeds some limit (55 m/ min). For increasing overall plan area of cell, troughs are preferred in place of square cells. Though height can also increase the volume of pulp held, this situation will result in an inefficient mechanical design in terms of pumping action. Mechanical cells with shallowest volume of pulp are favoured in terms of power consumption. Similarly, long troughs may not be preferred to handle large tonnage of ore, since most of the separating work is done in the first 1-2 m and the remaining length of the trough is to insure against by-passing of the values which can be better controlled in a bank of individual square cells than in a long tank. Thus, for a specific case of plant design, a compromise is made for different elements. Further, in case of individual square cells, adjustment for new reagents developed is more convenient.

13.3.5 DETERMINATION OF SIZE AND NUMBER OF CELLS, AMOUNT OF PULP REQUIRED, AND ITS DENSITIY

The basic criteria required for calculating the size and number of cells is the optimum flotation time for each operation, which is determined in each individual case by carrying out the specific experiments under laboratory and semi-commercial conditions.

The number of flotation chambers (N) can be calculated by the following general formula:

$$N = \frac{V_a\, t}{V_b\, K} \tag{13.2}$$

where, V_a = quantity of pulp entering a certain flotation operation (in m³/ min)
t = time required for the given flotation (in minutes)
V_b = chamber capacity (in m³)
and K = ratio of pulp volume in the chamber to the chamber's geometric volume (usually varies between 0.65 and 0.75).

For trough type, the following formula may be used:

$$L = \frac{V_a\, t}{A \cdot K} \tag{13.3}$$

where, L = length of trough (in m)
A = cross-sectional area occupied by the pulp (in m²).
t, K, and V_a are same as in Eq. (13.2).
However, the maximum length of the trough is limited to 10 m.

If the amount of ore required for flotation (Q tonnes/day), the density of the ore (δ), liquid to solid ratio in the pulp by weight (R) are known, then

$$V_a = \frac{Q}{1440}\left(R + \frac{1}{\delta}\right) \text{ m}^3/\text{min} \tag{13.4}$$

Then the values of Q and R can be calculated from Eq. (13.4).

$$Q = \frac{V_a \times 1{,}440}{\delta R + 1} \text{ tonnes/day} \qquad (13.5)$$

and

$$R = \frac{V_a \delta \times 1{,}440 - Q}{Q \delta} \qquad (13.6)$$

The following example will illustrate the calculation of various parameters:

Example: If an ore having a density of 3.5 is to be floated at the rate of 3,000 tonnes/day, with a pulp density of 25 per cent of solids and flotation time 10 min, find the number of chambers required for mechanical type machines. Capacity of chamber may be considered to be 0.75 m³.

Solution: Liquid–Solid ratio, $R = \dfrac{L}{S} = \dfrac{75}{25} = 3$. Quantity of pulp for flotation (per minute)

$$V_a = \frac{3{,}000}{1{,}440}\left(3 + \frac{1}{3.5}\right) \text{ m}^3/\text{min}.$$

$$= 6.845 \text{ m}^3/\text{min}.$$

Now the number of chambers required (N)

$$N = \frac{6.845 \times 10}{0.75 \times 0.7} \text{ (from Eq. (13.2) } K = 0.7)$$

$$= 130.4$$

Since number of chambers must be a multiple of 2, the result may be rounded-up to give $N = 132$.

13.3.6 FACTORS AFFECTING THE OPERATION OF FLOTATION MACHINE

a) *Peripheral Speed of the Impeller*

Higher peripheral speed causes the greater amount of air drawn in with higher power consumption. Excessively higher speeds result in excessive pulp mixing, leading to detachment of large mineral particles from air-bubbles.

b) *Pulp Density*

Increase in pulp density results into reduction in amount of air drawn in and increased power consumption. However, very diluted pulps cannot be used in flotation, an average pulp density is most suitable.

c) *Amount of Pulp Reaching the Impeller and Its Point of Feed*

These two factors work together and affect the operation of impeller. The best practice is to feed the pulp at the impeller central zone in approximately optimum amounts, since increased amount chokes the impeller due to restricted amount of air drawn in. A part of the circulating pulp should also be fed directly to the impeller blade periphery, since air consumption increases with increase in circulating pulp. The pulp fed to the impeller periphery is ejected into the chamber more efficiently, without clogging the impeller.

d) *Pulp Level Height Above the Impeller*

This should be maintained to an optimum as it affects considerably the air consumption and power requirements. With increase of this height, air consumption falls, and power requirement increases. Excessive increase in this height will cause the deterioration of flotation.

e) *Frother Concentration*

Decreasing concentration of frother to its optimum point results in reduction of atmospheric air drawn in by the impeller due to more intensive circulation of fine air-bubbles and increased evolution of air in the impeller chambers (since frother reduces surface tension on the air–water interface).

f) *Condition Under which the Pulp–Air Mixture Leaves the Impeller Chamber*

An area of turbulence holding the pulp is formed nearest to the impeller, as a result of which the speed of pulp passage from impeller to chamber decreases considerably. This effect may be overcome by fitting guide vanes, which steer the pulp away from the impeller without turbulence.

g) *Clearance between Impeller and the Stator Blades*

This clearance greatly affects the efficiency of impeller. By increasing this clearance to more than 8–10 mm, impeller efficiency drops considerably.

h) *Impeller Diameter and Its Angular Velocity*

These factors affect the power requirement as well as air consumption. Power consumption is approximately proportional to the cube of impeller diameter and proportional to square of angular velocity. The air consumption can be approximately represented by the following relation

$$Q_{air} = 135\, D^3 \left(n^2 - \frac{100}{D} \right) \tag{13.7}$$

where Q_{air} = amount of air in litres/min,

$\quad\quad D$ = diameter of impeller in metres, and

$\quad\quad n$ = impeller speed in rpm.

13.4 Auxiliary Flotation Equipments

There is a great variety of auxiliary equipments required in flotation plants, but the important ones having direct relation to flotation are reagent feeders and conditioning tanks.

13.4.1 Reagent Feeders

Special devices are needed to have an accurate and uniform addition of reagents to the flotation circuit. The design and their operation depend main-ly on the characteristics of reagents. In general, the reagents may be dry powders such as lime, liquids such as solutions of xanthates, soaps, acids, etc., or viscous liquids such as oils. Accordingly, reagent feeders may be classified into three separate groups as follows:

a) *Dry Reagent Feeders*

Dry reagents can be fed by spreading them upon a belt or band feeder, which consists of a small conveyor band and is driven slowly by means of a worm drive. As the belt moves, a layer of powdered reagent in controlled width and height is fed from the hopper. Another device is portable feeder and is known as rotary feeder, which consists of a cylindrical hopper with an outlet hole at the bottom and a round disc mounted below the hopper. The hopper delivers the reagent to the disc which rotates at an adjustable speed and delivers the reagent by turning of the plate by means of a special drive. These were used earlier for feeding lime to ball mills. Presently these feeders are not in direct use, as lime is being added in the form of milk of lime, but can be used in manufacture of milk of lime in small installations. Supply of milk of lime can be made by tapping from ring mains through adjustable nozzles.

b) *Liquid Reagent Feeders*

The reagents which are almost insoluble in water (pine oil), are either fed directly at 100 per cent strength, or sometimes emulsified to a suitable dilu-tion before use. Aqueous solutions such as of xanthates, dithiophosphates, oleic acid, etc., are made in varying solution strengths (2, 5, 10 per cent etc.). In case of feeding highly corrosive reagents, the parts coming in contact with them (tanks, pipes, channels, pumps, etc.) should be made of suitable mate-rials (e.g. lead lined for H_2SO_4, wood tanks for $CuSO_4$, etc).

The most common feeders for liquid reagents are skip and bucket type. The skip feeder consists of a rectangular tank filled with liquid reagent and equipped with skip-bucket with an extended spout. The skip is periodically raised to tilt and discharge the reagent and lowered to fill the reagent with the help of a connecting rod. Bucket feeder (Fig. 13.7) consists of a tank for liquid reagent and a vertical revolving disc carrying small buckets at varying horizontal distances. The disc is partially submerged in the reagent and carries a number of buckets, free to turn about their mounting pivots. The disc is rotated slowly to submerge the buckets and get them filled where-as rising buckets discharge the reagent. The rotating buckets keep the solu-tion stirred. Several parallel discs can be driven from a common shaft for feeding reagent to a number of points or from separate compartments in the tank.

c) *Viscous Reagent Feeders*

Viscous reagents such as oils, are usually fed by drum-type feeder (Fig.

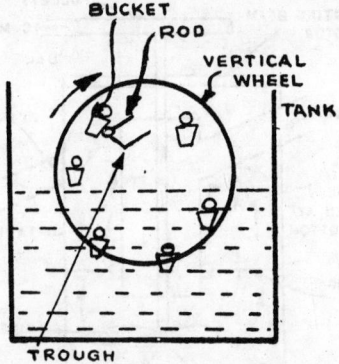

Fig. 13.7. Bucket feeder.

13.8). The reagent is filled into the tank, in which a drum is partially submerged and rotates slowly. On rotating, the drum surface picks up a thin layer of reagent, which is removed from the drum by a scraper. The reagent flows to a trough. A constant level of reagent in the tank is maintained with the help of a float valve. The rate of reagent feed can be controlled by altering the width of the scraper, or by varying the number of scrapers in contact with the drum.

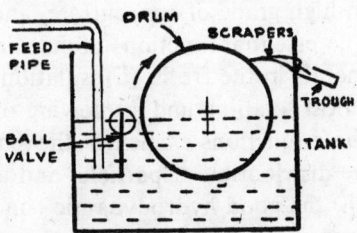

Fig. 13.8. Drum feeder.

13.4.2 CONDITIONING TANKS OR CONTACT VATS

Contact vats or conditioning tanks are used to bring mineral particles into contact with reagents for required periods of time and thus produce a homogenous pulp with uniform reagent concentration, density and particle size throughout. The contact vats (Fig. 13.9) are cylindrical tanks having an internal central pipe which is open at the top and at the bottom. The pulp enters the upper part of the pipe, having a number of holes in its sides. A shaft is mounted along the vertical axis of the vat, which is driven by a V-belt. A mixer (impeller) is mounted on the lower end of the shaft, but about one-third up the height of the tank from its bottom to prevent silting of pulp. Rotation of impeller causes a very high rate of pulp mixing due to (i) vertical circulation of the pulp and (ii) rotation of pulp in horizontal plane. The capacity of tank is usually 0.25–0.5 m³ depending on the time required for

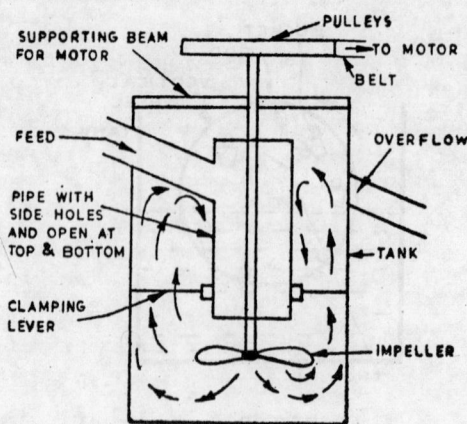

Fig. 13.9 Contact vat.

reagent pulp contact (established experimentally). The method of calculating the capacity of conditioning tank is the same as for the capacity of flotation machine.

13.5 Design of Flotation Machines

Due to continuous efforts made for better economy and efficiency, and simplicity of operation, it could be possible to treat low-grade ores profitably by improved designs of flotation cells. The objectives of flotation cell design are high recovery with high grade of concentrate, and economy of installation and operation. The essential functions of machine are (i) selective separation of required minerals in the froth, (ii) isolation of froth from tailing by overflow or mechanical scraping and (c) flowing of froth out of the cell. In order to perform these functions efficiently, the required conditions in the cells are (i) an effective distribution of particles and air-bubbles throughout the cell volume, and (ii) the good hydrodynamic conditions for the attachment of air-bubbles to particles.

In general, the design criteria are different for different modes of mechanisms of flotation, but owing to the complexity of designing based on the mechanisms, most of the designs have been developed on the basis of gross requirements of agitation and aeration. The design of a flotation machine (or cell) is largely empirical and aims at maintaining high throughout subject to the previously stated criteria.

13.5.1 Design Evaluation

If a cell of a given design (manufacture) is selected, the size can be fixed through the diameter of impeller (D) which fixes the rest of the geometry of the system. The rpm of impeller (N) will determine the maximum possible aeration for a self-aerating cell. The control of aeration can be achieved by the use of a throttling device. The smaller sized self-aerating cells oftenly provide aeration close to the flooding value, i.e. additional air provided by blower does not get dispersed, but escapes from the impeller zone in slugs. Whereas, in case of larger cells, aeration is usually below the required level and thus blower air will be needed. The design should provide the repeated circulation of particles through air swept zone. This can be obtained either by means of recirculation within the single cell or by multiplication of cells in series. Other variables to be optimised for a selected design include size and number of cells, impeller speed, and aeration.

Economy of installation and operation requires high capacity, lower power consumption, ease of operation, and low maintenance. Further, coarser the feed that can be used, the less is the amount of energy required for grinding. Following features of design are common to all types of flotation machines.

a) *Selection Zone and Transportation of Particles to It*

Selection zone is the bubble column supported laterally at the bottom of the machine. Its capacity depends on the interfacial area that it contains and is determined by its overall volume and by the size of the bubble constituents. Height of column and rate of bubble rise determine the time available for separation of any given mass of pulp presented to a given horizontal area at the bottom. Particles which get attached to a bubble tend to be carried to the top of the column. The depth of column affects only the grade of concentrate, whereas horizontal column area determines the volume of pulp to be treated per unit volume of column for a given depth with a given feed rate. In general, high-grade feeds require less air than low-grade. Therefore, more air is needed in scavengers than in roughers.

Transport of feed to the column is influenced by sweeping of pulp upward from the body into the column having fine particles held around the bubble by viscosity. In general, pneumatic cells maintain the greatest proportion of pulp in sweeping condition. However, sweeping effect gets reduced by large sized bubbles, clogging of holes, and sedimentation, and eddy currents producing actual downward movement of bubbles relative to the cell walls in some parts of the cell.

In air-lift cells, the use of volume is proportionately less for air sweeping compared to other machines. But in this case, the air is more finely and completely dispersed and the air-swept zone is more viscous than in any other type of bubble column cell. A given mass of air remains in the deepcell air-lift for a period at least twice as long as in the standard-cell, with the same degree of agitation. Hence the dispersion of air in the pulp cascaded to the air-swept zone is much finer in the deep cell and the volume held is also larger. As a result, capacity of the deep cell per unit depth is five to six times more than that of a shallow cell of the same width requiring almost equal power consumption.

b) *Tailing Flow*

Tailing flow is normally transverse to froth flow. In trough-type cells without transverse partitions (usually pneumatic and cascade type), pulp flow is often in the form of a confused double spiral around parallel horizontal axes, which are symmetrically spaced on both sides of the longitudinal centre plane of the trough. The tighter the spiral and faster the peripheral rate, more frequently is the pulp brought into the air-swept zone.

c) *Tank Design*

Tank should be so designed that settled material returns to the air-swept zone as rapidly as possible. This can be accomplished by sloping the tank bottom toward the point of air introduction.

d) *Mechanical Features*

Mechanical features should be considered in the light of performing the primary functions, i.e. (i) making proper recovery and grade of concentrate, (ii) continuity and ease of operation, and (iii) labour costs involved in maintenance. In addition to the provision for air sweeping, protection of bubble column, and direction of separating flows, provision should be made to prevent sanding up. This involves elimination of dead spots in the pulp path, arrangement of pulp passages without allowing the pulp to settle down, and provision of sand bleeders which short-circuit the settled material directly to a zone of greater agitation. The cell should be so designed that it can be restarted after an unexpected shut down, without emptying. It is easy to restrict the machines operating with extraneous air supply. For other machines, piping should be provided for high pressure flushing water to impeller zones, closed pulp passages, sand bleeders and similar other areas. The pipes used should be as short as possible, it is more important particularly for heavy ores.

Impellers operate in areas of fully developed turbulence, with Reynold's numbers more than 10. The impellers can be characterised by their power numbers (power number $n_p = Hg/\rho\ ND$, where H = horse power, g = acceleration due to gravity, ρ = pulp density, D = diameter of impeller, and N = number of blades). The objective in designing flotation impellers is to obtain a high power number, which can permit the same power intensity (a measure of suspension and dispersion of particles, and mixing and dispersion of air) with lower rotational speed, and low wear and maintenance costs.

13.5.2 Design Considerations for Effective Suspension and Distribution of Particles

The major problem is to reduce the tendency of coarser sizes to segregate and their settling toward the bottom of the cell, which would otherwise decrease the chances of their flotation. Effective suspension is the result of velocity patterns of pulp, which must sweep out all areas of the cell. The finest particles remain in suspension in all regions, whereas heavier particles, except in the immediate vicinity of the impeller, will tend to move downward from the pulp. Thus, stable suspension is a dynamic state and requires flowline loops of varying velocities. The coarser particles predominating regions will require higher velocity flow-lines.

For effective suspension and distribution of particles, the quality as well as quantity of aeration should be considered. Quality of aeration refers to the distribution of bubble diameters, which depends on the presence of a frother and the extent of agitation. Aeration quantity may be considered as the volumetric air-flow rate (ratio of flow rate to area or volume of cell) giving a nominal air retention time.

Air-flow rate and impeller speed (air-flow number) have a critical effect on flotation of coarse particles. This can be related to the stability of the particle–bubble aggregates with respect to turbulence and also to particle suspension effectiveness. Particle suspension is favoured by higher impeller speed and lower air-flow. However, particle–bubble stability gets reduced by turbulence caused by higher impeller speeds. On the other hand, lower impeller speed and higher air-flow favour reduced turbulence and stability of particle-bubble aggregates, and thus it creates the optimum conditions. But further increase in air-flow may cause less effective suspension due to reduced pulp flow and these conditions will not favour recovery of coarser sizes. Combined effect of aeration and impeller speed may be summarised as follows:

a) There is a specific hydrodynamic condition with any cell expressed by the flow number for optimum recovery and grade of concentrate, particularly for the coarser size fractions.

b) The coarser feed requires better aeration–impeller speed combination. Finer feeds may require more air, whereas for coarser feeds, throttling of air supply may be needed.

c) In case of flotation of coarse particles, problem of cell effectiveness and cell design is more critical, whereas with finer feeds (−100 microns) selection and design of cell is not critical.

13.5.3 Design Problems for Larger Cells

There is an increasing trend to use coarser floation feed, for which larger cells should be employed. Therefore, the design criteria include (a) optimum coarse particle suspension, and (b) provision of test facilities for flow number suitable to each grind and ore density combination. This will require provisions for variable air supply and impeller speeds even with self-aerating cells. Other design considerations may include the following:

i) *Multi spindle type cell*: This avoids the scale-up problem but the cost savings are not obtained as with the single large tank.

ii) *Single large tank with single mechanism*: It can be designed on the basis of models with principles of similitude. However, in this case hydrodynamic and flotation behaviour of each size becomes unpredictable.

In larger cells, the problem is to reduce turbulence directly under the froth while retaining the required flow conditions for suspension below it. For this, multibladed turbine may be most effective.

13.6 Flotation Schemes

In practice, it is generally not possible to obtain a useable high-grade concentrate and tailing containing no useful mineral in one single stage due to the reasons : (a) Minerals to be separated may have identical properties, which will reduce the selectivity of minerals, (b) in many cases, more than two products are needed, and (c) in case of complex ores, it is sometimes, useful to crush the individual products and process them further to obtain more complete separation. Therefore, processing of certain ores may need different flotation schemes consisting of several flotation operations, and transporting of individual products from one operation to another. The flotation schemes are extremely varied, and are dependent on (i) flotation properties of the ores to be concentrated (ii) grade of concentration products required, and (iii) economics of the process. Various flotation operations may be termed as follows :

a) *Basic flotation*: This refers to first flotation operation to separate specific groups of minerals. Several basic flotations may be included in a flotation scheme, particularly when treating polymetallic ores, basic lead and basic zinc flotations are carried out.

b) *Repurification*: This refers to repetition of flotation operation to further improve the concentrate quality.

c) *Control of tailing flotation*: This refers to the operation in which the flotation tailings are reprocessed to reduce the valuables float mineral going into tailings.

The flotation schemes may be classified depending on their primary features (i.e. number of stages of concentration, number of concentration cycles, and the purpose of the individual concentration stages and cycles), and secondary features (i.e. number of repurifying and control flotation operations in the various stages and cycles and the return points for intermediate products).

13.6.1 FLOTATION SCHEME BASED ON NUMBER OF STAGES

A stage in flotation may be referred to one ore crushing operation followed by a group of flotation operations. Based on number of stages of concentration, the flotation schemes may be classified as single-stage, two stage, and multi-stage schemes which are shown in Fig. 13.10.

a) *Single-stage Flotation Scheme*

This scheme [Fig. 13.10(a)] is generally provided for the flotation of ground and classified ore. If the intermediate products need grinding, they are returned to the initial grinding mill. If fine slimes affect the flotation operation considerably, ground-ore pulp is classified into sands and slimes and then flotated separately [Fig. 13.10(b)].

b) *Two-stage Flotation Scheme*

This is provided for the repeat grinding of concentrate [Fig. 13.10(c)], tailings [Fig. 13.10(d)] or middlings [Fig. 13.10(e)] obtained in the first flotation cycle.

c) *Multi-stage Flotation Scheme*

This scheme [Fig. 13.10(f)] may be quite varied, comprising of different combinations of single- and two-stage flotation schemes. Some schemes may even have more than three stages, e.g. some cases of graphite flotation.

13.6.2 FLOTATION SCHEMES BASED ON NUMBER OF FLOTATION
CYCLES

These are referred to a group of flotation operations, in which one or more finished products (not requiring further flotation) are obtained. Each stage of flotation scheme may have several cycles. Flotation schemes based on number of cycles are shown in Fig. 13.11, which include single cycle, double cycle and double cycle-two-stage flotation schemes.

13.6.3 FLOTATION SCHEMES BASED ON NUMBER OF PRODUCTS

In flotation of ores when several concentrates are to be obtained, flotation schemes can be distinguished as collective, sequence selective and collective-selective flotation schemes.

a) *Collective Flotation Scheme*

In this a bulk concentrate containing several valuable components is produced, e.g. in flotation of gold, both gold and sulphide minerals pass into the concentrate.

b) *Sequence Selective Flotation*

In this scheme [Fig. 13.12 (a)] the individual valuable components are obtained in flotation of ore one-by-one, in sequence. The most water repellent minerals are removed first, which is followed by removal of minerals more difficult to separate. These schemes are most commonly used in flotation of sulphide minerals. These schemes are specifically suited when either large gangue inclusions are present in the mineral or coarse inclusions of mineral are present in the gangue.

c) *Collective–Selective Flotation*

In this scheme [Fig. 13.12(b)] several valuable minerals are first separated into a froth product and subsequently separated into individuals from each other. The collective concentrate is generally subjected to repeat grinding before further separation into individual concentrates. This scheme may be employed when minerals are present in each other in finely divided form, having their concentrations fairly large and also occur in gangue. In such ores, first the ore is crushed fairly coarse and the concretions are floated into a collective concentrate. The collective concentrate is then subjected to repeat grinding and the individual minerals separated from each other.

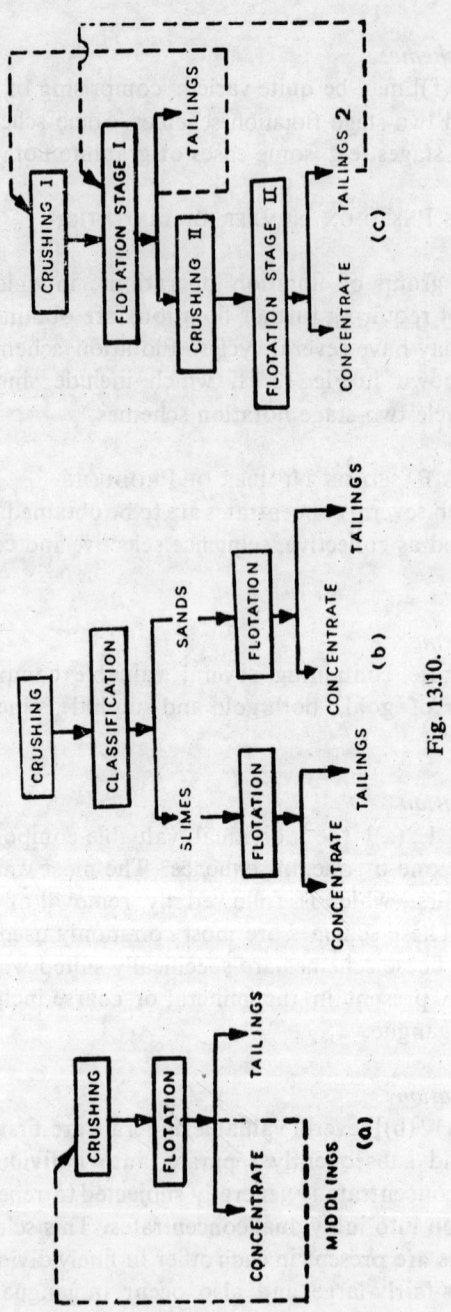

Fig. 13.10.

(Contd.)

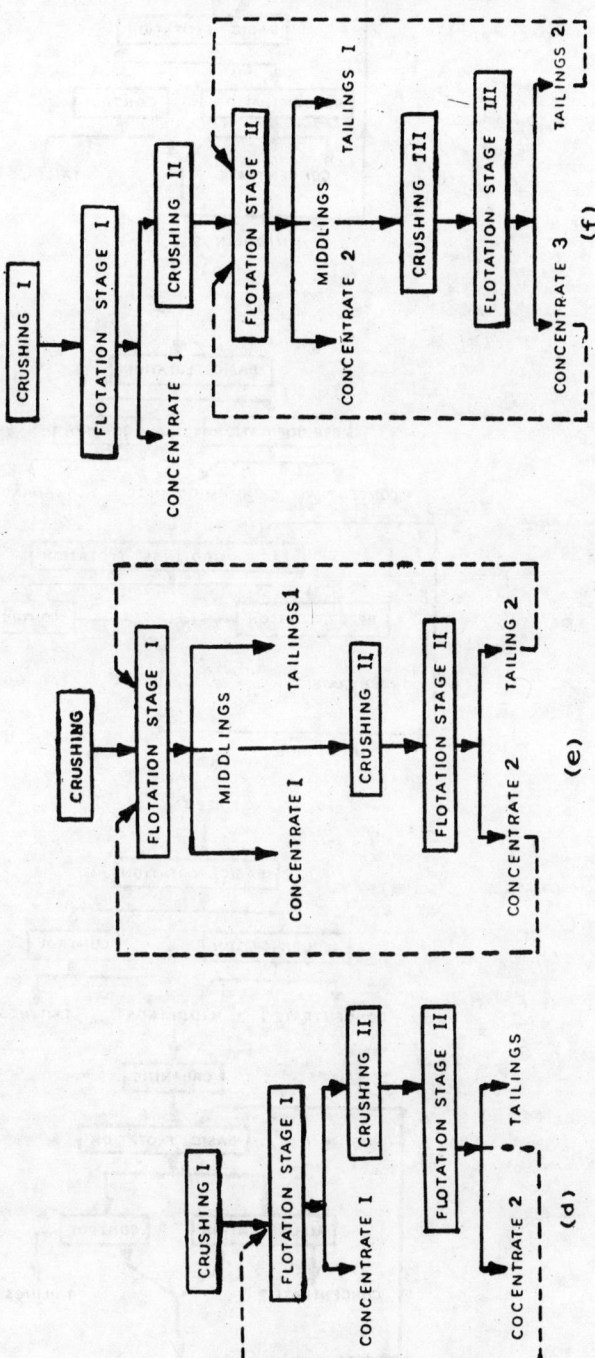

Fig. 13.10. Flotation schemes based on number of stages.

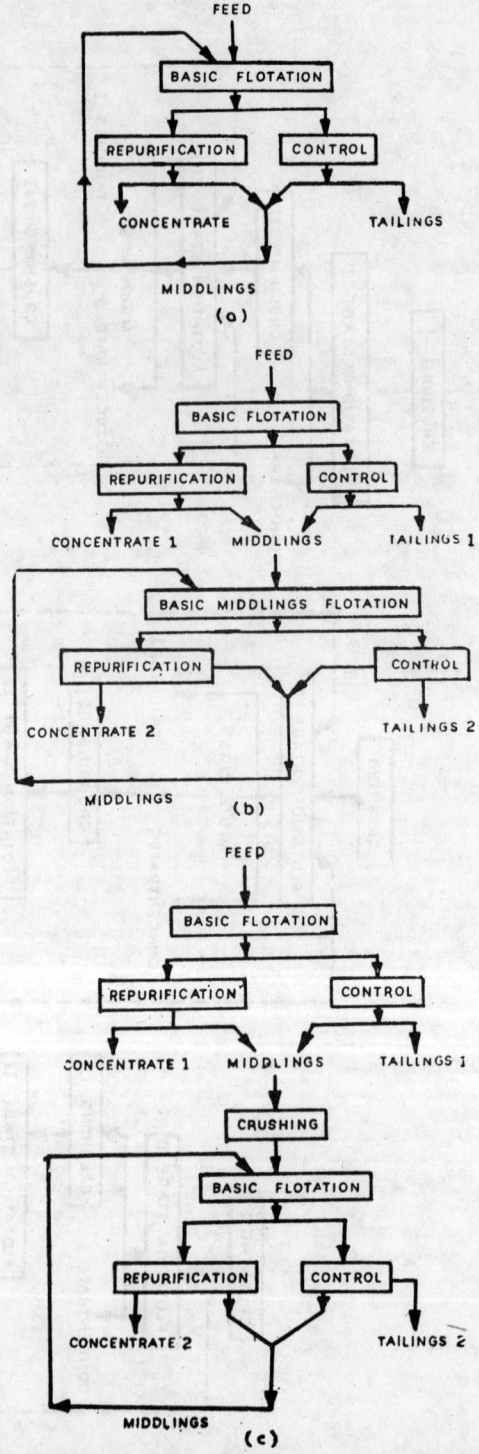

Fig. 13.11. Flotation schemes based on number of flotation cycles.

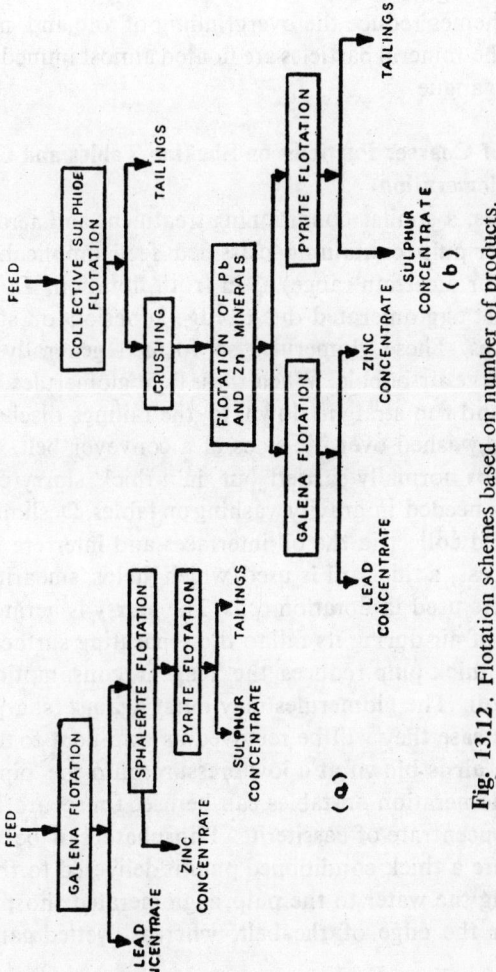

Fig. 13.12. Flotation schemes based on number of products.

Out of the above various schemes, stage flotation schemes possess various advantages as follows:

i) These schemes consist of alternating grinding and flotation operations, which extract the valuable minerals into the concentrates as they are liberated from gangue.

ii) These schemes are advantageous, where complete mineral separation requires fine grinding of an ore.

iii) These schemes reduce the overgrinding of ore and mineral losses in the tailings, as the mineral particles are floated almost immediately after their liberation from gangue.

13.7 Flotation of Coarser Particles on Shaking Tables and Conveyor Belts (Agglomeration)

In this process, a similar conditioning treatment and aeration are applied to a much thicker pulp containing classified feed (applicable for 3 mm to 70 microns but for a certain range) as in froth flotation. In this case, aerophilic particles get agglomerated due to edge-adhesion or sticking together by fine air-bubbles. These glomerules so formed generally arrange themselves round a large air-bubble. When these light glomerules reach a shaking table, they rise and run straight down to the tailings discharge. Alternatively, they can be washed over the sides of a conveyor belt.

Conditioning is normally carried out in a thick slurry of deslimed and classified feed as needed in gravity washing on tables. Desliming is necessary since slimes would collect in the oil interfaces and interfere with conditioning. In this process, a thick oil is used which helps smearing of the solids instead of frothers used in flotation cells. The slurry is aerated by cascading and entrapment of air during its fall to the separating surface.

The use of a thick pulp reduces the reagent consumption and helps in smearing of the oil. The glomerules may burst against sharp riffle-edges of table and in this case they will be removed as skin-float to the tailings launder. In practice, air is blown at a low pressure, into the pipes crossing the table deck. Agglomeration on tables can be used for separation of sulphides from a rough concentrate of cassiterite. Phosphate can be treated on conveyor belts, where a thick conditioned pulp is delivered to the middle of the belt. On spraying the water to the pulp, agglomerated phosphate aggregates get washed over the edge of the belt, whereas wetted gangue goes to the discharge end.

13.8 Flotation Practice

For the recovery of the same mineral from different ore bodies, conditions for flotation may be quite different, since small differences in the crystal structure and chemical constituents of minerals in different ore bodies may affect the properties of mineral to be recovered by flotation. Therefore, in plant practice, different schemes are usually adopted for a specific ore and not to the mineral in general. Though, a general scheme of liberation, activation, gangue depression, optimum froth texture, etc., may be adopted for the separation of required mineral values, the specific details are modified for various reasons, such as, physical, chemical, economic, etc. The various steps involved in flotation can be summarised as follows:

a) Grinding of ore in water to required mesh size (usually 200 microns).

b) Dilution of pulp formed to required pulp density (25–45 per cent solids).

c) Addition of small quantities of reagents to modify the surfaces of required minerals.

d) Addition of collector in optimum amounts.

e) Addition of frother to stabilise the froth.

f) Aeration of pulp in a suitable flotation cell.

g) Removal of mineral-bearing froth from the aerated pulp.

The possible objectives of flotation are discussed below:

i) *Flotation of one or more minerals simultaneously*: This is a general procedure and can be employed to obtain a mixed bulk concentrate of valuable minerals. For example, copper minerals can be concentrated in one single concentrate, which does not require further separation before smelting. When two sulphides such as of copper and lead readily float together, they can be bulk floated and further separated by depressing one in a further treatment.

ii) *Depression of one or more minerals from the bulk concentrate*: This procedure is used to clean a bulk concentrate by depressing undesired constituents. For example, removal of the last amounts of lime, silica, and iron from a fluorspar to obtain the acid-grade product. This procedure can also be used to separate two values by specific depression of one from the bulk float, e.g. separation of copper and lead sulphides.

iii) *Flotation of one or more minerals to obtain a concentrate having value intimately associated with others*: This is an indirect method employed to float auriferous pyrite or a fraction of an uranium bearing ore by floating the minerals most associated with the required values.

iv) *Flotation of one or more gangue minerals*: This is an alternative to procedure (ii) and can be used in cases such as the final upgrading of a dirty phosphate float, where silica is removed by flotation.

A general reagent scheme employed in flotation of single mineral is given in Table 13.2. Higher alcohols mentioned under frothers are usually employed to improve the frothing action of the other reagents. However, in practice, the scheme of flotation reagents, sequence of different operations, and number of total steps will vary depending on the individual case. Some important examples of flotation concentration are described in the following paragraphs.

13.8.1 FLOTATION OF SULPHIDES

Treatment of sulphide ores is the most important subject to flotation concentration, owing to the excellent flotation properties of sulphide

Table 13.2. General reagent practice for flotation of important single minerals

S. No.	Mineral/chemical formula	Principal constituents per cent	Dispersants	pH regulators	Selectivity differentiating agents	Depressants	Activators	Collectors	Frothers	Associated minerals
1	2	3	4	5	6	7	8	9	10	11
1.	Bauxite, $Al_2O_3 \cdot 2H_2O$	Al 39.13	Polyphosphate	NaOH	Paraffin	—	—	Oleic acid/cyanamid reagents	Higher alcohols	Clay, kaolinite, lime stone, Fe, dolomite, Ti
2.	Corundum, Al_2O_3	Al 53.4	Orthotoluidine	NaOH	—	Excess acid	—	Oleic acid and its salts	Higher alcohols, pine oil	—
3.	Cryolite, Na_3AlF_6	F 54.3 Na 32.9	—	—	—	—	$CuSO_4$	Oleic acid and its salts	Higher alcohols	—
4.	Feldspars, $(AlSi_3O_8)$ Na, K, Ca	Na, K, Ca About 18	HF	—	—	—	HF	Amines	Pine oil, higher alcohols	—
5.	Stibnite, Sb_2S_3	Sb 71.7 S 28.3	Na_2SiO_3	Na_2CO_3	—	Excess alkali, NaCN	$CuSO_4$	Ethyl and sec. butyl xanthates	Pine oil	Quartz, barite, galena, gold
6.	Arsenopyrite, FeAsS	As 46 Fe 34.3 S 19.7	Na_2SiO_3	Na_2CO_3	—	NaCN, lime	$CuSO_4$	Ethyl to amyl xanthates	Pine oil	Au and sulphides of Cu, Fe, Pb,
7.	Barytes, $BaSO_4$	Ba 58.1	Na_2SiO_3	Na_2CO_3, citric acid	Aerosol	$AlCl_3$, $FeCl_2$	$Pb(NO_3)_2$	Oleic acid, and fatty acid soaps, higher alcohol sulphates	Pine oil, cresylic acid	Fe, Mn, PbS, ZnS, $CuFeS_2$
8.	Beryl, $Be_3Al_2Si_6O_{18}$	Be 14.1 Si 66.8 Al 19.1	—	HF	—	H_2SO_4	$Pb(NO_3)_2$	Amines, oleic and fatty acid salts, cyanamide	—	Granite, mica, clay, slate

9.	Borax Na₂B₄O₇· 10H₂O	B < 11.4	Aniline, Starch, Dextrin, Quebracho	—	—	—	—	Fatty acids	Aniline, xylidine, pyridine	—
10.	Calcite, CaCO₃	Ca < 40	—	Na₂CO₃	—	Quebracho, Na₂SiO₃, K₂Cr₂O₇	—	Oleic acid, sulphonated oleic fatty acid salts	Pine oil	—
11.	Coal	C < 95	Na₂SiO₃	—	—	Tannin, quebracho	—	Fuel oil paraffin, cresyls	Pine oil, cresylic acid	—
12.	Diamond	C < 100	—	—	—	—	—	Natural grease, petroleum jelly on tables	—	Shale, clay, quartz iron, silicates of Fe and Mg
13.	Graphite	C < 99	Na₂SiO₃	Na₂CO₃	—	Starch	—	Paraffin fuel oil	Pine oil, long chain alcohol	Spinel, shale, mica, silica, schist
14.	Monozite (CeLaYt)-PO₄	—	Na₂SiO₃	Na₂CO₃	—	Strong acids	—	Oleic acid, fatty acid salts	Pine oil	—
15.	Chromite FeO·Cr₂O₃	Cr 46.5	NaOH	Na₂CO₃	Fluo-silicate	—	Phospho-molybdic and phospho-tungstic acids	R 800 series fatty acids, alkylo-amines	Pine oil fuel oil	Serpentine, black sands
16.	Azurite	Cu 53.3	Na₂SiO₃	Na₂CO₃	Na₂S	Quebracho	Poly-	Fatty acids	Pine oil	—

(Contd.)

Table 13.2. (*Contd.*)

1	2	3	4	5	6	7	8	9	10	11
	$2CuCO_3$-$Cu(OH)_2$						sul-phides	and salts, xanthates, aerofloat 25	gas oil cresylic	—
17.	Bornite Cu_5FeS_4	Cu 63.3 Fe 11.1 S 25.6	Na_2SiO_3	$Ca(OH)_2$	—	NaCN	—	Xanthates, aerofloats	Pine oil	—
18.	Chalcocite Cu_2S	Cu 79.9 S 20.1	Na_2SiO_3	$Ca(OH)_2$	—	NaCN	—	Xanthates, aerofloats	Pine oil	—
19.	Chalcopy-rite $CuFeS_2$	Cu 34.6 Fe 30.4 S 35.0	Na_2SiO_3	—	—	NaCN	—	Xanthates, aerofloats	Pine oil	—
20.	Fluorite, CaF_2	F 48.7 Ca 51.3	Na_2SiO_3	H_2SO_4	—	Quebracho $K_2Cr_2O_7$	—	Oleic acid, sulphonated fatty acids	Pine oil	—
21.	Gold, Au	Au < 100 Ag	—	Na_2CO_3	—	Lime, NaCN	R 404	Xanthates, aerofloats	Cresylic pine oil	—
22.	Petzite Au Ag_3Te_2	Au 25.5 Ag 42 Te 32.5	—	—	—	NaCN	—	Xanthates	Pine oil	—
23.	Auriferous pyrite/iron pyrite, FeS_2	Fe 46.6 S 53.4	—	H_2SO_4	—	NaCN, lime	$CuSO_4$	Xanthate	Pine oil	—
24.	Hematite, Fe_2O_3	Fe < 70	—	HS_2O_4	—	Tannin phosphate	De-slime	R 801, R 825, tall oil, oleic acid	Pine oil, Fuel oil	—

25.	Cerussite, PbCO$_3$	Pb 77.5	—	Na$_2$CO$_3$	—	—	Na$_2$S, CuSO$_4$	Xanthate, R 404, aerofloat	Pine oil cresylic	—
26.	Galena, PbS	Pb 86.8	Na$_2$SiO$_3$	Na$_2$CO$_3$	—	K$_2$Cr$_2$O$_7$	—	Aerofloat, xanthate	Pine oil	—
27.	Magnesite MgCO$_3$	Mg 28.8	—	—	—	Tannin	—	Oleic acid and oleates	Pine oil	—
28.	Talc, H$_2$Mg$_3$Si$_4$O$_{12}$	Mg 2.8	Polyphosphates	—	—	Starch, glue	—	Short chain amines R 825	Pine oil paraffin	—
29.	Pyrolusite, MnO$_2$	Mn 63.2	Na$_2$SiO$_3$	Na$_2$CO$_3$	—	Phosphate, quebracho	—	Fatty acids and salts	Pine oil paraffin, eucalyptus oil	Silica
30.	Cinnabar, HgS	Hg 86.2	—	—	—	Na$_2$SiO$_3$	—	Amyl xanthate	Pine oil	—
31.	Molybdenite, MoS$_2$	—	—	—	—	—	—	Min. oil, xanthate, aerofloat	Pine oil	—
32.	Pentlandite, (FeNi)S	—	Na$_2$SiO$_3$	Alkalies	—	Lime	CuSO$_4$	Pentasol, xanthate	Pine oil	—
33.	Apatite, Ca$_5$(FCl$_2$)-(PO$_4$)$_3$	P$_2$O$_5$ 42	—	NaOH, Na$_2$CO$_3$	—	HF	—	Fatty acids and salts	Pine oil	—

(Contd.)

Table 13.2. (*Contd.*)

1	2	3	4	5	6	7	8	9	10	11
34.	Phosphate rock $Ca_3(PO_4)_2$	—	—	NaOH	—	—	—	Fatty acids	Pine oil	—
35.	Scheelite $CaWO_4$	W 63.9	Na_2SiO_3	Na_2CO_3	Acrosol	Excess Na_2SiO_3 Quebracho	—	Oleic and fatty acid salts	Pine oil	—
36.	Wolframite (FeMn) WO_4	W 61	—	—	—	—	—	Oleic acid R 708, 710	Pine oil, fuel oil	—
37.	Carnotite, $K_2O \cdot 2UO_4 \cdot V_2O_5 \cdot 8H_2O$	—	Na_2SiO_3	Na_2CO_3	—	—	$Pb(NO_3)_2$	Fatty acids	Pine oil, cresylic	—
38.	Pitchblende, $x UO_2 \cdot y UO_3$	—	—	—	—	—	$FeCl_3$	Fatty acids, amines	Pine oil	—
39.	Smithsonite $ZnCO_3$	Zn 52.1	Na_9SiO_3	Na_2CO_3	—	—	Na_2S	Fatty acids, amine acetate	Pine oil	—
40.	Sphalerite, ZnS	Zn 58.7	—	Lime	—	$ZnSO_4$, $NaHSO_3$, SO_2	$CuSO_4$	Xanthates	Pine oil	—

minerals. For processing of various sulphide ores, sequence (direct) selective flotation or collective–selective flotation or the combination of two can be employed. Usually xanthates are used as collectors and pine oil as frother. Some examples of sulphide flotation are given below:

i) *Flotation of Lead–Zinc Ore*

These ores contain galena (PbS) and sphalerite (ZnS) as principal valuable constituents, along with certain gangue minerals (quartz, silicate, dolomite etc.). Xanthates (usually butyl or amyl) are used as collector but do not have selective action on individual sulphide. Therefore, NaCN is used first to create the conditions under which xanthate will render only the galena water repellent, whereas sphalerite will remain unattacked by xanthate. Thus NaCN works as an activator for galena. $ZnSO_4$ is used as a depressor for sphalerite. After selective flotation of galena under suitable pH conditions (pH 6.5 to 8), the pulp is brought to a pH of 9–11 by adding lime. It is followed by addition of a suitable quantity of $CuSO_4$ in the form of a saturated solution. Most of the copper ions react with residual cyanide and precipitates of hydrate or carbonate are formed. These salts along with some free copper ions plate the zinc lattice ions with copper ions and thus activate the sphalerite. After adjusting the concentration of collector and frother, sphalerite is selectively floated leaving pyrite in the pulp.

ii) *Selective Flotation of Copper Sulphide Ores*

Most copper sulphide minerals are associated with worthless pyrite, or pyrrhotite and also arsenopyrite, which should be separated and rejected. If other valuable sulphides such as sphalerite, galena, pentlandite, molybdenite, etc., are present, their separation will also be advantageous. Further, many copper sulphide ores also contain appreciable quantity of precious metals and it is desirable to include them in copper concentrates, if possible separate them by amalgamation.

For flotation of copper-bearing sulphide minerals, xanthates are the common collectors, preferably ethyl or propyl xanthate is used. In order to reject pyrite, a higher pH 8.5–12 (about 12 for chalcocite-bornite ores, and 8.5 to 10 for chalcopyrite ores) is employed. The value of pH is adjusted by addition of lime, and a small amount of NaCN may also be employed to control the rejection of pyrite.

13.8.2 FLOTATION OF NON-POLAR, NON-SULPHIDE MINERALS

The minerals under this group are coals, graphite, talc, etc. These minerals possess high natural water repellent properties (contact angle being 50–90°). Other favourable properties for flotation are relatively low density and low hardness. Due to the favourable flotation properties, these minerals can be floated with the following advantages:

a) Comparatively simple and cheap reagents such as hydrocarbon oils can be used as collectors.

b) Frothers are rarely used.

c) Relatively coarser particles can be floated.

The problems associated in flotation of these non-ionising minerals are :

a) Formation of large quantities of fine slime.

b) Wetting down and dispersion of the gangue is difficult.

c) Sulphur acidifies the pulps causing the corrosion of metal surfaces.

13.8.3 FLOTATION OF OXIDISED AND MIXED NON-FERROUS ORES

This group's oxidised ores are carbonates, sulphates, hydrates, and silicates of Cu, Pb, and Zn, and mixed ores having sulphides of these metals along with oxidised ores. The important minerals among oxidised ores are malachite [$CuCO_3 \cdot Cu(OH)_2$], azurite [$2CuCO_3 \cdot Cu(OH)_2$] chrysocolla (Cu $SiO_3 \cdot H_2O$), cerussite ($PbCO_3$), anglesite ($PbSO_4$), smithsonite ($ZnCO_3$), and other oxidised minerals such as arsenates, vanadates, phosphates, etc., are also of industrial importance. These ores can be floated at high pH (9.5–11) by the following methods:

a) *Flotation by Sulphidisation*

This is the most common method employed, which consists of sulphidisation of minerals by Na_2S or $(NH_4)_2S$ followed by flotation with collectors and frothers employed for flotation of sulphides.

b) *Direct Flotation Employing Carboxyl Collectors*

Though this method gives better extraction of minerals, it does not find much application due to the low selectivity of these collectors.

c) *Flotation Employing Xanthates of Higher Alcohols with a High Reagent Consumption*

This method involves high cost of reagents and does not provide any advantage over the above methods.

d) *Combined Method*

This is applicable to oxidised and mixed copper ores and consists of initial treatment of ground ore with H_2SO_4 to dissolve the copper. It is then followed by cementation of dissolved copper with iron and finally flotation of the cemented copper with undissolved sulphides. This method gives high copper extraction from oxidised ores, which are difficult to concentrate.

e) *Direct Flotation Employing Mercaptans*

This method does not find much use due to the extremely unpleasant odour of mercaptans.

13.8.4 FLOTATION OF POLAR NON-SULPHIDE MINERALS

This group includes minerals having alkali earth cations (Ca, Mg, Sr, and Ba), such as calcite, fluorspar, apatite, barite, magnesite, phosphates, scheelite, dolomite, etc. These minerals have high chemical activity towards anion type collectors and high susceptibility to hydration due to their pronounced ionic character of the crystal lattice bond. Xanthates and dithiophosphates cannot be used as collectors for these minerals, since they do not show any collector properties to these minerals due to the high solubility of alkali earth metal xanthates and dithiophosphates in water. Anion collectors having carboxyl solidophil group (soaps and organic acids) form compounds with alkali earth cations, which are practically insoluble, work as effective collectors for these minerals. Flotation of these minerals can be carried out without preliminary activation. However, the selective separation of these minerals is extremely difficult owing to their similar flotation properties. Selective separation of minerals having identical cations (e.g. calcite and scheelite, calcite and fluorite, scheelite and fluorite, etc.) is particularly more difficult.

Minerals of this group can be separated from each other and from other non-sulphides by use of selective regulators (electrolytes, Na_2SiO_3, tannin, dextrin, etc.) and carrying out the flotation at an optimum value of pulp pH, which is of great importance for selective separation of these minerals.

13.8.5 FLOTATION OF PRECIOUS METALS

Precious metals are usually found in their native form as well as in the form of chemical compounds, such as tellurides of gold and silver and sulphides of silver and platinum group metals. They are also found as finely disseminated inclusions in sulphides as well as in the form of solid solution of precious metal sulphides with another sulphide, e.g. palladium sulphide in pentlandite (Ni-mineral). Usually flotation is employed prior to extraction of precious metals by metallurgical treatment. These minerals are extracted in concentrates by flotation of Pb-Zn, Cu, and Cu-Ni ores. For recovery of precious metals, following types of ores/materials can be processed by flotation:

a) Gold-pyrite ores by collective flotation.

b) Cu, As, and Pb sulphide ores by selective or collective–selective flotation.

c) Pd-Pt sulphide ores by collective flotation.

d) Flotation of cyanidation tailings containing tellurides.

e) Carbonaceous and graphite ores containing precious metals.

Gold bearing pyrites are treated by flotation to extract the free gold and gold associated with pyrite. These ores can be readily floated with xanthates as collectors, pine oil as frother, and a suitable depressant for the gangue.

13.8.6 FLOTATION OF SILICATES

This group includes quartz, feldspar, sillimenite, garnet, kyanite, mica, etc. These minerals are water-avid due to the presence of oxygen and their polar nature. The flotation of these minerals can be carried out with fatty-acid type anion collectors having long hydrocarbon chain after activation with heavy metal ions. Most of these minerals cannot be sulphidised. Their flotation requires a close control of pH and depression of gangue by a wetting agent. Cation reagents may be effective collectors for some minerals of this group, but in this case water should be soft and pulp fairly free from slime. However, the selective separation of these minerals is a complex problem.

13.8.7 Flotation of Soluble Salts

The separation of soluble salts by flotation is relatively a new field and is becoming more important in chemical industries. This process principally differs in conditions and the reactions between reagents and the salts to be floated from the usual flotation of minerals practically insoluble in water.

Flotation of soluble salts is carried out in saturated aqueous solutions of the salts present in the material to be concentrated. The collector usually employed is fatty acid and frother is not required due to strong frother properties of concentrated salt solutions. Sometimes an activator (usually lead nitrate) may be needed. For example, KNO_3 and $NaCl$ can be separated by floating out $NaCl$, employing fatty acids.

13.8.8 Flotation of Precipitates and Ions

Precipitate flotation is based on the same principles as mineral flotation. For example, in the preparation of potable water, the adjustment of pH to precipitate ferric ions as hydroxides eliminates most of the iron and colloidal matter by a subsequent flotation step. In this case, the water contaminants render the hydrolysis product sufficiently hydrophobic for bubble-particle adhesion to permit flotation under quiescent conditions.

Alternatively, dissolved ions can be electrostatically attached to selective cations, which in turn can be collected in a foam layer. The recovery of Au,Pd,Pt and Ir by such foam fractionation methods, as well as the removal of uranium from aqueous solutions by the use of amines, has been made.

13.9 Process Control and Measurements

However efficient a plant is running, small changes occur in performance of the process continuously, due to various variable factors such as size of feed, pH of pulp, mineral content, etc. Therefore, it becomes necessary for the mineral engineer to keep a constant watch on the variables of the process and their effect. The important variables requiring measurement and control are discussed as follows:

a) Size of Feed and Products

Beaker decantation for sizing is quite simple to use. Tests are made on a composite sample taken from the routine sample, in proportion to the tonnage represented by each of these samples over the period concerned. Sometimes, even the sense of touch serves to detect a change in coarseness of the feed and products.

b) QUALITY OF FROTH

From the colour and appearance of froth, a trained operator can easily detect at a glance the quality of froth. This may be aided by the provision of fluorescent or monochromatic light. or the exclusion of day light from the flotation section.

c) ASSAYING OF PRODUCTS

Simple spot assay can be used to check the chemical composition of products. A colorimetric method for copper, nickel, manganese, etc., can determine their contents within 10 per cent of their values in a few minutes. In case of complex mixture, such as ZnS and PbS, determination for associated iron may provide rapid information. Oxidised copper can be tested by shaking a few grams of ore in a test tube with ammonia, filtering and observing the colour of filtrate (blue or colourless). Tailings in flotation of cerussite can be tested after panning by adding few drops of Na_2S. The presence of cerussite will be shown by brown colour. In all cases the aim is to detect the trouble making conditions, to avoid the loss of values. The analysis laboratory is usually several hours behind the operation with the information and thus any measure taken to reduce the time-lag is important. Continuous fluorescence analysis of samples drawn from required points gives rapid indication of any change in composition. If it is coupled to corrective devices, necessary corrections can be applied rapidly.

d) MICROSCOPIC EXAMINATION OF PRODUCTS

Microscopic studies of tailings and scavenger froths may give the guide to control of process. Details of microscopic examination in process control are given in Ch. 17. Changes in reagent or machine setting may be carried out depending upon the changes. Microscopic examination of samples would establish the reasons for the failure of a particular mineral to float, and thus corrective measure can be taken for it.

e) PULP pH

Use of the correct pH is very critical for separation of minerals, their optimum recovery and grade of concentrate. The value of pH can be measured rapidly with a portable pH meter, but it is likely to be damaged. A comparator can be used with suitable indicator with less chances of getting damaged.

f) HEIGHT OF FROTHING HEAD

It determines the amount of sorting work done on the floated particles. Height of frothing head can be measured by a floating gauge. For an efficient operation, frothing head should decrease slowly from head to tail. The last scavenger cell should show only a mineralised fugitive froth. The weirs of these final cells are set high to facilitate the removal of the last froth into the froth launder.

g) PULP DENSITY

Scavening operation needs higher pulp density (30–45 per cent solids) whereas cleaning requires low pulp density (15–20 per cent solids). The pulp density can be measured by hydrometer and necessary corrections are applied. Pulp density can also be measured by taking the pulp streams from different points, followed by filtration and drying of the solids. Pulp density can be found by knowing the weight of dried solids and volume of pulp.

h) TEMPERATURE OF PULP

In general, conditioning and flotation are carried at the plant temperature, but in some specific cases it may become necessary to maintain the pulp temperature higher than the plant temperature (e.g. in case of using fatty acids). Temperature can be measured with the help of a thermometer and necessary changes can be made.

13.10 Recent Developments in Flotation Process

Developments in flotation technology may be considered in various respects, such as new reagents resulting into improved flotation flow sheets, equipment design, increasing size of flotation cell, new methods in process control of flotation process, etc. Some notable advances and developments in the field of flotation technology are as follows:

13.10.1 INCREASED SIZE OF FLOTATION CELLS

Main efforts have been to construct larger machines of conventional design or modified to meet specific needs. In the mid-sixties a number of designs of flotation impellers (Denver DR, Minmet Cross barred rotor, Boliden FR rotor, etc.) have been developed. At that time the usual size of a cell was about 2.8 m^3. After the development of these impellers, cells of conventional design could be constructed up to 17 m^3 in size to treat very high tonnage of low-grade ores. Use of these large cells reduces the total number of units with their ancillary motors, switches, power lines, walkways, etc. These result in savings of space and energy. Presently, the cells as large as 36 m^3 in volume (Denver 36 m^3 DR) are in use.

In order to meet the new economic requirements for treatment of very low-grade ores, increasing interest in the design of pneumatic mechanical flotation machine (a subaerated machine with air supplied) has been observed during the last three decades. I–Z type pneumo-mechanical flotation machines developed in Poland, are used for the concentration of coal and nonferrous metal ores. Presently, single impeller cells having up to 30 m^3 are in use.

13.10.2 DEVELOPMENTS IN DESIGN OF FLOTATION MACHINES

An injection flotation cell known as Davcra cell has been developed in Australia. This cell has the form of a tall rectangular tank. The pulp and air are admitted through an injector at the base of the cell (Fig. 13.13). This arrangement provides more efficient mixing of air–pulp compared to the mixing in conventional machines.

Another new design of flotation cell known as Nagahm cell, has been developed in Japan particularly for the recovery of ions, precipitates, and ultrafine particles arising in overflow from concentrate thickeners. This cell incorporates various favourable aspects such as (a) large volume of bubbles generated, (b) evenly distributed small bubbles, (c) non-turbulent condition in cell, and (d) provision for discharging large volumes of froth. Several

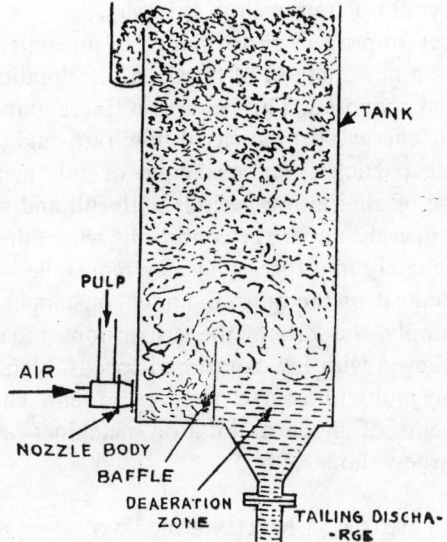

Fig. 13.13 Davcra flotation cell.

other methods are also being developed for floating fine particles. Electrolytic action may be used to generate small bubbles capable of floating ultrafine particles.

A modified technique using froth column as a filter has been developed in Russia. In this case the same range of reagents may be used. The conditioned pulp is fed on to the pulp of a prepared froth column of about 1 m depth. Hydrophobic particles get trapped in the froth while the hydrophilic particles get washed through the bubble wall fluid to join the water phase below the froth column. Froth is skimmed from the top and the froth column working as bed is continuously regenerated from below. This design has the advantages of retaining coarser material in the froth column, which cannot be floated by a conventional cell. However, fine gangue particles may also be retained. Therefore, this new process should be operated as a complement to conventional one, on a fraction separated by tabling or classification prior to flotation.

13.10.3 DEVELOPMENTS IN TECHNOLOGY

Flocculation of −2 micron slimes can be effected on a selective basis. In this case small amounts of polymers are added which are adsorbed on to specific minerals causing them to form large flocs. These flocs can readily be separated from the remaining dispersed mineral species.

A new technique called ultra-flotation has been developed to float −5 micron material. This technique can be used to float contaminants away from minerals. Some carrier minerals (such as barite, calcite, fluorite, silica, etc.) are ground to a slime and conditioned to cause their attachment to the unwanted contaminants and finally the coated particles are floated off. For example, anatase could be removed in this way.

In order to get improved performance of pneumatic and mechanical flotation machines, a new method of flotation, i.e. flotation with intermittent air supply has been proposed. This possesses three importan: advantages, i.e. (a) savings in energy, (b) increased flow rate, and (c) improvement in flotation of coarse particles. The principle of this method consists of an adjustment of time of alternating periods with full and reduced air supply. This system can operate under the conditions where the solid particles are not dispersed adequately in a three-phase system (solid–water–air) and there is a tendency to deposit on the bottom, under constant air supply. By the use of intermittent supply, the state of the suspension in the flotation machine can be maintained even with low rotational speeds, high pulp density, considerable air supply and with coarse particles present. This new method fulfills the requirements of modern flotation machines and has led to the development of large-volume cells.

13.10.4 Use of Collector as Mixture of Two

Suitable addition of amino-acid along with xanthate to sulphide ore flotation results in the promotion of xanthate adsorption on sulphide mineral. Simultaneously it also acts as a dispersing agent for slimes.

13.10.5 Magnetic Treatment Applied to Flotation

In the last 25 years, many possibilities of considerable improvements in the flotation, thickening and filtration of pulp, and reagent solutions after the magnetic treatment of water have been discovered. It has been established that magnetic treatment carried out under optimum conditions improves flotation rate and recovery.

13.10.6 Electrochemical Treatment of Water

Electrochemical treatment of process water may provide the control of pH, redox potential, hardness and ion content of the medium. By using electrolysed water, an increase in selection and recovery of minerals may result by a shift in the redox potential, a reduction of hardness and the presence of finely dispersed gas bubbles in the aqueous phase.

The use of catholyte in copper–tin, zinc, and barite flotation, and anolyte in pyrite flotation makes it possible to increase the recovery of copper in its concentrate by 3 per cent, the tin in its concentrate by 2 per cent, the zinc in its concentrate by 5 per cent and the barite by 3 per cent. It also allows a 40 per cent reduction in soda and H_2SO_4 consumption.

13.11 Future Trends

The recent decline in the mining and metallurgical industries has shifted emphasis away from conventional flotation technology. Nonetheless, understanding of the fundamentals of flotation is by no means complete and development of sophisticated instrumentation is leading to combined research on materials science and on surface fundamentals of flotation.

As new applications, such as control of the so-called pulp potential, control of redox conditions, E_h, use of critical surface tension of wetting, and flotation of ions and dissolved species, such as toxic and hazardous materials, become more widespread, froth flotation will be employed more and more.

Computer technology and instrumental control now make larger plants possible with higher-grade products and more complete recoveries. Furthermore, pyrometallurgy and hydrometallurgy are now used together with flotation for recovery of metals from slags or melts.

The flotation of precious metal-bearing ores and native gold is widely applied despite the apparent fate of sulfide mineral flotation. Less familiar systems such as rare-earth minerals are finding increased interest. Column flotation promises wider application. The custom-design of flotation chemicals is also under development and wider application of flotation in recycling industries is at its early stages.

CHAPTER 14

Magnetic and Electrical Separation

The practical significance of magnetic separation was recognised in 1792 for separating iron ore by magnetic attraction. However, it could be possible to make its commercial use after 57 years, i.e. in 1849, when a patent for a commercial magnetic separator was issued in the U.S.A. Since that time, the science and technology of applied magnetism have developed rapidly. The scientific understanding of the nature of magnetic force started only in the early 20th century. By combination of various forces of magnets with gravitational and frictional forces, it is possible to separate various mineral fractions. As in other processes of concentration, magnetic separation also produces three products, i.e. concentrates, middlings, and tailings.

14.1 Magnetism

The field around a magnet exerts a force called *magnetic field* which resembles a gravitational or electric field of force. When a material is placed in a magnetic field, it is subjected to either attraction or repulsion. In magnetic separation, the magnetic susceptibility of certain minerals is taken advantage of. Some minerals are strongly magnetic, whereas some other are weakly or feebly magnetic. At low intensity, strongly magnetic minerals can be separated from weakly or non-magnetic and by variation in the intensity, weakly and feebly magnetic minerals can be separated. Depending on the attraction or repulsion, the magnetic materials can be classified into two categories, i.e. (a) paramagnetic, and (b) diamagnetic.

Paramagnetic materials are those which are attracted along the lines of force of a magnetic field to the point of greater magnetic field intensity. Ferromagnetism is considered to be a branch of paramagnetism. Ferromagnetic bodies are differentiated from other paramagnetic bodies by the magnitude of their reaction with a magnetic field. In case of paramagnetic materials, the value of magnetic susceptibility (k) will be small and positive, i.e. slight tendency to line-up parallel to the lines of force of a field, whereas in case of ferromagnetic materials the value of magnetic susceptibility will be large and positive, i.e. strong tendency to line-up parallel to the lines of force. Before 1939, only few ferromagnetic materials (mainly iron and magnetite) were available for industrial application. Today, a large number of ferromagnetic materials are available. The most important are ferrites (natural or synthetic).

Diamagnetic materials are repelled from the regions of greater field intensity along the lines of force of magnetic field to points of smaller magnetic field intensity. Diamagnetic forces are too weak to be of any use in processing of minerals. The minerals responding to these with many feebly paramagnetic ones are termed as non-magnetics.

14.2 Magnetic Materials and Their Separation

The oldest known magnetic material is magnetite (lodestone) which is natural ferrite (Fe_3O_4) or ($Fe^{++} Fe_2^{+++}O_4^{--}$). All ferrites contain their trivalent Fe-ions and their oxygen anions in 2-to-4 ratio. In synthetic ferrites, the divalent can be provided by any metallic atom, small enough to fit into the crystal lattice. The general formula of ferrites can be represented as $X^{++}Fe_2^{+++}O_4^{--}$. In this case the resistance to flow of current is high, as the crystals are ionic and the lower electron shells of their constituent atoms are filled. The spin of the electrons round their atomic nuclei produces magnetic effect. Pair electrons in most materials spin in opposite directions and therefore, cancel out, but the atoms of transition elements (Mn, Fe, Ni, and Co) contain at least one unpaired electron and thus these are magnetic in nature.

Magnetic separation can be effected to those materials in which a natural or induced degree of polarity can be sustained during their passage through a field of magnetic flux. In order to use ferromagnetism, this field should be steady and is produced either by the use of permanent magnets or electromagnets energised by direct current or alternating current (in case of ferromagnetism). For the successful magnetic separation, the particles exposed to the magnetic flux should respond with adequate strength to overcome frictional, gravitational and inertial constraints. The magnetic separation, thus should be aided by size control, controlled rate of feed, and by the use of appliances giving the gravity effects on dry or pulped ore.

14.3 Mechanism of Magnetic Separation

The degrees of susceptibility differentiate strongly and weakly magnetic materials into separate products. The attracted particles should be deflected from the others in the moving stream of material. The other factors influencing the separation are size and specific gravity of particles, and freedom of movement of particles relative to moving stream.

Magnetisation per unit mass (σ) may be defined by the following equation:

$$\sigma = \frac{B-H}{4\pi\rho} \tag{14.1}$$

where, B = induced flux per unit area measured in gauss (1 gauss = 1 maxwell/sq. cm),

H = flux density of the existing field (in gauss),

and ρ = specific gravity of the particle.

The susceptibility (ratio of magnetisation intensity to magnetic field strength) may be defined by the following equation as X.

$$X = \frac{\sigma}{H} \tag{14.2}$$

The value of X is measured in electromagnetic units (emu's). The unit magnetic pole (m) may be defined as the following:

$$m = r \cdot F \cdot \mu \tag{14.3}$$

where, r = distance of magnetic pole from a like pole,
F = repulsion, and
μ = magnetic permeability compared with air (unity).
The unit of H may be defined as the following:

$$H = F/m \tag{14.4}$$

On achieving the maximum magnetisation, a particle is said to be saturated. At the field strengths usually employed in processing of ores, all ferromagnetic materials (magnetite, pyrrhotite, magnetopyrite, and ferro-silicon) become saturated. The value of X for magnetite is 93 emu/g. Between this and non-magnetic materials, there is a large number of weakly and conditionally magnetic materials. In addition, many iron minerals acquire ferromagnetic properties after reduction roasting. Although at the lower end of the scale, susceptibilities of paramagnetic materials vary from 10^{-4} to 10^{-6} emu/g, it is possible to treat such materials by the use of high intensity separators. Strong fields (up to 20,000 gauss) are produced by powerful electromagnets for this purpose. The pole pieces should be properly designed to give the appropriate field gradient in the flux path. The field of permanent magnets does not exceed 7,000 gauss, and thus are not suitable for the separation of feebly magnetic materials.

In order to produce a steep flux gradient, the attracting pole is given a wedge-shape, whereas counterpole is made flat (Fig. 14.1). The lines of force

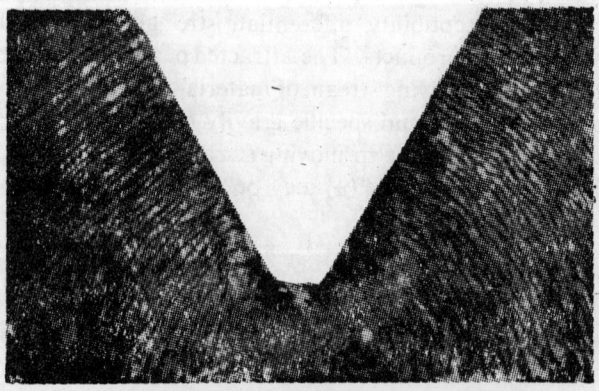

Fig. 14.1. Lines of magnetic flux.

spring from the poles normal to the surface. Therefore, the greatest attraction occurs along the sharp edge of a wedge-shaped pole, and practically there is no attractive tendency just adjacent to a flat pole surface. The attraction force is inversely proportional to the square of the distance between the attracting pole and the attracted particle. In a number of applications, the magnetic alloys of high gauss have replaced electromagnets. These magnetic alloys are described not only according to composition, but also in terms of their *coercivity* (H_c), remanence (B_r) and maximum gauss (BH_{max}). Coercivity may be defined as the magnetic force necessary to demagnetise a substance which has been magnetised to saturation. The remanence is the residual magnetism in a ferromagnetic substance (its hysteresis) after withdrawing the external magnetic force. The properties of some important magnetic alloys are given in Table 14.1.

Table 14.1. Properties of some magnetic alloys

Alloy	B_r gauss	H_c oersted	HB_{max} M gauss-oersted
35 % Cobalt steel	9,000	250	0.95
Alni (Nial)	5,600	580	1.25
Alnico (normal)	7,250	560	1.7
Alcomax II	13,000	580	5.4
Hycomax I	9,000	825	3.2
Hynico II	6,000	900	1.8
Columax	13,500	740	7.5
6 % Tungsten steel	10,500	65	0.3

Electromagnets are employed for heavy-duty work, as in this case it is easier to control the field flux by varying direct current as well as the number of turns in the coils surrounding the high permeability steel core. A further control is obtained in operation by adjusting the air-gap between the particles and the attracting magnet.

14.4 Minerals Responding to Magnetic Separation and Factors Affecting the Separation

Magnetite, franklinite, ilmenite, and wolframite can be directly concentrated by magnetism. Many other minerals may also respond to magnetic separation depending on the distinctiveness of reaction of magnetic minerals at a given mesh size. The feebly magnetic minerals such as biotite, garnet, basalt, zeolites, pyrochlore, muscovite, and chlorite can be concentrated by magnetic separators.

Good data on the magnetic susceptibility and permeability of minerals are not available, since many factors influence the measurements. The important factors influencing the magnetic properties of minerals are:

a) Mechanically held impurities in the grains.

b) Dissolved impurities in the mineral of interest.

c) Grain size and crystal aggregation of the crystal complex.

The difficulty is also experienced in making the measurements of magnetic properties on fine particles and due to the effect of intensity of the magnetic field, in which the determination is being carried out. However, an idea of the magnetic permeability of various minerals may be had from Table 14.2.

Table 14.2. Magnetic permeability of various minerals

Material	Permeability
Iron	2.16
Magnetite	1.47
Franklinite	1.41
Ilmenite	1.28
Pyrrhotite	1.078
Siderite	1.022
Hematite	1.008 to 1.024
Zircon	1.002 to 1.029
Limonite	1.0088 to 1.0099
Corundum	1.0018 to 1.025
Pyrolusite	1.0078 to 1.0088
Manganite	1.0061
Garnet	1.0047
Quartz	1.0022 to 1.0055
Rutile	1.0030 to 1.0053
Pyrite	1.0007 to 1.0064
Sphalerite	1.0007 to 1.0057
Dolomite	1.0015 to 1.0056
Apatite	1.0026
Willemite	1.0024
Talc	1.0008
Arsenopyrite	1.0017
Chalcopyrite	1.0016
Gypsum	1.0005 to 1.0033
Fluorite	1.0010 to 1.0017
Zincite	1.0012
Orthoclase	1.0001 to 1.0011
Calcite	1.0004

A tiny inclusion of magnetite in a relatively large particle of a non-magnetic particle may affect considerably its response to a magnetic field and may present it as a feebly magnetic particle. The effect of dissolved impurities is large. An example is the influence of dissolved iron on the magnetic properties of ferruginous sphalerite (marmatite). Variations in magnetic susceptibility have been observed to result from variation in particle size also, however, the variation is small and can be neglected for approximation.

For industrial concentration by magnetism, following requirements should be met for an efficient operation:

a) A thin layer of particles should move through consecutive magnetic fields with successive reversal of their polarity. This arrangement causes the ferromagnetic particle to respond by turning through 180° and thus any entrained gangue is freed.

b) Successive magnetic fields should be increasingly strong or the air-gap should decrease progressively to effect the removal of most strongly magnetic particles first and the weaker ones later.

c) The feed should be presented in such a way that a gangue particle should not pin down a ferromagnetic particle on which it may be lying.

d) The machine used should provide converging field at each point of separation, a means of regulating flux intensity, speed control through the field, scrubbing by pole-reversal to free the entrained gangue, and separate discharge for concentrate, middlings (weakly magnetic) and tailings (non-magnetic).

e) The machine employed should suit to the optimum liberation size of the ore.

f) If the treatment is dry, moisture content should not exceed 0.5 per cent.

g) Feed should be provided in a specified size range.

14.5 Application of Magnetic Separation

Initially, magnetic separation was employed to separate strongly magnetic iron ores (magnetite) from gangue or other less magnetic minerals. With the advancement of technology and design of machines, it was adopted in separation of ores containing iron or manganese, which are only feebly magnetic. For example, hemetite, limonite, and siderite can be separated from their gangue by application of high intensity fields. Magnetic separation finds its application in the direct separation of many other minerals which are very near to one another in specific gravity. The important separations carried out include (a) sphalerite from pyrite, (b) sphalerite from garnet and rhodonite, (c) rhodonite from garnet, (d) separation of various silicates of manganese from one another and also their separation from zincite, willemite and calcite, (e) separation of tin–tungsten concentrates, (f) rutile from apatite, (g) rutile, garnet, and monazite from one–another, (h) garnet and other iron minerals from corundum, (i) siderite from cryolite, (j) emery from gangue, (k) biotite from gangue, (l) hornblende from valuable minerals, (m) chromite from other heavy minerals, and (n) rutile from ilmenite.

Other applications include (a) removal of deleterious magnetic material from a product such as china clay, silica sand, etc., (b) cleaning of magnetic material used in heavy media separation, and (c) removal of tramp iron from ore to safeguard machines.

14.6 Magnetic Separators

In the earlier times the only industrial magnetic separators available were using electromagnets, and the permanent magnets were weak and ineffective on industrial scale. However, with the advent of alnicos and ceramic permanent magnets, it could be possible to use permanent magnets industrially. Improvements have also been made in use of electromagnets. Many of the earlier problems associated with magnetic separators have been overcome due to the development of better magnetic materials and scientific designs. Various magnetic separators can be classified on the basis of medium employed in separation (dry or wet), the mode of presentation of feed, the mode of disposal of the products, and whether the magnets are moving or stationary. The various magnetic separators employed industrially are discussed in the following sections.

14.6.1 Dry Magnetic Separator

Dry magnetic separation can be practised at low as well as high intensity. The machines with low intensity are more common in concentration of lump ores and coarse sand (to remove strongly magnetic particles). A rough concentrate obtained from this process may need further comminution and magnetic treatment. Dry magnetic separator possesses the following advantages:

a) More economic due to absence of water requirement which may be 2,500–5,000 litres/tonne in wet magnetic separator.

b) The process is attractive in the field of dry autogenous grinding and climates subject to freezing, if wet method is used.

Various types of dry magnetic separators are discussed below:

i) Dry Belt Magnetic Separators

One of the oldest and most successful magnetic separator of belt type is Wetherill magnetic separator (Fig. 14.2). This is a high powered machine. The material flows from the hopper to the feed roller, which discharges the feed in a uniform layer over the entire width of the conveyor belt, passing between the magnets (electromagnets). The poles are arranged one above the other. The poles of the upper magnets have the shape of a sharp wedge, while the lower poles are flat. This arrangement provides the jump of magnetic particles toward the upper poles (as soon as they are brought by con-

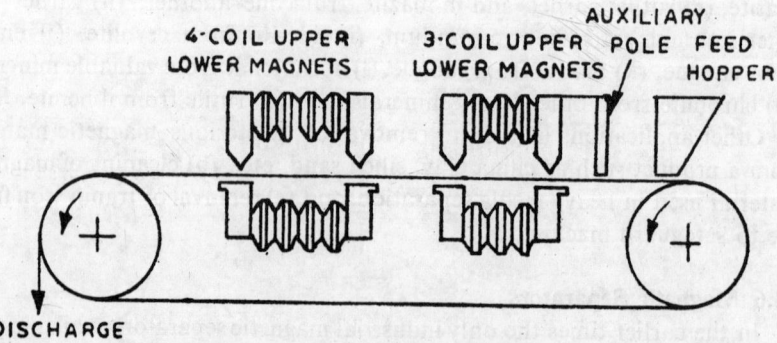

Fig. 14.2. Belt type dry magnetic separator.

veyor belt into the magnetic field) due to concentration of magnetic force on the wedge-shaped edges. The cross-belts move between feed-belt and wedge poles and prevent the adhering of magnetic particles to the poles. The cross-belts convey the magnetic particles out of the system. The same principle has been used in many other belt magnetic separators.

This type of machine can be constructed with one, two or three pairs of magnets. Each pair of pole is provided with the control for current strength. This machine has been widely used earlier in concentration of zinc–lead–iron sulphides (zinc blende as magnetic product), magnetite ores (magnetite as magnetic product), and many other applications.

ii) Dry-drum Separator

This is an early machine employing low intensity and is still used in various forms. It consists of a series of oppositely charged or same polarity magnets in a semicircular form as shown in Fig. 14.3. The magnet is held stationary and surrounded by a renewable non-magnetic rotating cylinder (e.g. made of brass). The feed is introduced at the top of the drum. The non-magnetic particles rebound off the drum or are carried around by the drum rotation until they fall by gravity. The ferromagnetics are gripped by magnets to the drum and roll with each change in polarity, till they are dropped off below into suitable receivers.

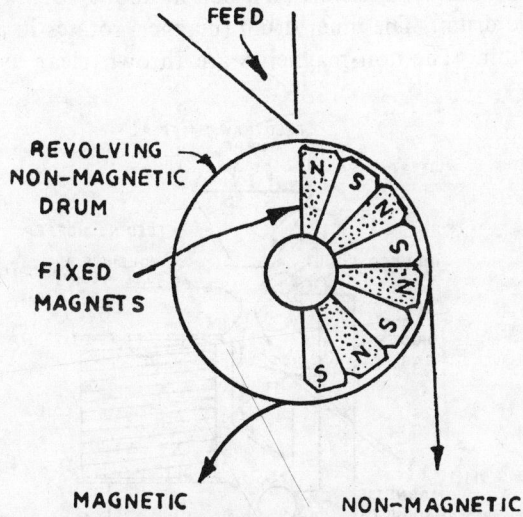

Fig. 14.3. Schematic diagram of dry-drum magnetic separator.

In this case, the strong electromagnets cannot be used and thus these are usually low-intensity magnetic separators (600 gauss field strength). Though it has become obsolescent but can be successfully used for coarse feeds (+ 6 mm). With the development of strong magnetic alloys, a large variety of dry-drum separators are developed to treat the wide range of materials.

iii) *High Intensity Dry Magnetic Separator*

The present day work mainly requires the treatment of weakly magnetic sand and fine powders which require the use of high-intensity magnetic separator. The removal of trapped gangue particles from magnetic aggregates can be facilitated by a rotating field achieved by setting permanent magnets with a pole distance of only a few centimetres. Centrifugal force of the rotating drum can be used to discharge the more weakly attached feebly magnetic particles. The magnets used in these drums can be close packed. In high-intensity dry magnetic separator, an efficient separation can be achieved by having a single layer of particles, high operating speeds, making magnetic field in circular, spiral, or helical shape, making use of gravity force for separation, and using an opposing air stream.

Basically the separator consists of permanent magnets on a drum which rotates at a different speed inside an independent stainless steel drum. The magnetic particles are deflected during their fall through air. Various types of machines have been developed to handle 3 mm to 50 micron size materials. The main difference is between suitable dynamics for easily settled material and that easily airborne. One such separator (Erzbergan dry magnetic concentrator) is shown in Fig. 14.4, which is used for the treatment of weakly magnetic sands (2 mm to 70 microns). The system can use up to seven laminated rolls with progressively smaller gaps.

In one system the feed travels on a belt at about 125 m/min and passes over the double drum. The inner drum (magnet) rotates in a reverse direction at 125 m/min. The non-magnetics are thrown clear as the belt turns

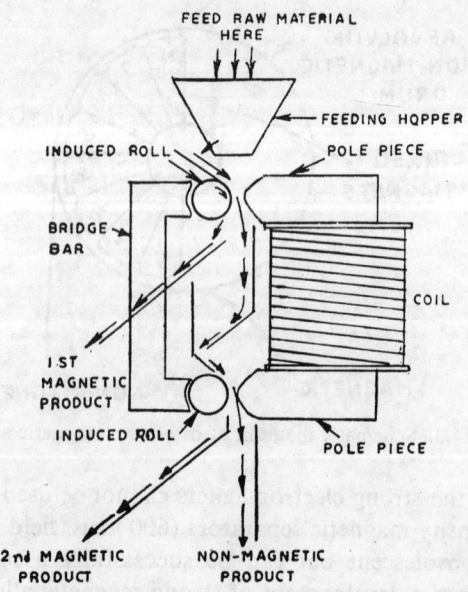

Fig. 14.4. An arrangement for the treatment of weakly magnetic sands.

over the drum, whereas the magnetic particles cling as they are accelerated by the effect of field flux, until the magnetic particles are separately discharged when belt leaves the drum. In another system, the magnet drum is mounted in a second unit above the feed belt and travels in the opposite direction. The flux density and gap are adjusted in such a way that the magnetic particles are picked up more than once and dropped before their final adherence to the upper belt. The speed of the upper belt is two times of the feed belt, and the magnet drum rotates in the opposite direction as before. The separation of clinging particles is further aided by an air blast.

A relatively finer material (250–75 microns) is treated in a different manner. The material is fed from a hopper on to a rotating stainless steel drum. The inner magnetic drum rotates inside the stainless steel drum, but at a different speed. The feeding is aided by the stirring of material caused by the influence of the reversing flux. The feed is stratified with non-magnetic gangue on top.

iv) Magnetic Precipitator

A device called the magnetic precipitator (Fig. 14.5) has been developed to treat very fine dust in an airborne stream. The entering air travels in a helical path and the dust is pressed against the outer rotating wall of a cylinder. The outer system of stationary or rotating magnets induces a helical magnetic field. The magnetic particles travel upward against an air draught and at the top enter a weaker magnetic field, where they are thrown off. The non-magnetics are blown down to a control discharge. These machines have been successful on Aerofall mill doing the dry grinding.

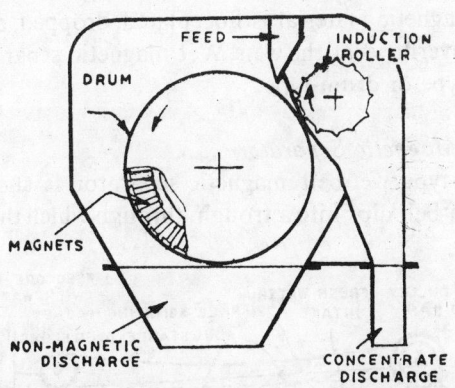

Fig. 14.5. Simplified diagram of a magnetic precipitator.

v) Pick-up Type of Drum Separator

This is shown in Fig. 14.6. This is employed for the ores containing large proportion of magnetic materials and to obtain a high-grade concentrate. The magnetic particles are lifted from the passing stream and carried forward proportionally to their tenacity, to separate middlings and fully mag-

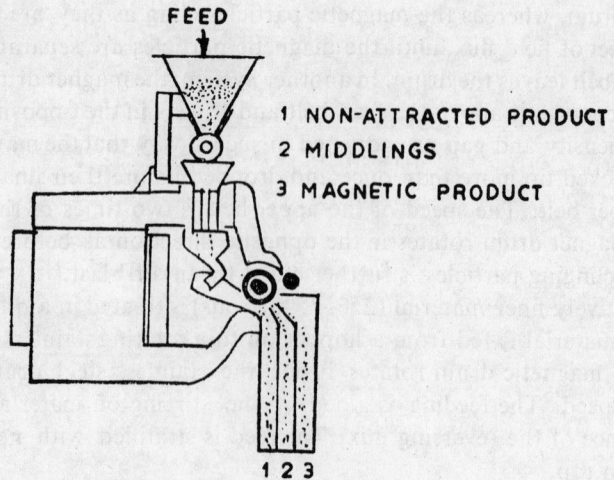

FEED

1 NON-ATTRACTED PRODUCT
2 MIDDLINGS
3 MAGNETIC PRODUCT

1 2 3

Fig. 14.6. Illustration of pick-up type of drum separator.

netic drop-off points. The non-magnetic particles are thrown from the drum, whereas seized particles are carried down till they leave the field and fall clear.

14.6.2 WET MAGNETIC SEPARATORS

Wet magnetic separators are commonly used in concentration of ferro-magnetic sands and purification of magnetic media (magnetite or ferro-silicon) used in heavy-media separation. One of the oldest devices known as magnetic log washer is a spiral classifier with magnets below its trough, in which the ferromagnetic material is flocculated, dropped, and dragged, while the barren pulp overflows at the weir. Wet magnetic separators are also basically either belt-type or drum-type.

a) Belt-type Wet Magnetic Separator

The Crockett-type wet-belt magnetic separator is shown in Fig. 14.7. In this machine, a belt dips into a trough, through which the pulp is fed under

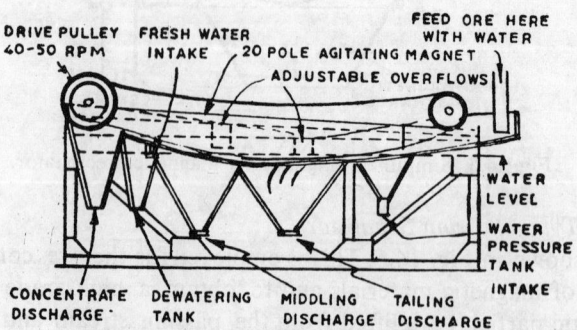

DRIVE PULLEY   FRESH WATER   FEED ORE HERE
40-50 RPM   INTAKE   WITH WATER
                     20 POLE AGITATING MAGNET
                     ADJUSTABLE OVERFLOWS

WATER LEVEL

WATER PRESSURE TANK INTAKE

CONCENTRATE   DEWATERING   MIDDLING   TAILING
DISCHARGE   TANK   DISCHARGE   DISCHARGE

Fig. 14.7. Belt-type wet magnetic separator.

the belt. The magnets are provided above the belt which lift the magnetic particles. The non-magnetics (gangue) not attracted to the first magnets above the moving belt, are dropped immediately and eliminated in the first hopper whereas the magnetic concentrate is carried further to a dewatering tank and is discharged there. Fresh water is added near the concentrate end and flows counter-current to the movement of belt. The magnet assembly is made in alternate polarity to facilitate turning over and agitation of the magnetic material held to the belt. The belt-type magnetic separators are now replaced either by drum-type or a special type.

b) Drum-type Wet Magnetic Separator

A single drum wet magnetic separator (Fig. 14.8) is based on pick-up principle. Magnetic particles are lifted from the pulp, while the non-magnetic particles (tailings) pass through. The hydraulic current of water keeps the pulp in suspension, and the magnetic fraction passes through a series of north and south fields before the concentrate is removed by sprays. The drum dips into a receiving tank and transports the magnetic fraction to the discharge point. As soon as the feed leaves the feed box, it is spread over the whole width of the receiving tank.

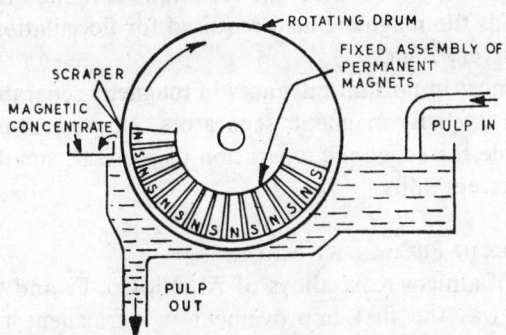

Fig. 14.8. Drum-type wet magnetic separator.

A system comprising of two drums can also be used, in which the tails from the first drum are treated (scavanged) in the second drum. However, the use of second drum can now be omitted due to the development of ceramic magnets. Presently, drum separators using permanent magnets are preferred since they are more compact.

The diameter of drum and width occupied by the magnets depend on the amount of pulp treated per unit time, solid content of pulp, percentage of magnetics in ore, desired operating efficiency, and the required grade of concentrate. In single drum separator, feed is usually below 400 litre/min containing up to 25 per cent solids, whereas double drum system can handle the feed up to 600–800 litre/min with up to 50 per cent solids. The usual width of drum range from 75 to 100 cm giving a discharge rate of magnetic 3 or 5 tonne/hr with a single and double drum system, respectively.

14.7 Advances in Magnetic Separation

The current design of wet magnetic separators for treating weakly magnetic minerals is based on the presence of ferromagnetic bodies of varying shapes in magnetic field (grooved plates, balls, rods, steel wool, etc.) which are known as matrices. A high intensity magnetic field with high gradient can be induced on the surface of these matrices, and as a result a very strong magnetic force develops which can attract even weakly magnetic particles and thus effecting their separation from non-magnetic particles.

The usual disadvantage of these designs is the problem of keeping the matrix clean and long enough for industrial conditions as the industrial pulps or slurries usually contain wood, plastics, tramp iron, strongly magnetic minerals, oversized grains, etc. The filtration of pulp to remove these unwanted materials is not much effective and thus a frequent cleaning or replacement of matrix will be needed to restore the efficiency. In order to solve this problem, special arrangements of magnet coils (e.g. quadrupole) have been developed, which are capable of producing the desired magnetic field heterogeneity.

One interesting possibility of matrixless system is the application of magnetic flocculation to the separation of magnetic and non-magnetic constituents. However, the application of this technique is limited to strongly magnetic materials, as the magnetic field required for flocculation of paramagnetic material is very high.

One of the most important advances in magnetic separators is the development of high intensity magnetic separators. In most attempts, the aim has been to enable the economic separation of minerals small in size and of low magnetic susceptibility.

14.7.1 ADVANCES IN PERMANENT MAGNETS

The advent of alnicos (cast alloys of Al, Ni, Co, Fe and Cu) magnets in the early forties was the first improvement in permanent magnets. These magnets are much stronger than earlier permanent magnets and have much greater resistance to demagnetisation. In fact, the strength of alnicos is well comparable with the strength of powerful electromagnets. However, the use of alnicos could not be possible at that time, due to their cost being very high. Introduction of ceramic magnet material in the 50s was the major breakthrough. Ceramic magnets were made from barium ferric oxide powder by cold pressing the slurry, followed by magnetic orientation, sintered, ground, and finally cut with diamond saws into the required shapes and sizes. At present, almost all the magnetic separators (except high intensity machines) are fitted with ceramic magnets.

14.7.2 NEW CIRCUITS

The development of improved magnetic materials led to improvements and new developments in magnetic circuits as well as designs of separators, which resulted in improved quality of product with better efficiency of concentration. Perhaps, the most important developments have been in wet-drum separators, in which new circuits have been introduced for both alnico and ceramic magnets. The major parameters in the design and applications of magnetic separators affecting the performance are (a) flux density, and (b) gradient of the flux density. These two factors should be given equal weightage for the optimisation of the magnetic force of attraction. As a result, higher gradient, wet and dry, and permanent magnetic drum-type separators have been developed. In many cases, these new separators may equal or exceed the performance of electromagnet separators and thus old electromagnet drums are vanishing from the industries. The newly developed magnetic separators are associated with the following advantages:

a) Saving in quantities of magnetic materials.

b) Greater capacity.

c) Elimination of the problems associated with electromagnets (such as coil failure, maintenance, and the need for constant power sources).

14.7.3 HIGH SPEED DRUMS

The high-speed dry drum separators have been developed with various options (with respect to strength, capacity, etc.) which can treat a wide range of particle sizes of iron ore, slag, and other ferromagnetic materials which are difficult to treat. Newly developed drums employ high strength magnetic elements and high speed of shell rotation (100–500 m/min), which introduce a third factor, i.e. centrifugal force (in addition to magnetic attraction and gravity in old type drums) which is directly proportional to the diameter of drum. In order to take maximum advantage of these forces, powerful magnetic circuits are designed, which permit high rotational speeds of the drum shell. Use of balanced magnetic attraction, field depth, and shell speed results in higher efficiency and higher capacity of the machine.

14.7.4 HIGH INTENSITY WET MAGNETIC SEPARATORS

High intensity magnetic fields are required to separate weakly magnetic minerals (such as hematite, siderite, ilmenite, ores of Cr, Mn, W, Ni, Mo, etc.) from non-magnetics (China clay, silica sand, sillimenite, zircon, etc.). When very high intensity fields are required, the use of electromagnetic coils is preferred. However, when relatively less intensities are sufficient, permanent magnets may be employed.

The basic elements of a continuous high intensity wet magnetic separator are: (a) an electromagnetic coil to generate the magnetic field, (b) a rotating ring containing flux converging elements (known as magnetic matrix), (c) an entry for the feed and wash water, and (d) discharge chutes for the non-magnetic and magnetic fractions below the ring. Magnetic particles are held in the matrix, whereas non-magnetics pass through the magnetised zone to the discharge chute. The trapped magnetic particles are released and flushed, as soon as they move out of magnetic zone. Table 14.3 represents the necessary magnetic intensities in gauss to extract some important minerals

**Table 14.3. Magnetic intensities required to extract various minerals
in high intensity wet magnetic separator**

Mineral	Magnetic intensity (gauss)
Apatite	14,000–18,000
Biotite	10,000–18,000
Braunite	14,000–18,000
Chromite	10,000–16,000
Columbite	12,000–16,000
Eipdote	14,000–20,000
Garnet	12,000–19,000
Goethite	15,000–18,000
Hematite	13,000–18,000
Ilmenite	8,000–16,000
Limonite	16,000–20,000
Magnetite	very small – 1,000
Monazite	14,000–20,000
Olivine	11,000–15,000
Pyrolusite	15,000–19,000
Pyrrhotite	1,000– 4,000
Siderite	10,000–18,000
Serpentine	3,500–18,000
Tantalite	12,000–16,000
Titaniferous-magnetite	500 – 3,000
Uraninite	18,000-24.000
Wolframite	11,000–16,000

employing a high intensity wet magnetic separation.

High intensity magnetic separators are built by several manufacturers, e.g. Eriez Magnetics, Sala HGMS, Jones WHIMS, Boxmag Rapid, etc. The main difference lies in the use of type of matrix. For example, Eriez used sandwich of steel wool and expanded metal, and Sala HGMS has employed a matrix ring made of sections of expanded metal sheets stacked one over the other. Most widely used Jones WHIMS separator (Fig. 14.9) employs one or two matrix rings fitted with grooved vertical plates. Grooves are provided to concentrate magnetic force at their peaks and provide valleys through which the non-magnetic material passes to the discharge funnel (at position A). Two magnet yokes at 180° with coils are enclosed in air cooled casings joined to the main frame.

The separator operates over cycles of about 4 sec and continuously. A high-intensity electromagnetic field is applied by the magnet through its poles to the grooved plates. The feed (-1 mm) is maintained, stirred in a hopper and introduced through the valve 1 for 2 sec. It is then cut-off and low velocity water is introduced through the valves 2 and 4. As the water flows, two rams give two short high pressure pulses which spread the retained magnetic particles and stir them on the grooved plates, flushing the middlings at the same time through the funnel, at delivery position B. After wash-

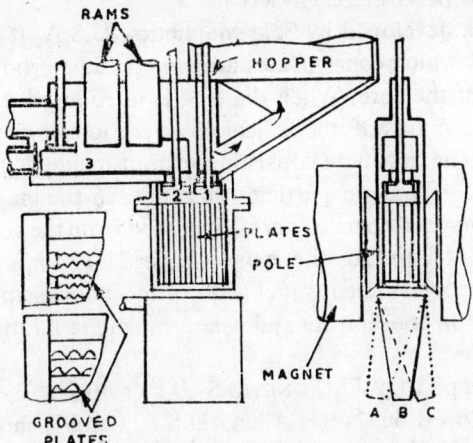

Fig. 14.9. Jones WHIMS high intensity magnetic separator.

ing for about 1.5 sec the magnetising current is cut-off and a short surge of high pressure water is introduced through the valves 2 and 3, and the delivery funnel comes to the position C at the same time. The whole cycle is then repeated. Jones separators are available in the range of 5-180 tonnes/hr throughput.

14.7.5 CRYOGENIC MAGNETIC SEPARATION

If the magnetic tractive force can be increased beyond the engineering and economic limits of the conventional magnetic systems, the use of wet magnetic separator can be further extended to finer particle size and feebly magnetic minerals. Superconducting magnets are capable of producing very intense and uniform magnetic fields up to 1,50,000 gauss (electromagnet can produce up to 20,000 gauss) with a field gradient of 7,000 gauss/cm or more.

At very low temperatures, some alloys show no resistance to electric currents. e.g. Nb–Ti at 4.2 K (temperature of liquid helium). It means that once an electric current starts to flow through a coil, it will continue without being connected to a power source, and, in effect, it will become a permanent magnet.

A pilot plant using such a magnet was built at the Department of Mining and Mineral Technology, Imperial College, London. The magnet consisted of a reverse pair of coils made of niobium–titanium alloy in a copper matrix, secured in reinforced glass fibre and vacuum impregnated with epoxy resins to keep a check on the forces between the windings and to prevent electrical breakdown. The external circuits are isolated from the magnet coils by a special switch made from superconducting wire. To prevent the heat ingress, the magnet is surrounded by liquid helium in a vacuum container. An efficient refrigerator maintains the liquid helium at 4.2 K with 1/10th of energy required to power a normal electromagnet with the same field strength. It may be possible to achieve capacities as high as 100 tonne/hr.

14.7.6 MAGNETO-DENSITY APPARATUS

This has been developed by Sala Magnetics, U.S.A. It consists of a helical core or cores which concentrates a magnetic field produced in an electric coil at one end of the core. When the feed is introduced, the magnetic fields concentrated at the tips of the helical cross-section divide the flow of feed into two paths. The first flow consisting of non-magnetic follows the helical path, whereas the magnetic particles flocculate in the magnetic field within the vessel. The magnetic particles are discharged from the vessel intermittently by periodically interrupting the magnetic field. The important features of this separator are simultaneous inclusion of mechanical and magnetic forces to yield a magnetic concentrate and a non-magnetic tailing.

14.7.7 CONTINUOUS FLOW MAGNETIC SEPARATOR

This is developed by James Allen, U.S.A., and is shown in Fig. 14.10. The feed is admitted in a helical coil mounted around a magnetic rotor having grooved edges which concentrate the revolving magnetic field. Direction of rotation is so designed that it causes the movement of magnetic and paramagnetic particles upward, toward the concentrate outlet. The non-magnetic particles move downward within the coil by gravity to the tailings outlet. An enclosed coil is the base of the unit which generates a strong magnetic field. This separator can be used for separating magnetic and paramagnetic minerals from the tailings.

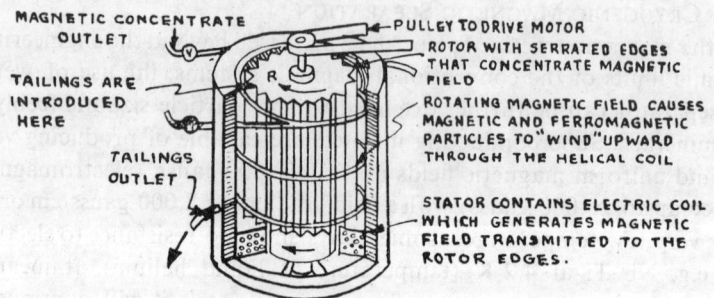

Fig. 14.10. Continuous flow magnetic separator.

14.7.8 WET–DRY MAGNETIC SEPARATOR

This separator, combining wet as well as dry processes has been developed by Aquafine, U.S.A. It consists of a metal cylinder within an annular high intensity magnetic coil. This unit is employed to purify the slurries of kaolin, magnesium carbonate, asbestos, zircon, bentonite, talc, etc. The feed is introduced at the bottom inlet which passes upward through internal hollow boxes containing fine iron particles (or ground steel wool) which trap the magnetic (iron) materials from the slurry in the boxes resulting into purified slurry. Once the boxes become loaded with magnetic impurities, the magnetic field is cut-off by demagnetisation of steel particles. The impurities are then flushed out with fresh water. The compressed air may also be used to aid the flushing out of magnetic impurities. The whole cycle is then repeated.

14.7.9 HORIZONTAL ROTARY WET MAGNETIC SEPARATOR

It is developed at Klockner Humboldt–Deutz, West Germany. In this separator (Fig. 14.11), a series of boxes containing slurry is rotated through a series of magnetised zones in a horizontal plane. Magnetisable particles are retained in the boxes by the attraction of magnetic particles to the ferro-magnetic plates in the boxes. As the boxes rotate from the magnetised zone to the neutral unmagnetised areas, the non-magnetics are flushed out of the boxes first. Later, the magnetic fraction is flushed off the plates in the boxes.

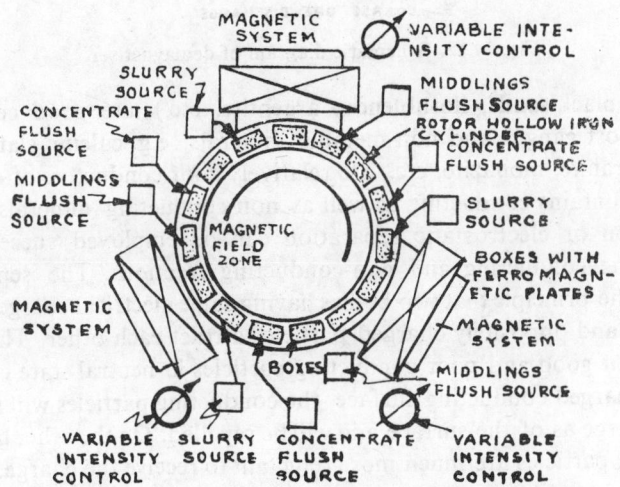

Fig. 14.11. Horizontal rotary wet magnetic separator.

14.8 Demagnetisers

After subjecting the minerals to magnetic field and withdrawal from the field, susceptible minerals retain a certain residual magnetism. In rougher concentrates and middlings, the residual magnetism is objectionable because it prevents the rearrangement of the particles and exclusion of mechanically entrapped gangue particles (non-magnetics). In order to remove this permanent magnetism, a demagnetiser is needed. One such system has been given by E.W. Davis, which is similar to that used to demagnetise watches. It consists of an a.c. coil through which the dry ore or the pulp is allowed to flow. The diameter of this coil reduces gradually to provide the gradual reducing intensity of a.c. magnetic field (as shown in Fig. 14.12). The alternation in the direction of the field as well as gradual reduction in the intensity of the field, is effective in removing the permanent magnetism from the particles.

14.9 Electrical Methods

Native metals (Au, Ag, Cu, etc.), many metal sulphides (chalcopyrite, galena, molybdenite, etc.) and other minerals (graphite, garnet, chalcocite,

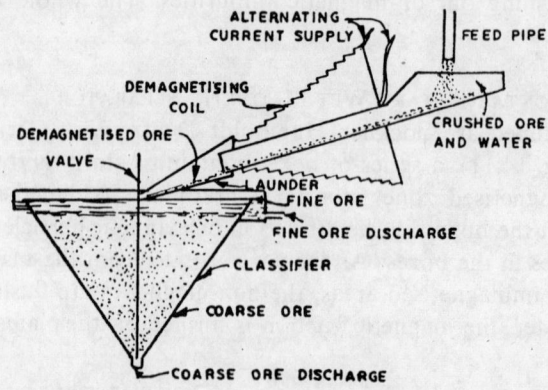

Fig. 14.12. Schematic diagram of demagnetiser.

tellurides, black sands, hornblende, argentite, etc.), are good conductors, whereas most gangue and non-metallic minerals, e.g. calcite, slates, spinel, gypsum, granite, monazite, etc., are relatively poor conductors of electricity. If an ore contains conducting as well as non-conducting minerals/particles, high-tension or electrostatic separation can be employed successfully to separate the conducting and non-conducting fractions. The separation is based on the principle that two bodies having same electrical charge will repel each other and oppositely charged particles attract each other. Therefore, if a mixture of good and poor conducting particles in neutral state is fed upon a highly charged conducting surface, the conducting particles will receive the similar charge as of the surface and will be repelled. On the other hand, non-conducting particles are much more reluctant to receive the charge, and thus, not repelled readily.

If a material charged to high potential with opposite charge to that of separating surface is fed, the good conductors will assume immediately the same charge as of the charged surface and will be repelled, whereas, non-conducting particles carrying a charge opposite to that of the surface will adhere to that surface, and thus this makes a sharper separation. However, theoretically, it is not necessary that one of two minerals should be a good conductor and the other be a poor conductor, but the difference in their conductivity will effect the separation.

14.10 Mechanism of Electrical Separation

Electrostatic separation is applicable in dry condition for small particles. The separation is based on the relative ability of particles to acquire and retain the electrical charge applied at a high voltage. When minerals are exposed to high voltage, they may capture or lose electrons and then be attracted to or repelled from, or neutralised by other bodies which are either grounded or charged.

The static electricity (employed in separation) is defined as the electrical charge which is temporarily fixed on the charged body. If there are more than one region having fixed charge in a system, a regional interaction will result. This interaction may be modified by the magnitude of charge, polarity, distance between the charged regions, and dielectric constants of the minerals concerned. All these effects are made use of in high tension separation, which is mainly based on electrostatic differentiation. The charge on particles may be acquired by conductance, ion bombardment (gaseous), friction, thermal strain, and light or radiation conductivity.

14.10.1 CHARGING OF PARTICLES BY CONDUCTANCE

When two types of particles are placed between a positive and negative electrodes and the particles rest on the positive electrode, the non-conducting particles will receive equal charges from both electrodes and thus will remain neutral. On the other hand, negative charge acquired by the conducting particles will be neutralised and thus will have entirely positive charge.

14.10.2 CHARGING OF PARTICLES BY ION BOMBARDMENT

When the mineral particles are allowed to pass through ionised gas, one of the ionic species of particles captures counter ions and as a result, a charge is acquired on the particles. If the resulting charged particles are contacted with an electrode or a ground surface, only non-conductors retain their charge and only in the areas which are not in physical contact with the surface.

14.10.3 CHARGING OF PARTICLES BY FRICTION

In this case the electrons are transferred and frictional charge is developed by rubbing together the two different compounds.

14.10.4 CHARGING OF PARTICLES BY THERMAL STRAINS

The thermal strains in crystals may give rise to local areas having opposed charge. This is also known as pyroelectric effect.

14.10.5 CHARGING OF PARTICLES BY LIGHT OR RADIATION CONDUCTIVITY

This is referred as a photoelectric effect. In this case, the incident light or X-rays cause the emitting of electrons as a result of which the particles acquire positive charges.

The forces employed in electrostatic separation are delicate and can be used to treat particles of low specific gravity having high ratio of surface to weight. Usually the upper limit or particle size is 3 mm. However, the shape also affects the practical size limit.

14.11 Conductivity of Minerals and Factors Affecting Them

The conductivity of minerals varies enormously from one mineral to another. The variations may also occur within the single ore, because of variation in solid solution or due to variable occlusion of moisture. The small impurities thus affect greatly their conductivity and their behaviour in electrostatic separators. The treatment of minerals with organic reagents may change their electrical response due to change in conductivity caused by different affinities for adsorbed water. A hydrophobic surface has a sharply decreased conductivity. Therefore, conditioning of minerals before separation by high tension can be looked as similar to the conditioning before flotation process. However, pre-treatment does not affect the electrical response of minerals possessing high conductivity.

The conductivity of minerals depends on the crystalline nature and the state of lattice. The relative conductivities (in ohm^{-1} cm^{-1}) of important minerals and metals are given in Table 14.4. The various factors affecting the conductivity of minerals and thereby separation are given below:

Table 14.4. Relative conductivities of important minerals and metals[6,8,9]

Metal/Mineral	Conductivity ohm^{-1} cm^{-1}
Copper	0.634×10^6
Gold	0.455×10^6
Covellite	8×10^3
Galena	3.35×10^3
Graphite	0.7×10^3
Pyrrhotite	119
Chalcocite	91
Pyrite	41.7
Magnetite	1.2
Chalcopyrite	0.98
Cuprite	25×10^{-3}
Siderite	0.14×10^{-3}
Marble	10^{-9} to 10^{-11}
Mica	10^{-12} to 10^{-17}
Quartz	10^{-14} to 10^{-19}
Sulphur	10^{-17}

14.11.1 SURFACE CONDITIONS OF MINERALS

If the mineral surfaces are contaminated, separation may be improved by wet attrition, chemical solution or de-dusting. However, the cost of retreatment and redrying should be justified. A superficial film of dust may affect the efficiency adversely. A selective conductivity can be produced by amines, barium chloride ($BaCl_2$), hydrogen fluoride (HF), sodium cyanide (NaCN) and copper sulphate ($CuSO_4$) due to the controlled humidity absorbed by these reagents.

14.11.2 TEMPERATURE

Differentiation in conductivity of most minerals can be improved by raising the feed temperature. Heating to a temperature as high as 600 °C has been used in some cases.

14.11.3 TYPE OF MATERIAL

From the electrostatic point, the materials may be (i) conductors (metallic or ionic) having their electrons highly mobile, (ii) insulators or dielectrics having the electron mobility very low, and (iii) semiconductors which are in between the two (higher electron mobility than insulators but less than conductors). Electron mobilities increase in all materials with the strength of the applied field. Thus by employing high field potentials, even the dielectrics behave as conductors. The semiconductors are highly sensitive to even small amounts of impurities.

14.11.4 SPECIFIC GRAVITY, SIZE AND SHAPE

The maximum treatable size depends on the specific gravity and shape. For example, heavy sulphides can be treated up to about 3 mm size, whereas coke can be treated up to a size of 25 mm. The minimum treatable size is about 50 microns, since smaller than this, do not flow freely through the separating field.

14.12 Electrical Separation Processes

In general, electrical separation processes may be classified as follows:

a) Processes depending on difference in contact electrification, either between particles or between the separating surfaces and particles.

b) Processes based on difference in conductivity.

c) Processes based on electric polarisation.

d) Processes based on difference in dielectric constants.

e) Processes utilising the principle of photoelectricity or photoconductance.

In industry, the processes based on electrical conductivity are applicable. These are termed as electrostatic and high tension separation methods.

In electrostatic separation, the mineral particles are charged by induction, conduction or friction from a charged surface. When the particles pass over a metal surface charged by induction, the particles are themselves charged by conduction from the charged surface. In frictional charging, the particles acquire a charge during their passage over a surface with an intimate contact. On completion of the charging process, the particles pass through a high intensity electric field and are deflected depending on the sign and magnitude of their charge.

In high tension separation, in addition to the electrostatic field (by conduction or friction), charging by convection (particles are sprayed with a discharge from a sharp edge points or fine wire) or ion bombardment is also used. In modern industrial practice, high tension separation using ion-bombardment is most important. The ions can be produced between an electrically charged wire and grounded or charged conducting body, separated from it by an air gap. Air around the wire becomes ionised and is attracted towards the grounded body, where the ions are discharged. Usually the field applied to the corona electrode is more than 30 KV and the ionised corona is visible as a luminous discharge. The mineral particles entering the electrified field are bombarded with gaseous ions from which they acquire negative charge.

14.12.1 INDUSTRIAL ELECTROSTATIC SEPARATORS

The electrostatic separator may be either roll–type or belt–type. The former one is common in practice, while the latter is used in some specific cases:

a) *Belt-type Electrostatic Separator*

It consists of a belt which may be horizontal or inclined at an appropriate angle. When an electrostatic field is applied, the conducting particles adhere to the belt and are carried over the upper end, while the non–conducting particles are not much influenced by the field and thus roll down the belt to fall off the lower end. The belt-type separator is suitable for the cleaning of many food products and the separation of scaly particles of ore and other materials from granular particles. The capacity of the separator may range from 1.5 to 2 tonne/hr.

b) *Roll-type Electrostatic Separator*

A simple electrostatic separator of this type is shown in Fig. 14.13, in which conducting surface is a slowly revolving roll. When the feed is drawn from a hopper over this charged revolving roll, the conducting particles are thrown clear of the path of free fall followed by the non-conducting particles. The path of poorer conductors will be determined mainly by the size and shape of the particles, and the speed of roll.

In another system (Fig. 14.14), the roll is grounded and the charged element is a fine wire placed at one side of the roll. In this case, the electrical spray emanating from the wire sprinkles charges on all particles. The conducting particles will lose their charges immediately by passing them to the grounded roll and will thus fall undeflected. On the other hand, the non-conducting particles acquire a charge and induce the opposite charge to the ground roll and resulting in their adhesion to the roll. The non-conducting

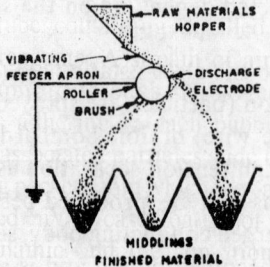

Fig. 14.13. Electrostatic separator using electrified roll.

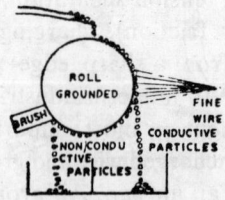

Fig. 14.14. Electrostatic separator using grounded roll.

particles drop off gradually due to leakage of charge or may be scrapped off by a nylon brush.

In practice, it may not be possible to obtain complete separation in one pass. The general practice is to re-pass either the residual fraction at lower potential, or to retreat both fractions.

14.12.2 HIGH TENSION SEPARATORS

A high tension separator is shown in Fig. 14.15. This is sometimes applicable to material which is too coarse for froth flotation, or where dryness is essential. It is also useful to recover small valuable concentrate from a gravity concentrate (e.g. recovery of diamond). In high tension separation, the surface potential of a material (its contact potential) is the measure of its amenability to separation. The potential between electrodes of the separator is limited by corona or by spark-over. The spark-over voltage is proportional to the density of the ambient gas, and hence its pressure and temperature. This is highest in nitrogen atmosphere, and diminishes in the order of air, CO_2, O_2, and H_2. The ionisation discharge also varies with humidity. Usually the electrodes used are in the form of fine wire, knife edge, or bars with projecting points.

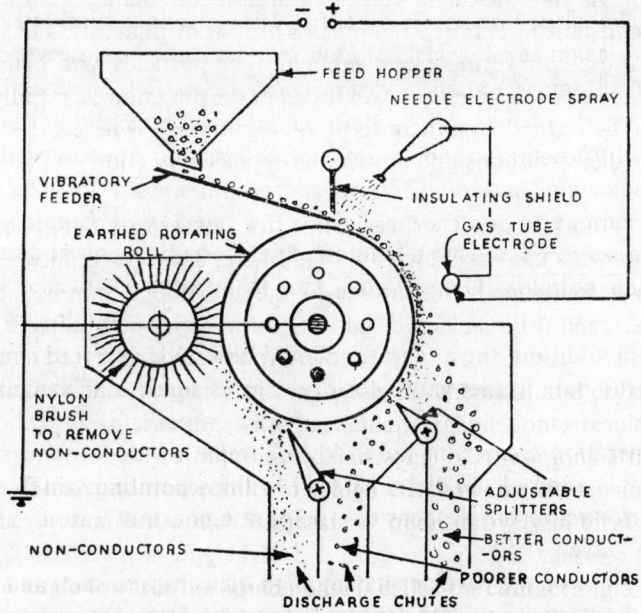

Fig. 14.15. High tension electrostatic separator.

In order to get an efficient separation by high tension separator, the feed should be closely sized (usully 1 to 0.1 mm), free from fines, and uniformly presented to the separator. In some cases, heating of material has been found to be beneficial. Presently, in most of the mineral processing plants, high tension separators are employed. Some important applications of high tension separator are given below:

a) Removal of fine dust and fumes from contaminated air or gas: The air is passed between negatively charged wire electrodes and earthed or positively charged plates (collecting electrodes). The particles are ionised, picked up by the plates, and periodically shaken down into hoppers by scrapping.

b) Separation of titanium beach sands containing ilmenite, rutile, zircon, and monazite together with lighter minerals has been successfully treated by high tension separation with high recoveries.

c) The process is applicable, where high-grade products are to be obtained from the mixed minerals having their densities quite close for effective gravitation separation to be used. For example, good response is obtained for chromite bearing sands, tungsten mixed minerals, microlite and tantalite, cassiterite and columbite, coarse pebble phosphate and feldspar, etc.

d) Separation of diamond from gravity concentrate.

14.13 Dielectric Separation System

This method is based on the difference between the dielectric constants (D) of different minerals. The process consists of suspending a mixture of minerals of varying dielectric constants in non-conducting fluid having the dielectric constant in between that of two groups of minerals, and subjecting the system to a converging electrical field. The particles with higher dielectric constants than the medium will travel in the direction of rapidly increasing electric field (tend to adhere to point electrodes), whereas the particles with lower dielectric constant move in the opposite direction and fall through.

14.13.1 Apparatus and Its Operation for Dielectric Separation

An apparatus used is shown in Fig. 14.16. This essentially consists of (a) two electrodes (one being earthed to a high voltage supply, i.e. 20 KV, 50 mA d.c.), and (b) a dielectric bed made of material with suitable dielectric constant. In addition, the apparatus should be provided a feed chute for introducing the pulp, outlet valves for drainage of liquid, and velocity regulation valve.

The operation consists of the following steps:

i) Application of the dielectric field to the separating vessel.

ii) Introducing the suspension (pulp) at a constant rate by use of the regulation valve.

iii) Capture of the particles having high dielectric constant and discharging away the particles of low dielectric constant from the system being in liquid.

iv) Removal of the electric field.

v) Recovery of particles of high dielectric constants by washing the dielectric bed.

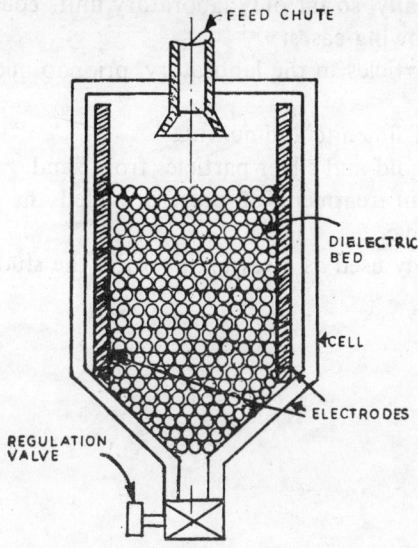

Fig. 14.16. Apparatus for dielectric separation.

14.13.2 DIELECTRIC CONSTANTS OF MEDIUM AND MINERALS

Dielectric constants of all the minerals are not known correctly due to the considerable variations in crystal orientation, porosity and moisture content of the same mineral. Usually, sulphides and minerals of high conductivity behave as having the dielectric constants higher than that of water (81). Other minerals have intermediate dielectric constants. The dielectric constants of some minerals and liquids are given in Table 14.5.

Table 14.5. Dielectric constants of some fluids and minerals[6]

Fluid	Dielectric constant	Mineral	Dielectric constant
Air	1	Cassiterite	27.7
Octane	1.9	Cuprite	16.2
Benzene	2.3	Quartz	3–6.5
Chloroform	5.2	Corundum	3–5.3
Phenol	9.7	Biotite	5–9.3
Pyridine	12.4	Calcite	3–8.5
Ethyl alcohol	26.8		
Nitrobenzene	36.5		
Dimethyl sulphate	55		
Water	81		

14.13.3 APPLICATIONS OF DIELECTRIC CONSTANT SEPARATION

The method is applicable to fine particles (finer than 0.2 mm). The process is very expensive. Though many attempts have been made to employ this process industrially, so far only laboratory units could be developed. It may be used in following cases:

a) Sorting of particles in the laboratory, prior to microscopic observation.

b) Separation of ilmenite and quartz.

c) Removal of gold and silver particles from sand.

d) Can be used in treatment of ores, provided the price of the liquid used is not prohibitive.

e) It is commonly used as a research tool in the study of minerals.

CHAPTER 15

Solid–Fluid Separation

In mineral processing, fluids (particularly water) are invariably used in many processes, such as gravity separation, heavy media separation, froth flotation, wet magnetic separation, etc. In order to make the product suitable in subsequent processing or use, as well as to regenerate and recycle the water/fluid, the liquid should be removed from the pulp as much as possible keeping the overall economics in view. Solid–fluid separation is also an equally important subject in water purification, in mining, metallurgical and almost all industrial operations. The separation of coarse solids is readily accomplished by settling, and it is not paid much attention. In treating of pulps with fine solids (froth flotation, sliming, flue dust, etc.), the importance of dewatering and settling dusts becomes considerable.

The mechanical separation of solids from liquids is accompanied by thickening and/or filtration. Final moisture is removed by drying. In some specific cases decantation may also be employed. Separation of solids from gases is also accompanied in a similar way, i.e. settling and filtration, which are almost counterparts of the corresponding procedures involving liquids.

15.1 Thickening (Sedimentation)

The English word *sedimentation* has been derived from the Latin word *sedere* meaning to sit down. As applied to mineral processing operations, sedimentation may be defined as 'separation of suspension into a supernatant clear fluid and relatively dense slurry containing higher proportion of solids'. The sedimentation unit operation as applied to mineral processing, has been termed as *thickening* which may have the primary purpose either (a) of increasing the solid content of the thickener products relative to the feed stream, or (b) clarification in which the solids are to be removed from the feed stream to produce a liquid product essentially free from particulate matter.

In thickening, the suspension of solids in liquids is allowed to settle until a clear liquid layer tops a mud layer. The clear liquid and mud are withdrawn from top and bottom respectively. The operation may be batch sedimentation or continuous sedimentation. Thickening consists of the following four steps:

i) Flocculation of very fine solid particles to form aggregates of flocs of many particles.
ii) Sedimentation of liquid–laden flocs, leaving clear supernatant liquid.
iii) Compaction of the sedimented flocs.
iv) Withdrawal of the clear fluid and thickened slurry.

The sedimentation may be either batch or continuous type. In the former type, the tank is filled with pulp and the solid particles are allowed to fall under gravity. The batch type requires longer time and has been used since older times. Continuous thickener appeared around the turn of the 19th century and is well accepted by the ore processing industry.

15.1.1 FLOCCULATION

A ground ore consists of particles finer than the finest, that can be settled in water within a reasonable time. If the pulp is so conditioned that the fine particles are grouped together to form flocs of a size large enough to be seen with the naked eye, the resulting flocs will settle with sufficient rate to provide a clear supernatant liquid and compacted sediment in a reasonable time. Therefore, the dispersed state may be considered as a *natural* condition, while that a flocculated state is obtained by addition of some suitable agents (electrolytes, etc.).

Mechanism of Flocculation

The dispersion or flocculation mainly depends on the conditions/forces present at solid–liquid interface. The mill water may be considered homogeneous in a broad sense, but is inhomogeneous with respect to its selective behaviour at a solid interface, since many chemicals and minerals are dissolved in mill water. The contribution from the minerals make the fluid phase still more complex. Thus, a pulp of finely divided mineral in water consists of mineral particles, water molecules in various stages of association, and a variety of ions (ions are mostly associated with one or more molecules of water). Ions and water molecules are constantly attaching themselves to the mineral particles, and conversely ions are detaching themselves from the mineral surfaces and entering into the liquid phase. At equilibrium, the two actions would be balancing each other. Following are the three basic phenomena to all flocculation and dispersion.

a) Selective sorption of ions.
b) Spontaneous and erratic motion of particles as a result of molecular bombardment.
c) Natural tendency for surface energy to change to kinetic energy.

In flocculation of discrete fine particles, the interfacial area is reduced and thereby surface energy also reduces. The process of flocculation proceeds according to second law of thermodynamics, (i.e. all naturally occurring processes are accompanied by an increase in the entropy of the system) and thus may be expected to occur spontaneously.

If the particles have ionised surfaces with similarly charged potential, a repulsive force will be exerted between approaching particles. On the other hand, its surface tension promotes cohesion which results in decrease of overall solid area of the cohesive particles, and thereby the total surface energy. If the repulsive force is dominant, the minute particles remain dispersed (peptised) in the pulp. When the mutual attraction is offered, the particles will flocculate. Therefore, the most favourable conditions for flocculation require electrical neutrality of particles. This neutral state is called the iso-electric point, or the zero potential, which refers to the pH value of 7. Near the iso-electric point, a colloid coagulates, while a pulp tends to flocculate.

The forces favouring flocculation are (a) magnetism (used in dense-media separation to manipulate ferro-silicon through its cleaning process), (b) ionic forces of attraction, and (c) secondary entropic forces. Each particle has an unmeasurable but adequate surface tension observed by reactions between the particles and the aqueous phase of the pulp. Further reduction in surface tension can be affected by adherence to other particles. If a number of particles can bind together so as to reduce the total free surface, the conditions are favourable for flocculation.

Factors Affecting Flocculation

The various factors affecting flocculation are discussed as follows:

a) *Concentration of particles in the pulp* (*pulp density*): If other conditions are favourable for flocculation, increasing of pulp density should result in rapid and complete flocculation. Since collision should precede adhesion, higher the pulp density (more population of particles), higher will be the rate of collision. Once the floccules are formed, they act as membranes and during their downward drift, they trap particles from the pulp.

b) *Temperature of pulp*: At higher temperature, the velocity of the pulp will be lower and the stirring effect will be stronger, caused due to ceaseless bombardment by moving molecules of water. Increase in temperature will provide the force to make more contacts of particles with each other, and if particles are already flocculated, their dispersion will be aided by increase of temperature.

c) *pH of pulp*: pH is the measure of ionisation or electrical charge. The flocculation is aided by the addition of an electrolyte capable of discharging the ions of opposite electrical charge to that on the colloidal particles. Further, flocculation increases with increase in ionic valency. When the dispersions are negatively charged, the cation of the electrolyte is more important, and conversely when dealing with positively charged dispersions.

d) *Size of particles*: If relatively coarse particles exist with fine particles, floccules of fine particles may entrap coarse particles, which would settle themselves without flocculating. This will give two effects, i.e. (i) coarse particles behave as if subject to flocculation (whereas they are not of themselves), and (ii) coarse particles considerably increase the weight of floccules without much increase in their volume. This weighting down is of importance in settling of floccules.

Flocculation Agents

When fine ores are processed, the final tailings or concentrate may be highly dispersed, slow to settle, and difficult to arrest on filter membrane. This problem can be overcome by adding more reagents without much increase in thickening area. Some proprietary reagents (known as flocculants) include glues, starch admixed with metal salts, cyanamid series of aeroflocs and surperflocs, and the nalcolytes. These flocculants have strong affinity for minerals and thus should be added carefully 'in stages into the pulp stream, otherwise the whole addition may be taken up by a small fraction of the pulp. Various other flocculating reagents are lime, H_2SO_4, alum, gypsum, $CuSO_4$, sodium aluminate, NH_4Cl, etc. The compounds having trivalent ions are more effective in flocculation.

15.1.2 SEDIMENTATION AND DEWATERING OF FLOCCULES

The floccules (flocs) are rounded aggregates of solid particles with interstitial fluid. Usually, the flocs consist of 10 or more volumes of fluid for each volume of solid, whereas tightly compressed and packed floccules (discharged as thickener underflow) rarely consist more than two to three volumes of fluid for each volume of solid.

The specific gravity of floccule (fluid and solid considered together) is very near to that of fluid, and thus settling velocity is much less even at a fairly large size. For example, settling velocity in water of spherical floccules of 0.5 mm in diameter and containing 90 per cent water can be calculated as follows:

According to Eq. (7.8)

$$V_{max} = \frac{\frac{2}{9} [(0.10)(3) + (0.90)(1) - 1](0.025)^2 \times 980}{0.01}$$

$$= 2.695 \text{ cm/sec.}$$

The above calculation is based on the assumption that the water within the floccule moves with the solid of the floccule. In fact, there is some movement of fluid with respect to solid within floccule, resulting in an increased friction and decreased velocity. In practice, flocs of 1 mm diameter or more are generally observed to settle as slowly as 0.1–0.2 cm/sec.

The overall dewatering of flocculated suspension in thickening involves the following processes:

a) Free-settling of flocs.
b) Hindered settling of flocs.
c) Exudation of water from the settled flocs under the influence of pulp pressure.

In case of thickening of very dilute suspensions, all three processes are operating. In the thickening of medium–thick suspensions, processes (a) and (c) are operating. In the thickening of heavy suspensions, only process (c) is operating.

a) *Free Settling of Flocs*

The flocs of a flocculated pulp behave as individual particles in respect of settling velocity. The free-settling flocs rapidly arrange themselves with the fast flocs at the bottom and the slow flocs at the top. The dewatering rate is controlled by the slowest flocs and thus most effective flocculants should be used to increase the settling rate of floccules. In practice, free-settling of flocs is only possible in water clarification and not in mineral pulps. However, for any individual flocs, the free-settling rate is constant.

b) *Hindered Settling of Flocs*

Under hindered settling conditions, the fast moving flocs travel through the slower settling flocs, or get trapped between them. Fast settling flocs are thus retarded, whereas slow-settling flocs accelerated by the blows of faster moving flocs resulting in movement of slow as well as fast moving flocs with the crowd. Therefore, under hindered settling conditions, the settling rate is practically uniform but goes on decreasing with time due to continuous increase in hindered character of settling. However, in spite of the generally slow settling of the mass as a whole, the overall settling rate may be faster than that of the slowest flocs under free-settling conditions.

c) *Dewatering of Flocs by Exudation*

After the flocs are settled, the intrafloccular liquid is exuded through the pores of the floccules. Exudation results in water oozing out at the top of the settled flocs and moving upward through the hindered settling column. The oozing of water continues with settling of hindered settling column until it vanishes. During this time, the settling of pulp is more or less at a constant rate.

Exudation is mainly effective after completion of hindered settling. With the continuation of the exudation process, the floccules become rigid enough and the exudation process is terminated. Figure 15.1 represents the settling of flocculated pulp with time. Further dewatering of settled flocs will require mechanical rupture of structures by the application of exuding forces or filtration.

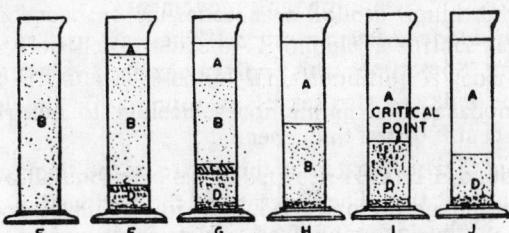

Fig. 15.1. Different stages of slime settling.

15.1.3 MECHANICAL WORKING OF THICKENER SLUDGE AND WITHDRAWAL OF CLEAR LIQUID AND COMPACTED SLUDGE

A considerable reduction in the water content of thickened sludge can be obtained by gentle stirring, to break the gel-like structures. The function of mechanical working of sludge is performed by the thickener rakes.

Clear liquid is discharged naturally by overflow, the rate of which, in turn, is controlled by the rate of feed of the settling tank.

The compacted flocs in the form of sludge can be removed by any one of the following two methods:

a) The simplest method of removing the sludge is by natural flow through a pipe provided with a goose neck discharge arrangement. The flow rate can be controlled by a valve as well as by the difference in head at the bottom of the tank due to the height of the pulp in the tank and the sludge within the goose neck outside the tank. However, there are operational difficulties, particularly due to variations in height of the pulp level and in the density of the thickened pulp.

b) A more appropriate system consists of diaphragm pump to discharge the sludge. This system is more steady in operation and is preferred. Further, it has the advantage of delivering the sludge sufficiently high to flow. readily into another thickener set-up at the same level.

15.1.4 THICKENING (SEDIMENTATION) EQUIPMENTS (THICKENERS)

Today, the sedimentation equipments, widely known as thickeners, are manufactured in many designs and shapes. For the convenience of discussion, these can be classified as conventional, high capacity, and other designs. These are discussed as follows:

a) *Conventional Thickeners*

These thickeners have the following common features (Fig. 15.2).

i) *Cylindrical shape*: Under portion of the thickener is cylindrical (diameter is much larger than height) which is attached to a shallow conical section having the apex oriented downward.

ii) *Annular flow launder*: As shown in Fig. 15.2, all thickeners are provided with an annular flow launder which may be located internal or external to the tank. If necessary, the launder may be equipped with a froth baffle. Overflow is controlled through a notched weir so that adjustments may be made to compensate for tank movement.

iii) *Walkway and feed-launder support*: These are used to service the centre drive mechanism. Sometimes, this walkway extends across the full diameter of the tank. The walkway also serves as support for the launder carrying the feed to the centre of the thickener.

iv) *Feed well*: This is located at the centre of the thickener and the feed launder terminates at this. The function of the feed well is to dissipate the kinetic energy of the incoming feed and to form a zone of quiescence for commencing the sedimentation. Feed wells may be made in different shapes and sizes. The method of introducing the feed and configuration of feed well, usually influence the sedimentation behaviour.

v) *Drive mechanism and rake arms*: The rakes are provided to give a

gentle movement to the sedimented solids from the periphery towards the
centre discharge point. The movement of the rakes is provided by a suitable
drive mechanism.

vi) *Underflow cone or trench*: This is located near the bottom centre. The
sedimented solids collected at the centre of the tank by the action of the
rake arms are removed from the underflow cone or trench.

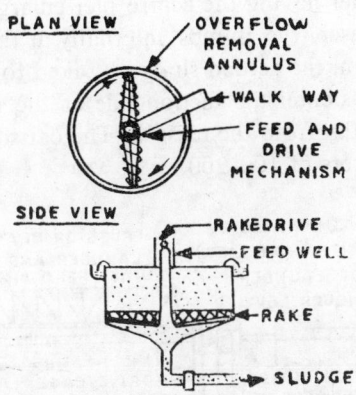

Fig. 15.2. Conventional thickener.

A typical thickener of conventional type is the Dorr thickener[37]. It consists of a cylindrical tank (about 3 m deep and relatively large diameter)
with a slowly revolving central shaft carrying the sweeping paddles. The
shaft is driven from a motor through speed reducers. The motor may be
mounted either at the centre on a permanent structure, or at the edge moving on a circular track and driving the shaft through a revolving truss. For
relatively small tanks, the centrally driven motor is preferred, whereas for
large tanks, the traction type is preferred.

Thickeners are usually installed outdoors to eliminate the cost of building. Thickener tanks may be made of wood, mild steel, stainless steel, elas-
tometers covered steel, or concrete. The larger tanks are made of concrete
and smaller tanks can be made of stainless steel, mild steel or wood depending on the chemical character of the pulp. The thickening cost is mainly a
capital cost, since power, attendance, maintenance, etc., are practically non-
existent.

The conventional thickeners may be classified on the basis of drive
mechanism and type of support for rakes. These include bridge or beam
type, column support type, cable support type and traction type thickeners[41].

In the bridge type thickener, the rakes and drive mechanism are supported by a truss which spans the whole diameter of the thickener tank. The
bridge can be supported from a tank well or from independent piers. In this
case, the discharge is generally drawn from the apex of a cone located at
the centre of the sloping bottom. The bridge type thickeners can be made
in sizes up to 50 m in diameter.

In the column type thickener (Fig. 15.3), the drive mechanism is supported by a column located in the centre of the thickener tank. The rake arms are attached to and supported from a drive cage which is connected to the drive machanism. The settled solids are discharged through an annular trench encircling the column. Thickeners of this type are commonly constructed in diameters up to 110 m. A variation of this arrangement is marketed as *caisson* thickener having the centre pier enlarged enough to provide the installation of discharge pumps internally at the bottom of the centre column. In this system, the settled sludge is raked to a trench surrounding the caisson. The solids enter the suction side of the underflow pumps and are discharged vertically from the caisson. The caisson thickeners have been constructed in diameters of 100–200 m.

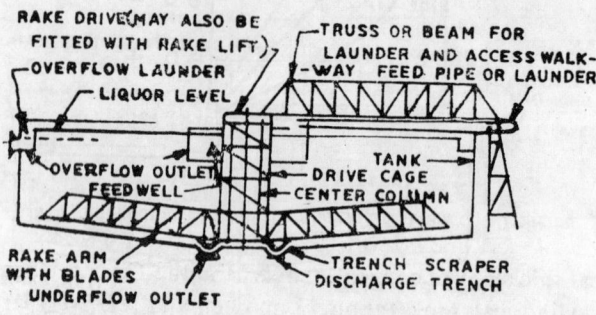

Fig. 15.3. Column support type thickener.

The cable support type thickener is based on the method of imparting the motion to the thickener arms. It consists of a hinged rake arm fastened to the bottom of the drive cage, or centre shaft. The hinge is so designed as to give simultaneous vertical and horizontal movement of the rake arm. The rake arm can be pulled by cables connected to drive arm structure, which is rigidly connected to the centre shaft or cage at a point below solution level. This design of rake provides an automatic lift when the torque developed in moving the rake through the sludge bed counterbalances the weight of rake. When torque required is less, the rake tends to move downward and settled sludge is moved toward discharge parts. In some cases, the drive arm can be arranged to be completely above the liquid level. This arrangement is generally employed when the drive arm suffers with scaling.

In traction thickener, the centre pier is required to partially support rake mechanism. However, the centre pier also serves as a pivot around which the rake mechanism rotates. The rake arm is moved by a drive mechanism or trolly connected to a single long arm extending from the centre pier to the tank periphery. The drive trolly is arranged in such a way that the motion is imparted to the rake arm by means of friction developed between one or more driven wheels and a curved peripheral rail. The addition of two or three short rake arms completes this arrangement. Traction thickeners are generally manufactured in the sizes of 60–130 m in diameter.

b) *High Capacity Thickeners*[37]

In recent years, various manufacturers have introduced thickeners known as *high capacity* or *high rate* thickeners. In general, these units are recognised by reduction of area requirement for installation. The area required for a conventional unit is 0.5–1 sq. m/tonne/day, whereas for high capacity thickeners the area required is as low as 0.03–0.06 sq. m/tonne/day. Several high capacity thickeners are being manufactured by various manufacturers. Few to mention are Eimco high capacity manufactured by Amster Corporation, Enviro-Clear thickeners manufactured by Envirotech Corporation, and Dynafloc thickener manufactured by Dorr-Oliver.

Enviro-Clear thickener consists of a feed pipe terminating in feed well which is of the inverted cone shape. The slurry entering the thickener is deflected downward by the feed well to an impingement plate where a horizontal velocity component is given to the solid and liquid. In operation, a circulating sludge bed is maintained by means of level detectors. Incoming particulate suspension is directed into the circulating sludge bed and flocs. The addition of flocculants, such as organic polymers is made to form the flocs which gain in mass by entrapment of incoming solid particles. After reaching a certain critical mass, the flocs tend to settle through the discharge cone. The liquid moves upward through the circulating sludge bed and passes to the overflow weirs. This thickener may be provided with radial as well as peripheral overflow launders. As the diameter of these units rarely exceeds 17–18 m, the drive mechanism is usually a bridge type. In this type of thickener, gravity plays only a small role in solid–liquid separation. The main effect of separation is achieved due to filtration of suspension through the agitated sludge.

In Eimco thickener, the feed enters through a hollow drive shaft in which flocculants are added and mixed by an internal mechanical mixer. The flocculated slurry from the mixing chamber is directed into a slurry blanket in which radially mounted and inclined plates are partially submerged. These plates improve the operation of the thickener. When the flocs grow to a critical size, they settle through the slurry bed and are transported by the rotating rake mechanism to centre cone. The height of the slurry blanket can be controlled automatically through the use of a level sensor.

In Dynafloc thickener, the design of high capacity is obtained by providing the effective mixing of the polymer flocculants intimately with the feed stream. The feed is split in two equal portions prior to entering the feed well. One-half of the feed is presented to the top plate or shelf and the other half is presented to the bottom shelf. The two portions flow in opposite directions to each other. In this design, a substantial reduction in thickening area is obtainable at the same or lower flocculant amounts, as employed in conventionally sized thickeners. The unit areas as low as 0.028 sq. m/tonne/day may be obtained with an equal concentration of underflow as in case of conventional units.

c) *Special Thickeners*

The Lamella thickener has been introduced by Parkson Corporation. It consists of inclined plates in close proximity to each other, so that the effective gravity settling area becomes the horizontal projected area of each plate. The pulp may be presented to the inclined lamella plates either directly through the feed box or into a flash mix and flocculation tank. The feed flows into the plates from the side and then upward, flowing out at the top of the tank through flow distribution orifices. The solids settle out on the surfaces of plates and slide downward into the sludge hopper. Additional thickening of the settled solids is obtained by compression with a low amplitude vibrator pack located in the sludge hopper. Another similar type of device is the tube settler in which the lamella plates are replaced by tube bundles.

Deep cone thickener concept has also been introduced by Denver Equipment Division, in which the density of flocculated solids is increased by the application of pressure in the form of a static head. The static head is the function of height of the solution above the discharge point.

15.1.5 OPERATION OF THICKENERS

In general, the operation is more or less identical in all the types of thickeners. The mill pulp carrying finely ground solids in suspension is fed centrally through a *trash screen*, which prevents the entry of any debris. The entering pulp displaces part of its volume as an overflow of moderately clean water. The rake arms provide a gentle radial drift of this overflowing water from centre to sides, and the solids fall slowly downward. The rakes revolve very slowly (one revolution in 2–8 min) through the compression zone, gathering and sweeping the settled slurry or sludge toward the central discharge well. During the revolution of rakes, channels through the flocs are formed, through which clear water can be squeezed upward. The thickening time depends on the dilution of the pulp, i.e. more dilute pulps require greater settling time, since floc formation is slower in dilute pulps.

The discharge rate from the well is controlled by means of a diaphragm pump run at a rate to allow about 60 cm of fully thickened slurry to be maintained in the compression zone. This layer holds back enough compressed pulp and ensures the removal of only completely settled slurry. If the zone becomes too thick, there will be a danger of overloading the rakes and thus the mechanism may be distorted. Alarm and trip mechanisms may be fitted to indicate such overloads. The overflow can be monitored by an electric-eye which gives warning when turbidity of overflow rises. Flushing points are also provided for the introduction of water and compressed air in the event of choking. In some large-size thickeners, bottom valves are used in place of pumping to allow the slurry to run-off.

The discharge of thickened slurry is usually led to a continuous filter. The filter is periodically shut down for servicing and during this period (may last for several hours), the thickener continues to receive feed and thus should be capable to store its slurry. Under such conditions, provision should be made for raising the rake mechanism to prevent overstrain. As soon as normal running has been restored and the loading has been reduced, the rakes should be lowered gently.

The main purpose of thickener is to reclaim water/fluid from muddy effluents, in adjustment of pulp density and in facilitating chemical reaction.

15.2 Solid–Liquid Separation by Decantation

This method is identical to thickening in principle. The pulp is allowed to settle in tanks and the clear supernatant solution is decanted by syphon or pumping. Decantation is usually subjected to fast settling pulps without causing any mechanical working of the pulp. In industry, counter-current decantation (C.C.D) is common in hydrometallurgical plants, where the leached solution is separated from the unreacted solids. This method employs upgrading or downgrading by stages (Fig. 15.4). Any number of retreatments required are used in line. The downgraded product from each separator provides the head feed of the next succeeding separator and being joined then by the upgraded product from the next separator lower down the line.

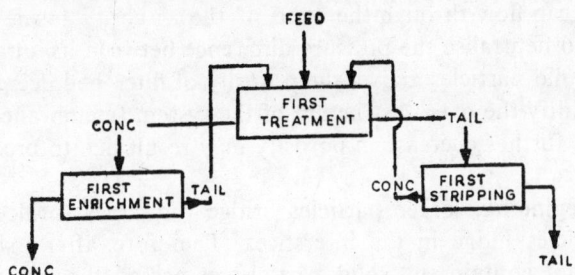

Fig. 15.4. Illustration of counter current decantation.

This method is mostly employed in separating the solids and liquids from pulps obtained in leaching the ores. For example, in gold cyanidation, barren solution (or last stage water) is worked from the tail end of the process up towards the feed end, while ore travels down by similar stages towards the tail end, losing part of its gold at each transfer.

15.3 Solid–Liquid Separation by Filtration

Filtration may be defined as the process of separating finely divided solid particles from a fluid accomplished by driving the pulp through a membrane, septum, cloth, or bed of natural or synthetic material, which is porous to permit the passage of fluid and retains the solids. The filter cloth should offer least resistance to the flow of fluid and should act mainly as a support for the deposited solids, which form the effective filter bed. The pulp is presented at one side of the filter at a higher pressure than that on the other side of the filter cloth/bed. The very fine material passing through the porous barrier forms a part of the filtrate. The pressure difference may be due to (a) simple gravity, (b) pressure applied to the pulp being fed, (c) vacuum on the filtrate side, or (d) effect of centrifugal force. After a certain layer of solid cake builds up on the filtering media, it should be removed, since beyond a certain thickness of the solid cake, the resistance to the flow of water becomes excessive.

15.3.1 MECHANISM OF FILTRATION

In the simplest form, the filter may be considered in the shape of a tube having small bore, through which the fluid is sucked, leaving the solid particles at the entrance of the tube. As the filtration commences, first the solids pass through the tube, but soon they form an arch or bridge across the opening, and thus allowing only clear liquid to pass. Therefore, in real sense, the filter bed consists of a layer of particles derived from the suspension subjected to filtration. The tube only acts as a framework to support the bed of solids. However, in order to bridge a pore by solids, the diameter of the coarsest particles must exceed a certain minimum size (it is generally one-third of the pore opening for coarse material, and is much smaller for fine particles).

A liquid can flow through the tube, if the viscosity inside the pore is insufficient to neutralise the pressure difference between its intake and discharge. As solid particles arrive, the porosity of filter bed decreases rapidly and thus modify the overall structure of the system (membrane plus arrested layer). A further decrease in porosity may result due to precipitation of calcium-salts.

In the beginning, larger particles bridge the tubes (orifices) and then smaller particles lodge in the interstices. Therefore, after a short run, the filtrate will not contain any solid particles as passed through in the beginning.

15.3.2 FACTORS AFFECTING THE RATE OF FILTRATION

In practice, the filtration rate and the ability of the filter barrier to arrest solids depends on the following factors:

a) *Filtering Area*

The rate of filtration increases directly with increase in filtering area. This is controlled by the device employed in filtration.

b) *Difference in Pressure Between the Two Sides of the Filter*

The filtration rate is proportional to this pressure difference. This factor also depends on the device employed for filtration.

c) *Average Cross-section of the Pores within the Filter Cake*

The filter surface may be considered as made of many capillaries and the flow through each capillary is proportional to the fourth power of its radius. This factor is dependent on the character of the pulp.

d) *Number of Pores Per Unit Area of Filter Cloth*

The filtration rate is proportional to the number of pores per unit area. For a given proportion of pore space in a filter cake, the number of capillaries per unit area is inversely proportional to the square of the radius of capillaries. This is controlled by the character of the pulp.

e) *Thickness of Filter Cake*

The filtration rate is inversely proportional to the thickness of filter cake. This is dependent on the way in which the filtering equipment is operated.

f) *Pulp Temperature*

High pulp temperature favours filtration rate due to decrease in viscosity of fluid and more coagulation of particles.

g) *Size-range of Particles*

The filtration rate is adversely affected by a wider range of particles, since small particles lodge themselves in the interstices of the coarser particles, clogging the pores.

h) *Specific Surface of Solids*

Filtration rate is inversely proportional to the specific surface of solids, i.e. finer the particles, slower will be the rate of filtration.

i) *Degree of Flocculation*

If the pulp is well dispersed, the particles will have a firm packing in the cake and thus the pores will be extremely small, since each large pore will be clogged rapidly by finer particles. On the other hand, if the pulp is flocculated, the cake is porous, having relatively large pores and filtration will be fast. For example, filtration of dispersed slime may be 50–100 times slower than filtration of the same slime after flocculation. An overflocculated pulp will be filtered readily, yielding relatively a wet cake, while an underflocculated pulp will be filtered slowly yielding relatively a drier cake.

15.3.3 FILTERING EQUIPMENTS

The filters can be classified on the basis of the type of force of action applied to the filtering operation. The various types of filters include (a) gravity filters, (b) pressure filters, (c) suction or vacuum filters, and (d) centrifugal filters. These are described as follows.

a) *Gravity Filters*

In gravity filters, the main function of the septum is to act as a support for the bed of material. The leaching tanks employed in cyanidation may be considered as gravity filters. Their thick beds are maintained by exclusion of most of the fine sand, which would otherwise clog the interstices between the coarse particles and prevent filtration (drainage). For clarification, a thinner sand bed is used and the fine material is allowed to clog the interstices up to a certain extent. The control can be maintained by skimming out some of the uppermost layer of sand. Thickeners can be used for this purpose which are called filter-thickeners (devices which combine filtration with thickening).

An example of this type is the Hardinge super thickener (Fig. 15.5), in which filtration is used on a sand bed. This gives two clarified liquors, one overflowing the rim of the tank and the other as a filtrate from the porous sand bed.

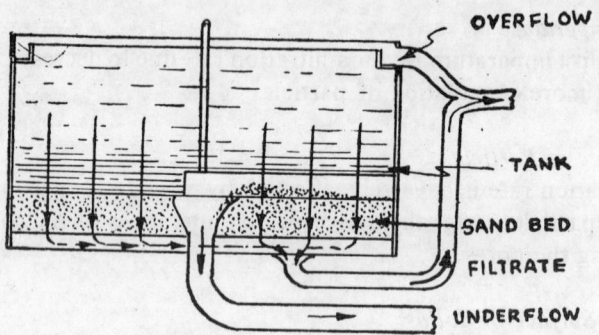

Fig. 15.5. Illustration of Hardinge filter thickener.

b) *Pressure Filters*

These are big vessels containing porous filtering sections. The feed is admitted in the vessel and the air pressure is applied. These are known as plate-and-frame filters. These are intermittent in operation. The septum is stretched over a frame which is provided with channels for pulp feed, wash liquor, and filtrate discharge. A number of such frames are assembled in a press, and pulp is pumped in; when the pressure reaches a determined value, the entry of pulp is stopped. The solid cake filling each frame is then washed with water under pressure, which can be discharged separately. The press is then opened and the cake is removed. These devices, also known as filter presses are generally used in the chemical or hydrometallurgical industry, where interest centres on filtrate, i.e. in clarification of cyanide leach liquor. Their use in mineral processing plants does not find much place, since here the interest centres on cake.

c) *Vacuum Filters*

The vacuum filters employ suction or vacuum at the filtrate side. These may be operated either intermittently or continuously. The intermittent type is generally used in cases where the amount of solids involved is small, e.g. in clarifying pregnant solution or in catching gold precipitate. In this case a number of frames covered individually by canvas envelope, are suspended in the tank receiving the feed. The frames are connected through pipes to a common *leader* attached to a vacuum pump. The liquid is drawn through the frames. The frames are taken out periodically and cleaned.

An intermediate between the intermittent and the truly continuous filter is the Genter filter which is semicontinuous in operation and also incorporates the thickening process. This consists of a cylindrical tank which receives the pulp. The canvas bags hang from each arm and are submerged in the tank. The canvas bags are kept apart by wooden blocks. The liquid is withdrawn by vacuum through the canvas bags. At regular intervals, the vacuum is cut off and the liquid flow is reversed for few seconds to detach the sludge collected on the outside of each canvas bag. The sludge slides down to the bottom of the tank and is raked to a centre discharge. The displacement of sludge can be aided by low-pressure air. The sludge is intermediate in consistency between the flowing slurry of a thickener discharge and the solid cake of a drum filter.

Today, most widely used filters are continuous vacuum filters. These are usually drum or disc filters. Both the types have several variants in design to meet special requirements. Developmental work is also active to improve the working of existing designs as well as to create new types. As a result, some new filters have emerged out, e.g., belt-drum filters, top feed filters, magnetic filters, and disc filters. The principle of operation is same in all these types.

A vacuum filter installation (Fig. 15.6) comprises a tank containing a cloth covered drum or disc and an agitator to maintain the slurry in suspension. The drum or disc is divided into three sections, i.e. (i) cake building zone, (ii) dewatering zone, and (iii) discharge zone. The first two sections are connected to a vacuum system so that the slurry is drawn against the cloth. The liquid passes through the cloth, while the solids build up on the surface. As the drum or disc revolves, the cake emerges from the slurry and the residual liquid is removed by the vacuum. In a third stage, cake is discharged.

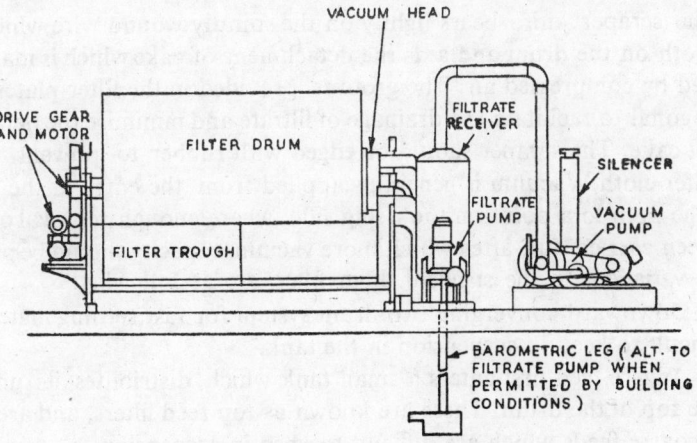

Fig. 15.6. Vacuum filter installation.

Drum filter: The filter unit consists of a metal drum having filter plates (of wood or metal) on its circumference. The filter plates are grooved to permit flow of filtrate drawn through the cloth or a membrane wrapped round the drum and bound on by wire so as to rest on the plates. The drum rotates at the speed of 2–12 minutes per revolution depending upon the characteristics of the pulp. The trough made of metal structure, receives the pulp and settling is prevented by agitation. Height of slurry in trough is controlled between certain maximum and minimum levels. Vacuum during submergence is produced by vacuum pump through a valve head and distributing ports, each port serving its own plates. As plate leaves the pick-up zone, it passes progressively through the drainage, washing, drying, and blown-back zones, and then finally reaching the scrapper knife, where the cake and cloth bellies outward under the influence of compressed air. The valve parts can be connected in various ways through the valve head depending on the desired series of operation. A sectioned drawing of drum filter is shown in Fig. 15.7.

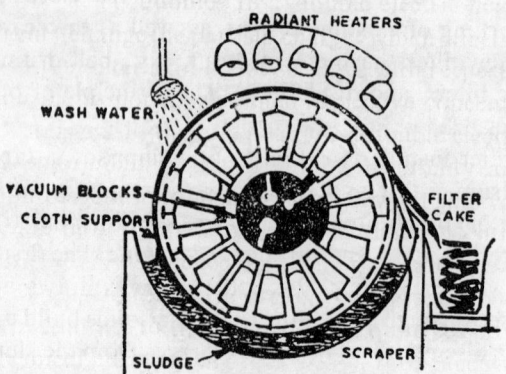

Fig. 15.7. Cross-section of rotary filter.

The scraper knife bears lightly on the spirally wound wire which holds the cloth on the drum and aids the detachment of cake which is mainly performed by compressed air. The grooves provided in the filter plates should be diagonal to facilitate the drainage of filtrate and minimise the detachment of wet cake. The scraper should be edged with rubber to prevent wear of the filter cloth. Vacuum is generally applied from the entry of the segment into the pulp to a point on the rising side, where enough removal of filtrate has been carried out, after which more vacuum capacity should be provided. Some variations in the standard drum filter, are as following:

i) Downward converging two-drum system for fast settling material that is difficult to keep in suspension in the tank.

ii) Where feed first enters a small tank which distributes its underflow on the top of the drum. These are known as top feed filters, and are used to treat coarse feeds which are difficult to keep in suspension.

iii) In one design, the pulp floods down the rising side of the drum, sealing the cracks in the cake. When the pregnant solution should be displaced before discarding the filter cake, e.g. in the cyanide process, the vacuum maintained should be strong in most of the arc of the drum, through which cake moves after emerging from the slurry tank. This arrangement provides the use of enough wash water to displace and recover solution entrapped in the pores of cake. In such working, cracks in cake would endanger the vacuum system.

iv) The belt-drum filter, in which cloth passes over a series of rollers after leaving the drum. It is equipped with a brake-roller discharge or with an air-knife system, by which even very thin and sticky cakes can be removed. The cloth can be washed from both sides in each revolution. The belt-drum filter is ideal for handling very fine grained pulps or other pulps that tend to blind the filter cloth.

v) Magnetic filters, which are used for dewatering coarse magnetic concentrates. The principle is the same as for standard drum and top feed filters, except that permanent magnets are installed inside the drums. Under the influence of magnetic field, the particles are classified such that the larger particles settle first. The charge in magnetic field causes the particles to turn along their axes, facilitating rapid drainage and resulting in greater capacity.

Disc filter: In place of drum, it consists of a tank and a number of independent discs (Fig. 15.8), each composed of detachable segments and each running in its own compartment of the filter tank. Suction and blowing of air are provided as in drum filter. It has two attractive features, specially for the smaller plants, i.e. (i) the ease of replacing the cloth, and (ii) possibility of handling more than one product by using an appropriate number of compartments for each type of product. All segments are connected to large diameter external drainage tubes.

The filter surface used may be of canvas, cotton cloth, porous rubber, nylon, terylene, woollen, etc. For some special uses in chemical engineering,

Fig. 15.8. A view of disc filter plant.

glass fibre and stainless steel mesh cloth have been employed, however their use in ore processing is rare. Among the various materials, nylon and terylene are commonly used due to its superior resistance to abrasion and clogging, easy removal of cake from its smooth surface, and longer life. Terylene can also withstand relatively higher temperatures (100°C). The disc filters are suitable for handling large quantities of fine pulp, since these give high performance of filtration.

d) *Centrifugal filters*

These consist of a horizontal or vertical basket of cylindrical or cylindro-conical shape which rotates at high speed (Fig. 15.9). A centrifugal force generated due to high speed of rotation can be used for separating the solids from the liquids. The pulp is introduced at the centre of the drum. The cylindrical surface is lined with screen or other filtering surface. The centrifugal force causes the pulp to press on the filtering cylinder and the liquid to pass through the filtering surface (cloth, membrane, etc.). The centrifugal filters are limited by their relatively small size and thus their use in filtering fine pulps is possible only on small-scale or chemical plants and thus find a

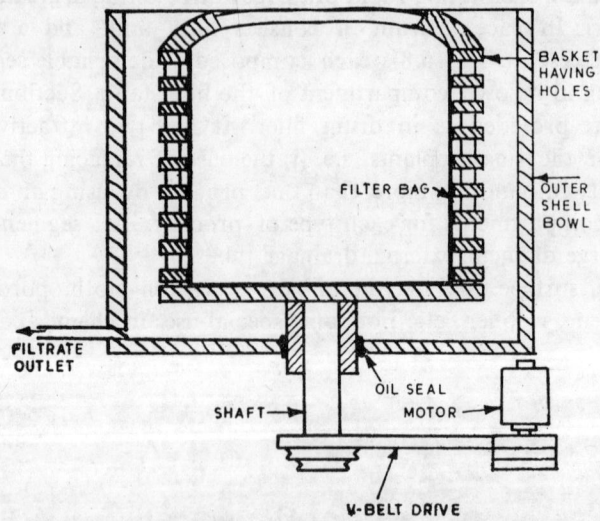

Fig. 15.9. Simplified diagram of centrifugal filter.

limited application in ore processing industry. In centrifugal filters, the cake of solid separates to the side and the clear liquid flows through. The deposited solids can be removed by scraping. These are essentially intermittent in operation. Principally, these filters have been used in dewatering granular coal to a low moisture content and to remove solutions from metal powders produced by electrolytic processes.

e) *Electrophoretic Filters*

These are based on the fact that the clays suspended in a suitable electrolyte can be made to migrate under the influence of an electric current. In this case, the suspension of clay in water having suitable electrolyte is subjected to d.c. current. The process may be carried out in a tank with suitable electrodes. The kaolin and other clay particles migrate to one of the electrodes, where they are deposited in the form of a layer. The filter cake is removed either intermittently as in filter presses or continuously as in vacuum filters. When some fine silica is also present in suspension of clays, some preferential settling of quartz may take place, which will result in somewhat purer dewatered clay.

15.4 Dust Elimination from Air/Gases

Dust elimination is a counterpart of clarification from fluids and the methods used to eliminate dust from air/gases are similar to those employed in clarification of liquids, but with some differences. Though dust elimination has little relevance with respect to economics, it is becoming of more and more concern with the view of creating better working conditions for workers, since the dust creates health hazards to the persons working in the industry.

The viscosity of air/gas is approximately only one-sixth of that of water and the specific gravity of air/gas is negligible. Thus the particles remaining suspended in air are much finer (one-eighth to one-tenth of size) than the particles remaining suspended in water for any given time. Practically, the dusts are made up of particles usually finer than 10 microns. The particles in the size range of 0.5–2 microns are the worst from a physiological standpoint. The various processes employed in dust elimination are analogous to clarification (by settling), thickening (flocculating and settling), and process of filtration (passage through bags). The various methods are discussed below:

15.4.1 Dust Elimination by Settling

This can be accomplished by simply passing off the dust laden air/gases through an enlarged settling chamber. It permits removal of the coarse particles but does not remove most of the fine particles and thus this method is not efficient from a physiological stand-point. The improved means of dust removal by settling is one which uses the centrifugal force along with gravity. When centrifugal force of the order of about 100 times of gravity is applied, dusts even finer than 1 micron can be made to settle. This method is an efficient one to remove very fine particles which are physiologically harmful. The devices employed are known as *cyclones* which are discussed in detail under Ch. 7. For effective elimination of dust, large velocities of air and small diameters of cyclones are required.

15.4.2 Dust Elimination by Flocculation and Settling

The Cottrell treater is an important example of dust eliminations by flocculation and settling. The dust laden air/gas is passed in a large chamber, where an electrical field of steep gradient is set up. The flying ions have the effect of flocculating the dust, which then settles quickly to the bottom of the chamber or on the electrodes towards which the dust particles are carried by ionic charges. Flocculation may be obtained by a.c., continuous d.c., or intermittent d.c. The results are reported to be better with d.c. The periodic shaking of the electrodes results in settlement of major portion of the dust. A typical installation is shown in Fig. 15.10. This is also known as electrical precipitator. The efficiency of precipitation/separation depends on (a) time for which the dust laden air/gas remains in the electrical field, (b) distance between electrodes, and (c) electrical field gradient. For a given installation, the efficiency, E may be expressed as the following:

$$E = 1 - Kt$$

where, K = precipitation factor, depending on the nature of gas and suspended particles, and on the type of precipitator,

and t = time for which gas/air remains in the electrical field.

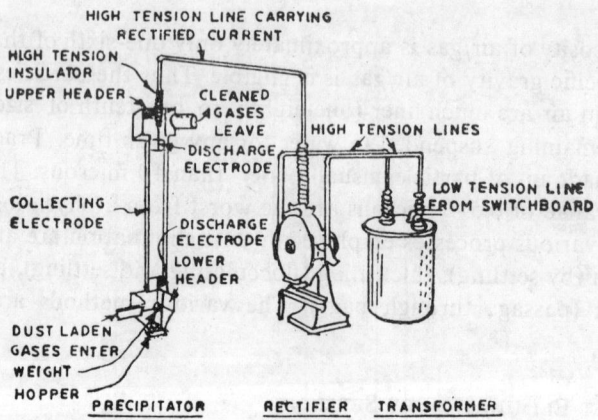

Fig. 15.10. Schematic diagram of electrical precipitator.

A new method of eliminating dust by flocculation has been proposed in which ultrasonic waves are utilised. The waves concentrate the dust at their nodes and bring the particles in their close proximity. As a result, flocculation takes place.

15.4.3 DUST ELIMINATION BY SYNTHETIC FOGS AND RAINS

This is an imitation of the natural way of eliminating dust and is very attractive. The water acts as a bond between particles similar to oil acting as a bond in agglomeration. The practical problems are (a) to create a curtain of rain across the dust laden gas stream, (b) to collect the muddy water, (c) to separate the water from solids, and (d) to return the clean water to the rain maker. A simple device to effect this system consists of a revolving drum which picks up water and sprays it across the gas stream.

Another device is a hydrocaptor consisting of a hollow drum fitted with screen surfaces and revolving in such a way as to dip and pick up water which drops as a curtain across the dust laden air/gas.

15.4.4 DUST ELIMINATION BY FILTERS

The filter assembly employed for dust removal from gases/air is known as bag houses. The bag houses are bulky in installation and expensive in operation, since the handling of product is a problem. A widely known unit is Birtley filter which is a cyclone operating to remove the coarser particles. In the upper portion of the cyclone, a number of filter bags are placed which retain the fines. After certain periods (every 2 hr or so), the suction is cutoff and the upper part of the filter is connected to atmosphere. This reversal of pressure clears the clogged pores, which is equivalent to the blow-back of pulp in vacuum filters. Discharge of the settled dust is intermittent, i.e. a valve is opened (manually or automatically) when the weight of the dust becomes excessive.

15.5 Removal of Water by Drying

This is the method of separating liquids from solids by evaporation of liquid and removal of the vapours. Drying is an expensive process in comparison to thickening and filtration (the cost incurred per unit volume of fluid removed). However, drying has its own place, since filtration cannot remove the moisture content below certain minimum level, which may be too high for the next operation or marketing purposes.

Some important methods of drying the products of mineral processing plants are as follows:

a) FLASH DRYING

This is accomplished by dropping the wet material in a tower against a rising current of hot gases.

b) ROTARY DRYING

This is carried out in a horizontal cylinder in which the material is turned over and over against or with a current of hot gases.

c) RABBLE HEARTH DRYING

In this method, the drying is affected by mechanically turning over of the material on a horizontal hearth in a current of hot gases.

15.6 Floto-Flocculation Method for Clarification of Sewage and Effluents[18]

The floto-flocculation method is regarded as the flotation aggregation for intensification of floc formation in suspensions treated by long-chain organic reagents before settling. This method has been widely used for the treatment of large volumes of waste water and recycled water from ore-processing and metallurgical plants.

The method takes into account the appropriate choice of reagent and creation of suitable conditions for particle hydrophobisation. The flocculation of particles in dilute suspensions by surface-active hydrophobic agents can be divided into the following stages:

a) Introduction of a hydrophobic reagent into the suspension. The reagent is strongly adsorbed on the surface of particles and is evenly distributed on the particle surface.

b) Introduction of air-bubbles into the hydrophobic particle suspension for the floto-aggregation of the particles on the surface of gas-bubbles and in the froth.

c) Transportation of the flocculated suspension into the thickeners.

This mechanism of aggregation of suspended particles provides sufficiently higher rate of clarification and greater purity of the clarified solution than the general conditioning of suspensions with a hydrophobic agent.

The device employed in floto-flocculation is a floto-flocculator (Fig. 15.11) which operates in the same way as coagulation chambers. The floto-flocculator may be installed by the side or inside of a thickener. The retention time depends on the impurities and flocculating reagents. In most of the cases, employing polymer flocculants, the retention time is 30–60 sec, while it is more than even 3 min in case of using carboxylic acids. In fact, the floto-flocculators may be regarded as flowing devices, in which air dispersion is provided through some perforated rubber tubes. The intensity of air flow may be 50–70 m³/hr. The clarification and formation of larger flocs is favoured by increasing the aeration time up to a certain limit, as well as by an increase in the thickeners of aerated water column.

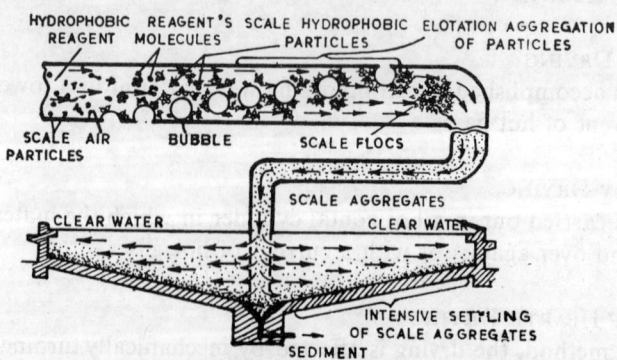

Fig. 15.11. Simplified sketch of floto-flocculator.

The method of floto-flocculation has made possible (i) to intensify clarification with organic flocculants, and (ii) to use a number of surface active agents of low toxicity with little capital cost. This method has resulted in a greatly improved ecology in the vicinity of metallurgical and ore processing plants. Presently, this method is employed for the treatment of waste water from metallurgical (hot rolling, blast furnaces, electrofiltration of dusty air and gas) and chemical processing of ores (leach solutions of various ores/minerals, such as copper, tungsten, gold, etc.). The floto-flocculation method can be effectively used to treat the majority of the industrial waters and slurries in metallurgical and ore processing plants.

CHAPTER 16

Chemical Processing of Ores

The extraction of specific values from their ores by dissolution or simple chemical treatment such as roasting, calcination, and fusion. etc., may be regarded as the area of chemical metallurgy, rather than mineral engineering. In the last four decades, there has been increasing trend to treat low-grade ores due to the fast depletion of rich ores. The treatment of low-grade and difficult to treat, involve handling of huge tonnage and combined skill of chemist and mineral engineer. Therefore, the area of chemical processing of minerals/ores has now become of more concern to mineral engineer. particularly when there is no other better alternative to treat the specific ores for the recovery of valuables.

In gravity concentration, only physical structure of minerals is altered. In flotation, only minute changes are made at the surfaces of specific minerals, and there is a very minor effect on chemical composition. In aqueous chemical processing (leaching) of ores, either the desired metal or compound is separated from the finely ground ore by dissolution or the gangue and undesired compounds are dissolved out from the ore. In dry chemical processes, chemical reactions may be used even at high temperatures in solid–gaseous or solid–solid phases (roasting, sintering, segregation, etc.). The values are then recovered from the respective solutions or residue and likewise products by suitable techniques.

The most important and oldest method involving chemical treatment is processing of gold ores by cyanide leaching. The other treatments include bacterial leaching of ores, leaching of uranium and thorium ores, beneficiation of ilmenite by selective leaching, roasting, segregation roasting, sintering, etc. These are discussed in the following pages.

16.1 Leaching of Ores

A large tonnage of metals is produced by hydrometallurgical techniques for which *leaching* is the first and essential step. Leaching may be defined as the treatment of crushed/ground ore (with or without pretreatment) with water or other cheap reagent (H_2SO_4, HCl, NaOH, Na_2CO_3, $FeCl_3$, etc.) to effect either the preferential dissolution of valuables or undesired constituents. Leaching may be carried out in several ways, such as *leaching in situ, heap leaching, percolation leaching, agitation leaching, pressure leaching, and bacterial leaching.*

16.1.1 LEACHING IN SITU

This refers to the leaching in mine itself. When it is not economic to mine lean ores, this type of leaching is adopted. Since, *in situ* leaching of undisturbed ore bodies is not likely to be successful, the rock is broken and a suitable permeable texture and suitable shape of ore body is obtained to facilitate the percolation of solution. Sometimes surface water may be diverted to percolate through old mine workings which takes away the values in solution down into a sump, from where the solution can be pumped for the recovery of values. In another technique, a number of holes are drilled into the upper surface of an ore body and an additional hole is made at another point which is at its lowest level. Water is fed through the shorter holes and pregnant solution is withdrawn from the deepest point.

16.1.2. HEAP LEACHING

As the name implies, this is carried by stacking the ore in heaps and then spraying the solution/water over the heap at different points to give proper distribution of solution. The run-of-mine ore (<200 mm) is stacked in the open on impermeable rocks or clay pads with drainage culverts built into them. The heaps of ore are sprayed regularly with the oxidising and acid solution (obtained from precipitation plant) which is rich in ferric sulphate. The heaps of ore should be sufficiently ventilated to provide plenty of air to keep down the temperature required to inhibit the oxidation and sulphation of FeS_2. The sprayed solution seeps into the ore lumps, and reacts with minerals (such as copper) and dissolves some of the metals. The drained solution is sent to the recovery plant. This mechanism is faster than diffusion but still quite slow. A heap of 1,00,000 tonnes of sulphide ore at Rio Tinto in Spain took two years to leach it.

16.1.3 PERCOLATION LEACHING

This refers to the leaching performed in large concrete tanks by percolating the suitable solution through the crushed ore (-6 mm) stored in tanks. These tanks can hold as large as 10,000 tonnes of ore. Leaching liquor is fed into each tank which is held for several hours and drained out. Usually each tank is percolated by a number of solutions and each time with more concentrated solution than the preceding one. In the last tank, the fresh solution is employed. Similarly, the solution is passed to fresher ore at each move until it is strong enough to be sent to the recovery plant. In this way a counterflow leaching is carried out discontinuously. A tank load of about 10,000 tonnes will require 5–10 days for leaching. The most successful application of percolation leaching has been to oxidised ores which do not require oxidation.

16.1.4 AGITATION LEACHING

In this the ore is suspended in leaching solution in a vat/tank and stirred mechanically or with compressed air jets. When fine grinding is required to expose the mineral particles, the ore becomes too fine for the percolation method. Flotation concentrates and gold ores can be leached efficiently with agitation. In this type, the contact time is reduced to about two days. In this leaching, the pregnant solution does not get separated at its own. The separation of leach liquor is effected by classifiers, thickeners and filters.

16.1.5 PRESSURE LEACHING

Usually the leaching is carried out at atmospheric pressure and low temperature. However, in certain cases pressure leaching can be applied when it is not possible to dissolve the required constituents under normal conditions. Leaching under pressure makes possible the use of elevated temperatures and thus rendering the dissolution much faster. The high pressure also permits the use of gaseous reagents (O_2, NH_3, CO_2 etc.) to be used at higher activities, than in normal pressures and temperatures. The use of high pressure and temperature accelerates the dissolution and also modifies the reaction equilibrium position favourably. Pressure leaching can be used in the cyanidation of gold under oxygen atmosphere, bauxite leaching with NaOH, leaching of mixed Ni–Cu–Co sulphide ores with ammonia and oxygen, leaching of lateritic nickel ores, etc.

Many batch and continuous systems have been developed to process various ores/concentrates under pressure (2–40 kg/cm^2) and high temperatures (up to 250°C). The usual equipment used for pressure leaching is an autoclave provided with suitable lining to withstand temperature, pressure, and chemical attack of the reagents.

16.1.6 BACTERIAL LEACHING

In the presence of certain bacteria such as *Thiobacillus ferro-oxidans*, *Thiobacillus thioxidans*, and *Ferrobacillus sulphoxidans* (often found in mine water) the leaching rate of ores is increased considerably, particularly sulphide ores respond to this. It is believed that pyrite present in the ore is oxidised to H_2SO_4 and $FeSO_4$. The bacteria obtain their energy by the oxidation of ferrous iron to ferric salts or by oxidising sulphur, metal sulphides or the lower oxyacids of the sulphur. Therefore, either the metal sulphides are oxidised directly or an acid ferric sulphate solution is produced which is capable to oxidise most of the metal sulphides and render the metal ions soluble and reduction of ferric sulphate to ferrous sulphate which again provides the food for bacterial oxidation. The bacteria also exhibit a self-bufferring action, which enables the maintenance of uniform pH. In case of ores/minerals of low iron content, iron sulphate may be added for maintenance of pH. After leaching, the metal like copper can be precipitated by hydrogen under pressure, or iron. The various reactions involved in bacterial leaching are:

$$2FeS_2 + 7O_2 + 2H_2O = 2FeSO_4 + 2H_2SO_4 \qquad (16.1)$$

$$4FeSO_4 + 2H_2SO_4 + O_2 + \text{bacteria} = 2Fe_2(SO_4)_3 + 2H_2O \quad (16.2)$$

$$7Fe_2(SO_4)_3 + FeS_2 + 8H_2O = 15FeSO_4 + 8H_2SO_4 \qquad (16.3)$$

or $$Cu_2S + 2Fe_2(SO_4)_3 = 2CuSO_4 + 4FeSO_4 + S \qquad (16.4)$$

$$2S + 3O_2 + 2H_2O + \text{bacteria} = 2H_2SO_4 \qquad (16.5)$$

Leaching of ore may be carried out by stirring the crushed ore in a Pachuka vessel (tall cylindrical tanks having conical bottom) with aerated solution or by circulating the solution through the ore (underground or piled under a plastic sheet) with proper species of bacteria.

Bacterial leaching is being employed successfully to leach lean ores. In U.S.A. about 15 per cent of the total production of primary copper is obtained through bacterial leaching. The use of bacterial leaching is becoming more important, since there is considerable decline in the general grades of the ores. Bacterial leaching of ores offers the following advantages:

a) Leaching rate is greatly increased.

b) Very lean ores or tailings can be treated by bacterial leaching.

c) It may be possible to avoid the step of mining, since *in situ leaching* is possible.

d) The process is very cheap, since only a small quantity of H_2SO_4 is required to maintain the desired concentration of ferric iron and a little power is required for aeration and circulation of leaching solution.

e) It does not involve practically any investment in equipment and building.

In addition to copper ores, the bacterial leaching has been tried successfully for treating of various sulphide ores (nickel sulphide, molybdenite, ZnS, etc.) as well as oxide ores (uranium ores).

16.2 Ion Exchange

The sudden emergence of uranium as a major strategic element during the years of second world war (1939–44) led to urgent development of new methods for the extraction of uranium. An intensive study and world-wide use of ion-exchanger resins at an intermediate stage of chemical extraction and/or concentration was made. In ion exchange (IX) reaction, ions are exchanged at the boundary between two phases. A solid ion exchanger is an insoluble resin possessing accessible exchange cations, anions, or amphoteric ions. It differs from sorbing solids in respect of stoichiometric action in ion exchange. The ions captured by the exchanger from the solution can be replaced by an equivalent amount of similarly charged ions. The selective capture depends on relative ion size, valences, and concentrations of the ions displaced.

A liquid ion exchanger is a compound which is dissolved in a liquid immiscible with the solution carrying the ions to be exchanged. The immiscible phase may be kerosene, tetra-chlorethylene, chloroform, etc. A high flash point kerosene is widely used in industries. The ion exchanger compounds used are aliphatic amines, dialkyl phosphates and fatty acids. However, industrially the ion-exchange technique is limited to use of solid exchangers and the liquid ion exchangers are covered under solvent (liquid–liquid) extraction.

16.2.1 SOLID EXCHANGERS

These are solid electrolytes having inert matrices (R), special polar groups (G), and charged ions (X) available for exchange. Therefore, a solid ion exchanger can be represented by $R(GX)_n$. If G is negative, X is cation and the solid is *cation exchanger*, while G being positive, X is an anion and the material is *anion exchanger*. Inorganic ion exchangers are zeolite, glauconite, natural montmorillonite, and synthetic sodium alumino-silicates. Some coals also develop ion exchange properties on sulphonation and oxidation. The usual ion exchange materials used industrially are based on synthetic resins, such as phenol formaldehyde and styrene-divinyl benzene.

The polar groups in cation exchangers are sulphonic acid ($-SO_3H$), carboxylic acid ($-COOH$), sulphonium ($-SR_2OH$) and phosphonium ($-PR_3OH$). The reaction in ion exchange process can be represented as follows:

$$R(GX)_n + nM^{\pm} \rightarrow R(GM)_n + nX^{\pm} \qquad (16.6)$$

The rate of reaction depends mainly on the diffusion of exchanging ions to the resin particle and across the film of solution at its surface. This is aided by increase in active area, temperature rise, concentration of ions and their relative size. The ions leaving the ion exchange media are replaced by an equivalent amount of other counter-ions to maintain electronegativity.

In order to use ion exchange process, the ion species captured by solid ion exchanger should be at least partly dissociated and mobile. The solid ion-exchanger should also not be soluble and destructible in the solvent. The dielectric constant of solvent should be high (for water 81).

The resins may be considered as gels with an irregular macromolecular matrix of hydrocarbon chains in a three-dimensional network. This matrix carries $-SO_3$, $-COO^-$, $-PO_3$—groups in case of cation exchanger, whereas $-NH_3^+$, $>NH_2^+$, $>N^+$, $-S^{3+}$ groups in anion exchanger. The capacity of ion exchanger depends on the nature of the groups. A weak acid ($-COO^-$) is only ionised at high pH, while a strong acid ($-SO_3^-$) remains ionised even at low pH.

In most cases, the ion-exchange resin is used in bead form, but resins can be formed in many other shapes such as discs, plugs, mats, belts, ribbons, etc. The solid ion exchangers employed in mineral processing are selected for their stability (resistance to swelling), toughness to abrasion and breakage, loading capacity (capacity to retain the number of iogenic groups per specific volume/weight), and ionic sign. The usual cation exchangers are sulphonated coals, and sulphonated carboxylic and phosphoric resins. Anion exchangers include a range of weak to strong basic resin matrices.

16.2.2 OPERATION OF ION-EXCHANGE PROCESS

The process consists of three steps, i.e. loading, back washing, and elution.

a) *Loading of Resin*

The operation of ion exchange starts with the loading of the exchange sites by ions drawn from the pregnant aqueous solution. An ion-exchange column used in operation is shown in Fig. 16.1. The pregnant liquor with its load of A-type ions is allowed to percolate downward in the ion-exchange column which is packed with resin beads loaded with ions of B-type. As the solution comes in contact with resin, A-ions from solution deposit in a narrow zone near the entry point and takes up the equivalent number of B-ions from the resin. With the continuous downward percolation of solution, the resin becomes loaded and the zone of exchange shifts through the column until the resin beads become saturated. This stage is called *breakthrough point*, and A-ions then begin to appear in the effluent, even when the lowest zone is not fully loaded. Thus, in order to avoid loss and to use full loading

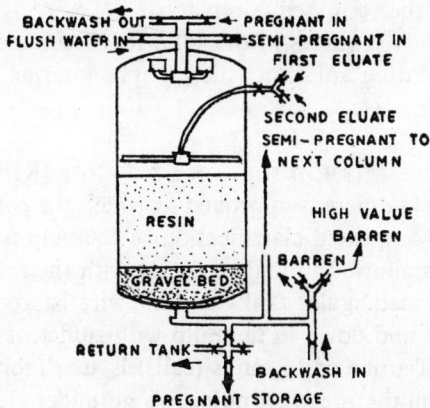

Fig. 16.1 Illustrative diagram of an ion exchange column.

capacity of resin, a series of three ion-exchange columns can be kept on stream. When breakthrough point is reached, the first column is switched out. New pregnant solution is then fed to the second column, and the initial third one now becomes second. A fresh column is put after the new second one, which works as the third column. The arrangement allows the use of three columns being loaded continuously, while the fourth one is being unloaded and reactivated.

b) *Back Washing*

When loading of the first column is stopped, its resin may not be saturated due to some channelling effect in the bed. The channelling results into incomplete contact between beads and solution. Further, if the liquor is not fully clarified, some surfaces of beads may be masked by slime. There-fore, the next operation consists of stirring of the bed by sufficient water rising upward to scrub the beads without flushing them out. This operation is called *back washing*, which removes deposited slime and any residual un-drained pregnant solution. In some cases, this step may not be necessary.

c) *Elution*

This is the displacement of captured ions from the sites of resin by flush-ing of a suitable chemical solution (eluant) through the column. When this operation is complete, the resin sites are restored to their original condition and the beads are regenerated. The pregnant eluate is run off either to storage tank or for precipitation treatment. The column is again given a back wash to remove the residual eluant. The column then becomes ready for loading.

During the working life, the beads become progressively loaded with undesired ions picked up from pregnant solution and physical poisons (such as polymerising collodial silica and organic materials). This requires perio-dic regeneration of resins and when the situation becomes too far for an efficient operation, the resin bed is replaced.

Final recovery of the values in the solid form is made by suitable chemi-cal treatment of eluting solution (such as preciptation, filtration, drying, calcination, reduction, etc.).

16.2.3 Resin in Pulp Method for Ion Exchange (RIP)

This is an entirely different approach in which the coarser sand particles are first removed by repeated classification of the pulp to about 10 per cent solids and then it is allowed to make contact with the resin. The pulp flows through a series of rectangular tanks and the wire baskets loaded with resin beads are jigged up and down in the pulp with sufficient force to dilate and contract its load. If a line of 14 tanks (cells) is used, seven are provided to adsorb the ions from the pregnant pulp, five go under elution, one receives preliminary back wash and one receives post-eluating back wash.

16.2.4 Examples of Use of Ion Exchange

The well known example is the softening of water by zeolite. The most important application of ion exchange in mineral processing is recovery of uranium which may occur in leach liquor as complex anion in H_2SO_4 solu-tion. The reaction with RNO_3 will be

$$UO_2(SO_4)_3^{4-} + 4\,RNO_3 = R_4UO_2(SO_4)_3 + 4\,NO_3^- \qquad (16.7)$$

In this particular case, the separation of uranium from other metals as cations is of great importance. The application of ion exchange to other metals is governed by the cost of operation.

16.3 Liquid–Liquid (Solvent) Extraction

In liquid–liquid (solvent) extraction, instead of using a solid resin and an aqueous solution in ion exchange, the former is replaced by an immiscible organic liquid which extracts the metal species from the pregnant solution. Some diluents such as kerosene and xylene are generally used as a carrier of organic solvent. The organic liquid and the pregnant solution are mechanically stirred as they flow through a series of mixing tanks or columns, The conditions of stirring should be such that a temporary emulsification should be produced and separated out. The metal values get transferred from aqueous phase to organic phase. The loaded organic phase is then treated with suitable chemicals to extract back the values either in solution or precipitate form. This step is known as stripping.

The main attractions of solvent extraction are its simplicity and continuity of operation. The immiscible phase (organic liquid) should be highly selective for the species to be transferred from the aqueous phase. Both phases must be quick to de-emulsify and separate after stopping the stirring.

16.3.1 Mechanism of Solvent Extraction

An extractant (organic compound which extracts/values from aqueous solution) should fulfill the requirements such as (a) selectivity, (b) high extraction capacity, (c) ability of being stripped easily, (d) ability to separate from aqueous phase, i.e. having low viscosity, high surface tension and low density, (e) non-toxic, (f) non-flammable, (g) stable, (h) non-volatile, and (i) cheap. However, it may not be possible to have all these requirements in a single extractant, and thus a compromise is made. The mechanism of extracting the species by organic solvent may be physical or chemical. In physical transfer, absorption of species occurs in extraction, e.g. extraction of As, Sb, Ge, Hg, etc., from HCl acid solutions by hydrocarbons or chlorinated hydrocarbons (benzene, and chloroform). In chemical interaction, species in aqueous phase interact with the extractant and the product (organometallic complex) is soluble in the organic phase. Following are the three types of chemical interactions:

a) Ion-Pair Transfer

In this case, electrically neutral species interact with the extractant to form an additional compound which is soluble in the organic phase. For example, uranium in nitric acid solution and trybutyl phosphate (TBP):

$$UO_2(NO_3)_2 + 2\ TBP \rightarrow (TBP)_2 \cdot UO_2(NO_3)_2 \qquad (16.8)$$

$$TBP = \begin{matrix} RO \\ RO-P = O\ (R = C_4H_9) \\ RO \end{matrix}$$

b) Ion Exchange

In this case, extractable species are charged positively or negatively (cationic or anionic exchange, respectively) and are stochiometrically exchanged with the ion of similar charge from organic phase. For example, extraction of uranium from monoalkyl phosphoric acid:

$$2\ RO{-}\underset{\underset{OH}{|}}{\overset{\overset{O}{\|}}{P}}{-}OH + UO_2^{++} \rightarrow RO{-}\underset{\underset{OH}{|}}{\overset{\overset{O}{\|}}{P}}{-}O\cdot UO_2{-}O{-}\underset{\underset{OH}{|}}{\overset{\overset{O}{\|}}{P}}{-}OR + 2\ H^+$$

(16.9)

$$\underset{\text{(Org)}}{(R_3NH)_2\ SO_4} + \underset{\text{(aq.)}}{UO_2(SO_4)_2^{--}} = \underset{\text{(Org)}}{(R_3NH)_2\ UO_2(SO_4)_2} + \underset{\text{(aq.)}}{SO_4^{--}}$$

(16.10)

c) *Chelate Extraction*

In this case, extractant molecule containing both acidic and basic functions combines with a metallic ion with both groups operating. As a result a chelate or inner salt is formed which is soluble in diluent. For example, extraction of Be^{++} by acetyl acetone:

$$2\underset{\underset{CH_3}{|}}{\underset{C=O}{\overset{\overset{CH_3}{|}}{\underset{\|}{HC}}{\overset{|}{C}}{-}OH}} + Be^{++} \longrightarrow 2H^+ + \underset{\underset{CH_3}{|}}{\underset{C=O}{\overset{\overset{CH_3}{|}}{\underset{\|}{HC}}{C}{-}O}}\ \ Be\ \ \underset{\underset{CH_3}{|}}{\underset{C=O}{\overset{\overset{CH_3}{|}}{O{-}C}{\underset{\|}{CH}}}} \quad (16.11)$$

16.3.2 SOLVENT EXTRACTION TECHNIQUE

A general flowsheet of solvent extraction process is shown in Fig. 16.2. For a considerable recovery of values in one single batch extraction process, the value of distribution coefficient, D (ratio of concentration of a metal species in the organic phase to that in aqueous phase under equilibrium

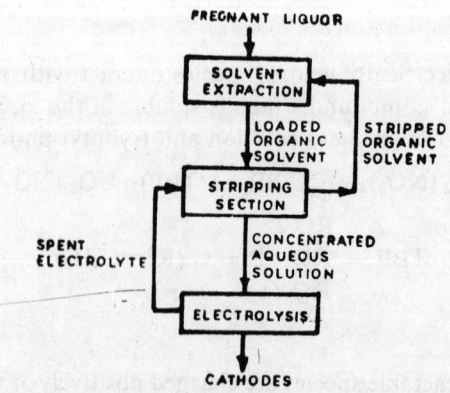

Fig. 16.2. General scheme of solvent extraction.

conditions) and ratio of organic phase/aqueous phase volumes should be large enough. Since D is generally not high and organic/aqueous phase ratio can also not be maintained very high for economic reasons, a number of successive extractions may be employed. Various techniques used are multiple batch extraction, counter-current batch extraction and continuous counter-current practice.

a) Multiple Batch Extraction

In this technique a number of successive extractions of metal values are carried with fresh solvent. However, in order to recover almost the entire value, a large number of stages will be required, which is not convenient in practice. This is limited to laboratory studies only.

b) Counter-Current Batch Extraction

In this technique, the portion of aqueous phase containing the solute (to be extracted) is extracted with successive portions of the organic phase in such a way that the fresh solvent always extracts from the weakest aqueous phase and the most loaded solvent extracts from the solute-rich aqueous feed.

c) Continuous Counter-Current Technique

In this technique two phases pass continuously in opposite directions as in a column. In this case the difference in density between the organic and aqueous phases provides the driving force for counterflow.

16.3.3 SOLVENT EXTRACTION PRACTICE

Large-scale mixer/settler as well as column units are available which operate continuously. The immiscible phases pass in opposite directions. Different types of column include spray columns, packed columns, perforated plate columns, rotary annular columns, pulse columns, rotating shaft columns, etc.

Solvent extraction finds its use in recovery of uranium, thorium, beryllium, cesium, boron and in separation of zirconium from hafnium, cobalt from nickel, copper, nickel, zinc and cobalt, and many similar applications.

The most popular solvents are marketed under proprietary names, LIX and KELEX with a number appended. LIX reagents are hydroxy-oximes and KELEX are hydroxy-quinolines.

16.4 Roasting

Roasting may be defined as a process which involves chemical changes at somewhat higher temperatures (200–800°C) other than decomposition and fusion. Generally, the reactions take place with furnace atmosphere. Roasting may effect drying and calcination also. The main purposes of roasting are the following:

a) Oxidising roast to burn out sulphur partly or completely from sulphides, e. g. copper concentrates (partly), zinc concentrate (fully), and auriferous sulphides (partly).

b) Volatilising roast to separate the elements with volatile oxides, such as As_2O_3, Sb_2O_3, or ZnO which can be recovered as fumes in roasting of auriferous sulphide concentrate.

c) Chloridising roast for conversion of certain metals (from oxides or sulphides) to chlorides. This may be carried out under oxidising or reducing conditions, e.g. roasting of ilmenite with HCl or chlorine in presence of carbon.

d) Sulphating roast to convert certain metal sulphides to sulphates, usually prior to leaching, e.g. roasting of chalcopyrite and zinc concentrate under controlled conditions.

e) Magnetising roast, usually a controlled reduction of hematite to magnetite to facilitate the magnetic separation, e.g. roasting of low-grade iron ores other than magnetite.

f) Reduction roast of oxide to metal, e.g. reduction roast of lateritic nickel ores to render them suitable for the ammonical treatment.

g) Carburising roast to prepare calcine for chlorination, e.g. formation of silicon carbide for chlorination.

The primitive roasting was carried out in *heaps* and *stalls*, which was replaced by the use of shaft furnaces and simple air furnaces. Later on, roasting kilns and multiple hearth roaster were developed. Presently, the roasters used are mainly suspension roaster, flash roaster and fluid-bed roaster.

16.4.1 SUSPENSION ROASTING

The suspension roaster is shown in Fig. 16.3. In this case the ore is dried and preheated on two top hearths and then falls through the central section of the furnace against the oxidising furnace gases.

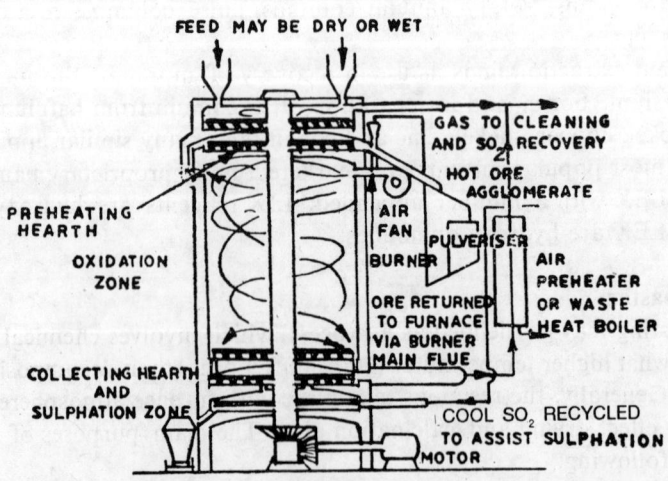

Fig. 16.3 Schematic representation of suspension roaster.

16.4.2 FLASH ROASTING

This is a modification of suspension roasting in which the pre-heated ore is injected through a burner with pre-heated air instead of pulverised fuel. The flash roasting is most suitable for the roasting of sulphides which oxidise exothermally and no additional fuel is needed. This is also known as autogeneous roasting.

16.4.3 FLUID-BED ROASTING

The use of fluidised bed in roasting of fine concentrates has become most attractive due to faster rate of reaction. When a gas is passed upward through a bed of fine articles (2–0.02 mm), the bed will expand and fluidised bed condition is obtained, in which individual particle is exposed to oxidising atmosphere resulting into instantaneous roasting. A fluosolid roaster is shown in Fig. 16.4. The roasting may be carried out autogenously if the reactions are adequately exothermic, otherwise, some fuel such as pulverised coal, or gaseous fuel should be introduced with air along with concentrate.

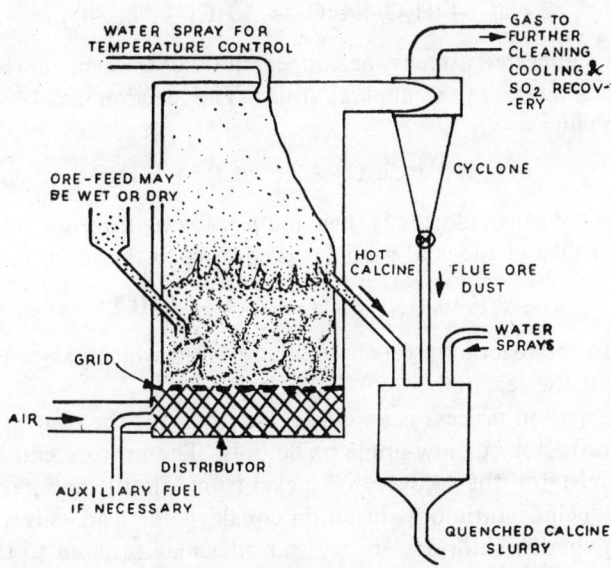

Fig. 16.4. Fluidised bed roaster.

In case of magnetising roasting of hematite or reduction roasting of nickel ores, the atmosphere of the furnace should be reducing and thus it should have sufficient CO or H_2 or the mixture of the two.

16.4.4 SEGREGATION ROASTING[50-52]

The segregation roasting, also known as TORCO (treatment of refractory copper ores) process was first developed to produce copper concentrates from refractory copper ores (aluminate-silicate ores found in Mountania and Zambia) containing about 2.5 per cent copper. These ores do not respond to normal flotation or leaching processes. The segregation process may be regarded as chloridising roasting process under suitable reducing conditions. The segregation process principally consists of the following steps:

a) Pre-heating of fine ore (60 per cent–50 microns) in a coal fired fluid-bed reactor.

b) Reaction of hot coal–ore mixture with NaCl at about 800°C in a vertical kiln. The product obtained from the vertical kiln is a particulate coal–ore mixture having metallic copper particles segregated on the coal surface. The segregation process is believed to proceed by the formation of volatile cuprous chloride, followed by reduction of chloride by hydrocarbons at the surfaces of coal.

c) Flotation of kiln product to obtain a concentrate containing 30–50 per cent copper which can be sent for the recovery of copper by standard methods (smelting or leaching followed by electrolysis.).

In the process, it is generally recognised that the actual chloridising agent is HCl, which may be generated according to the following reaction:

$$2\ NaCl + H_2O + SiO_2 \rightarrow 2\ HCl + Na_2SiO_3 \qquad (16.12)$$

HCl thus liberated attacks the copper minerals to form cuprous chloride which diffuses out from the mineral grains. The reaction may be represented as the following:

$$6\ HCl + 3\ Cu_2O \rightarrow 2\ (CuCl)_3 + 3\ H_2O \qquad (16.13)$$

The gaseous cuprous chloride is then reduced by hydrogen to metallic copper at the surface of the coal particles according to the reaction:

$$2\ (CuCl)_3 + 3\ H_2 \rightarrow 2\ Cu + 6\ HCl \qquad (16.14)$$

According to the reaction (16.14), HCl is released which passes through the same cycle of the reaction.

The segregation process is an efficient method for the recovery of metals from their refractory or low-grade oxide ores. The process can be successfully employed for the recovery of nickel from lateritic and garnierite ore. The metals such as antimony, bismuth, cobalt, gold, lead, silver, tin, etc., which form volatile chlorides or oxy-chlorides may respond to the segregation roasting.

16.5 Sintering

The process of sintering involves the heating of ore/mixture of ore, flux and reducing agent below the melting point, but certain amount of fusion of low-melting constituents occurs. It involves important chemical reactions and formation of new chemical compounds (generally low-melting) which act as bonding agent (bridge) between ore particles. Generally, sintering is employed for roasting as well as agglomeration to reduce or eliminate particularly sulphide and carbonate by oxidation, reduction, and/or dissociation. Sometimes sintering is referred to as *blast roasting*. The sintering may also be applied for blending in fluxes required during smelting or for agglomerating returned sinter fines, flue dust, filter cake, etc.

Sintering may be carried out either in continuous or in a batch process. In the batch process, each grate rests on its own suction box. The simple batch process does not involve much capital cost and thus can be used for irregular small production units. The continuous machine (Dwight–Lloyd type) shown in Fig. 16.5 is more common, particularly in a large plant. The Dwight–Lloyd machine consists of grates in sections of pallets travelling in an endless belt over a series of suction boxes, which are controlled individually. The first box remains under the ignition hood and the last one works only as a discharge point.

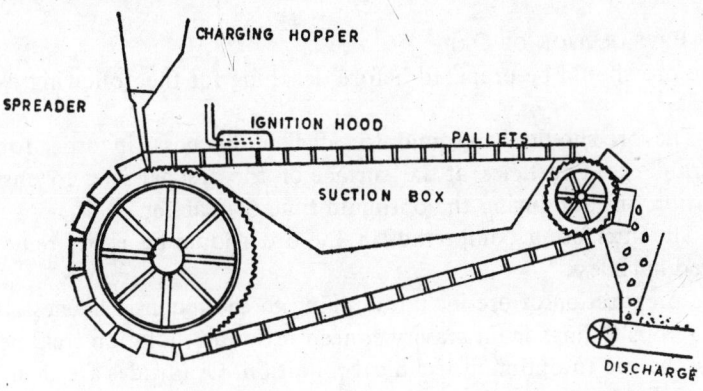

Fig. 16.5. Schematic diagram of Dwight/Lloyd sintering machine.

Sintering is employed for iron ores, zinc concentrate and lead concentrate. In iron-ore sintering, the ore should be in the size range of 15 mm to 150 microns, whereas coke breeze and limestone added should be —3 mm in size. The sulphide ores (zinc and lead ores) are sintered without adding any fuel, since sulphur acts as a fuel. In fact, in order to control the temperature (not to allow to rise too high) it is essential to recycle a large amount of returned fines. The bed may be laid in two layers of different characteristics, i.e. bottom layer may contain less fuel. Ignition is carried out by use of gas or oil burners under the hood, over the first suction box.

Gases drawn through the bed should be cooled and scrubbed free of dust and SO_2 before their release to atmosphere. The temperature of the sinter bed is controlled through (a) fuel content of the feed, (b) suction rate, (c) dilution of charge with inactive returned fines in pyritic sintering, and (d) permeability of bed.

In some special cases such as lead concentrates, updraught sintering using ignited bottom layer covered with 20–30 cm of sinter mixture can be applied. In this case, the compounds formed by the interaction of oxide and sulphide becomes solid and remain in the sinter.

16.6 Cyanide Process for Recovery of Gold

This process was first used in 1894 to recover the gold from the ores containing gold in the form of sulphide, which did not respond to amalgamation. This has been used later to recover gold which escapes from gravity concentration or flotation concentration, as well as to treat flotation concentrates. The process consists of the following stages:

a) Preparation of ore to expose its gold values.
b) Leaching of gold ores to dissolve gold values.
c) Separation of solids from gold-rich solution.
d) Clarification and deaeration of solution.
e) Precipitation of gold from solution.
f) Treatment of precipitated gold.

16.6.1 PREPARATION OF ORE

The ore should be prepared before leaching for the following two reasons:

a) The ore should be ground to sufficient fineness, in order to expose recoverable gold particles at the surface of the ore particles to ensure the dissolution of gold during the optimum time of reaction.

b) The interfering compounds in the ore should be either removed or rendered harmless.

The preparation of ore includes milling to the required fineness, mixing of auriferous tailings from gravity concentration or flotation treatment and making the ore to optimum technical condition. Cyanicides are then removed, neutralised or reduced to tolerable limits. Comminution of ore is carried out to the point where the maximum economical recovery can be made and the increased cost in further comminution is not compensated by increased recovery of gold. The extent of grinding depends on the gold content of the ore, and the mode of occurrence of gold.

16.6.2 LEACHING OF GOLD ORES

The dissolution of gold in cyanide is slow and the rate of dissolution is a function of total surface exposed to attack. Large particles of gold will require lengthy period for dissolution and thus requiring excessively large volume of pulp holding vessels to treat huge tonnage of gold ore. Fortunately, due to high specific gravity of gold, the large particles of gold are easily trapped and the holding capacity is required to treat only finer particles escaping from gravity concentration. The dissolution time is usually 8–30 hr and during this period, a large tonnage of gold ore along with much greater weight of solution should be kept in moving and thoroughly agitated condition. All this is done to extract few grams of gold from each tonne of ground ore, and thus operation should be designed to consume minimum power.

Coarse crushed ore can be leached by percolating the cyanide bearing solution through moderately thick beds. Use of vacuum to suck the leach liquor down through the bed, and stirring the liquor in the upper layers of the bed improve the dissolution rate. Sand or finer-size ore can be leached directly into collecting tanks, but to treat the large tonnage, the material is transferred to leachings vats via bottom discharge openings of collecting tanks. The leaching vats are large round tanks (up to 18 m in diameter and 5 m in depth) made of steel, concrete, or wood. In order to have efficient leaching, the sand should be distributed uniformly and the channel formation should be avoided, otherwise the solution will run without making the searching contact with all the grains as required for dissolution.

The other requirements in leaching are the presence of sufficient lime and oxygen/air to prevent the destruction of cyanide. Some air comes with new solution and during drainage, whereas additional air can be introduced by top raking, use of compressed air blowers, or applying suction from below. Sometimes, leaching tanks can be built in pairs, being one above the other to facilitate the transfer of material and to provide economy of space.

Leaching is carried out with strong cyanide (0.05–0.1 per cent NaCN) with a solid–liquid ratio of 70 : 30 to 70 : 25. A cover of about 30 cm height of liquor is maintained above the sand to obtain continuous percolation. The strong solution so obtained is sent for precipitation and from time to time charge is drained out. Nearing the completion of dissolution, weaker cyanide (0.02–0.03 per cent NaCN) is employed to complete the reaction and displace pregnant solution (containing gold values). This solution is used for making up strong solution and not for precipitation. Final washing is done with water or waste liquor from the slime plant. The whole cycle of treatment takes 4–7 days. The tailings are then removed to dump. This method of leaching has been successfully employed to treat clayey ores and retreatment of old dumps.

Milling of the ore can also be carried out in cyanide solution or water. When cyanide solution is not employed in milling, maximum possible water is removed by decantation and thickening before the leaching of slime. The pulp is adjusted for its lime content (0.002–0.025 per cent CaO of the pulp) in the mill. When the ore is ground in cyanide, the overflow is a valuable pregnant and thus goes to precipitation section. The settled slurry is then diluted with barren cyanide solution as it is being transferred into agitators. For agitation, the pulp is thinned to a liquid–solid ratio of about 1.2 : 1 with adequate dissolved cyanide to give a strength of 0.01 per cent NaCN or KCN and protective lime 0.0005 per cent CaO.

In leaching, the reaction involved can be represented as the following:

$$2Au + 4NaCN + 1/2 \, O_2 + H_2O = 2Na \, Au \, (CN)_2 + 2NaOH \quad (16.15)$$

In many flow sheets, washing treatment may be given prior to leaching to remove undesired constituents such as soluble salts and clayey fine material (which interfere with cyanidation), cyanicides' and other substances (upset gold–cyanide reaction), partly oxidised sulphides (mineral acids are produced), some organic matters such as graphite and charcoal (precipitate the dissolved gold prematurely), and chromium (which interferes reprecipitation of gold) at the suitable stage of the process.

16.6.3 SOLID-LIQUID SEPARATION

After the dissolution of gold to an optimum level the pregnant solution is displaced. The first wash of new feed with cyanide solution dissolves most of the gold and this solution is directly sent to the precipitation section. The remaining solid consists of pregnant solution in the form of film held by each particle by capillary action in the spaces between particles. This attached cyanide solution should be removed by washing with barren cyanide, low-value cyanide, or water. During various stages of leaching, the cyanide solution is available, but the amount of new wash water used should not exceed the volume of discarded foul liquor. Therefore, the various washings and precipitation section produce cyanide solution which is either of the following:

a) Rich enough in gold to be sent for precipitation.

b) Mildly auriferous and capable to pickup more gold before precipitation.

c) Foul enough requiring regeneration or to be discarded.

d) Barren solution returning from the precipitation section.

The primary separation can be effected by two methods, i.e. (i) counter-current decantation, and (ii) filtration. A combination of the two may also be used.

Counter-current decantation may be used through a series of thickeners and filters with intermediate repulping of the filter cake. Basically, removal of solution from pulp involves two steps, i.e. (i) concentration of the solids into the smallest possible bulk by removal of the liquid, and (ii) mixing of the solids with more dilute liquid which is flowing counter-current to them. The latter step is the displacement of liquid film held around each particle. The filters employed are usually of continuous type due to its various advantages (better control, compact housing, and less operating time). Filtration is generally applied to slurry discharged from thickeners. The cake obtained can be repulped in water or barren cyanide and then refiltered.

16.6.4 CLARIFICATION AND DE-AERATION

The cyanide solutions of different concentrations and varying degrees of contamination, aeration, and gold contents are circulating through the plant. The richest of these solutions is diverted continuously to the precipitation section for recovery of gold. The filtrate obtained is usually cloudy due to the presence of very fine material not trapped on the filter cloth. If the precipitation is carried out from this solution, the process will be interfered partly due to formation of slime coating on precipitating agent (usually zinc) and partly by contamination of the gold with slime (this will flocculate and choke the filters used to retain gold slime). Therefore, it is a standard practice to clarify the filtrate before precipitation to obtain a clear liquid. The clarification can be accompanied by passing the liquor through a tank containing a bed of sand. The colloids are trapped at the top of the sand bed in the form of a film, which can be skimmed periodically. Small thickeners can also be employed for clarification. During the clarification, the solution gets de-oxygenated to some extent. After the clarification, the solution is deaerated by vacuum to remove oxygen as much as possible.

16.6.5 PRECIPITATION OF GOLD FROM SOLUTION

After clarification and deaeration, the gold is precipitated by adding zinc dust/turnings to the pregnant solution. The gold is precipitated according to the following reaction:

$$2 \, NaAu \, (CN)_2 + Zn \rightleftharpoons Na_2Zn \, (CN)_4 + 2 \, Au \qquad (16.16)$$

In old practice, lead coated zinc turnings or wire (freshly produced by dipping into 10 per cent lead nitrate or acetate solution) contained in a wire basket and into steel boxes were used. The pregnant solution flows upward through each box, over a weir, and then down to the bottom of the next box in the series through which it is to flow upward. Finally, -850 micron material is periodically withdrawn from the head boxes and sent to bullion treatment.

In modern practice, zinc is used in the form of dust. A typical scheme is shown in Fig. 16.6. The pregnant solution runs into the clarifying tank, which is kept to a constant level by means of a float valve. The solution from the clarifying tank is drawn through filtering leaves (made of canvas and covered with diatomaceous silica and hung in the tank) by vacuum. The solution flows to a deaeration tower through a manifold. In the deaeration tower, most of the dissolved oxygen and CO_2 are removed. A small quantity of sodium phosphate is added to retard the precipitation of lime.

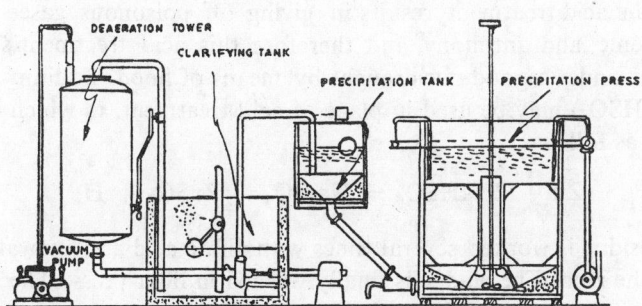

Fig. 16.6. A precipitation unit for gold precipitation by zinc.

Clarified and deaerated pregnant solution is withdrawn from the bottom of the deaeration tower by a special pump sealed against air leakage. A measured amount of lead activated zinc dust wetted with barren solution is introduced in the pregnant solution on its way to the precipitation tank. The zinc added presents a large active surface for the reaction and the precipitation takes place very fast. In case of excess time allowed for the reaction, zinc may be filmed over with calcium zincate according to the following reaction:

$$Ca(OH)_2 + Zn = CaZnO_2 + H_2 \tag{16.17}$$

In addition to this, $CaCO_3$ may also be formed on the zinc dust. Thus, zinc will be insulated and prevent the precipitation reaction.

From the precipitation tank, the solution flows to the precipitation press or filter, where the excess of zinc dust and gold slime are settled and held until the next clean-up, while the barren solution is pumped out to storage for reuse or to discard. The usual consumption of zinc is $0.01–0.03$ kg/m^3 of pregnant solution, but when silver is present, the amount may increase to $0.1–0.3$ kg/m^3. If the cyanide solution contains some silver and copper, they will activate the zinc, and lead need not be used. If copper is present in considerable amount, trouble may be caused due to plating of copper on zinc. However, this problem is not likely while using zinc dust due to large surface area available. If the zinc is added in insufficient quantity for complete precipitation of metals, preferential precipitation of gold will occur. This tendency can be utilised to obtain gold-rich and copper-rich slimes separately by adding zinc dust into the solution in two successive additions (the first addition of zinc is made in starving quantity).

From pregnant solution, gold can also be extracted by carbon in the form of charcoal and some variety of graphite. Activated charcoal can retain up to 50 kg of gold per tonne of carbon.

16.6.6 TREATMENT OF PRECIPITATED GOLD

The slime obtained after filtration consists of gold/silver with zinc content below 10 per cent. In order to remove zinc, the slime is first treated with H_2SO_4. The acid treatment results in giving off poisonous gases carrying HCN, arsenic, and antimony, and therefore this acid treatment should be carried out under a good out-draught by means of hood. Sodium acid sulphate ($NaHSO_4$) may be used in place of acid treatment, in which case zinc is reacted as follows:

$$Zn + 2NaHSO_4 = Na_2SO_4 + ZnSO_4 + H_2 \tag{16.18}$$

The residue is worked several times with dilute acid and hot water to remove all the zinc. The sludge is finally washed in filter press. The moisture is then removed by squeezing and subsequent drying on iron trays coated with bone ash or white enamel. The dried mass is then calcined at 500–550°C which is then mixed with a suitable amount of flux (consisting of silica sand 25–40 per cent, borax 40–60 per cent, soda ash 10 per cent, fluorspar 5 per cent, and MnO_2 0–45 per cent) and melted in a suitable furnace (oil fired tilting, or electrical). A part of borax may be replaced by soda ash (Na_2CO_3) or fluorspar, and a part of MnO_2 may be replaced by $NaNO_3$.

The molten metal is poured into a conical mould and later remelted into a bullion bar. A sample may be taken out before pouring by dipping a small ladle into the molten gold. The slag, ash, and linings usually contain some values, which may be recovered by grinding and retreatment.

16.6.7 TREATMENT OF COMPLEX GOLD ORES

Many gold ores present difficulties in cyanidation due to the presence of substances which destroy the cyanide. For example, graphite leads to premature precipitation of the gold from the leach solution. Other sulphide minerals particularly pyrrhotite consume oxygen which is supplied during aeration of the pulp resulting into interfering the dissolution of the gold. Oxidised copper ores may react with cyanide to form stable complexes not capable of dissolving gold. Small amounts of nickel or chromium ions would interfere with the precipitation of gold. When tellurides and arsenopyrites are present, special methods may be required to ensure sufficient extraction in a reasonable time. In one of the older treatments, the ore may be ground extremely fine and agitated with cyanogen bromide (Br CN)-cyanide solution and adding a cyanogen bromide periodically. In this case the solution is held almost neutral with least alkali. Presence of arsenic may not cause a problem, but the usual association of antimony or partly oxidised iron minerals may present problems and reduce efficiency of gold extraction.

The difficulties of treating these complex or difficult ores may be overcome by two methods. One is the use of froth flotation to collect auriferous sulphides in a small bulk, which can be given an extremely fine grinding and special treatment. Another one is roasting, which can be employed to destroy the sulphides almost completely and thereby getting rid of the associated problems. *Dead roasting* of whole of the ore to remove all tellurium, sulphur, and graphite is seldom used for being expensive and difficult to control within desired limits of temperature.

In case of antimonial ores, Sb_2S_3 (stibnite) is soluble in alkalis (alkali is always needed to protect cyanide) and the compound formed decomposes cyanide to thioantimonate. The possible reactions may be

$$2Sb_2S_3 + 6Ca(OH)_2 = Ca_3(SbO_3)_2 + Ca_3(SbS_3)_2 + 6H_2O \quad (16.19)$$

$$Ca_3(SbS_3)_2 + 6NaCN + 3O_2 = 6NaCNS + Ca_3(SbO_3)_2 \quad (16.20)$$

The various methods to overcome this problem include pre-leaching of such ores with 2–4 per cent NaOH, oxidation roasting, and prolonged weathering.

Carbon is usually present in the form of graphite and absorbs the gold from its solution in cyanide. This can be removed by pre-flotation. Sometimes graphite will rise as removable scum on the classifier by adding a small amount of paraffin or diesel oil. In some cases, graphite sorbs mineral oils preferentially and form a oily coating which prevents the pick-up of gold by the graphite.

In cyanide leaching of gold, it is essential to use oxygen and thus any mineral being oxygen-avid will reduce the efficiency of extraction. Among the various minerals, pyrrhotite is the most active. These ores can be preaerated at low alkalinity and then leached with cyanide. In addition, the ore can be milled with low-strength cyanide. In this case, freshly sheared sulphide mineral surfaces will get minimum opportunity to decompose the cyanide.

16.6.8 CYANIDATION OF FLOTATION CONCENTRATES

The flotation concentrate in addition to the higher gold content, carries most of the cyanicides and other sulphide minerals (pyrite, pyrrhotite, chalcopyrite, galena, arseno-pyrite, etc.) and graphite. All these are likely to cause problems in cyanidation. If the tailing from this flotation still carries enough gold to repay for straight cyanidation, it can be processed without any difficulty as most of the interfering minerals have been removed.

All the minerals of the flotation concentrate do not carry gold. If graphite and non-auriferous sulphides can be removed from the flotation concentration, the treatment would be simplified. The important points in cyaniding of flotation concentrate are:

a) High gold content of the concentrate.

b) Small volume of material in which gold value is concentrated.

c) Rapid oxidation of sulphides during aeration and its effect on cyanidation reaction.

d) Gold is finely disseminated in particles of pyrite.

The first two effects are the economical ones. The effect (c) will generate H_2SO_4 during aeration and thus the alkalinity of the pulp has to be watched very closely to prevent the destruction of cyanide. When the pulp containing sulphide is aerated in presence of lime, hydroxyl ions react with pyrite and liberate sulphide ions into the pulp. These sulphide ions react with oxygen from the aeration and form sulphate ions via the formation of thiosulphate and sulphite ions. In the presence of cyanide, this reaction may be accompanied by the formation of thiosalt, which does not dissolve gold. Therefore, pre-aeration prior to cyanide leaching is carried out (with or without lime addition) which is followed by decantation of water carrying the sulphur salts.

Therefore, efficient cyanidation of flotation concentrate will involve the following steps:

a) Sufficient regrinding followed by aeration to stabilise the interfering surfaces.

b) Removal of fouled water.

c) Finally, cyanidation with sufficient aeration and close watching of the protective alkalinity of the pulp.

These controls are more easily applied in batch treatment than continuous treatment. The leaching is usually carried out in Pachuka agitators capable of holding about 5 tonnes of solids. The tank is charged for about 12 hr, the solids are settled and the overflow water is returned to the grinding mill. The thickened pulp is then aerated without lime for 18–20 hr and allowed to settle. The clear foul water is siphoned to waste. It is followed by treatment with weak cyanide (0.02–0.05 per cent or 0.1–0.8 per cent for silver ores) with lime and agitated for 12 hr. The pregnant solution is decanted and a subsequent agitation for 12 hr is given. Pregnant solutions obtained in leaching of flotation concentrates are rich and thus efficiency of precipitation should be checked carefully.

Flotation can also be employed to the tailings from the cyanidation plant. In this case the ore may be ground in cyanide and subjected to usual extraction treatment. The final tailing is treated by flotation. The auriferous sulphide concentrate thus obtained is given suitable treatment to recover the values.

16.7 Recovery of Gold by Amalgamation

Amalgamation is the most important process used earlier in recovery of gold. This process involves the treatment of crushed gold ore with liquid mercury, as a result, gold is dissolved in mercury forming a gold–mercury alloy known as amalgam (alloys of mercury with various elemental metals such as Au, Ag, etc.). Amalgamation can also be considered as adhesion process similar to oil–mineral and air–mineral adhesion process. The resulting amalgam can be trapped and removed from the ore pulp. The mercury is then removed and the residual gold is refined.

Gold and mercury are known to form two compounds, i.e. Au_2Hg, and $AuHg_2$, both melting with a peritectic reaction at 400–420 °C and 310 °C, respectively. Gold is not soluble in solid mercury, but dissolves to some extent in liquid mercury. The solubility of gold in liquid mercury increases with temperature and is around 0.06 per cent at ordinary working temperature. The amalgam formed is however, a complex mixture which is not at equilibrium. It consists of gold, many compounds of gold and mercury, and solution of gold in mercury. Similar to gold, other metals such as silver can also be amalgamated. However, the conditions required may be quite different. The amalgamability of native gold depends on the surface condition of mercury. This is identical to the condition responsible in flotation and agglomeration, where the process is controlled by the character of mineral surface and air–water interface. In amalgamation also, various reagents may be effective in favouring or preventing the wetting of gold by mercury.

16.7.1 AMALGAMATION INHIBITORS

These are the substances which interfere (inhibit) the amalgamation reaction. The various inhibitors fall into following categories:

a) *Substances Dissolved in Water*

These are usually alkali sulphides and some flotation agents. Their effect is probably due to the formation of surface coatings absorbed, or chemically bonded on either gold or mercury or both. These substances tend to flour the mercury into fine droplets and tarnishing of gold.

b) *Undissolved or Suspended Substances*

These substances tend to spread at metal surface. These are some sulphide minerals, particularly containing antimony and arsenic, and organic matters (oils, grease and other contaminants). The effect of these undissolved inhibitors is identical to the dissolved substances.

c) *Substances Driven Mechanically into the Gold Surface*

These may be associated due to grinding devices or weathering agencies. The gold covered with adherent iron oxide is an example of this type of inhibitors.

Many methods have been proposed to overcome the action of inhibitors. However, only few could be used in practice. The important ones include the use of alkalies or alkali cyanides in the pulp, use of sodium or zinc amalgam in place of mercury, and the auxiliary use of electric current passing from pulp to amalgam. In all the cases, the aim is to provide better wetting of gold by mercury.

16.7.2 AMALGAMATION PRACTICE

The mercury used in amalgamation may be free liquid moving around in a mill or drum. Mercury may also be used in the form of a film at the surface of copper plate. Usually sodium amalgam is preferred over mercury for amalgamation due to two reasons, i.e. (a) clear mercury is too fluid, whereas amalgam can be easily applied to the plate, and (b) sodium amalgam improves the recovery of gold. The various methods used to bring the liberated gold particles into contact with mercury are (i) plate amalgamation, (ii) immersion contact, and (iii) grinding contact.

16.7.3 AMALGAM TREATMENT

From the various points, i.e. plates, mercury traps, riffles, etc., amalgam is obtained in the form of viscous liquid. Many undesired materials such as sand, iron minerals, ore fragments, skimmings of fouled mercury, etc., may be carried along with amalgam. This mixture can be first purified by thinning it with fresh mercury, agitating the mixture with hot wash-water and skimming off the rising impurities. The mixture may be given a special clean-up in pestle and mortar, washing and grinding. After giving the clean-up, the amalgam is squeezed through soft leather or canvas in an amalgam press. Practically the gold remains as a putty-like grey amalgam containing 20–45 per cent gold. However, a small amount of gold may be squeezed through and thus this filterate should be used in clean-up work to avoid the gold losses.

The next operation in treatment of amalgam is distilling off the mercury from it. The amalgam is placed in a mercury retort and heated to drive off the mercury completely: The remaining residue consists of more or less clean gold which is taken out and melted with flux (silica, soda ash, and borax), to remove the remaining impurities in slag. The discarded slag, crucibles, furnace linings and other auriferous materials are periodically ground and retreated.

16.8 Treatment of Oxide Nickel Ores[53–56]

Nickel ores are chiefly sulphides and oxides. The sulphide ores can be concentrated by the conventional froth flotation, while the oxide ores cannot be upgraded by physical methods. On the other hand, the position of nickel oxide ore reserves is more favourable (75 per cent) compared to the reserves of sulphide ores (25 per cent). Due to the difficulty of treating oxide ore, a major portion of nickel (75 per cent) is extracted from sulphide ores, and 25 per cent from oxide ores. In order to recover nickel values from its oxide ores, chemical treatment is inevitable due to the following reasons:

a) Nickel is finely disseminated in the iron matrix.

b) Ore contains higher amount of moisture (more than 20 per cent).

c) Association of other high melting constituents.

d) Requirement of large amount of energy and reducing agent for pyro-metallurgical treatment.

The various methods of treating the nickel oxide ores include furnace smelting, ammonia leaching, and sulphuric acid leaching. These are discussed below:

16.8.1 SMELTING OF NICKEL OXIDE ORE

This consists of smelting the ore in blast furnace or arc furnace along with pyrite (FeS) or gypsum ($CaSO_4$) and reducing agent. It yields a nickel matte which is further processed by metallurgical methods to obtain nickel metal.

16.8.2 AMMONIA LEACHING PROCESS

This process is used to extract nickel from Cuban nickeliferous ores at Nicaro (Cuba). The process consists of preferential reduction of nickel oxide with reducing agent (CO, CH_4, C, etc.) to metallic state and subsequent leaching of reduced mass with ammonical solution of ammonium carbonate. The reactions involved may be written as

$$Ni + \frac{n}{2}(NH_4)_2 CO_3 + \tfrac{1}{2}O_2 = Ni(NH_3)_n CO_3 + \frac{n}{2}H_2O +$$

$$\left(\frac{n}{2} - 1\right) CO_2 \qquad (16.21)$$

where n is usually 4 or 6 (coordination number of the complex formed).

The iron present in the form of an alloy (as a result of reduction) in reduced product gets oxidised at higher pH in the presence of carbonates to hydroxides. If leaching is carried out only with ammonium carbonate, carbonic acid will be formed and some ammonia will be needed to neutralise the same. Therefore, ammonia–ammonium carbonate leaching is preferred and the reaction involved can be written as

$$2Fe-Ni + \frac{5}{2}O_2 + 12NH_3 + 2CO_2 + 3H_2O =$$

$$2Ni(NH_3)_6 CO_3 + 2Fe(OH)_3 \qquad (16.22)$$

After solid–liquid separation (by decantation/thickener and filtration), the solution containing nickel–amine complex is heated with steam to decompose the complex into insoluble basic nickel carbonate and ammonia according to the following reaction:

$$5Ni\,(NH_3)_6\,CO_3 + 6H_2O = 3Ni\,(OH)_2.\,2NiCO_3 +$$
$$30NH_3 + 3H_2CO_3. \qquad (16.23)$$

The filtered cake of basic nickel carbonate is calcined at a temperature of about 800 °C in a rotary kiln. The resulting nickel oxide (NiO) is obtained according to the following reaction:

$$3Ni\,(OH)_2.\,2NiCO_3 = 5NiO + 3H_2O + 2CO_2 \qquad (16.24)$$

The resulting nickel oxide can be reduced with H_2 or CO or a mixture of the two. Alternatively, nickel oxide can be dissolved in H_2SO_4 and the solution is then electrolysed to obtain metallic nickel.

The reduction can be carried out in a fluidised roaster or a rotary kiln. Leaching can be carried out in pressure tight stainless steel vessels with stirring arrangements. For decomposition of complex amines, large-size distillation units are employed.

Alternatively, reduced ore can be leached with ammonia–ammonium sulphate solution, in which case $Ni\,(NH_3)_n SO_4$ complex is formed. The resulting solution can be treated by solvent extraction and electrolysis to obtain metallic nickel. In this case, ammonium sulphate will be produced in excess as by-product and thus limiting the use of this method by the demand of ammonium sulphate.

16.8.3 Sulphuric Acid Leaching

Under high pressure and high temperature, nickel from the ore dissolves preferentially, while dissolution of iron is suppressed due to the formation of ferric iron. Direct H_2SO_4 leaching of oxide nickel ore at high pressure and temperature has been developed to treat Moa–Bay deposits in Cuba. This ore is of limonitic type having low magnesia (acid consuming) and relatively high cobalt. This process has an advantage over the ammonia leach process, that it is a simpler one-step process, i.e. the ore can be wet screened and slurried with water to the desired solid–liquid ratio and then used directly for pressure leaching. The formation of nickel sulphate by the attack of H_2SO_4 on nickel oxide is formed according to the following reaction:

$$NiO + H_2SO_4 = NiSO_4 + H_2O \qquad (16.25)$$

The principal factor controlling the leaching reaction is the ratio of acid to ore. A rough division of acid consumption is one-third to free acid, one-third to tailing, and one-third to soluble salts. The leaching is carried out in stainless steel high-pressure autoclaves. The solution obtained after leaching and washing is quite dilute, which cannot be used directly for the recovery of metals. The recovery of metals from the resulting dilute solution can be affected as follows:

The leach solution is first neutralised by $CaCO_3$, and precipitated $CaSO_4 \cdot 2H_2O$ (gypsum) is removed as underflow in a thickener. It is followed by precipitation of Ni and Co as their sulphides by adding H_2S or Na_2S to the clear neutral solution at a suitable pH. If any Cu, Pb, and Zn are present, they also get precipitated as sulphides and join with precipitate of nickel and cobalt sulphides. The precipitate so obtained can be treated by various conventional methods. Alternatively, the leach liquor can be treated by solvent extraction followed by electrolysis and this technique is gaining more importance.

16.9 Treatment of Low-Grade Uranium Ores[44]

Some low-grade ores such as Indian Ores of Jaduguda cannot be beneficiated by physical methods (jigging, tabling, etc.) and the recovery of uranium from these ore necessitates direct chemical treatment. The important processes used commercially are (a) acid leaching, and (b) carbonate leaching, which are discussed as follows:

16.9.1 ACID LEACHING OF URANIUM ORES

In this case a fairly thick pulp (60 per cent solids) is agitated with H_2SO_4 and oxidising agents (MnO_2 or, $NaClO_3$, and ferric sulphate). The presence of oxidising agent is necessary to maintain the uranium in its reactive state, i.e. U_3O_8. The pregnant solution is then removed and stripped. The reactions involved are

$$6H_2SO_4 + 3MnO_2 + 3UO_2 = 3UO_2SO_4 + 3MnSO_4 + 6H_2O \qquad (16.26)$$

$$3H_2SO_4 + NaClO_3 + 3UO_2 = 3UO_2SO_4 + NaCl + 3H_2O \qquad (16.27)$$

$$3Fe_2(SO_4)_3 + 3UO_2 = 3UO_2SO_4 + 6FeSO_4 \qquad (16.28)$$

Ferrous sulphate formed in reaction (16.28) may be oxidised by MnO_2 as follows:

$$2FeSO_4 + MnO_2 + 2H_2SO_4 = Fe_2(SO_4)_3 + MnSO_4 + 2H_2O \qquad (16.29)$$

The reaction of partly oxidised uranium ore with ferric sulphate and sulphuric acid will be

$$UO_2 + 2UO_3 + Fe_2(SO_4)_3 + 2H_2SO_4 = 3UO_2SO_4 +$$
$$2FeSO_4 + 2H_2O \qquad (16.30)$$

Secondary uranium minerals readily dissolve in dilute H_2SO_4, but the tetravalent oxides require oxidising conditions. When suitable pyrite mineral in adequate amount occurs in the ore, it is the source of sulphate and no oxidising agent will be required in such a case.

16.9.2 CARBONATE LEACHING OF URANIUM ORES

When the ore contains excessive carbonates, the alkaline leach with Na_2CO_3 and $NaHCO_3$ can be employed. In this case oxidation is carried out either by preaeration or the use of permanganate and pulps may require heating to 70 °C or more. Carbonate leaching does not require costly materials for the construction of equipments as required to prevent corrosion in an acid leach plant. The treatment of pregnant solution is also simpler.

However, both the processes (acid leach and carbonate leach) have certain operating problems.

The reaction in carbonate leaching in alkaline solution is

$$U_3O_8 + \tfrac{1}{2} O_2 + 3Na_2CO_3 + 6NaHCO_3 =$$

$$3Na_4UO_2 (CO_3)_3 + 3H_2O \qquad (16.31)$$

or $\qquad U_3O_8 + \tfrac{1}{2} O_2 + 3CO_3^{--} + 6HCO_3^{--} =$

$$3UO_2 (CO_3)_3^{4-} + 3H_2O \qquad (16.32)$$

16.9.3 RECOVERY OF URANIUM FROM LEACH LIQUORS

Recovery of uranium from pregnant solution can be effected by chemical precipitation, static-bed ion exchange, fluid-bed ion exchange, or by the use of organic solvents. The chemical precipitation method based on the use of H_2O_2, fluoride, phosphate, or carbonate in an alkaline liquor is simple but costly. Ionic exchange using anionic resins is usually employed for acid-leach liquor. The uranium is seized by the anions on the resins, and periodically removed by nitrates or nitric acid, and finally precipitated with ammonia, NaOH, or MgO.

16.10 Treatment of Thorium Ores

The different minerals of thorium are monazite (complex ore of thorium and other minerals), thorite ($ThSiO_4$), thorianite (ThO_2) and yttrocrasite (Y, Th, U, Ca)$_2O_{11}$. Monazite is the principal source which is recovered by separation of other associated minerals (rutile, ilmenite, zircon, sillimenite, quartz, etc.) by a number of operations (gravity, magnetic, electrostatic, etc.). The monazite mineral usually contains 5–10 per cent Th. There are two chief methods to break down the ore, i.e. H_2SO_4 leaching and alkali leaching.

16.10.1 H_2SO_4 LEACHING

Unground sand or coarsely ground rock is added slowly to concentrated H_2SO_4 at its fuming point in a high-silicon cast iron or tantalum–iron vessel. The thorium and rare earth phosphates rapidly react with H_2SO_4 to form sulphates and liberate phosphoric acid according to the following reactions:

$$Th_3 (PO_4)_4 + 6H_2SO_4 \rightarrow 3Th (SO_4)_2 + 4H_3PO_4 \qquad (16.33)$$

$$(RE)_3 (PO_4)_4 + 6H_2SO_4 \rightarrow 3RE (SO_4)_2 + 4H_3PO_4 \qquad (16.34)$$

$$ThSiO_4 + 2H_2SO_4 \rightarrow Th (SO_4)_2 + SiO_2 + 2H_2O \qquad (16.35)$$

During the reaction, the mixture becomes viscous paste having maximum viscosity at minimum amount of H_2SO_4. The reaction is exothermic and temperature should be controlled at about 225 °C. The reaction is completed by giving three to four stirrings. In order to carry away the radioactive gases liberated in reaction, good ventilation is necessary. Two tonnes of H_2SO_4 is usually employed for each tonne of monazite to convert more than 99 per cent of the pure monazite to water-soluble sulphate. The dilution of the paste with water is carried out in lead-lined vats with stirring and the liquor is cooled to about 30 °C. The amount of water added for dilution depends on subsequent treatment. Usually 10 tonnes of water per tonne of ore is adequate to dissolve thorium and rare earths completely. However, small amount may be employed when partial separation is needed, leaving a proportion of rare earths undissolved.

Rare earths are separated from the resulting solution by basicity separation. Thorium hydroxide or phosphate will precipitate at higher pH (pH 5.7 is required when the solution is almost free from phosphate). Thorium hydroxide formed is then reacted with nitric acid to get thorium nitrate, $Th(NO_3)_4$ which is evaporated to obtain crude thorium nitrate. This is further subjected to purification by solvent extraction.

16.10.2 ALKALI LEACHING

In alkali leaching, NaOH is used either in molten condition (m.p. 318 °C) or as a concentrated solution at about 140 °C to break down monazite. This process is employed for treating Indian monazite obtained from sea beach sand of South-West Coast. Monazite of −50 microns is treated with NaOH, to convert rare earths and thorium phosphate into oxides with the formation of sodium phosphate according to the following reactions:

$$Th_3 (PO_4)_4 + 12\, NaOH \rightarrow 3ThO_2 + 4Na_3 PO_4 + 6H_2O \quad (16.36)$$

In practice, ThO_2 tends to be hydrated.

$$2\,(RE)_3 (PO_4)_4 + 24\, NaOH \rightarrow 8\, Na_3PO_4 + 3\,(RE)_2O_3 +$$
$$12\, H_2O + 3/2\, O_2 \qquad (16.37)$$

The reaction is carried out in a stainless steel vessel fitted with a low-speed stirrer, heated jacket and reflux condenser. The concentrated NaOH solution is first heated in the reaction vessel to about 130 °C, and then ground monazite (−50 microns) is added gradually over a period of about 30 min. Reaction is then carried out for about 4 hr under reflux at 140–150 °C. As in case of H_2SO_4 breakdown, in this case also, good ventilation is required to remove radioactive thoron gas under safe conditions.

Control of reaction temperature is very important. If the temperature exceeds 145 °C by more than a small margin or fused NaOH is used in place of aqueous solution, the resulting oxides will be unreactive and will not dissolve completely in HNO_3 which is required for further treatment.

Filtration of oxides is preferably carried out at higher temperatures (up to about 110 °C) in order to keep the trisodium phosphate in solution. Digestion at this temperature for some time facilitates the filtration by increasing the particle size. The filter cake is then washed with hot water (by decantation and finally by filtration) to remove phosphate content. A large proportion of phosphate may be recovered in the form of large crystals obtainable by evaporating the filtrate to NaOH concentration of about 47 per cent. The residual liquor is quite pure NaOH, which can be recycled to the breakdown process.

The mixed oxide cake is dissolved in nitric acid (H_2O_2 may be added to promote dissolution) and the solution is subjected to further purification by solvent extraction to separate uranium and other rare earths.

16.11 Chemical Beneficiation of Ilmenite

In ilmenite, iron oxide and titanium dioxide are held in spinel form and it is not possible to remove iron oxide therefrom by any physical means. Drastic chemical treatment has usually been applied to the mineral in order to yield rutile grade titania-rich product. Innumerable number of processes for chemical beneficiation of ilmenite have been reported in literature. However, the processes of commercial importance are only HCl based, i.e. either direct pressure leaching of ilmenite with HCl or partial reduction of ilmenite followed by HCl leaching.

Ilmenite can be pressure leached with HCl in glass lined autoclaves where iron oxide is preferentially dissolved forming iron chloride. The dissolved iron salt can be decanted off and remaining residue is washed with water. Usually concentrated acid (about 25 per cent), high temperature (130 °C) and a pressure of about 2–4 kg/cm² are employed for leaching. This process is associated with several drawbacks such as involvement of high pressure, high temperature and high concentration of acid. It is difficult to regenerate HCl of the required concentration in this case.

In the other method, iron oxide present in the ferric state is reduced to ferrous form by some reducing agent such as charcoal, coke, coal, H_2, CO, natural gas, etc., and the partially reduced ilmenite is leached with HCl (20 per cent) under atmospheric pressure and a temperature of about 104 °C. The iron oxide in ferrous form is more reactive with acid than in ferric form. The reaction can further be enhanced by pre-oxidation of ilmenite followed by reduction. This makes the ilmenite more porous which aids the leaching rate. The reactions involved may be represented as follows:

$$Fe_2O_3 \cdot x\ TiO_2 + C\ (H_2) = 2\ FeO + CO\ (H_2O) + xTiO_2 \qquad (16.38)$$

$$FeO + 2HCl = FeCl_2 + H_2O \qquad (16.39)$$

16.12 Removal of Phosphorus from High Phosphorus Manganese Ores

Manganese ores containing high phosphorus (more than 0.05 per cent) are not suitable for the production of iron, steel and ferroalloys. It is not possible to remove phosphorus from these ores by any physical method, but can be removed by leaching the ground ore with dilute HCl. The ore is ground to −70 microns and treated in rubber lined vessels fitted with stirrer and the reaction is carried out with 2 per cent HCl. Finally the slurry is filtered and washed to remove phosphoric acid generated, which can be used for fertilisers. In this way manganese ores could be rendered suitable for their use in iron and steel industry.

In India some manganese ores are of very high grade but contain high phosphorus. These ores can be used after removal of phosphorus by HCl leaching.

16.13 Purification of Graphite

Graphite concentrates obtained from flotation technique are usually 95–97 per cent pure and 2–3 per cent silica still remains in the concentrate. In order to make high-grade graphite suitable for electrical brushes, etc., this remaining silica should be removed. This can be accomplished by leaching the graphite concentrate with dilute hydrofluoric acid (HF) in a vessel lined with teflon (PTFE). The vessels lined with cashew-nut resin based polymers may also be employed. HF treatment of graphite concentrate yields graphite purity 99.5 per cent or more.

CHAPTER 17

Application of Ore Microscopy in Processing of Ores[61-66]

The main application of microscopy in the field of ore processing is to identify mineral constituents, their intergrowth, liberation size of valuables, composition of the gangue minerals, and thus in selecting and identifying the suitable process to be adopted. The natural state of mineral recognisable only by microscope is fundamentally important for the selection of suitable ore-processing method. These microscopic studies can evalute an operating process and help in improvement in the existing process or to develop a new process. A general view of the various physical methods of ore processing applicable to the range of grain size is represented in Fig. 17.1. This classification is only conditionally valid since the efficiency of a specific method is dependent on many natural properties of the ore.

17.1 Evaluation of Mineralogical Composition and Association of Various Minerals

The correct knowledge of mineralogical composition can be obtained only by microscopic examination. The microscopic results may also be complemented by other methods of investigation such as chemical and microchemical analysis, spectrographic and X-ray analysis, blow-pipe tests, etc. However, these methods cannot replace the microscopic method, as the natural form of various minerals is not recognisable by any other method except microscopic examination.

The microscopic investigation may be carried out in hand specimen, in polished section, or on grain sample. In case of hand specimen, a large number of specimen are examined for a reasonably correct assessment. Since most of the non-ferrous minerals are opaque, the microscopic examination of polished sections is of much greater importance than that of hand specimens. The grain samples are investigated on a microscope glass plate with or without immersion liquid, or in mounted thin polished sections. On grain samples, the determination of opaque minerals is quite difficult due to their similarity. The recognition of minerals becomes extremely difficult when a surface of mineral loses its characteristic appearance due to oxidation which

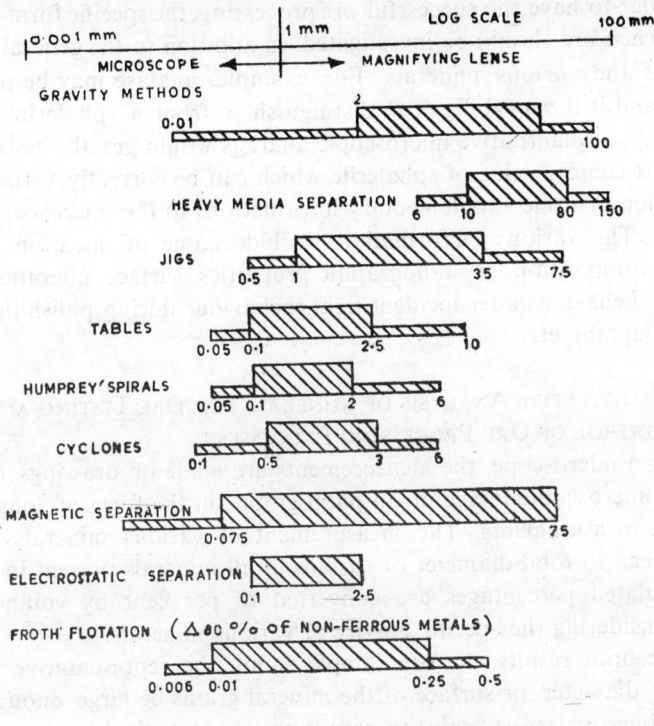

Fig. 17.1. Diagrammatic view of the fields of various ore dressing processes.

may occur sometimes even within few hours. For quantitative information, the studies are carried out on large polished sections. The microscopic studies may be divided into the following three categories:

a) Determination of minerals.

b) Quantitative analysis of various mineral constituents.

c) Study of textures.

17.1.1 DETERMINATION OF MINERALS

The determination of all minerals in the ore as completely as possible is essential, since for the efficient ore processing, the establishment of mineralogical source of the metal/element content of the ore (determined by chemical analysis) is required. A rapid development of ore-processing technique is always required when economic or technical requirements demand such development. For example, the ore deposits considered uninteresting 25 years ago due to their low metal content or complexity are the largest metal sources in the world today. This has happened mainly due to advanced methods of microscopic investigations.

In order to have the successful ore processing, the specific form of occurrence in each ore should be investigated, in addition to the general determination of the various minerals. For example, anatase may be present in zinc ore and it is very difficult to distinguish it from a sphalerite. In such an instance, a quantitative microscopic analysis would greatly assist in obtaining the correct value of sphalerite which can be correctly estimated by the addition of some supplementary informations to the microscopic investigations. The various informations include name of location of ore, chemical composition, crystallographic properties, surface alteration, etching effect, behavior under incident light, behaviour during polishing, X-ray powder diagram, etc.

17.1.2 Quantitative Analysis of Minerals for Ore Testing and Control of Ore-Processing Processes

Under a microscope, the measurements are made on drawings obtained from the microscopic image or on photographs in the form of longitudinal or surface measurements. The measurement of various minerals is made with respect to total diameter or surface of all minerals present in the ore. The calculated percentages are converted to per cent by volume or, by weight considering the specific gravity of various minerals.

For accurate results, the ore sample should be representative and the measured diameter or surface of the mineral grains be large enough (usually 100 times of largest grain) to permit proper statistical evaluation. The possibility of errors is enhanced by non-homogeneous distribution of the minerals, change in grain-size, and texture.

For the purpose of ore beneficiation, the grain size of a homogeneous mineral is not of much interest except in case of some specific ore-processing method, e.g. flotation. Therefore, considerable comminution of the ore prior to such a quantitative analysis eliminates many sources of errors, particularly arising due to non-homogeneous distribution, rapid change in grain-size, etc. Polished section of grain specimens simultaneously reveals the shape, composition and degree of the various mineral grains, which in turn predicts the degree of liberation.

17.1.3 Study of Textures of Ore Minerals

The identification of various minerals usually does not indicate much about their appearance and relative or absolute size on which their technical recovery is dependent. The study of texture of ore minerals is aimed to obtain these informations. The observed texture of minerals may prove helpful in the determination of certain minerals as type of texture is usually a characteristic of a specific mineral. The mineralogical composition and texture of the ore particles after comminution and their behaviour are important in course of ore processing. Texture of minerals may be observed during their identification.

Similar to mineral identification procedure, determination of texture of ore minerals also requires some generally acceptale consistent pattern. For ore beneficiation purpose, the most suitable pattern of investigations is given by Ramdohr and Schneiderhohn. The general classification of textures of minerals according to requirement of ore processing may be represented as follows:

a) *Texture According to Properties of Mineral Grain*

In ore processing, the term *mineral grain* is usually referred to a single mineral grain of any physical and chemical composition separated from its surroundings. In the case of beneficiation process, it is difficult to maintain the behaviour of various grains as a homogeneous one, as the behaviour largely depends upon the ore beneficiation method used. For example, certain galena–quartz grains behave homogeneously in jigging operation and produce a homogeneous product at a certain grain size, whereas the same ore may not produce homogeneous product in flotation. Sphalerite having finely disseminated chalcopyrite may be considered as homogeneous for flotation. Similarly, from the ore beneficiation point of view, ilmenite (titano–magnetite) can be considered as a homogeneous compound, where the separation of the two compounds is impossible. However, this is not true in the case of fine inclusions of gold in sulphides.

The texture depending on the properties of mineral grain may be studied as internal texture of mineral grain (homogeneous grain, heterogeneous grain, and inclusions) and external texture of mineral grain (grain shape, grain size and grain bonding).

Internal texture and its effect on ore processing: In ores, a mineral grain is rarely homogeneous and inhomogeneity may be either physical or chemical. Purely physical inhomogeneity generally affects the comminution whereas the chemical inhomogeneity would affect mainly floatability. A sharp separation of two properties (physical and chemical) is not possible and usually both act together.

The physical inhomogeneity includes irregularities in the growth of minerals, change in porosity, deformation of mineral grains, and intergranular films of other minerals between the grains of parent mineral. The physical inhomogeneity affects the size of liberation and behaviour during crushing and grinding.

The chemical inhomogeneity is generally associated with physical inhomogeneity and it is evidenced by isomorphous replacements, solid solutions (e.g. invisible gold in pyrite) and absorptive contamination (e.g. pyrolusite containing varying amounts of As, Co, Cu, etc.). The chemical inhomogeneity may either be zonal or equally distributed. Great difficulties may be faced in beneficiation of such ores due to their chemical instability, rapid oxidation and pyrophoric properties. It is not possible to separate isomorphic compositions and solid solutions into homogeneous phases.

In ore beneficiation, the most important form of inhomogeneity is the internal zonal texture which is recognised by the change of optical properties, variation in colour, hardness, etc. The internal zoning of the minerals may be considered of great significance as the concentrates produced will not have a uniform composition. Physical zoning caused due to deformation is sensitive in varying degrees to chemical alteration and replacement, and influences the floatability of the minerals. The important example of such a case is the alteration of galena to cerussite in certain zones of the mineral. The behaviour of such galena in flotation may be quite different from normal galena. This feature cannot be explained without the microscopic investigation.

Certain zones of mineralisation may form zones of weakness. These zones disintegrate preferentially during comminution. The surface properties (chemical and physical) of the disintegrated particles may not be same as those of original minerals, and thus affect the floatability of the minerals considerably.

Mineral inclusions may be present due to physical and/or chemical inhomogeneity of minerals. These inclusions in the form of rounds or droplets, laminates, etc., are generally represented by well defined minerals, e.g. copper in zinc mineral, and valuable metals (Au, Pt, Ag, etc.) in sulphides. From the technical point of view of ore processing, usually it is not easy to differentiate the behaviours of inclusions of ore minerals and intergrowth of two minerals. This is particularly true when the inclusions to be liberated by comminution are considerably large. If the inclusions are quite small, their liberation is difficult.

External textures of mineral grain and its effect on ore processing: After establishing the physical and chemical properties and internal texture of a single mineral grain, its external texture may be investigated, which is also quite important. The external texture may be distinguished in the form of grain shape, grain size and grain bond.

Grain shape may be distinguished as cubic, tabular or lamellar. Well developed crystals possessing the smooth surfaces show a tendency to break out easily from their surroundings, whereas the minerals showing stronger intergrowth with neighbouring mineral render the separation more difficult.

Grain size described as *coarse* or *fine* is not valid for the purpose of ore processing, since in many cases finely crystallised aggregates may be easily separated at coarse grinding (e.g. finely crystallised galena) and in some cases even coarsely crystallised may require fine grinding for liberation (e.g. bismuth skeleton of several centimetres). Visual inspection or simple tests may be adequate to decide the process in case of coarsely mineralised massive ore, whereas microscopic examination should be carried out in case of finely intergrown parts of the ore body requiring additional comminution for liberation of ore minerals.

Grain bond is considered as the power of adhesion between two or more minerals at their mutual grain boundaries. In other words it may be considered as the resistance to the separation of individual constituents by comminution along the grain boundaries. In most of the cases, quantitative measurements are not possible. All transitions between the two extremes, i.e. *loose sand* and *quartzite sand stone* are possible. In a crude ore, the bond strength of the various grains of a homogeneous mineral (e.g. magnetite aggregate) is of very little importance compared with the mutual bond strength of various sulphide minerals and of these sulphide minerals to the gangue minerals (e.g. sphalerite, chalcopyrite, pyrrhotite, etc.). Very fine grained intergrowths may prevent the separation of two minerals along the grain boundary.

b) *Textures Depending on the Properties of Mineral Aggregates*

This group may concern monomineral as well as polymineral aggregates. In most cases, description of a particular texture of a mineral may not be required. It is usually sufficient to note the arrangement of a particular valuable mineral such as sulphide within the aggregate. Depending on the properties of mineral aggregates, the various textures observed may be described as shape, grain size and space arrangement of the mineral constituents.

The shape of mineral constituent may be either simple or intergrown. Simple grain textures can many times be disintegrated easily into their constituents by the application of weak stresses. Intergranular textures of ore minerals having interspaces between the mineral crystals filled with other minerals (such as sulphides) present difficulties in the process of comminution, whereas bond of limonite will facilitate the separation of sulphide grains.

In a polymineral aggregate, various grain sizes are common which may conveniently be determined for the purpose of ore processing. In a crude ore, a particular mineral may be present in various shapes and sizes, e.g. chalcopyrite occurs in lead–zinc ores in the interstices of the various minerals in addition to its occurrence in sphalerite as ex-solution particles. The former may be liberated at reasonable size by comminution and recovered, whereas the liberation of finest ex-solution particles will not be possible and this portion of copper is considered to be non-recoverable.

The space lattice arrangement is of main importance for ore beneficiation, particularly when employing gravity methods. Pronounced porosity of mineral may affect the specific gravity to an uncontrollable degree. For example, the oxidised ore from ferruginous outcrops, usually is not amenable to gravity separation, as the porous minerals may be both heavier and lighter than the compact mineral. However, in flotation process, the porosity of the material is not much important.

17.2 Optical Instruments Employed in Microscopic Studies

The simplest instrument of magnification is the hand lens which is quite often used for rough work. The hand lens may be a simple biconvex lens (magnification up to 6), aplantic lens (magnification up to 20), anastigmatic magnifying glass consisting of four lenses, of which two are cemented (most perfect and magnification up to 40), etc. The usual magnifications of hand lenses are 8–10 times and in a very few cases 20 times or more.

For higher magnification, a compound microscope is employed. In this case, a real magnified and reversed primary image is formed by the objective, which is further magnified by the eyepiece (total magnification = magnification of objective × magnification of eyepiece). The basic principle and construction of the microscope are the same as of reflected light microscope. The characteristic feature of an ore dressing microscope is the illumination of the object by the reflected light falling from above but outside of the objective. In this microscope only the diffused light reflected from the irregular surfaces of the individual grain is observed. Hence, the observations are frequently made under natural light conditions obtained by employing a day-light filter (light blue) in artificial light.

Examination under transmitted light is also carried out in the usual way. A further characteristic feature of ore-dressing microscope is the possibility of simultaneous use of reflected and transmitted light. For special purpose, polarisation may be used for the reflected or transmitted light.

For the studies of coarser and medium grain products (products of wet gravity concentration) binocular microscopes up to 30 × magnification and stereo ore-dressing microscopes up to 150 × magnification are employed. For the fine and finest grain sizes of the flotation process, ore-dressing microscopes with magnification up to 1,250 ×, are employed. The characteristic features of binocular and stereo microscopes are, (a) large working distances, (b) upright and true microscopic image, and (c) large and bright field of observation. Stereoscopic binocular microscope is illustrated in Fig. 17.2, which is widely available.

For the finest products of flotation process (grain sizes are usually below 100 microns), microscopes of higher magnification should be employed, in which the principle of stereo-observation is not possible. This type of ore-dressing microscope is manufactured by Leitz Company (Germany). It is known as Ultropak—Ore-Dressing Microscope (shown in Fig. 17.3) and satisfies all the practical requirements of an ore-dressing study with respect to rapid determination of the various minerals according to their amount, size, etc.

17.3 Methods and Procedure of Investigations

The information such as mineralogical composition, minerals present, grain size, form of association, etc., required for the proper planning of the ore-dressing operations are obtained by the systematic microscopic investigations of polished and thin sections of the ore employing reflected or transmitted light. The polished sections are of no significance for the fundamental investigations of various products at different stages of operations or performance of a specific equipment. Further, the polished sections cannot be prepared by classical methods and some special methods are needed which are time consuming. Therefore, for the routine work and control of the per-

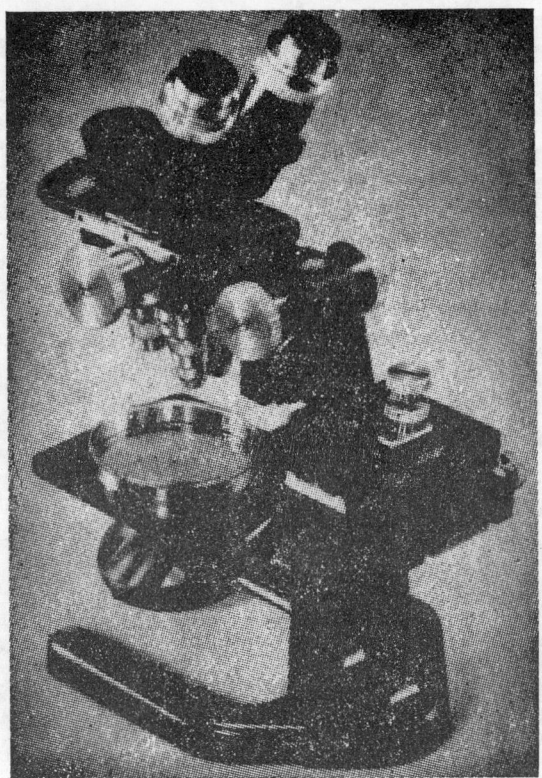

Fig. 17.2. Stereoscopic ore-dressing microscope.

formance of various equipments during the operation of ore-dressing plant, a rapid procedure of investigation is required, in order to correct the performance of a specific equipment, if required. This requires the examination of dispersed grain sample by special ore-dressing microscope.

17.3.1 SAMPLING AND SCREENING

The sampling of mill products is done with well established procedures. For quantitative evaluation of the results of microscopic investigations and for the preparation of a balanced flow sheet of the various minerals, shift samples or monthly composite samples should be used. However, in case of studying the performance of a specific mill equipment, samples may be collected without regard to timing. The mill samples are dried, thoroughly mixed and split to a range of 50–100 g. The flotation reagents that may be present, are removed by washing the sample with alchohol or other suitable solvents such as xylene.

The microscopic studies of the mineral content of ore dressing product are greatly facilitated by screening the sample into various fractions. An unscreened sample may be used for microscopic determination of the mine-

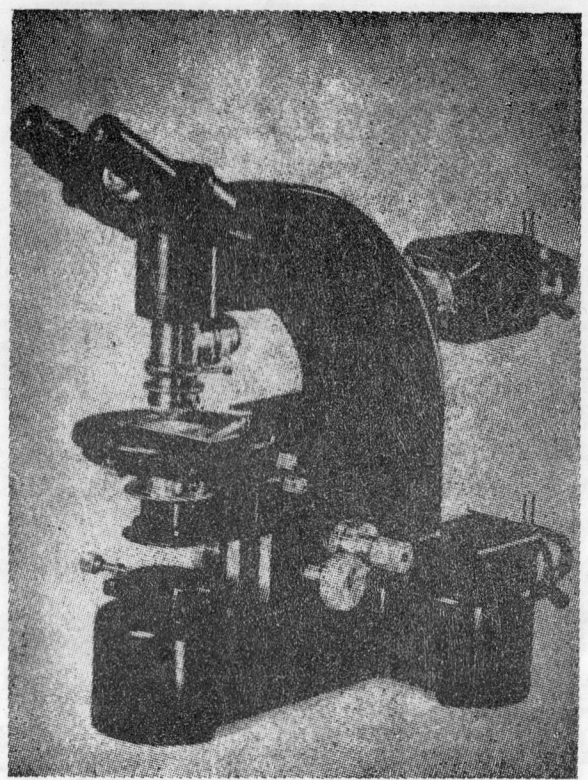

Fig. 17.3. Ultropak ore-dressing microscope.

rals and their grain size. A screen analysis of mill product separates the product into grain fractions, that are weighed separately. Considering the classifier overflow and the returned classifier sands as separate products, the balance-sheet of grain distribution can be prepared, from which the efficiency of a mill operation can be established.

The screen analysis and classification down to 40 microns are time consuming and delay the actual investigation, but it is generally unavoidable. The infrasized fractions (below 40 microns) of a mill product are individually investigated microscopically and the various mineral particles counted.

17.3.2 Preparation of Samples

The preparation of samples for microscopic studies is quite simple. With coarsely grained material, the sample is placed in a small flat glass dish and covered by an immersion liquid (water, glycerine, nitrobenzene, etc.). A binocular stereo-microscope may be used for observation. In case of Ultropak—Ore-dressing Microscope, low magnifying objectives can be used with special dipping caps. Investigations without an immersion liquid are rarely carried out (with only relatively coarse grain sizes of 2 to 0.5 mm).

For medium-fine or fine grains, the grains are dispersed into a few drops of immersion liquid placed on an ordinary microscope glass slide, and then covered by a cover glass. The preparation of such a sample does not require more than a minute. The grains in the dispersed sample have the advantage of being held by the cover glass in approximately one plane. This type of sample permits the exact examination, determination, and counting of the individual grains in only one plane.

Methylene iodide may be employed for studying some specific minerals. This gives an advantage of rapid determination of specific ore minerals in a mixture of gangue minerals due to their separation in the heavy media liquid (methylene iodide) according to their specific gravities. By providing a thick layer of the immersion liquid, both layers of minerals (heavy in bottom of the slide and lighter on the top of the slide) can be brought separately into focus and investigated without disturbing one another. The application of methylene iodide is extremely important in various studies such as ore-dressing products of the pyrite–sphalerite ore, pyrite–chalco-pyrite ores, etc.

17.3.3 Microscopic Counting

Microscopic counting of grains in the coarser grain fraction is carried out with the binocular stereo-microscope, using a net ruled in a flat glass dish or with the micrometer net of the eyepiece. The counting of the finer grains with the Ultropak–Ore-Dressing Microscope is carried out with the micrometer-net in the eyepiece.

The accuracy of counting can be increased considerably by increasing the number of grains to be counted, provided the sample is a true representative of that fraction. The counting is greatly facilitated by employing the classified material. Based on the average grain size of each grain fraction, the total volume of a particular mineral and its per cent of total weight according to its density can be calculated. By knowing the chemical composition of the mineral, the total amount of a particular element present in a specific ore-dressing product can be determined. The results are summarised in tables and graphs, which disclose the degree of liberation of the various minerals in the various grain size fractions (efficiency of liberation) and the degree of concentration during the separation and beneficiation.

17.4 Microscopic Studies in Ore Processing

The important characteristic features leading to the proper identification of minerals include colour, lustre, translucency, reflectivity, cleavability, fracturing, grain size, form of the crystal, inclusions, intergrowths, refractive index, etc. In studies of ore processing methods, it is assumed that the minerals present and their properties in the crude ore are known.

The microscopic examination data may be obtained from (a) grinding operation which reveals the degree of liberation, (b) classification processes which should provide properly classified grain sizes, and (c) separation and concentration processes which govern the extent of separation and concentration for each mineral constituent. These stages are usually repeated several times in an ore-processing operation and thus studies are required for individual stage.

17.4.1 MICROSCOPIC INVESTIGATION OF GRINDING CIRCUIT PRODUCTS

From the screen analyses, the information regarding the grain size distribution is obtained, whereas microscopic investigations provide the additional information about the mineralogical and physical composition of the individual grains in each of the size fractions. This is an important information as the behaviour of minerals during comminution largely depends on the nature of the comminution forces (stressing, striking, straining, etc.). The various minerals respond in different manner to these forces, depending on their cohesiveness. For example, sphalerite and galena will be broken in ball mills in larger amounts of fines due to their excellent cleavability as compared to the hard and tough pyrite with practically no cleavability. Grains composed of several minerals will show a very different behaviour on comminution than the individual minerals and thus a possibility of selective comminution can be determined microscopically, but not by screen analysis.

The degree of liberation of an ore after its comminution can be assessed by microscopic examination of ore-processing products. Only by microscopic studies it is possible to determine the prossibility and extent of separation of various minerals of the ore cleanly from each other. Further, the fine attachment of one mineral to another as well as grain bond between the various minerals can be observed under microscope. Determination of degree of liberation by microscopic studies cannot be replaced either by chemical analysis or by a screen analysis. It provides the information for exact determination of comminution efficiency for a particular grinding unit or for the entire grinding operation. The microscopic examination of the different fractions obtained in screen analysis provides an additional information about the presence of over or undersizes in each of the screened fractions.

17.4.2 MICROSCOPIC STUDIES OF CLASSIFIER PRODUCTS

The microscopic examination of classifier products is required to obtain information about the distribution of liberated or of still locked mineral particles according to grain size. In a classifier, overflow and the returning sand may have been already liberated, and hence these should be transported for the separation instead of returning to the mill for further grinding. A decision about this is dependent mainly on the proportion of liberated particles being returned to the mill, which can be determined only microscopically. On the other side, the classifier overflow may contain excess of middlings in the coarser grain sizes and this will require the change in conditions of classifications. In this case if chemical analysis alone is made, it will give only the total amount of various elements present in the classifier overflow or sand, whereas the information about the degree of interlocking of the various minerals can be obtained by microscopic investigation.

17.4.3 MICROSCOPIC STUDIES FOR THE EFFICIENCY OF SEPARATION AND CONCENTRATION

From the microscopic investigations of ore-processing products with respect to their mineralogical composition, it is possible to get a rapid estimate of the degree of concentration and the efficiency of ore-processing operation. The proportions of the various minerals identified according to their physical properties are determined and are compared with the initial amount in the ore. The relation of the two values represents the efficiency of separation and concentration.

17.5 Microscopic Studies of Important Minerals and Their Role in Beneficiation Process

17.5.1 COPPER

High grade copper ores (more than 6 per cent Cu) containing some precious metals can be directly smelted. Direct smelting may also be adopted for self-fluxing ores not amenable for beneficiation due to the presence of copper minerals in finely distributed form.

Native copper and sulphide copper ores are concentrated either by gravity method or preferably by flotation, depending on the fineness of the copper minerals. Though the oxide and carbonate ores may be floated in coarser grain sizes, leaching is preferred. Silicate ores may be processed by chemical segregation roasting or used as fluxing material in the smelting operation.

In flotation of copper ore, the most important consideration is whether the copper minerals are associated with unfloatable gangue minerals such as quartz, calcite, etc., or greater amounts of undesired floatable minerals such as pyrite, pyrrhotite, etc., are present. The general requirement of the beneficiation is to get the high-grade concentrate. The possibilities for beneficiation and processing of an ore mainly depend on the mode of occurrence and the association of the various copper minerals. In case of flotation process microscopic studies may be carried out at various points such as crude ore, classifier overflow, feed copper concentrate, middlings, and tailing. By microscopic examinations, the amount of chalcopyrite associated with pyrite, pyrrhotite magnetite, and gangue as well as free chalcopyrite can be established in various products. This will help in further improving the flotation process for better recovery of copper.

17.5.2 GOLD

The microscopic investigations of gold ores and various mill products are quite difficult owing to the small quantities of gold in the crude ore. In addition to this, high specific gravity of gold makes it impracticable to obtain a representative sample due to segregation tendency. However, with regard to application of a particular ore-processing method, the form of occurrence of gold is most important and this can be found only by microscopic studies.

From the view point of ore processing, the native gold may be classified as *clear gold, tarnished gold* and *coated gold.* Clear gold is of the placer or primary deposits and occurs with a clean metallic surface. Such gold is easily amenable to cyanidation or amalgamation. Tarnished gold carries a cover of thin film (less than 1 micron) of gold-free material. Coated gold carries a thick layer of other material such as iron oxide. Coated gold has to be specially treated to make it amenable for cyanidation or amalgamation. The coatings may be removed either by grinding or acid leaching or both.

Gold may also occur in considerable amounts as solid solutions in sulphides, particularly pyrite and arsenopyrite. This type of gold is not visible and thus cannot be liberated by grinding and therefore, the gold recovery by usual ore-processing methods is not feasible.

Mineralised gold may occur in the form of tellurides. These are neither amenable to cyanidation nor to amalgamation. The gold can be concentrated usually by flotation and recovered only by pyrometallurgical techniques from the concentrates.

17.5.3 Lead and Zinc

Usually lead and zinc minerals occur closely associated or intergrown. The separation of lead and zinc minerals by selective flotation is one of the most important work of ore-dressing processes. The most important mineral of lead is galena (PbS) which may alter into cerussite ($PbCO_3$), anglesite ($PbSO_4$) and pyromorphite ($Pb_5Cl [PO_4]_3$). These altered forms may play an important role in processing of the ore. Zinc usually occurs as zinc sulphide (ZnS) which may be present in two modifications, i.e., sphalerite or wortzite. A smaller or larger portion of zinc ions may be replaced by Fe-ions in solid solution. Sometimes iron content may go up to 20 per cent (marmatite). In the oxidation zones, carbonates of zinc as smithsonite ($ZnCO_3$) hydrozincite ($2 ZnCO_3. 3 Zn (OH)_2$) or silicates as hemimorphite ($Zn_4 [OH]_2 Si_2O_7]. H_2O$) are generally formed. These minerals usually enrich the zinc content of the ore and under the general name *calamine* an earthy and iron bearing outcrop is mined.

In primary ore deposits, galena and sphalerite are usually associated with many other sulphides as well as lead sulphosalts and depending on the genesis, the ore may be simple or complex in nature. Galena is usually associated with silver content which is of special significance. Silver may be present in the form of silver bearing minerals which are determinable only by microscopic examination. In coarse comminution of the ore, the silver minerals remain within the larger grains of galena, whereas, fine grinding leads in their abrasion and thereby resulting their loss into slime sizes.

In ore processing, it is important to know the form in which lead is present, i.e. as sulphide, or cerussite or anglesite. It may also be possible that a part of the lead content is present as lead sulphosalts, which would act as carriers of valuable silver as well as undesirable elements (As, Sb, Bi, etc.). The floatability of these lead sulphosalts is generally quite different from that of galena. The removal of these detrimental constituents is most essential but usually not achieved without heavy loss of lead as well as precious metals.

Quite often copper occurs as ex-solution particle within the sphalerite or as individual chalcopyrite grains. The latter usually carries large amounts of silver and small amounts of gold recoverable with lead concentrate, whereas copper ex-solved in sphalerite cannot be recovered. A substantial amount of cobalt and nickel may be present in hydrothermal lead–zinc ore bodies. The cobalt and nickel contents are usually accumulated on the grain boundaries of chalcopyrite or are present within the chalcopyrite grain itself. These cobalt and nickel contents are accumulated in the zinc concentrate and adversely affect the electrolytic production of zinc.

In processing of lead–zinc ores, large quantities of free sphalerite lead concentrates may often be mistaken for grains of sphalerite covered by a very thin layer of galena. Fractures of galena preferably take place according to cleavages and not according to the grain boundaries. This type of galena–sphalerite association is usually a characteristic of recrystallised or metamorphosed ore bodies. In determination of floatability of grains, the extent of surface coating of galena on sphalerite is of main importance. If a coating of galena is complete, then even a submicroscopically fine coating will alter the sphalerite grain to a galena grain with respect to its floatability. This behaviour is true for all lead–zinc ore beneficiation operations, and therefore, it is necessary to carry out the microscopic investigations of mill products by employing polished sections. The grain boundaries of an intergrown particle should also be studied with respect to its mineralogical composition to assess the possibility of further comminution.

In flotation of galena, the formation of cerussite and its form of occurrence are of major importance. Anglesite is also formed frequently ahead of cerussite, which prevents the flotation of galena. If the cerussite is present in the form of newly broken surfaces, difficulties in flotation will not be experienced after sulphidisation. However, in case the particles of ground galena are coated by a thin layer of cerussite with its natural surface of formation, the galena particles will not float at all, as the sulphidisation of naturally formed surface of cerussite is not possible. Therefore, an abundance of cerussite in coarsely mineralised lead ore is not harmful compared to the new surfaces of cerussite formed by comminution.

17.5.4 NICKEL

Based on the possibility of beneficiation, the nickel minerals are (a) oxidised minerals originating from gels and altered nickel-bearing serpentine, and (b) sulphides and arsenides of nickel. The sulphides occur in liquid-magnetic ore formation and found as pentlandite [(Fe, Ni) S] along with chalcopyrite and pyrrhotite. These ores frequently contain some Au, Ag and Pt. The arsenides of nickel belong to the Ag–Co–Ni–Bi–U type of deposits occurring in the form of veins. If pyrrhotite and pentlandite are present in intergrown form, they cannot be completely liberated due to the friability of pentlandite even at very fine grinding. Magnetite–pyrrhotite cannot be separated from pentlandite–chalcopyrite by magnetic separation, due to the reason that pentlandite is always present in pyrrhotite. In spite of this fact, the magnetic separation has been often tried and applied. The presence of larger amounts of magnetic cubunite makes the process of beneficiation still more complicated.

Depending on the form of association of nickel minerals and metallurgical process employed, selective flotation of nickel-concentrate poor in copper and a copper-concentrate poor in nickel is carried out. If hydrometallurgical process is to be applied, bulk flotation may be carried out to recover all sulphides in a single concentrate. Oxide ores are not amenable to any beneficiation process, and hence chemical treatment is required for recovery of nickel.

17.5.5 MOLYBDENUM

Molybdenum is mainly mined as molybdenite (MoS_2) and wulfenite ($PbMoO_4$). The latter can be concentrated easily by gravity methods owing to the high specific gravity (6.7–6.9) of the mineral, whereas the former can only be recovered by flotation. The fine distribution and the perfect cleavage of molybdenite lead to the formation of fine and easily floatable folia as a result of which high-grade concentrates are obtainable from low-grade ores.

17.5.6 TIN

Secondary placer deposits represent the major portion of the world production of tin. The microscopic studies in fine sizes may reveal interesting information with regard to proper designing of suitable ore-dressing methods such as Humphery spirals, centrifuge, etc. High-temperature vein deposits of cassiterite usually contain small amounts of sulphides, arsenides, native bismuth, molybdenite, etc., along with frequent presence of wolframite. Low-temperature cassiterite deposits occur with larger amounts of various sulphides, arsenides, silver minerals, etc. These ores require special methods of beneficiation.

17.5.7 TUNGSTEN

The gradational ferberite ($FeWO_4$) and huebnerite ($MnWO_4$) are the main sources of tungsten and these are known under the common name of wolframite. Pure huebnerite and pure ferberite are rarely found. From the viewpoint of ore beneficiation, huebnerite is not magnetic and thus cannot be separated by magnetic separation from the associated cassiterite, whereas ferberite responds to magnetic separation. Another important mineral of tungsten is scheelite ($CaWO_4$)

17.5.8 URANIUM

From the viewpoint of ore dressing, the uranium ores may be classified as (a) pitchblende, (b) uraninite and samerskite, (c) oxidation products of uranium, and (d) carnotite.

Pitchblende occurs in the veins of the cobalt–nickel–silver–bismuth–uranium deposits. The comminution of such ores produces mineral particles of all intermediate specific gravities due to the irregular distribution of both major constituents of the ore, i.e. pitchblende and gangue minerals. Therefore, the liberation of these minerals and recovery by employing physical methods is not possible. However, when fractures are filled by secondary pitchblende, galena, chalcopyrite, etc., it is possible to liberate various sulphides and arsenides by cosiderable comminution. In earlier years the pitchblende was concentrated by gravity methods, which resulted in high losses of pitchblende due to complex mineralisation. This could be observed only microscopically.

Uraninite and samarskite occur in pegmatites, usually in the irregularly distributed form. The low-grade ores can only be beneficiated by hand picking. Oxidation products of uranium are formed by weathering of pitchblende. Any physical beneficiation is not possible due to their fine earthy form and variable composition. These are generally hand picked and subjected to chemical treatment (leaching). Carnotite occurs as an impregnation of sand stones (formerly mined as vanadium ore) which contain carnotite, cementing the quartz grains. Carnotite could not be beneficiated due to its high friability.

Microscopic investigations of the ore reveal the problems of separation of the individual constituents of the ore by comminution, particularly in obtaining the grains of similar physical properties necessary in gravity concentration. However, with the development of leaching procedure, along with ion-exchange and/or solvent extraction, all former methods have become obsolete. Microscopical examinations determine the mineralogical composition of the crude ore, which is of great significance in selection of a suitable ore-processing method.

17.6 Some Practical Examples

17.6.1 DETERMINATION OF THE SEPARATION EFFICIENCY OF RAKE CLASSIFIER FOR PYRITE-SPHALERITE

The screen analysis has been carried out on returned sands and overflow. It revealed that the ratio of the returned sands to the amout of overflow is $4,210 : 4,790 = 87.9 : 100$. The calculated values of the mineralogical compositions of classifier sands and overflow in each grain fraction are summarised in Table 17.1, from which the cumulative values can be plotted as shown

Table 17.1. Mineralogical composition of various grain fractions in weight per cent in overflow and returned classifier sand

Grain fraction in microns	Wt. %	Mineralogical composition in wt.%					
		ZnS	ZnS/FeS$_2$	FeS$_2$/ZnS	FeS$_2$	Gangue	Total
Overflow 53.2%							
100	11.7	—	—	—	—	—	—
100–75	11.9	4.4	9.1	74.5	5.8	6.1	99.9
75–60	2.9	3.8	4.3	77.0	9.6	5.4	100.1
60–40	6.5	3.4	9.9	65.0	16.2	5.5	100
< 40	67.0	—	—	—	—	—	—
Returned Classifier sand							
100	87.4	—	—	—	—	—	—
100–75	5.35	1.2	4.6	88.5	4.6	1.1	100
75–60	0.64	4.7	4.5	79.3	9.8	1.7	100
60–40	0.6	4.6	5.4	72.6	15.9	1.3	99.8
< 40	6.0	—	—	—	—	—	—

Based on data from Ref. 45.

in Fig. 17.4. Because of the mineralogically very fine and intimate association of the various minerals in the crude ore (observed microscopically as in Fig. 17.5), the grain fraction of 100 to 40 microns are of special interest. A sample of dispersed grains immersed in methylene iodide (specific gravity 3.33) consisted of grain sizes ranging from 100 to 40 microns. Because of the separation of minerals according to their density, the gangue minerals floated underneath the cover glass and were easily distinguished from the heavy minerals. The results of counting and calculation of weight per cent of each grain fraction are summarised in Table 17.1.

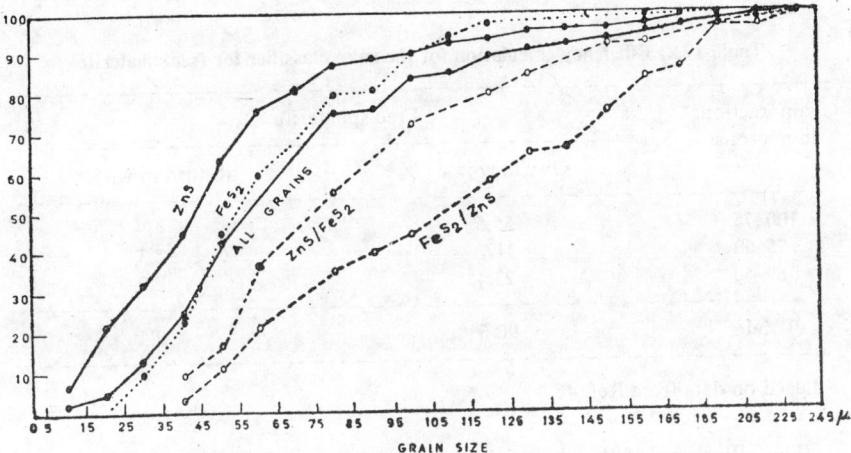

Fig. 17.4. Cumulative values of each mineral or its composite grains in the individual grain fraction.

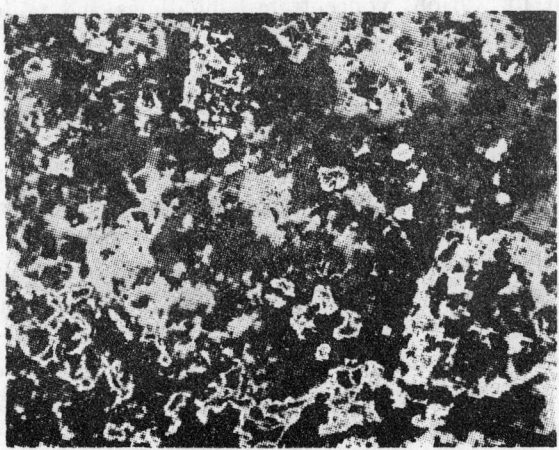

Fig. 17.5. Association of pyrite/sphalerite minerals as observed under microscope. Sphalerite (light-grey), pyrite (white), and calcite (dark-grey).

Mineral grain counts have not been made in grain fractions over 100 microns, because most of the grains still consisted almost completely of intergrown minerals. The grain fractions below 40 micron would have required further classification by elutriation or infrasizing.

The microscopically determined sphalerite values show for ZnS, the grade and efficiency of the classifier for the grain fractions of 100 to 40 microns. This holds, if (a) the weight percentages of the various grain fractions are taken as factors (Table 17.1, column—2) of the ratio of overflow to return, and (b) the total amount of free sphalerite of these fractions is equal to 100 (Table 17.2).

Table 17.2. Efficiency calculation for the rake classifier for free sphalerite

Grain fractions in microns	Free sphalerite	
	Overflow in wt.%	Return in wt.%
100–75	55.5	4.9
75–60	11.7	2.3
69–40	23.4	2.2
Total	90.6	9.4

Based on data from Ref. 45.

The efficiency of the rake-classifier with regard to free sphalerite in the grain fractions of 100 to 40 microns is 90.6 per cent. Only 9.4 per cent of total free sphalerite is returned to the ball mill. For other minerals, the efficiency of classifier can be determined in a similar way. Similar studies can be carried out for various concentrates, middlings and tailings, and the efficiency of the process may be calculated.

17.6.2 MICROSCOPICAL ANALYSIS OF FLOTATION CONCENTRATE AND TAILINGS OF CHALCOPYRITE

The screen analysis of concentrate and tailings is carried out in an identical manner as discussed in the previous section. It revealed that the ratio of concentrate and tailing was 670 : 9330. The microscopic examination was carried out of individual grain fractions. The results are presented in Tables 17.3 and 17.4. In very-fine size fractions, chalcopyrite was calculated on the basis of assays.

The microscopically determined values of chalcopyrite show the grade and efficiency of flotation as a whole operation, as the final concentrate and the final tailings are considered for examination. This holds under the conditions as mentioned in previous section. The efficiency may be calculated, as shown in Table 17.5.

The efficiency of the flotation with regard to the free chalcopyrite in the

Table 17.3. Results of microscopical examination of copper concentrate[45]

Grain fractions in microns	Wt. %	Cu Assay %	Mineral content present (%)			
			Chalcopyrite	Pyrite	Pyrrhotite	Magnetite
+ 147	0.99	21.6	62.61	10.99	5.07	1.38
−147 to + 104	2.44	28.6	82.90	5.92	2.72	2.01
−104 to + 74	8.04	25.5	73.92	11.46	6.50	1.77
− 74 to + 52	13.63	20.0	57.98	21.52	18.70	1.67
− 52 to + 37	12.51	18.1	52.47	13.74	28.75	0.74
− 37 to + 26	8.03	17.2	49.86	7.13	27.32	1.88
− 26	54.36	16.0	46.40			
			Chalcopyrite calculated by chemical analysis			

Table 17.4. Results of microscopical examination of copper tailing[45]

Grain fraction in micron	Wt. %	Cu assay %	Free chalco-pyrite %	Chalcopyrite associated with in %			Chalcopyrite associated with gangue in %	Total chalcopyrite
				Pyrite	Pyrrhotite	Magnetite		
+ 147	8.32	0.374	0.644	0.053	0.038	0.008	0.342	1.085
−147 to + 104	8.94	0.300	0.435	0.030	0.142	0.016	0.246	0.869
−104 to + 74	9.04	0.226	0.381	0.049	0.056	0.015	0.155	0.856
− 74 to + 52	11.64	0.186	0.370	0.045	0.089	0.006	0.029	0.536
− 52 to + 37	8.82	0.124	0.250	0.010	0.020	0.005	0.074	0.359
− 37 to + 26	7.18	0.107	0.244	0.0	0.0	0.014	0.053	0.311
− 26	37.24	0.109	0.306					
				Chalcopyrite calculated by chemical analysis				

Table 17.5. Calculation of efficiency of flotation for chalcopyrite[46]

Grain fraction in microns	Free chalcopyrite in %	
	Concentrate	Tailings
+ 147	1.10	1.32
−147 to + 104	3.58	0.93
−104 to + 74	10.53	0.83
− 74 to + 52	14.01	1.06
− 52 to + 37	11.06	0.54
− 37 to + 26	7.10	0.43
− 26	44.71	2.80
Total	92.09	7.91

grain fractions of 147 to −26 microns is 92.09 per cent. Only 7.91 per cent of total free chalcopyrite goes to the tailing. The results obtained can also be presented graphically on the basis of cumulative weight per cent versus grain-size fraction.

CHAPTER 18

Processing and Flowsheets for Important Ores

The ores/minerals of different places (even having the same mineral) differ so widely in their physical properties, distribution, gangue association, association of other valuable minerals, etc., that a general process flowsheet cannot be adopted. Therefore, some representative industrial processing of ores is briefly illustrated in this chapter. Since this subject is very wide, only the important cases are taken in this chapter.

18.1 Processing of Aluminium Ores[8,30,47]

18.1.1 PROCESSING OF BAUXITE ($Al_2O_3 \cdot 2H_2O$)

It is the predominant ore used in the production of aluminium. The main impurities associated are iron oxide, phosphate, titania, and silica, together with small amounts of zirconium, chromium, vanadium, gallium, etc. Bauxites from different areas differ in their basic characteristics and thus the actual conditions and operations employed also differ for the treatment of bauxite to obtain alumina for reduction to aluminium.

Preliminary treatment usually consists of a wash scrubbing which removes considerable amount of alumina-rich tailings. After relatively coarse grind and desliming, it can be subjected to froth flotation by using suitable collector (800 series) in acid pulp (using H_2SO_4) with fuel oil (to stabilise froth). If the concentrate thus produced contains iron or titanium minerals, tabling or high intensity magnetic separation can be used. Sometimes, the bauxite may be roasted to convert ferrous oxide into ferric oxide. The most widely and accepted method for processing of bauxite is Bayer process, which consists of NaOH leaching of ground bauxite in autoclaves. The Indian bauxites are generally not given any pre-treatment to the ore before its leaching. The basic principle of Bayer process in all the flowsheets remains the same. The flowsheets may differ in respect of type of equipments and number of operations carried out with some modifications depending upon the ore and place.

A schematic representation of bauxite processing as practised at Hindustan Aluminium Company (HINDALCO), Renukoot is shown in Fig. 18.1. The ore assaying about 50 per cent Al_2O_3 with 13–15 per cent Fe_2O_3, 3–4 per cent SiO_2 and 8–12 per cent TiO_2 is crushed to -50 mm and fed to ball mills where it is ground with spent liquor. The discharge slurry at -0.85 mm and about 45 per cent solids is digested at 150–240°C under a pressure of 40 kg/cm². After the leaching is complete, the pressure is released in stages and the hot slurry is discharged and allowed to settle in thickeners. The residue known as red-mud (Al_2O_3 15-20 per cent, SiO_2 7–10 per cent, Fe_2O_3 24-26 per cent, TiO_2 20–25 per cent, CaO 3–4 per cent, and Na_2O 8–11 per cent) is washed and discarded. The overflow is filtered, diluted, cooled and sent to precipitators with seeded alumina. The solution is agitated by

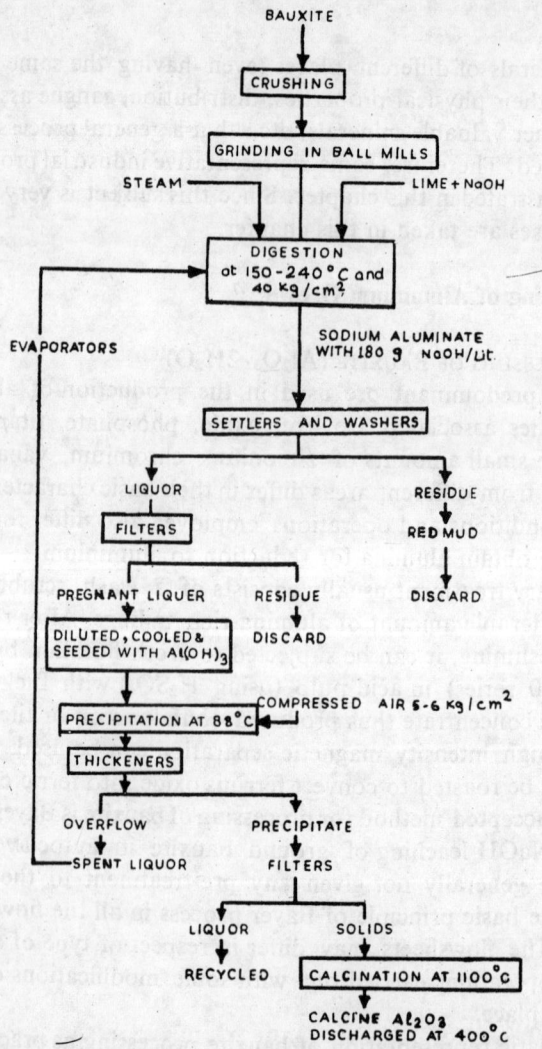

Fig. 18.1 Flowsheet of bauxite processing of HINDALCO, Renukoot.

compressed air for 30 hr, during which alumina precipitates on seeds. The precipitate is then settled, washed and filtered. The precipitate is calcined at 1100°C to yield anhydrous Al_2O_3. The aluminous liquor containing about 50 per cent of alumina, is recycled with sodium hydroxide.

18.1.2 PROCESSING OF FELDSPAR, (Na, K, Ca) $AlSi_3O_8$

Feldspar is processed by froth floatation. The process uses hydrofluoric acid as a pH regulator, as an activator for feldspar, and also as silica depressant. Amines may be used to promote flotation. The frothers employed are generally pine oil, fuel oil and higher alcohols. The ceramic industry requires the product as slime free and granular (0.85 mm to 70 microns). The minerals to be removed are usually garnet, mica, quartz and particularly ferromagnesians (dark silicates containing iron which affect the colour of fired ceramics).

18.1.3 PROCESSING OF KYANITE (Al_2SiO_5)

A pre-treatment of scrubbing with NaOH solution and desliming may be given before the flotation of kyanite. Oleic acid and fatty-acid salts are used as collectors, whereas frothers used are pine oil and higher alcohols. H_2SO_4 is used in cleaning stage.

18.2 Processing of Antimony and Arsenic Ores[25]

Antimony sulphide (Sb_2S_3) and arsenopyrite (FeAsS) are the chief minerals of antimony and arsenic, respectively. The flotation of these minerals with xanthogenate collectors requires activation with $CuSO_4$ or other heavy metal salts in an alkaline pulp (with soda ash). Cyanides depress these minerals. The specific dosage, reaction time, pH, etc., should be worked out for individual cases.

A standard plan requires a large circulating load in the grinding section to minimise the overgrinding (stibnite is very friable). The pH of pulp is adjusted to 7.6 with lime and activated with copper sulphate or lead acetate (3 kg/tonne). Arsenopyrite or pyrite are depressed in the flotation cleaning stage by the use of 0.02 kg NaCN per tonne. If talc is present, it is also depressed in this stage by yellow dextrin (0.22 kg/tonne). The collector used is generally sodium isopropyl xanthate (0.02 kg/tonne). The grade of concentrate is usually above 50 per cent antimony.

18.3 Treatment of Asbestos[8]

The objectives in processing of asbestos are (i) to free the potential fibres from the slabby gangue (b) to obtain the maximum length of the final product, and (c) to remove dust and grit at each stage of release. The treatment of asbestos involves special problems of dust catching and disposal, since about 10 tonnes of air is drawn through the passing material per tonne of final product. The comminution may start with impact crushing or the use of Aerofall mills, cone crushers, impactors, ball mills or hammer mill with decreasing feed size and increasing fluffiness. After rough drying, the broken ore is screened, and recrushed. Screening and crushing may be repeated several times depending from ore to ore. During the passage along the screen the ore is stratified. The undersize is either waste or middlings having short fibres and thus further treatment would be required. The oversize discharge is then comminuted to finer size and freed fibre is lifted by air elutriation using aspirators placed over the screens and trapping the lifted fibre in cyclones.

The flowsheets are quite different for different ores due mainly to the variations in the length, strength and milling properties of the fibres. As long fibres are desired, gradual fiberisation is carried out with repeated aspirations to remove finished product at the earliest liberation and to shake down the entrained gangue.

18.4 Barite Concentration[8]

The objective of processing barite is to produce barium sulphate suitable for paint, paper, pharmaceutical and other industries. When colour is important, the rigid specifications should be maintained, whereas discolouration can be tolerated in low-grade concentrates. The lower grade barite concentrates are used to blend drilling muds in oil wells and must have a density of minimum 4.25, 98 per cent, -75 microns, and a viscosity of at least 60 centipoise. The ores of barite are generally associated with iron carbonates, iron oxides, quartz, calcite and fluorite. A general treatment of barite consists of gravity concentration carried out at the coarser liberation sizes, followed by froth flotation. The pH of the pulp in flotation is maintained between 8 and 10 with soda ash and/or NaOH. In the circuit, sodium silicate is used as dispersant. The collectors for the iron minerals include R 824 or R 825, and tall oil or sodium oleate, with pine oil as frother. If the concentrate is to be used solid as drilling mud, heating during the drying stage may be used to expel undesirable reagents.

18.5 Concentration of Beryllium Ore

The pure mineral beryl ($3 BeO \cdot Al_2O_3 \cdot 6SiO_2$) contains 14 per cent BeO, and the marketable grade should contain a minimum of 10 per cent BeO. The major gangue minerals are quartz, albite, and muscovite, while the associated heavy minerals are cassiterite, columbite and tantalite which are rarely in economic amounts. Sometimes, spodumene (a lithium mica) may occur in recoverable amounts. After floating off muscovite, the pulp is conditioned at pH 5 with a tallow amine acetate and the bulk flotation of beryl and feldspar is carried out using an alcohol as a frother. The separation of beryl and feldspar is carried out by conditioning the pulp with calcium hypochlorite followed by flotation of beryl with petroleum sulphonate. Beryl can also be separated from quartz with an anionic collector, because it has zero charge at pH 3.7–4.6, when the other minerals have negative charge. Spodumene may be recovered with beryl in bulk float.

18.6 Processing of Chromite Ores

Earlier, only selective mining of high-grade chromite ores was practised with little or no beneficiation (hand-picking was sufficient). The increased demand of the ore necessitated bulk mining of the ore and thus demanding beneficiation of low-grade chromite ores. The processing of chromite ores usually involves chemical treatment (roast reduction and/or leaching). The attainable grades of mineral depend on (a) the amounts and types of gangue associated and (b) chemical composition of the chromite mineral. The chromite ores can be classified as simple ores, ferruginous ores, and chrome-spinels.

The industrial mineral chromite ($FeO \cdot Cr_2O_3$) is marketed either in terms of its chromium content or its Cr/Fe ratio, where metallurgical grade is concerned, whereas for bricks, the refractory character of the ore is important. The simple ores contain light minerals such as serpentine, olivine, talc, chlorite, magnesite, calcite, dolomite, etc. The chromite mineral can be liberated from these ores at coarser sizes (about 1 mm). The beneficiation of simple ores involves specific gravity methods, such as heavy media separation, jigging, tabling, and spirals. In India, the examples of these simple ores are Nausahi ores from Orissa and Kittaburu ore from Bihar.

The ferruginous ores contain iron minerals (generally hydrates) as the main gangue and thus a reduction roast treatment (at temperatures 500–600°C) will be required to convert the non-magnetic iron to magnetic form. This is then followed by wet or dry magnetic separation at sand sizes to yield a chromite concentrate and another iron concentrate as a by-product. This type of ore is called Gujang ore in Orissa.

The chrome-spinels contain chromium in finely disseminated form, not recoverable by simple physical methods, or in chemical combination with associated impurities, particularly iron. These ores, thus, will require chemical treatment to yield high-grade concentrates. The treatment may be reduction roasting followed by acid leaching to remove iron values selectively.

18.7 Treatment of Clay[8]

China clay (kaolinite) is a hydrous aluminium silicate ($Al_2O_3 . SiO_2 . nH_2O$). It is used mainly as inert filler, and a constituent of ceramics and for coating in the paper industry. If the end-product is of −2 microns, it is the best for paper, provided it can be bleached. The coarser particles (up to 50 microns) are used as fillers. China stone (undecomposed granite) can be used in the manufacture of clay slip after fluorine is removed by flotation of part of its mica.

China clay is either dug out or slurried from the quarries by high pressure jets and washed through sand traps, sluices and settlement pits to remove most of the undesired sand. The rest of the sand is removed by mechanical classifiers, hydroseparators, hydrocyclones, or centrifugal classifiers. Crude kaoline is sometimes classified into two grades, i.e. fine (for coating), and coarse (for filler). Bleaching is performed at a pH 3.5–4 with zinc or sodium hydrosulphite with addition of H_2SO_4 to slurry. Dewatering of fine grade clay, starts with thickening through a high speed centrifuge and then it is followed by filtration (or drum filters). Sometimes, heating may be required to reduce the viscosity and speed up filtration. Finally, the cake is dried and bagged.

18.8 Processing of Coal[9,30,46]

Processing of coal may be accompanied to achieve the following objectives:

a) Improvement of technical performance.
b) Grading into sizes for sale.
c) Separation of undesirable constituents.

The coal obtained from mechanised mining is a mixture of various sizes from dust upward, with random inclusions of chalk, clay, wall rock, pyrite, etc. The coal contains two kinds of ash forming dirt, i.e. fixed and free. The fixed ash is derived from the inorganic matter which has grown in the tissues of the original coal forming plants together with fine silt entrained during deposition of the seams. The fixed ash is not removable by standard methods. The free dirt is extraneous to the true seam, coming from roof, floor, or seat measures. It includes clay, black shales intercalated with the coal and coal locked to calcareous and other mined dirt. This free dirt can usually be removed by suitable methods.

The coal preparation consists mainly of the following four operations:

i) Screening or sizing.
ii) Mixing, blending.
iii) Cleaning of coal (removal of incombustibles).
iv) De-dusting or de-watering.

All the coal cleaning plants invariably employ heavy media separation (using ferro-silicon or magnetite, or sand as suspension in water), with or without Baumjig. The recently installed washeries have incorporated the provision for the treatment of fine coals either by cyclones or flotation. Figures 18.2 and 18.3 represent two typical coal-washing flowsheets. The former is for simple heavy media separation, whereas the latter is for a central composite washery.

18.9 Processing of Copper Ores[8,25,46]

The copper bearing ores are available in such a wide range that industrial processing of ores uses nearly every processing technique. Copper minerals range from straight sulphides (copper–iron sulphides) to oxides, carbonates, silicates and chlorides. The associated valuable minerals (gold, silver, cobalt, molybdenum, nickel, germanium, etc.) should also be recovered separately. The important methods of processing include chemical extraction (leaching) and its modification in leach precipitation—float (L.P.F.) process, straight flotation of typical copper ores, differential flotation of mixed ores, and pyrometallurgical treatment in segregation process.

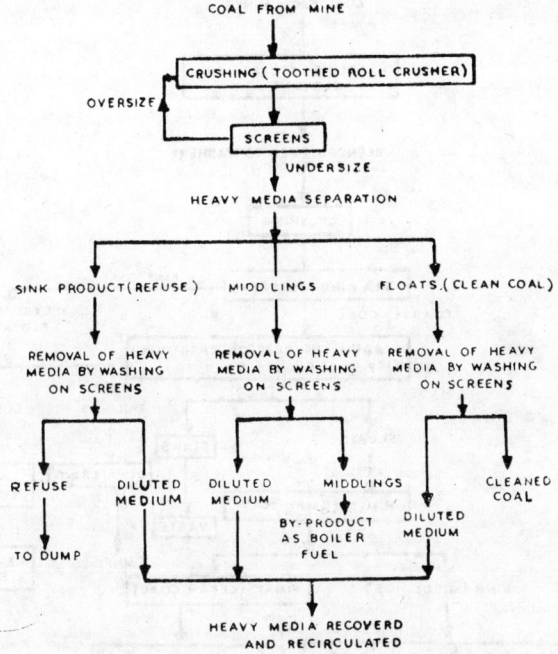

Fig. 18.2. Coal washing by heavy media separation.

18.9.1 CHEMICAL PROCESSING

The oxide ores, such as azurite ($Cu_3(OH)_2(CO_3)_2$), brochanite ($Cu_4(OH)_6 \cdot SO_4$), cuprite (Cu_2O), malachite ($Cu_2(OH)_2CO_3$), etc., are leached with H_2SO_4, dissolving copper values as copper sulphates, whereas the sulphide ores are leached with ferric salts or ammonia (since sulphide is insoluble in acids).

Heap roasting and leaching have been practised since the 16th century. Even presently, dump-leaching is an important source of copper. Pit-leaching has been practised for many years and the methods are now mechanised and accelerated. Where H_2SO_4 is used for leaching, copper from its solution can be recovered either by precipitation or by electrolysis.

If an ore is mixed, i.e. oxide and sulphide, the oxide fraction can be leached with sulphuric acid and the copper is recovered from solution, whereas the remaining ore is subjected to flotation to recover copper sulphide. A typical flowsheet for this process as adopted at Hayden is shown in Fig. 18.4.

The sulphide ores can be directly floated to yield a concentrate containing 12–24 per cent copper which can be leached directly with ferric salt or ammonia to yield copper in solution. From solution, copper can be recovered by electrolysis or hydrogen reduction. Segregation process for the recovery of copper from refractory ores is discussed in Ch. 16.

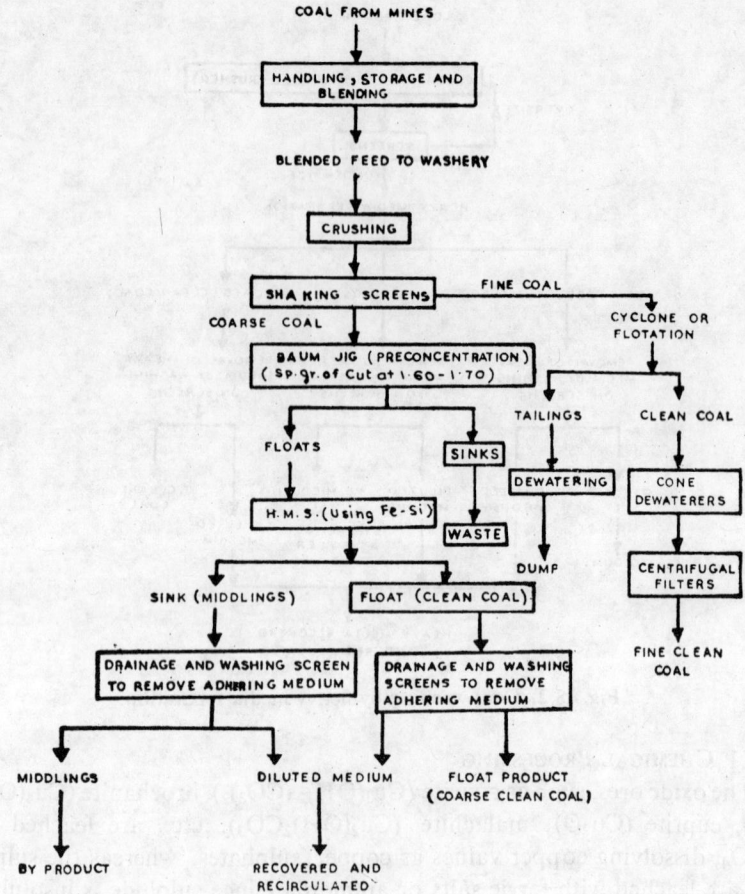

COAL FROM MINES

HANDLING , STORAGE AND BLENDING

BLENDED FEED TO WASHERY

CRUSHING

SHAKING SCREENS — FINE COAL

COARSE COAL — CYCLONE OR FLOTATION

BAUM JIG (PRECONCENTRATION)
(Sp. gr. of Cut at 1·60–1·70)

FLOATS — SINKS — TAILINGS — CLEAN COAL

H. M. S. (Using Fe-Si) — DEWATERING — CONE DEWATERERS

SINKS — WASTE

SINK (MIDDLINGS) — FLOAT (CLEAN COAL) — DUMP — CENTRIFUGAL FILTERS

DRAINAGE AND WASHING SCREEN TO REMOVE ADHERING MEDIUM

DRAINAGE AND WASHING SCREENS TO REMOVE ADHERING MEDIUM

FINE CLEAN COAL

MIDDLINGS — DILUTED MEDIUM — FLOAT PRODUCT (COARSE CLEAN COAL)

BY PRODUCT — RECOVERED AND RECIRCULATED

Fig. 18.3. Simplified flowsheet for a composite scheme of coal cleaning.

18.9.2 FLOTATION OF COPPER ORES

The sulphides containing native copper, selenides, tellurides, arsenides, and antimonides respond readily to xanthate collectors, notably potassium ethyl xanthate. Selectivity can be improved by taking care in grinding (avoiding unnecessary oxidation) and by using suitable polysulphides where oxidation is unavoidable. The main sulphides are chalcocite (Cu_2S), covellite (CuS), chalcopyrite ($CuFeS_2$), and bornite (Cu_2FeS_4). In general, greater the copper-iron ratio, higher is the upper limit of pH and lower the risk of depression through the use of lime (used to depress pyrite). Copper, the most readily floatable of the major sulphides (except molybdenite), is taken first in a differential flotation. Cu_2S floats at pH up to 14, and chalcopyrite (with increasing iron) floats at pH 11.8. Generally, the purpose of flotation is to float maximum possible copper for a specific grade of concentrate, and a suitable pH is maintained. Pyrite is the usual constituent in the ore and it is depressed at pH above 11 by addition of controlled lime.

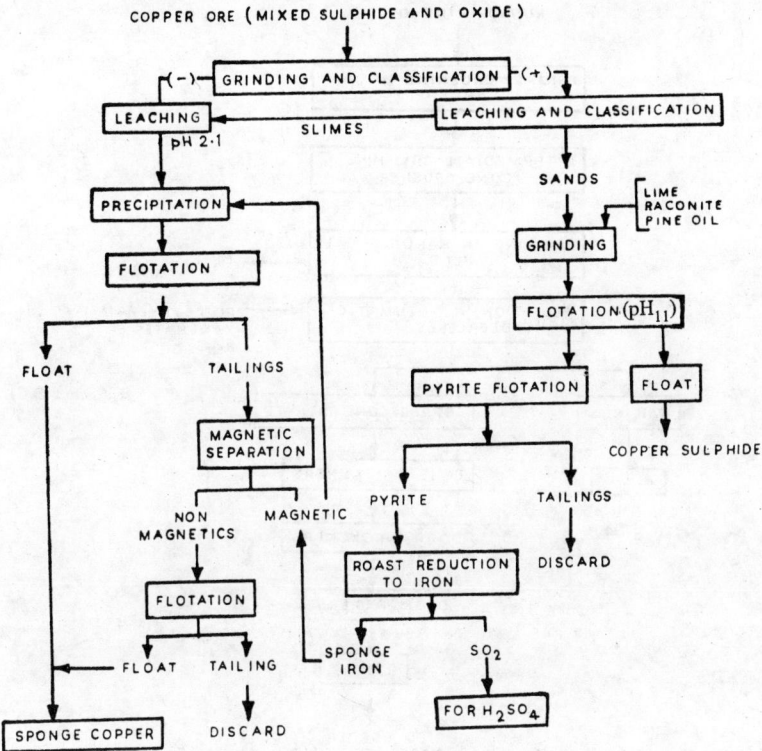

Fig. 18.4. Leach/precipitation/float process for mixed sulphide and oxide ore of copper.

A simplified flowsheet for flotation of copper ore as practised at Hindustan Copper Ltd., Ghatshila (Bihar) is shown in Fig. 18.5. The flowsheet used at Khetri (Rajasthan) is also identical in operations.

A special case is the reflotation of copper sulphide from a copper–nickel matte. In this case, the pH is held between 10 and 12.4, since a high OH⁻ concentration is needed to depress nickel sulphide. Though, xanthate may be used, the preferred collector is diphenyl guanidine (0.22 kg/tonne) or di-ortho tolyl guanidine which is added during grinding. Pine oil may be added as frother and lime is used to control the pH. Usually 3–6 rougher-cleaner stages are used with intermediate regrind.

The main non-sulphide ores to be floated are azurite, cuprite, and malachite. The general treatment plan for floating of non-sulphide ores consists of conditioning with collectors based on fatty acids (e.g. palm oil or cottonseed oil). The collector is emulsified with hot solution of sodium carbonate. Sodium sulphide is added in the conditioners and along the flotation line. The use of tall oil aids frothing. There are numerous flowsheets for processing of copper ores by flotation depending upon the grade and type of ore in a particular area.

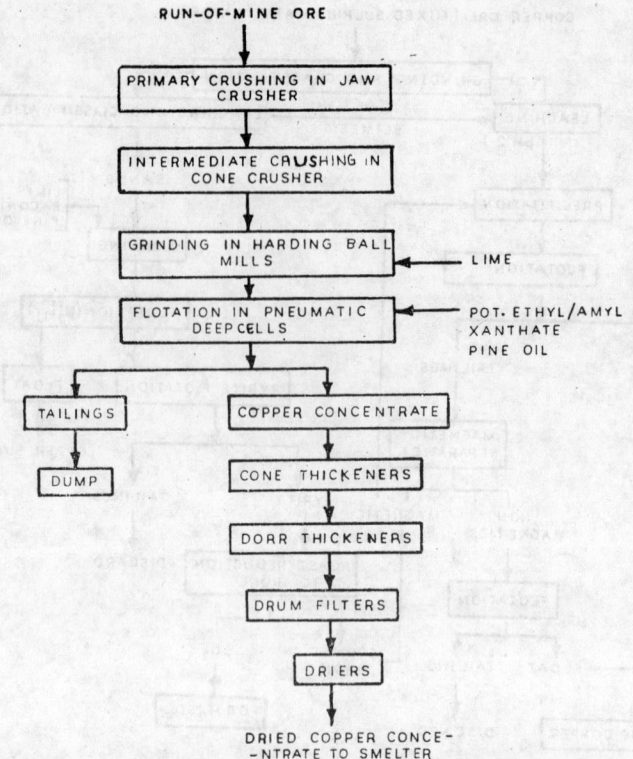

Fig. 18.5. Simplified flowsheet for flotation of copper ore at Ghatshila.

18.10 Diamond Concentration[8]

The dominant factors in diamond concentration are (a) concentration ratio is of the order of 20 million to 1, (b) the diamonds should be recovered intact and undamaged, and (c) a high percentage of recovery is required. Three main types of deposits exploited for the recovery of diamonds are (a) kimberlite known as blue ground (ore of volcanic pipes and fissures), (b) marine terraces, and (c) alluvial gravels including off-shore deposits. Kimberlite consists mostly of serpentine with minor amounts of calcite, ilmenite, pyrite, etc., and rarely, a diamond. Marine terraces and alluvial gravel require only screening before treatment, whereas kimberlite has to be disintegrated. Concentration of kimberlite starts with washing pans worked at a ratio of 32 to 1, which is followed by heavy media separation at a ratio of 6–25 to 1. The sinking fraction containing diamonds and heavy minerals is passed over a shaking table smeared with heavy grease. The diamonds adhere to the grease whereas other minerals do not adhere. After certain interval, the grease with diamonds can be scrapped off as a much purer concentrate (50,000–80,000 : 1 ratio of concentration). Final work is completed with hand sorting at a ratio of 2 : 1.

The alluvial and marine deposits do not respond to grease tabling (Indian diamonds of Panna do not respond to grease table). In this case, the surface is restored to its water-repelling state by gentle scrubbing of the HMS product and conditioned with a fatty acid (if grease tabling is to be carried out). The HMS concentrate after drying at 130°C can also be subjected to electrostatic separation to recover the diamonds.

18.11 Processing of Gold Ores[43,46]

In ores the gold is usually present as metal (in native form) alloyed with metallic silver and/or copper, as inclusions of gold (few microns in diameter) in metal sulphides (as pyrite, stibnite, arsenopyrite, galena, chalcopyrite, etc.), and as telluride or sulphotelluride. The process selected thus depends on the type of occurrence of gold in the ore. The various processes involved in treatment of gold ore are gravity concentration, amalgamation, flotation, cyanidation, and other chemical attack. The processing usually involves the use of more than one process. For example, the ore may be ground to −200 microns and subjected to gravity devices (jigging, tabling, etc.) and then may be followed by flotation, cyanidation, etc. The various methods for the treatment of gold ores are summarised in Fig. 18.6. The mixed methods are also possible. For example, the ore can be ground to −200 microns and treated with amalgamation alone or in combination with jigging or tabling and this can be followed by flotation. The tailings can then be treated by cyanidation, whereas the flotation concentrate can be roasted and then leached with cyanide.

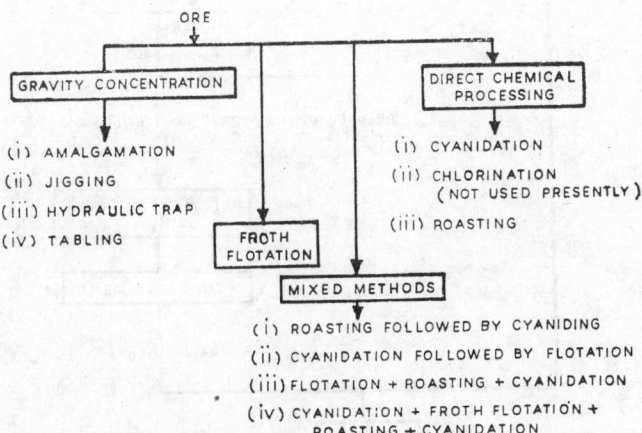

Fig. 18.6. Different methods for processing of gold ores.

The practice varies from place to place depending on the type of ore. At Kolar, native gold occurs in association with quartz. The ore is processed at different plants with slightly different flowsheets to meet the requirement of the ore. A simplified flowsheet is shown in Fig. 18.7.

One important example is processing of Canadian ores. These use the cyanidation after roasting, which is dictated by difference in structures of the

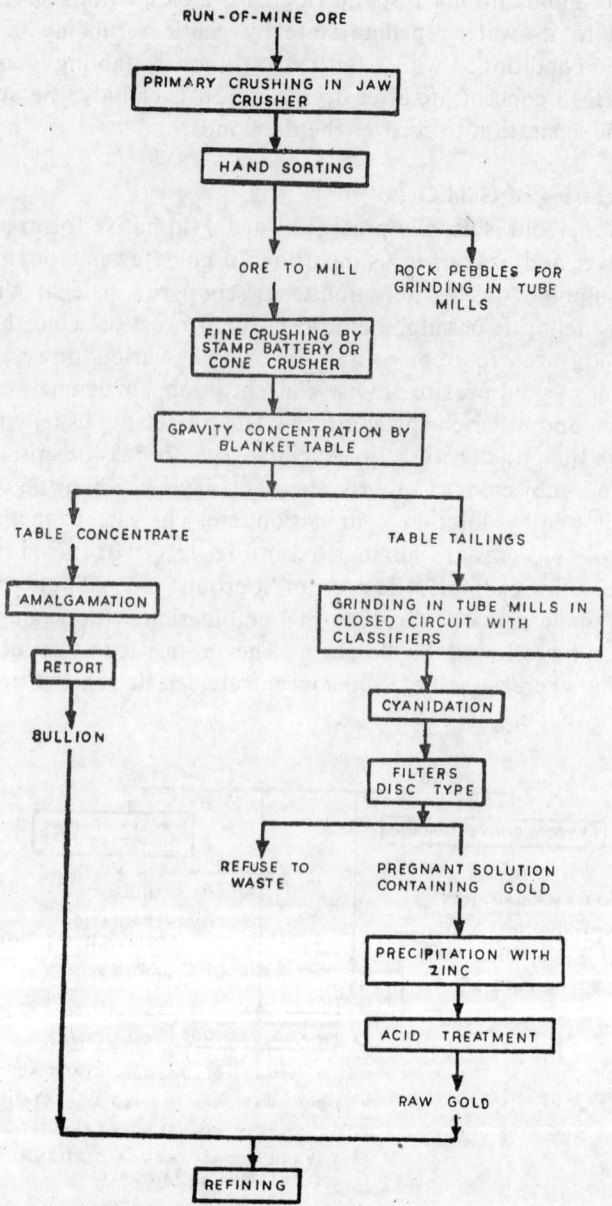

Fig. 18.7 Simplified flowsheet for treatment of gold ores at Kolar gold field.

ores. This ore consists of gold, arsenopyrite, pyrite, sphalerite, and stibnite in quartz. About 25 per cent of gold is recovered from the jig concentrate, 60 per cent by cyanidation and 15 per cent by cyanidation after roasting.

18.12 Processing of Graphite[8]

This form of carbon occurs in nature in coarsely crystalline as well as finely divided form with no clear dividing line. The associate minerals are mica, chlorite, quartz, hornblende, etc. Graphite is processed by flotation. The pulp is made of about pH 8 (alkaline) with soda ash. Sodium silicate is added for dispersion and to increase selectivity. The graphite is a readily floatable mineral. The collecting agents used are paraffin or diesel oil. Pine oil is used as a frother, and it may alone be sufficient with finely ground graphite. When it is possible to liberate clean coarse flakes by primary grinding in rod mills, separation is carried out by screening. In general, graphite can be upgraded by a series of operations as shown in Fig. 18.8. Since coarse flake fetches better price, care should be taken in avoiding overgrinding.

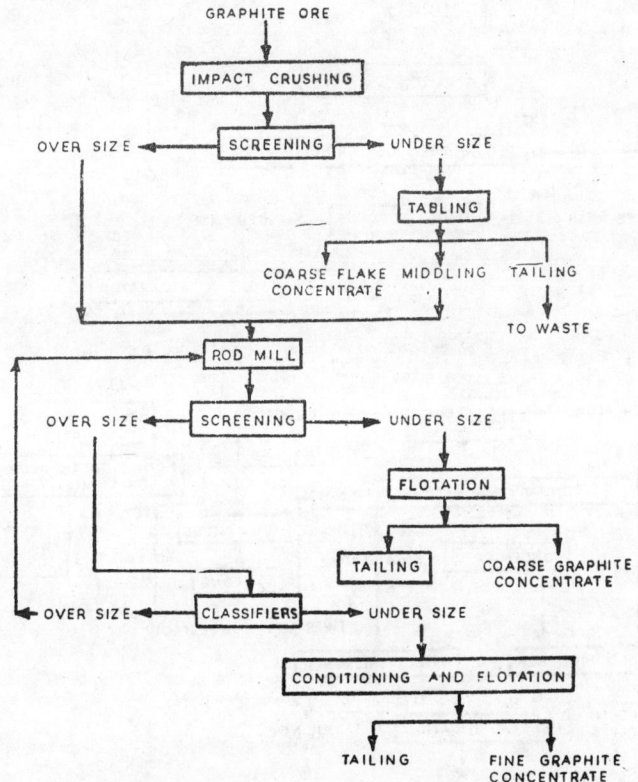

Fig. 18.8. Flowsheet for processing of graphite.

Impact crushing may be followed by screening, with tabling of the screen undersize to produce a coarse flake, tailings, and a middling. The middling together with the screen oversize goes to rod milling. The mill dischage can then be classified, undersize being treated by flotation and oversize by tabling. It may be desirable to table the concentrates, since graphite is ductile and tends to coat gangue minerals (reporting into float) which can be removed by gravity treatment.

Flotation yields a product of 90–95 per cent carbon content. This can be further purified by chemical treatment, e.g. leaching with hydrofluoric acid to remove silica.

18.13 Treatment of Iron Ores[30,47]

The main concentrating processes for iron ores are magnetic separation (directly or after reduction roasting), gravity concentration, and flotation. The problem associated with Indian iron ores is two-fold, i.e. (a) high content of gangue minerals particularly containing aluminium, and (b) relatively soft nature of the ore generating fines on handling. A simplified flowsheet for the processing of Kudremukh iron ore is shown in Fig. 18.9.

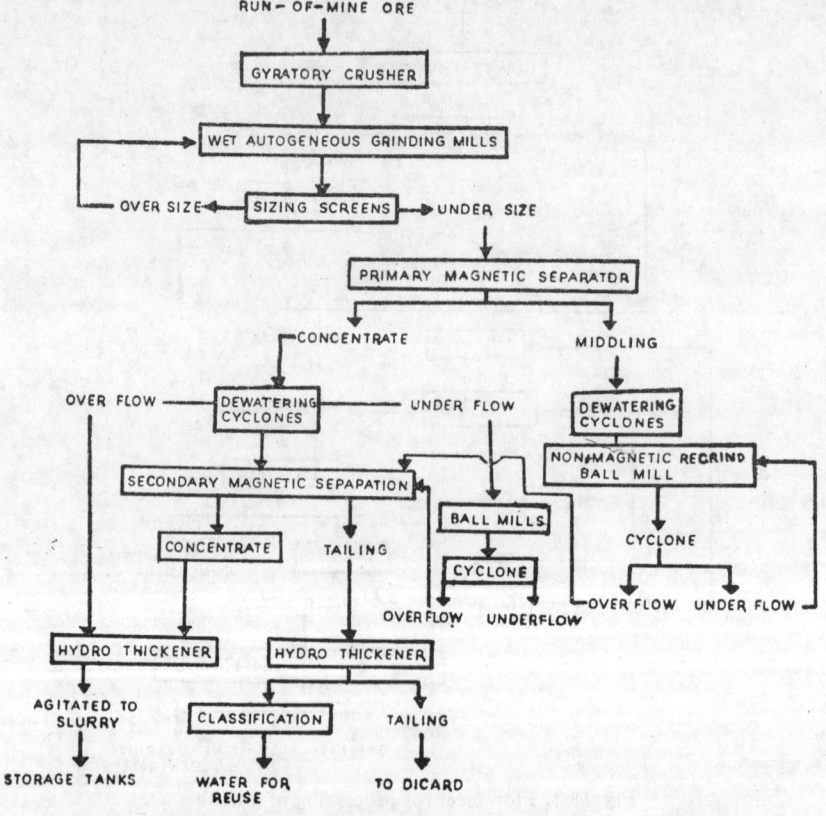

Fig. 18.9. Schematic flowsheet for the processing of Kudremukh iron ore.

High intensity magnetic separation can be used to concentrate weakly magnetic hematite ores. These can also be concentrated by froth flotation. Many laboratory and pilot plant investigations have been carried out, on flotation of various Indian ores of low grade. However, so far it could not be possible to adopt flotation technique in India due to high cost of operation.

Flotation of iron ore has been pioneered in Michigan in 1954. A —200 microns classifier overflow is deslimed and thickened by cyclones. The pulp is then conditioned at 70 per cent solids with 0.0025 kg/tonne of Aerosol, 0.7 kg of tall oil and then floated. Concentrates are filtered and stockpiled and drawn to the regrind section for reduction to at least 75 per cent — 50 microns, a size suitable for pelletising.

18.14 Concentration of Lead and Lead–Zinc Ores[8,25,43,46]

18.14.1 Concentration of Lead Ores

The main ores of lead are the galena (sulphide, PbS), partially oxidised sulphide, anglesite (sulphate, $PbSO_4$), and the cerussite (carbonate, $PbCO_3$). Galena usually occurs in association with sphalerite (ZnS) and other sulphides, from which it is differentially floated. The mineral galena floats readily with Aerofloat or xanthate, in a pulp made alkaline with Na_2CO_3. It may be depressed by lime at a pH exceeding 10.4. Potassium dichromate forms a non-reactive coating of lead chromate, and is sometimes used to depress lead from a bulk-float. If Aerofloat 25 or 31 is used, little or no pine oil is needed as a frother. Where galena has become tarnished, the use of cynamide (R. 404) may aid collection. When calcite gangue is present, sodium silicate is used to disperse the adhered slimed gangue.

Oxidised lead can be floated by making use of sodium sulphide, which produces a surface attractive to xanthate collectors. Since soluble sulphides depress clean lead and silver sulphides, these minerals must be removed in an earlier flotation operation. Since, excess Na_2S is in any case a depressant, starvation quantities should be employed. The stabilisation of newly sulphidised mineral particles is favoured by the use of copper sulphate. This is accompanied by two conditioning stages, starting with sulphidising treatment and followed by the use of the copper salt.

Most oxidised ores can first be treated by gravity concentration as far as possible to reduce the production of slimed values. Sulphidising agents are used just before the flotation, with sodium bicarbonate. Alkaline earths are liable to be harmful, since they tend to form coatings of insoluble carbonates on cerussite.

18.14.2 Concentration of Lead–Zinc Ores

In processing of lead–zinc ores (galena and sphalerite), the most important factor is the selection of galena and sphalerite, so that lead concentrates should be as free as possible from zinc, and zinc concentrate should be equally free from lead. Since, the zinc content of most ores is higher than its lead content, and the conditions of the process make it more convenient to float galena first, conditions arise in which lead concentrate picks up zinc. The lead concentrate can be kept zinc-free by using effective zinc blende depressants, by the selectivity of the collectors for lead flotation and by ensuring that no galena–sphalerite concretions are present in the ground ore. Generally, it is easier to produce a lead-free zinc concentrate. However, it

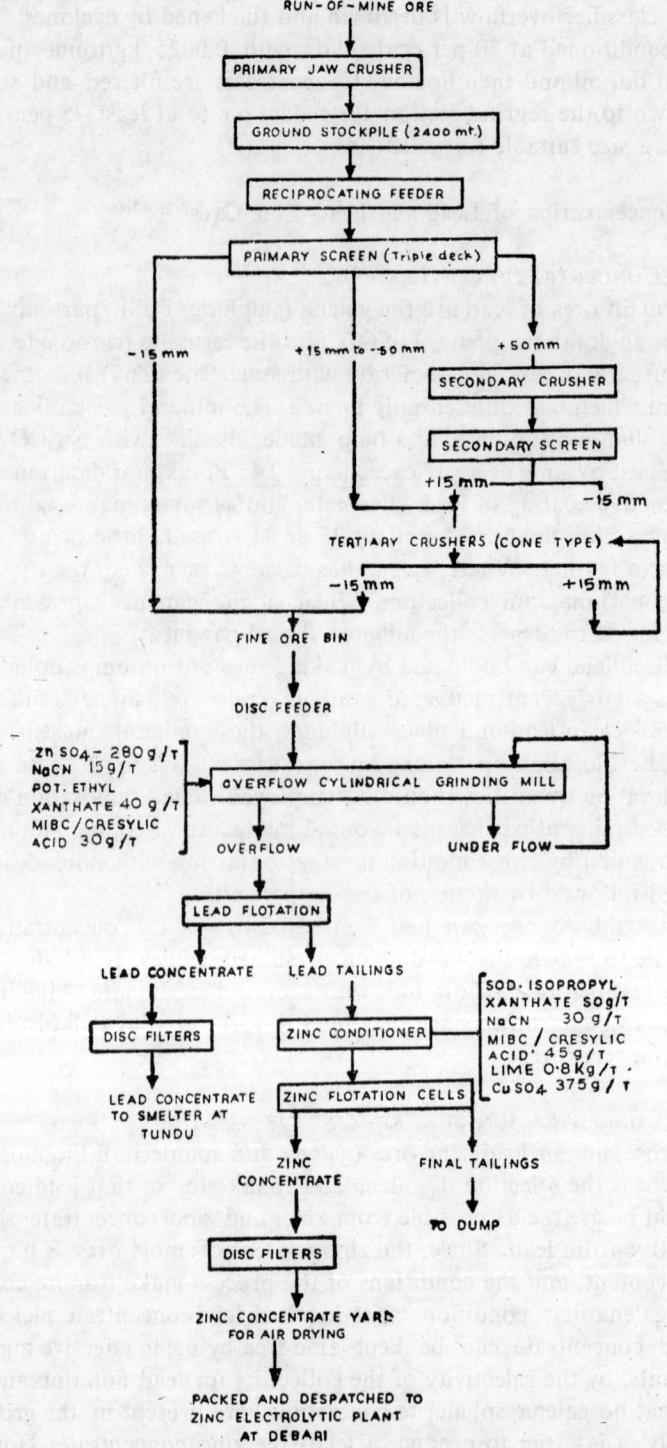

RUN-OF-MINE ORE

PRIMARY JAW CRUSHER

GROUND STOCKPILE (2400 mt.)

RECIPROCATING FEEDER

PRIMARY SCREEN (Triple deck)

−15 mm   +15 mm to −50 mm   +50 mm

SECONDARY CRUSHER

SECONDARY SCREEN

+15 mm   −15 mm

TERTIARY CRUSHERS (CONE TYPE)

−15 mm   +15 mm

FINE ORE BIN

DISC FEEDER

Zn SO₄ − 280 g/T
NaCN 15 g/T
POT. ETHYL
XANTHATE 40 g/T
MIBC / CRESYLIC
ACID 30 g/T

OVERFLOW CYLINDRICAL GRINDING MILL

OVERFLOW   UNDER FLOW

LEAD FLOTATION

LEAD CONCENTRATE   LEAD TAILINGS

DISC FILTERS   ZINC CONDITIONER

SOD. ISOPROPYL
XANTHATE 50 g/T
NaCN 30 g/T
MIBC / CRESYLIC
ACID 45 g/T
LIME 0.8 Kg/T
CuSO₄ 375 g/T

LEAD CONCENTRATE
TO SMELTER AT
TUNDU

ZINC FLOTATION CELLS

ZINC
CONCENTRATE   FINAL TAILINGS

TO DUMP

DISC FILTERS

ZINC CONCENTRATE YARD
FOR AIR DRYING

PACKED AND DESPATCHED TO
ZINC ELECTROLYTIC PLANT
AT DEBARI

Fig. 18.10. Simplified flowsheet of lead/zinc concentration at Zawar mines.

depends mainly on the amount of lead and zinc sulphide concretions present in the pulp. Selection of galena and sphalerite can be obtained by a number of methods. However, all the methods are based on galena being floated first, with depression of zinc blende. When galena flotation is complete, the sphalerite is activated with copper salts and zinc flotation is carried out. A simplified flowsheet for the ,treatment of lead–zinc ores at Zawar mines is shown in Fig. 18.10.

18.14.3. Concentration of Lead–Zinc–Copper Ores[43,46]

Concentration of lead–zinc–copper ores (containing galena, sphalerite, ane copper sulphide) yields three products, i.e. a lead, a zinc, and a copper concentrate. The ore generally shows remarkably fine dissemination. This type of ore is available in Rajpura Dariba (Rajasthan). A simplified flowsheet of processing Pb–Zn–Cu ore of Rajpura Dariba is shown in Fig. 18.11. The ore is ground to about −70 microns, Zinc blende is depressed with

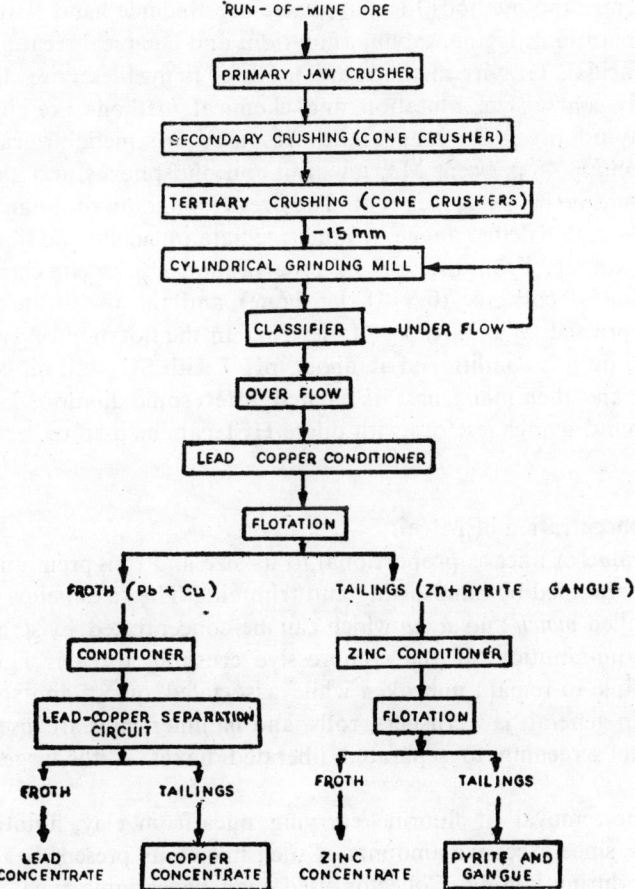

Fig. 18.11. Simplified flowsheet of Pb/Zn/Cu ore concentration at Rajpura Dariba mines.

cyanide and $ZnSO_4$. The collective lead–copper concentrate obtained is then mixed with Na_2S to desorb the collector, thickened, diluted with water and thickened for a second time to remove Na_2S and desorbed collector. The lead and copper are then separated by depressing the copper sulphides with cyanide and floating the lead. The lead concentrate so obtained is cleaned and the tailings which contain depressed copper sulphides, are copper concentrates. The tailings from bulk lead–copper flotation are transferred to basic zinc flotation after mixing with copper sulphate to activate the zinc blende. Lime is added in zinc flotation to depress pyrite. The reagents employed are usually butyl xanthate (0.16 kg/tonne), pine oil (0.09 kg/tonne), terpineol (0.01 kg/tonne), cyanide mixture (2.35 kg/tonne). $ZnSO_4$ (0.12 kg/ tonne), $CuSO_4$ (0.70 kg/tonne), Na_2S (2.0 kg/tonne); activated carbon (0.04 kg/tonne), and considerable amounts of lime.

18.15 Concentration of Manganese Ores[46]

Concentration methods for manganese ores include hand sorting, heavy media separation, jigging, tabling, flotation, and chemical treatment (leaching with acids). Gravity methods are used for manganese ores liberated at sufficiently coarse size. Flotation and chemical methods are applicable to finely ground ores. The manganese ores used for metallurgical purposes should contain 48 per cent Mn, low iron and phosphorus, and alumina plus silica should be below 11 per cent. The ores responding to flotation include (a) having high calcite gangue, (b) intermediate in calcite and silica, and (c) siliceous gangue. From the ores of type (a) and (b), calcite can be floated with oleic acid collector (0.5–1.8 kg/tonne), and the pyrolusite or manganite is depressed by 0.2–1.3 kg of dextrin. In the flotation of type (c), the thickened pulp is conditioned at about pH 7 with SO_2, tall oil is used as a collector, and then manganese is floated after some dilution. Leaching of finely ground manganese ore with dilute HCl can be used to remove phosphorus.

18.16 Concentration of Mica[8]

The value of mica is proportional to its size and thus premium grades of mica are obtained by handsorting and trimming. The sizes below 25 mm or so, are called *punch* and *scrap* which can be concentrated by screening and careful comminution. If the compressive crushing force is light, mica is quite flexible to remain unbroken while associated quartz and spar are detached. In general, jaw crushers, rolls, and hammer mills are used with intermediate screening to separated liberated flakes at the largest possible size.

For the removal of fluorine carrying mica from clay, flotation can be employed, since excessive amounts of such mica will present health hazard problems during kilning. Collector used may be cationic type, such as an amine acetate and the frother may be an alcohol. A pulp of pH up to 11 and moderately dilute is suitable for flotation of mica.

18.17 Concentration of Molybdenum Ores

The principal and most important mineral of molybdenum is molybdenite (MoS_2). The value of molybdenite is sometimes increased by the presence of rhenium (present up to 20 gm/tonne). Other minerals of molybdenum, such as powellite ($CuMoO_4$), wulfenite ($PbMoO_4$) and ferrimolybdite [$Fe_2 (MoO_4)_3 \cdot xH_2O$] are of less importance. The molybdenum ores may be divided into three groups, i.e. (a) purely molybdenum ores, (b) copper–molybdenum ores, and (c) molybdeno-tungsten (sometimes Cu–Mo–W) ores.

18.17.1 Flotation of Purely Molybdenum Ores

These ores contain other sulphide minerals (chalcopyrite, pyrite, etc.) in addition to molybdenite, but the other sulphides do not have any industrial value due to their small amounts and not extracted into separate concentrates. However, in extraction other sulphides should be depressed, since their presence is undesirable.

Molybdenite is one of the most easily floatable minerals (floatation properties are similar to graphite and other non-polar minerals). Thus most of the gangue can be discarded at a coarse grind. Molybdenite can be effectively floated with neutral hydrocarbon oils (kerosene, transformer oils, etc.). The flotation is aided by the use of pine oil as frother. Lime, cyanide, and Na_2S are used to keep down traces of pyrite and chalcopyrite in the cleaning stage at pH 8.3.

18.17.2 Flotation of Molybdenum–Copper Ores[25]

Molybdenite is bulk-floated with copper sulphides and then copper sulphides are depressed floating molybdenite. The bulk concentrate of molybdenum and copper is thickened and dewatered. The concentrate is repulped with fresh water and the pH is brought to 7.5 by adding H_2SO_4. The thiophosphate collector used is removed selectively from copper mineral by sodium ferro-cyanide (this has a short-lived effect). The pulp is slightly conditioned and floated to produce rough concentrate, which is further ground and cleaned. A small amount of Na_2S and NaCN is used to aid the further additions of ferro-cyanide. If xanthate collector is used in bulk flotation, starch may be used as a depressant. Copper minerals can also be depressed by phosphorus, arsenic, or antimony salts.

18.17.3 Concentration of Complex Molybdenum Ore[25]

The Climax Plant (Colorado, U.S.A.) is the largest molybdenum concentration plant in the world. The ore is of granite type, impregnated with quartz. In addition to the valuable molybdenite, other associated valuable minerals are wolframite, hubnerite, cassiterite, monazite, pyrite and chalcopyrite. The average content of Mo is 0.18–0.24 per cent, and the average content of tungsten trioxide is 0.002 per cent. The ore is processed to yield five concentrates, i.e., molybdenum, tin, monazite, tungsten, and pyrite concentrates. A simplified flowsheet of processing is shown in Fig. 18.12.

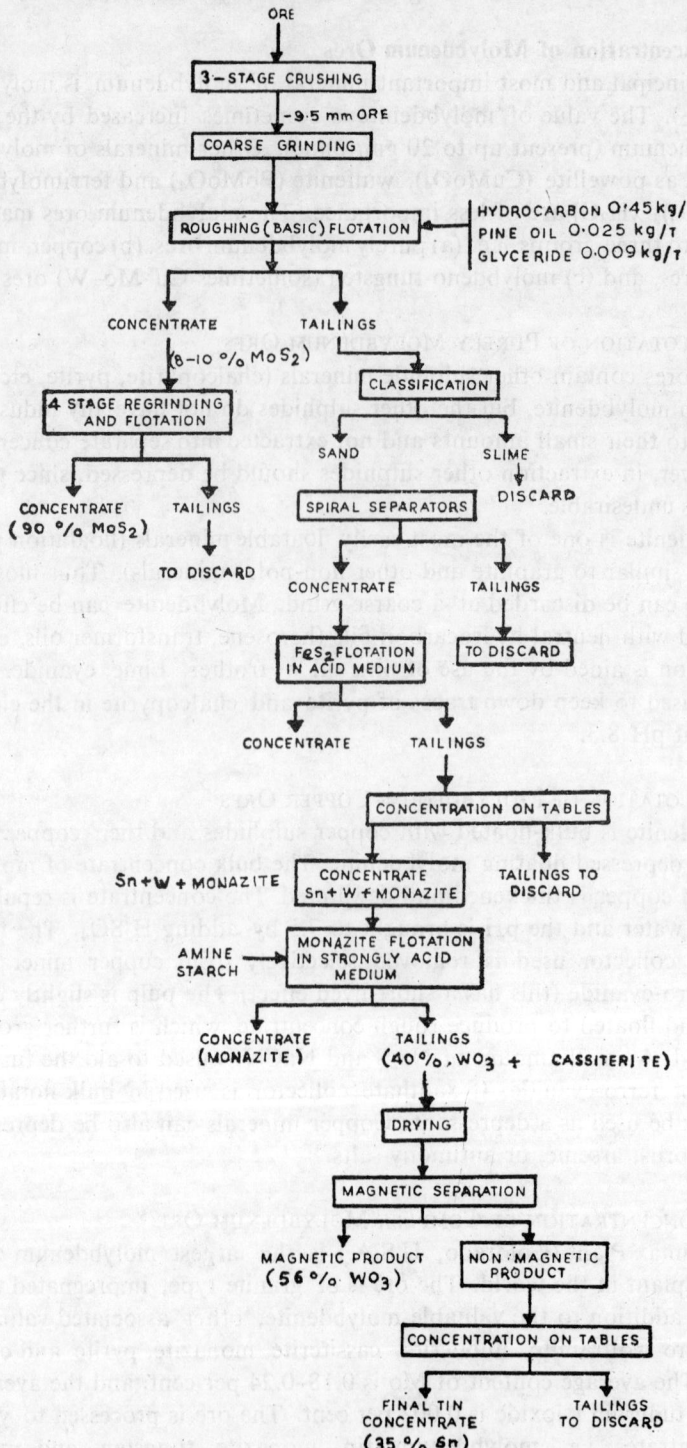

Fig. 18.12. Simplified flowsheet for processing of molybdenum ore at Climax Concentration Plant.

18.18 Processing of Nickel Ores[49]

Nickel ores are of two types, i.e. (a) sulphide ores, and (b) oxide ores. The former type can be concentrated by flotation whereas the latter types are not amenable to be concentrated by any simple method. The oxide ores are directly treated by chemical methods as already described under Ch. 16.

The main source of the world's nickel is the copper–nickel sulphide ores. The principal sulphide is pentlandite ($NiFeS_2$) usually associated with chalcopyrite and iron sulphides. International Nickel Company (Inco) and Sherritt Gordon use bulk copper–nickel float followed by differential flotation. The separation of copper and nickel can also be carried out after smelting the bulk concentrate. A simplified flowsheet for the treatment of nickel ore (carried out by Inco) is shown in Fig. 18.13.

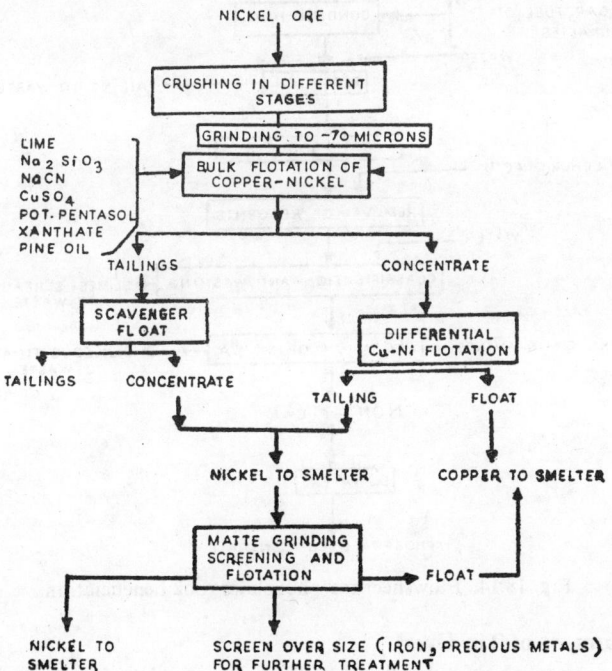

Fig. 18.13. Flowsheet for processing of nickel ores at Inco.

18.19 Beneficiation of Phosphate Rock[8,48]

The term phosphate rock includes chlorapatite ($Ca_5 (PO_4)_3 Cl$), fluorapatite ($Ca_5 (PO_4)_3 F$), and hydroxylapatite ($Ca_5 (PO_4)_3 (OH)$). In India, the rock phosphate is available in many areas, such as Hazaribagh, Singhbhoom, and Dalbhoom in Bihar, Tirichirapalli (Tamil Nadu), Pondicherry, Jamarkatra and Jaisalmer district in Rajasthan, and Mussoorie in Uttar Pradesh. The total reserves of these deposits are estimated to be about 40 m tonnes. Some of the rock phosphates are of a high grade of purity and are suitable as direct feed into the fertiliser plants without any beneficiation.

The mineralogical characteristics, such as minerals present, grain size, texture, and degree of intergrowth, etc., of different deposits vary greatly, thereby requiring use of different flowsheets for their beneficiation. A generalised flowsheet is shown in Fig. 18.14.

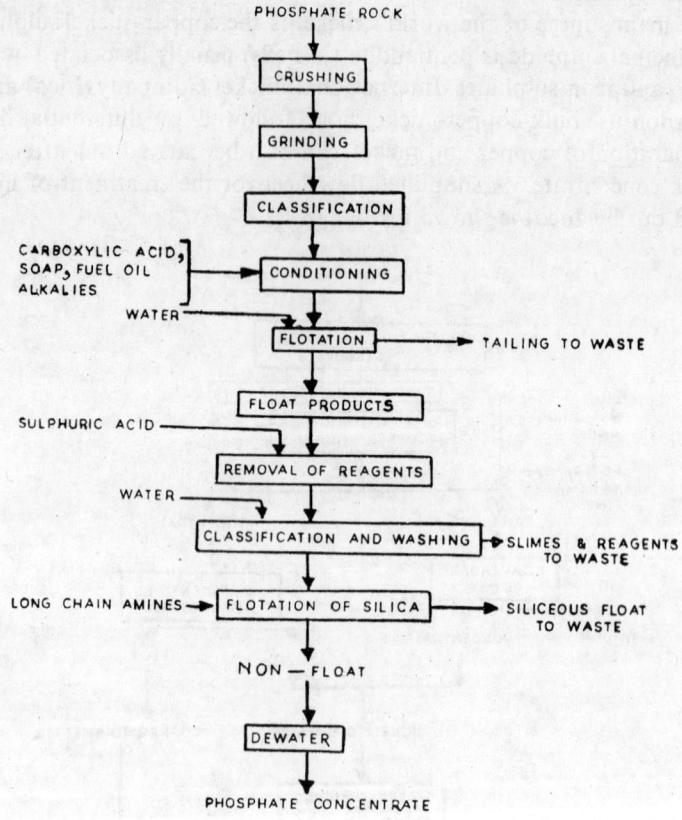

Fig. 18.14. Flowsheet for phosphate rock beneficiation.

18.20 Treatment of Tin Ores[8]

Gravity separation in treatment of cassiterite bearing ores is the most dominant. Where hard-rock mining is used, or a roughed gravity concentrate is to be upgraded after further grinding, overgrinding should be avoided since slimed cassiterite is difficult to recover. The usual methods employed are jigs, tables, spirals, etc., integrated with close screening and classification.

18.21 Processing of Titanium Bearing Beach Sands[44]

Beach sands available in India (Tamil Nadu and Kerala) are rich sources of many industrial minerals such as ilmenite, rutile, zircon, monazite, garnet, and sillimanite. Ilmenite and rutile are conducting minerals, whereas all the others are non-conducting. Similarly, ilmenite, monazite and garnet are magnetic and the remaining non-magnetic. In the recovery of economic minerals from beach sand deposits, no comminution process is needed but all other principal ore beneficiation processes are employed, e.g. screening, gravity separation (wet and dry), electrostatic separation, magnetic separation, flotation, etc. In India, processing of beach sands is carried out at two plants, i.e. Manavalakurichi (Tamil Nadu) and Chavara (Kerala). The flowsheets differ mainly in sequence of operations having the same basic principles. The flowsheet followed at Chavara (Kerala) is shown in Fig. 18.15.

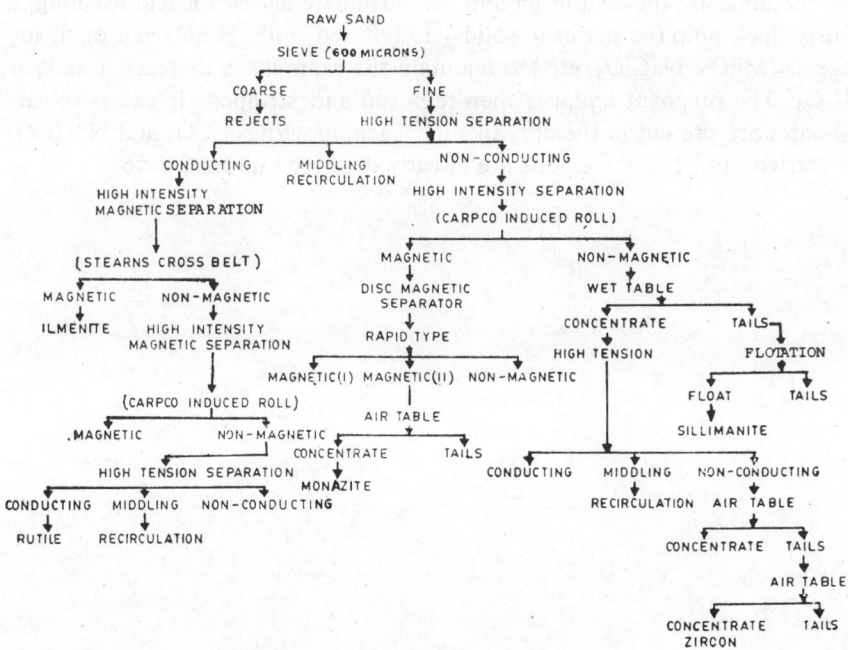

Fig. 18.15. Flowsheet for Chavara beach sands.

18.22 Concentration of Tungsten Ores[8]

The main ores of tungsten are (a) wolframite [(Fe, Mn) WO_4] having a specific gravity of 7.1–7.9 and it is feebly magnetic, and (b) scheelite ($CaWO_4$) having a specific gravity of 5.9–6.1. The dominating processes for concentration of tungsten ores are gravity methods. In gravity treatment, care should be taken to concentrate the value at the coarsest possible size, avoiding over-grinding. Jigs, tables, rag frames, and spirals are commonly used. Scheerlite can be concentrated by flotation also. It floats readily at a pH of 10.5 in softened water, using a carboxyl collector, sodium silicate as dispersant, and tannin as a depressant for associated calcite, fluorite, dolomite, and apatite. However, it is quite difficult to obtain economic grade as the scheelite itself is readily depressed.

18.23 Processing of Uranium Ores[44]

There is a large number of uranium minerals identified in nature, but the most important for economic recovery are the oxides, pitchblende and uraninite (x $UO_2 \cdot y UO_3$). Veins of uranium minerals are rarely of mining width, and usually are networks. This leads to dilution in mining by waste host rock. In general, the deposits are disseminated among other minerals. The uranium ores not amenable to physical treatment, and, thus, the concentration by physical methods is almost negligible. The typical uranium head value may be as low as 0.1 per cent (Indian ores at Jaduguda contain only 0.07 per cent). The ores require chemical processing. The dominant chemical methods are the acid leach and the carbonate leach. In acid leaching, a fairly thick pulp (60 per cent solids) is agitated with H_2SO_4 and oxidising agents (MnO_2, $NaClO_3$, etc.) to maintain the uranium in its reactive state as U_3O_8. The pregnant liquor is then removed and stripped. If excessive carbonates are present in the ore, alkaline leaching with Na_2CO_3 and $NaHCO_3$ is carried out. These methods are already described under Ch. 16.

CHAPTER 19

Sampling, Mill Control and Computer in Ore Processing

In order to obtain relevant informations for mill control and economic appraisal, routine sampling of various products at different stages is essential. The routine check is not limited to concentrates only, but it should cover the check at all possible stages, such as crushing, grinding, pulp preparation, conditioning, etc., and also the other products, such as tailings, middlings, waste water, etc. Further, check should be made whenever any additional control point is introduced. A metallurgical balance sheet should have the basic form as the following:

Units of values
   received in ore = Units of values in concentrate + units of values in transit + units of value in tailings + units of values unaccountable.

In addition to the information about recovery, the tests on samples give other technical informations, such as moisture content, crushed state, oxidation, etc., of the arriving ore, and the ways in which various processes have been conducted.

With the growing complexity of ore treatment, importance of maintaining optimum conditions at various stages of process has increased. These requirements can be met through removal of samples, their test, and application of necessary changes with the help of automatic control circuits. Recently, the use of computers in ore-processing plants has also been made successfully, resulting into fully automatic control of the process. Sampling and process control are associated together. Accuracy of control at various stages has resulted in high efficiency and smoothness of operation. Now it has become possible to have continuous check on important processing factors, and the use of automation resulting in a smooth operation.

19.1 Sampling and Control of Ore and Products at Different Stages of Process

Sampling may be considered as the operation of removing a part convenient in quantity for its tests from a bulk which is much greater in size. The samples should be drawn in such a way that the proportion and distribution of the quality to be tested are the same, in both the bulk and the sample. The reasons for sampling in ore processing can be listed as the following:

a) To obtain the information about the ore entering the plant for its treatment.

b) To inspect the conditions at selected points during the processing of ore, so that comparison can be made between the optimum requirements for efficient treatment and those existing in the process.

c) To disclose recovery and losses and to learn how to improve the recovery and reduce the losses.

Therefore, a mineral engineer should have the knowledge of the reasons for high tailing losses and poor concentrate, which can be obtained by taking the samples and their test. Since, quantity of sample tested is a very small fraction of the tonnage processed, single sample may give errors in tests. Use of appropriate method and sufficient number of samples reduce the overall error within tolerable limits. A sampling error is the divergence of the sampling information from the true value, which arises from defects in taking, handling, reducing, or treating the sample.

Sampling is a statistical technique based on the theory of probability. It should minimise the errors arising from various variables, such as surging, settlement from a pulp, segregation of the sizes in an ore bin, etc. Thus, the samples should be out at proper point and in suitable direction (e.g. in moving stream of material, the sample should be cut at a point of free fall and at right angle to the stream) at such intervals, which give a compromise between possible fluctuations in the quality being checked and problem of reducing a large sample to the size required for tests with sufficient accuracy. Generally, an automatic cutter is preferred, since it eliminates the personal errors of hand-sampling. The size of sample depends on aperture of the collecting device, speed and frequency of cutting the sample. A large sample may be reduced by secondary cutter, i.e. intermediate size reduction.

The sampling and control in the plant are concerned chiefly to assist the operators to maintain the status quo. Any change tried in plant could be fatal in the process line. When there are unaccountably poor or erratic results, some investigational work may be required to know the causes for poor performance.

The use of automatic controls to a process is justified when either it is not possible to control the process manually (e.g. nuclear power generation) or when by their use, economic gain is obtained.

19.1.1 SAMPLING ERRORS

These are differences between the observed or calculated values and the true ones. The errors fall into the following groups:

a) *Accidental errors*: These arise from special circumstances and cannot be predicted. These can be dealt only if they are observed to be occurring.

b) *Average error*: This is arithmetic mean of a series of samplings (plus or minus).

c) *Biassed, constant or systematic error*: This is a series of samplings which is always wrong in the same way, and thus it produces a cumulative error.

d) *Observation error*: This is due to misreading a signal or measuring index, or faulty recording.

e) *Personal error*: This may be random and may cancel out in a carefully observed series.

The adjustment of error is made by an *error band*, which is statistically a range of determined values (assumed to be experimentally valid), inside which the correct value is presumed to lie. A commonly used formula for adjustment of error is the root mean square formula, i.e.

$$V_m = \frac{\pm \sqrt{\sum d^2}}{(n-1)} \tag{19.1}$$

where, V_m = average deviation.

$\sum d$ = sum of the deviation of observed values from the mean value.

n = number of observations.

In statistical methodology, a term used is confidence interval defined as the limits of error of quantity obtainable from given data, when allowances have been made for the known chance variance in the collection of such data. A true value should lie within the confidence limits. The various considerations for the bulk sample are:

a) Representative weight.

b) The size to which the bulk sample should be reduced for a given purpose.

c) Tolerable limits of sampling error.

The solution to the above three problems can be obtained by the knowledge of (i) liberation size of the value, (ii) assay grade when pure, (iii) density of values and gangue, and (iv) maximum size of the particles in material to be sampled.

19.1.2 Sampling and Control at Grinding Circuits[13]

The various points of control in grinding circuit are (a) ball mill discharge (b) classifier returns, and (c) classifier overflow.

a) *Ball Mill Discharge*

The check of solid–liquid ratio is the most important control at this point, which is affected by collecting a canful of mill-discharge pulp at regular intervals and weighing it. If the pulp density is too high, the mill feed water should be increased, and vice versa. Sometimes, a sample is collected during the shift and the size analysis is performed. Sometimes the pH of the pulp is also checked. However, size analysis is performed on the products of classifier. The density of mill discharge is an important factor which affects the coating of balls and linears and thereby affecting the grinding action.

b) *Classifier Returns*

Hand samples are collected from the classifier returns at certain intervals and sizing analysis is performed. This makes a check on the efficiency of oversize returns and undersize released from the closed circuit. A check on quantity of circulating load is also possible.

c) *Classifier Underflow*

The sample of classifier underflow is of utmost importance in checking the efficiency of grinding. The sample should be cut automatically and subjected to assay analysis and sizing analysis. The assay analysis would check for the head value which should check roughly with that of the solid head sample, unless a concentrate is being withdrawn from the closed grinding circuit. The sizing analysis is performed to check the grinding efficiency which is essential, since the concentration efficiency depends on optimum liberation. In addition to the samples taken for assay and sizing analyses, the dip samples are taken regularly at their weir and tested for density and pH value. These running controls are employed to adjust water and alkali to maintain the correct mesh-of-grind and alkalinity of the pulp. The pulp density can be roughly measured with the help of a hydrometer. However, the usual method is to weigh a known volume of pulp. Balances are also available which give direct reading of specific gravity.

19.1.3 SAMPLING AND CONTROL OF PULPS

The process control depends to a certain extent on the testing and control of pulp streams. The best point of withdrawing the samples is at the launder overflow, where the pulp stream is falling freely in a flattish cross-section, e.g. the points where classifier overflow launders deliver the pulp to pump sumps or conditioners. The samples may be withdrawn either by a periodic diversion of the entire pulp stream, or by means of a cutter (a rectangular bottomless box with its length normal to the flow) and diverted to the sample bucket by means of a flexible hose.

The general purpose of sampling the pulp is to provide information, which will help in preparing a metallurgical balance sheet, which accounts for the units of value fed into the mill and those leaving as concentrate and tailings. Some specific informations are also obtained at each sampling point, which checks the state and suitability of pulp for its transfer to the next stage of treatment.

In case of pulps, reduction of the sample is performed very accurately and systematically. The whole sample is first made into suspension by stirring and then poured through a vessel which splits the pulp into two equal halves. One-half is rejected and the other half is split again after stirring. The process of stirring and splitting is repeated to obtain a convenient size for drying and further tests.

Pulp density (solid–liquid ratio) and flow rate are important control factors, since they represent throughput, when integrated together. Thus the pulp density and flow rate should be recorded continuously. The density of pulp having ore reasonably consistent in its specific gravity, can be observed by the absorption of γ-rays by the pulp stream. The results obtained give the accuracy within 2 per cent. In this method a capsule of cobalt-60 is placed on one side of the flow channel and a scintillation counter is placed on the other side. Variations occur depending on the amount of radiations absorbed by the solids in the flowing pulp. For the measurement of flow rate of pulp, a magnetic flowmeter can be used. The flowmeter is allowed to spin in the pulp at a point surrounded by a magnetic field. The induced voltage registered by immersed probe will give the flow rate.

19.1.4 SAMPLING AND CONTROL OF FLOTATION PRODUCTS[8,13]

In flotation line, several products are removed successively or diverted to other units and thus samples of each concentrate, rougher froth, and scavenger froth should be taken out. Sampling of pulp flowing through intermediate stages may also be necessary. However, the most important stage is the sampling of each final concentrate, which should be sampled periodically. The samples withdrawn over a certain period (e.g. one shift or one day) are mixed and assayed for the grade of the concentrate (total sulphur, metal values, and any other element of interest). Generally a simple assay for metal grade will be sufficient. A small quantity of concentrate may be retained from each shift to form a composite sample, which may be assayed for complete analysis at weekly or monthly intervals. In this case, a portion may be sized and assaying for grade may be carried out on each screen and subsieve size fractions. These informations on grade and size analysis are used to assess the work being done in the grinding circuit and the degree of liberation of each value.

The grade analysis of partially cleaned concentrates is performed only periodically on accumulated bulk samples. However, when the alterations in cells or circuits are planned, the analysis should be carried out on regular samples until the process is established. The routine tests of rough concentrates include their sizing analysis and microscopic examination. These give direct informations, which can be applied in maintaining the efficiency, or to improve any weak detail of the treatment.

The tailings from flotation circuits are tested for their grade on each shift's cumulative sample. The pH may be checked on each cut of sample to provide a cross-check on conditioning. The density of the pulp is checked at overflow from cells, which is generally required when over-dilution of the pulp lowers the flotation efficiency. This situation may arise when a heavy draft of concentrate is made or when the tailings from a series of cleaning sections dilute the new feed excessively. The samples are tested in the similar way as concentrate samples. In case of tailings also, a suitable quantity of each sample is retained and mixed to form a weekly or monthly sample, which is tested for size analysis (by screening and sedimentation down to 15 microns). Then each size fraction is assayed for the percentage of floatable product. Heavy and light fractions of each size are separated by panning and examined under the microscope to evaluate the reason for their failure to float, and thereby applying necessary corrections.

19.1.5 Sampling and Control in Heavy Media Separation[13]

The heavy media employed in separation are generally generated continuously. Sampling may be used to control the following factors:

a) Optimum viscosity.
b) Dilution of medium by ore slimes.
c) Rate of settlement of suspended heavy media particles.
d) Density of heavy media bath.

The density of media is a function of fluid density, viscosity of fluid, and the size-blend of the particles used to constitute the medium. During the separation, the particles become less angular due to wear, and the loss of weight is compensated to some extent by decrease in rubbing surface.

For tests and control, the samples are withdrawn periodically and tested for density, rate of settlement, and viscosity.

19.1.6 Sampling and Control in Cyanidation Products

The pulp samples are collected by cutting through the whole stream at an appropriate overflow point. Solutions may be sampled continuously by diversion of a steady drip or by drawing-off periodically. The testing of pulp sample is mainly concerned with losses of values in the tailings. The various reasons for the loss of gold are:

a) Gold losses may result due to incomplete washing of the discard, and allowing the pregnant solution to run to waste. This is checked by washing the filter cake and testing the filtrate for gold content.

b) Gold loss may occur due to formation of iron or aluminium oxide coating on gold particles, which prevents dissolution of gold.

c) Gold may be lost due to incomplete liberation, particularly that carried in auriferous pyrite.

d) Gold may not be completely dissolved due to short time given in agitators.

e) It may also be due to failure of the gravity and amalgamation sections to trap and hold coarse particles of gold.

f) Presence of graphite may also affect the recovery, and thus its presence should be watched.

The testing of cyanide solution is required to know about the following factors:

i) Oxygen content of the solution.
ii) Presence of fouling salts.
iii) Strength of the available cyanide.
iv) Adequacy of the protective alkali.
v) Efficiency of precipitation.

In order to assess the day's recovery accurately, the amounts of pulp, pregnant cyanide solution, stock solution, etc., held up in the plant should be known.

19.2 Automatic Measurements and Instrumentation[3,13,16]

19.2.1 AUTOMATIC MINERALOGICAL MEASUREMENTS AND LIBERATION ANALYSIS IN MINERAL PROCESSING

Earlier, a simple chemical analysis was considered to be sufficient to characterise a rock or an ore. However, at present, due to increased complexity and decreased grade of ores, sufficient mineralogical information is needed for the appropriate design of beneficiation processes. This mineralogical information is also important in optimising the existing ore processing operations. The various informations needed are mineral structure, mineral composition, particle and grain size distribution, particle shapes, textural characteristics, other properties of minerals (density, thermal, magnetic, etc.), variation of mineral properties with depth in deposit and time, and likewise other informations. This subject is already discussed in detail in Ch. 17.

In recent years, an automatic image analysis has been applied successfully for mineralogical measurements. The measurements are made on two-dimensional sections (thin or polished) by a method analogous to the traditional manual microscopic methods. An ideal automatic image analysis system would distinguish any particular mineral in a group of other minerals, with a spatial resolution of 1 micron or less. The image analysis devices suitable to meet many requirements are now available commercially. These devices can be mainly divided into two groups, i.e. (a) area measuring, and (b) line measuring devices. Most area measuring systems are based on television scanning techniques. An image of a specimen is viewed either under a beam of light or under an electron beam.

In mineral processing, the objective of most operations is the separation of minerals from the mixture of composite mineral particles. In grinding (an essential operation before beneficiation), the particles usually do not break at the boundary between mineral phases and thus composite (middlings), as well as liberated particles are produced. Therefore, it is of utmost importance to determine the proportion of liberated particles as well as the size distribution of minerals associated in composite minerals. Automatic image analysers are designed to achieve a quantitative characteristics of the geometrical properties as well as the relative relationships between the various constituents. The optical signals are used as the source of data processing. An image of sample observed under a microscope is formed in a television camera and is transferred for analysis and processing to various models. The apparent brightness is a function of various minerals in the sample. The image is decomposed automatically to a number of grey levels between the brightest level and pure black. In practice, the solution of a given mineral corresponds to a discrimination between these grey levels. For example, in the instrument Quantimet 720 of Imanco, there are 64 grey levels.

19.2.2 MEASUREMENTS AND INSTRUMENTATION IN ORE PREPARATIONS FOR FLOTATION

In flotation process, the cost of flotation reagents is a major factor in operating cost. Therefore, it is essential that the design and the system for preparation and distribution of reagents should be given the same importance as that given to the rest of the plant. The modern automatic analytical instruments of the titrator and absorptiometer type are capable of preparing reagent solutions automatically, while effective distribution and control of reagent addition can be exercised by automatic valves, positive displacement pumps and flowmeters. The efficiency of flotation and economic reagent utilisation mainly depend on the control of reagent addition in accordance with changes in the flotation feed, both in quality and quantity. This type of control can be achieved by suitable instrumentation analytical techniques.

The efficiency of flotation process is influenced by a large number of factors, and many of them interact and make the control of the process extremely difficult. In most plants, maintenance of constant pulp density will require variation in flow rate, contact time, froth level, and reagent addition rate. The contact time and froth level can be maintained by suitable automatic adjustment of inter-cell and overflow weirs. The rate of reagent addition is controlled in proportion to the rate of pulp flow. However, this arrangement does not take into account the change in grade of ores. Therefore, a means of detecting changes in grade is required before exercising the control. This problem may be solved by actual measurement of pulp feed grade by X-ray fluorescence techniques and adjusting the rate of reagent addition accordingly. Another possible solution is to measure the residual concentration of the reagents in the flotation pulp, since for an efficient flotation only a small excess of the reagent above the optimum is permissible.

19.2.3 AUTOMATIC WEIGHING ON CONVEYOR BELTS[3]

Conveyor-flowmeters can be employed for dry materials operating on the principle of weighing a section of the conveyor as shown in Fig. 19.1.

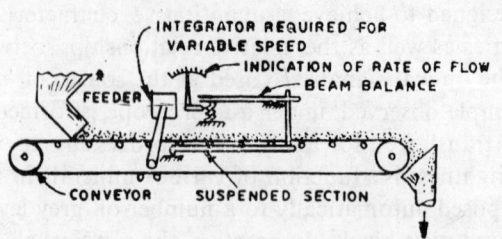

Fig. 19.1. Conveyor weighing for measuring flow of dry materials.

In this case, use of a spring balance is made and the flow is measured in weight units (tonnes/hr). If the conveyor belt is adequately flexible, the accuracy of measurement is usually within 1 per cent and sometimes 0.5 per cent. Integrators are used for determining total flow, and most such meters are compensated for the speed of conveyor. Other conveyor meters employ the automatic weighing scales of either the electric or pneumatic type, in place of spring balance shown in Fig. 19.1.

19.24 Pulp Density Recorder

A special pulp density recorder developed at Union Corporation, South Africa, is shown in Fig. 19.2. Two 50 mm steel tubes of equal length, each notched at a different depth below the pulp surface, are immersed in a rising stream of the pulp. The space between the notches is determined by the pulp density and the range of the recorder to be used. In each of the tubes, a column of water builds up to such a height that it just balances the pulp column above the notch. In each of the 50 mm steel tubes, bubbler tubes are placed with their open ends at the same level.

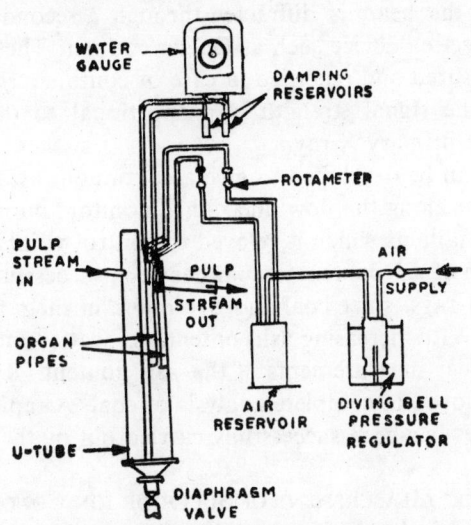

Fig. 19.2. Pulp density recorder.

In this instrument, when water alone is flowing, it will give a zero differential, and thus zero of instrument corresponds to specific gravity of 1.0. The bubbler tubes remain submerged in clear water, rather than pulp, and thus the build-up of pulp on them will be negligible even after a long time (several months). It has been found advantageous to supply a small drip of water (about 100 ml/min) to each of the 50 mm steel tubes.

The vessel in which the pulp density measuring element is placed, has the shape of a U-tube. The pulp enters the down-coming leg tangentially which assists the liberation of entrained air. The bend at the base of U-tube is drawn down to a 150 mm opening which is closed by an air operated valve, which operates automatically. This valve also serves as a convenient means to flush out the accumulated oversize material at regular intervals.

The general maintenance includes winding of the clock, inking of pen and changing of chart. Occasional adjustment of the rotameter (this supplies air to bubbler tubes) may be required. The unit should be cleaned after every six months.

19.2.5 Continuous Analysis of Pulp by X-ray Fluorescence

This technique of analysis is becoming popular for rapid and continuous analysis linked with plant control. This is based on a high intensity source of X-ray radiation or the use of radioactive isotopes. A simple system consists of an X-ray tube which directs the radiation on the pulp sample (static powder, or flowing pulp). The atoms in the sample are excited by the primary X-rays and emit secondary X-rays (fluorescent radiation) at characteristic wavelengths of the elements thus excited. This fluorescent beam then passes through a slit (collimator) adjusted in such a way that a selected band of radiation is directed upon an analysis crystal (LiF crystal). From the analysis crystal, the beam is diffracted through a second collimator and is scanned by a detecting device such as Geiger counter. The fluorescent energy can thus be measured and recorded in case of continuously flowing pulp on a strip chart. The signal strength is proportional to the selected atoms irradiated by the primary X-ray.

The system can be developed to give a continuous assay of flowing pulp at different points along the flow line. This monitors information on feed, concentrate and tailings which is relayed to control points.

Detemination of ash content in coal has been successfully done by means of either γ- or X-rays, since coal and ash differ in their reaction to short-wave radiation. With increasing ash content, the scatter diminishes sufficiently to permit direct measurements of the ash content. Radioactive source can also be employed in complete analysis of coal. A rapid analysis of lead and tungsten ores has been successfully carried out by the use of isotopes.

19.2.6 Automatic Measurement of Sulphide Ion Concentration

The high chemical activity of sulphide ions with respect to silver and extremely low value of the solubility product of silver sulphide (6.3×10^{-50}) ensure the preferential electrochemical reactions between the electrode and sulphide ions. This system also eliminates the influence of other anions on the measurement result and creates the necessary condition for the self-activation (self-sulphidisation) of the silver sulphide electrode in the solution being controlled (flotation pulp).

A standard calomel electrode is used as a comparative electrode along with silver sulphide electrode for determining the potential of the latter in the couple. This difference in the potentials of this electrode couple is measured with an automatic electronic potentiometer having a pneumatic output for controlling the pneumatic executive mechanism of the reagent feeder.

The silver sulphide electrode is placed directly in the flotation chamber. Depending on the e.m.f. of electrode, automatic potentiometer acts by means of a pneumatic executive mechanism of reagent feeder supplying sodium sulphide to conditioning tank.

19.3 Automatic Control in Mineral-processing Plants

With the growing complexity of ore treatment, rising cost of labour, power and material, it has become essential to maintain optimum conditions at key process stages. The optimisation of various operations in flow-sheet requires the use of automatic control which results in reduction in operating costs. Therefore, the aim of mechanisation and automation in ore-processing plant includes (a) to raise the productivity of labour, (b) to raise the productivity of process equipment, (c) to decrease the overall cost of ore processing, (d) to make the workers free from laborious operations, and (e) to maintain the highest technological and techno-economic performance of the plant.

Presently, most ore-processing plants are equipped with various devices at critically important points, either to regulate a process or to give automatic warning when something goes wrong.

The planning of automatic control can successfully be exercised, if the whole chain of operations is divided into sections, and the minimum number of control factors essential for an efficient working of each section is known. In principle, no ore should leave a section for the next in line, until it is in the correct condition to serve as feed to that section. In order to achieve the correct conditions in each section, a suitable automatic control system should be used.

Mechanisation and automation can be achieved in the following two main directions:

i) Automatically regulating the process with measuring and recording instruments.

ii) Stabilising the grade of raw and auxiliary materials and ensuring the uniform and continuous working of processes under perfect and strictly constant conditions.

In the first case, the ore processing plant requires fitting of large number of monitoring, measuring, and control devices along with other means of automation, which may be from simplest signalling units to complex computers and programming machines.

The second case includes the use of highly mechanised operations of raw material averaging, use of auxiliary materials of consistent grade (i.e. grinding media, reagents, water, etc.), and the use of durable and productive equipments, with the process being controlled on a centralised basis.

In order to employ complex mechanisation and automation to ore processing plants, adequate research and experimental designing work should be carried out. In general, this work concerns the following problems:

a) Investigation and study of various parameters such as statics and dynamics of processes.

b) Study of processes as objects of automation.

c) Determination of regularities of the processes.

d) Formulation of laws (algorithms) for the control of processes.

e) Development of transmitters, measuring and control devices, and auxiliary mechanisms suitable for use in ore-processing plants.

f) Development of new milling, concentration, and auxiliary equipments for ore-processing plants which satisfies complex mechanisation and automation raquirements.

g) Working out new grouping of solutions during the planning of plants, which facilitate the mechanisation of the main and auxiliary operations in them.

The various operations of an automatic control system include the following, which are performed in order:

a) Detection of change.

b) Transmission of warning signal.

c) Effective indication of variation.

d) Actuation of correcting mechanism.

There may be a time-lag network coupled with stage correction to prevent over-correction. For example, if pH is being measured in a conditioner 15 min beyond the point of control for addition, and the addition of lime is automatically adjusted with each measurement. In such a case, there should be a time-lag of at least 15 min between successive measurements, so that the changed pH of the pulp will be measured at the next measurement.

The automation can be further improved by incorporating the following measures for various adjustments:

i) Automatic analysers for the granulometric composition of the products from crushing or grinding section.

ii) Automatic and continuous analysers for the metal values in ores and products.

iii) Computer-information and controlling machines including systems of optimisation.

iv) Industrial television and other new engineering techniques.

Automatic control devices can monitor a detail in a process and can either give appropriate warning of variation, keep a record, or take necessary steps to correct a deviation from normal working.

Presently, it is possible to employ automatic control system in most of the processes in ore-processing plants. The important fields include feeding and blending, comminution and classification, gravity separation, chemical treatment, froth flotation, electromagnetism, etc.

19.3.1 PROTECTING CRUSHERS FROM TRAMP METAL IMPURITIES

Due to the metallic impurities caught in intermediate crusher, the stand-still time may account for 15–20 per cent of the total stand-still time. Several types of metal detectors have been developed to detect metallic objects in ore entering the crushers. A scheme for automatic removal of metallic impurities from the ore is shown in Fig. 19.3. If a ferromagnetic object comes on the conveyor, it is first detected as the belt moves along, by first metal detector, which gives impulse to the forced excitation of the windings of iron separator. The iron separator pulls the object and dumps it into the special bin. The second metal detector is a control device. If for any reason, the metallic object is not extracted by the iron separator, the motor of conveyor is switched off by means of the second metal detector.

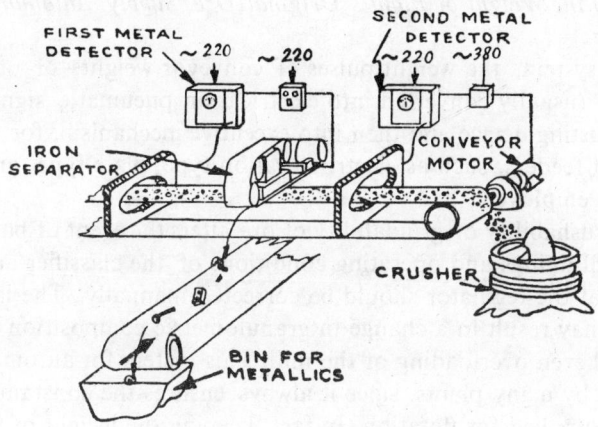

Fig. 19.3. Scheme for automatic removal of metallics.

19.3.2 AUTOMATION OF COMMINUTION PLANT

In comminution plants, the automatic control can deal with various matters, such as sequential starting and stopping, temperature rise in bearings, and loss of oil pressure. The transfer points can be scanned either by closed-circuit television or with a photoelectric cell. This receives a light beam arranged so as to be cut-off when ore piles up or conversely to emit a signal to stop the empty running conveyor belt. The various systems of crushing plant automation can be put into following three groups:

i) Systems of automatic stabilisation without any allowances for variation in the properties of ore and crushing conditions.

ii) Systems of automatic stabilisation relative to variations in the properties of ore and crushing conditions.

iii) Extreme systems for optimisation.

The first two groups take into account that optimum conditions (indices) have been attained by the stabilisation of different parameters of the process or their different combinations. In the third group, the aim is of seeking the possibility for the optimum plant operating conditions. The various factors affecting the choice of a particular system of automatic control are as the following:

i) Properties of ore.

ii) Technological scheme for treatment.
iii) The design of crushing and classifying equipment.
iv) Requirement of crushed product in respect of granulometric composition.
v) Permissible limits of volume variation of pulp.
vi) Economic considerations.

The various systems for automatic control in crushing plant are discussed below:

a) *Based on the Weight of Plant's Original Ore Supply Automatically Kept Constant*

In this system, the weight pulses of conveyor weights or other weight transmitters (usually converted into electrical or pneumatic signals) are fed into an adjusting device and then into executive mechanism, for which various types of feeders, such as electrovibration type, plate type, or pendulum type can be employed.

If the crushability or grindability of ore alters the event of ball charging, or if the mill linings and operating conditions of the classifier are changed, the setting of the regulator should be corrected manually. The lack of such correction may result in a change in granulometric composition of the final product and even overloading of the mill. This system for automatic control is preferred by many plants, since it always ensures the constant volume of pulp being supplied for flotation. In fact, keeping the weight of the original supply constant, while the properties of ore and crushing conditions change, the quality and the quantity of crushed product may change.

b) *System Based on Set of Mill Noise*

There is functional relation between the crushability of ore, and the mill productivity and noise. Many systems have been proposed for automatic regulation of input of ore into mills in relation to noise. These systems make use of acoustic regulators. A scheme for such a system is shown in Fig. 19.4.

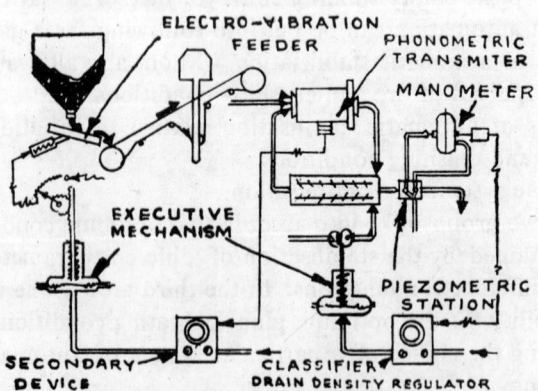

Fig. 19.4. Scheme for automatic regulation of changing a crushing plant based on set of mill noise.

The transmitter of this control system is a powerful loudspeaker mounted very near to the mill's drum (grinding mill), in the zone where the balls are falling. The transmitter produces an electrical signal which is proportional to the intensity of noise in the mill. This is picked up and averaged, and sent into the input of electronic potentiometer fitted with adjusting device. The command pulses from the electronic potentiometer actuate the executive mechanism controlling the productivity of feeder. The system does not require any circuit amplifiers.

c) *Systems Based on the Sum of Two Parameters*

The circulating load is a sensitive index of the performance of a comminution plant and has its optimum value, which results in maximum productivity. Thus a system based on controlling the constancy of the sum of two parameters, i.e. the weight of the original supply and the circulating load has been proposed. In this system,

$$Q + KS = \text{constant}$$

where Q = weight of original mill supply,
  S = weight of the circulating load in the classifer,
and K = factor expressing the ratio of original supply to the circulating load at an average crushability/grindability of the ore.

The system is suitable in those plants, where the original ore shows considerable variations in properties. The regulation system in such a case can be represented as shown in Fig. 19.5. The tonnage of the ore being transported to mill is perceived by the conveyor weight transmitter and recorded by a secondary device. The output of the secondary device is connected with one of the outputs of summation instrument, i.e. the regulator of the supply productivity. The output signal of regulator is proportional to the sum of the pulses being put into it. The regulator is connected to an executive mechanism, which alters the productivity of feeder. The weight of ore entering the mill is a function of the grindability of ore, because, for the ores difficult to crush/grind, the circulating load of classifier increases and

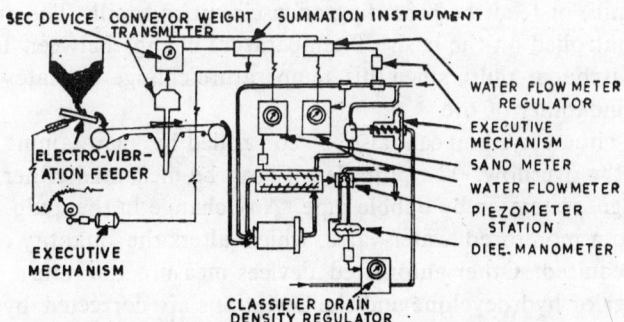

Fig. 19.5. Scheme for automatic regulation of charging a crushing plant - based on the sum of two parameters.

the regulator orders a certain decrease in the tonnage of the original supply. On the other hand, in case of readily grindable ores, the regulator orders a certain increase in the original supply.

This system can also work on the basis of stabilisation of the tonnage of the original supply. For this, the circuit of circulating load charge is switched off.

19.3.3 AUTOMATIC CONTROL IN WET GRINDING

In this case several types of signals can be used. These include changes in the following variables:

a) Grinding noise level in the ball/rod mill.
b) Circulating load in mill–classifier system.
c) Power draft to mill.
d) Pump loading.
e) Pulp density of mill discharge or classifier overflow.
f) Temperature rise through mill.
g) Feed water to mill or classifier pool.

Noise level can be controlled by means of a microphone placed close to the mill (as discussed in earlier section 19.3.2). The signal is used to vary the rate of new feed when the sound intensity deviates from the pre-set level.

The power draft to the mill should be maintained at its maximum. Decrease in power draft may result due to underload and overload of ore, wrong charge of grinding media, wear of liner, or incorrect solid–liquid ratio. The main use of power monitoring is to apply correction in ore loading.

The deviation in ratio of solid–liquid in the mill charge can be monitored by the use of radioactive source beamed on the mill discharge, or by semi-continuous weighing devices which check the density of discharge. The signal initiates any change necessary in the volume of head water-added, and in a flowmeter monitoring this water. It then signals any further adjustment to the feeder supplying new ore. This new ore may either enter the closed circuit (ball mill) or feed the preceding open-circuit rod mill. The process can also be controlled on the basis of temperature change between feed water and mill discharge pulp, since this temperature change indicates the variations in grindability of ore.

Closed-circuit system can also be controlled by maintaining the pulp density in the overflow. The pulp density can be measured either by radioactive gauge system or by bubble pipe. Any change in the pulp density is signalled to a monitored water-valve, which alters the quantity of diluting water as required. Other automated devices measure the return load from the classifier or hydrocyclone and the deviations are corrected by the regulation of new feed. The vacuum in the vortex of the hydrocyclone can be probed and monitored to control the return load from the classifier.

19.3.4 AUTOMATIC CONTROL IN FEEDING AND BLENDING

For an efficient smooth running of the process, it is necessary to have the control over the rate and quality of the process feed. In addition to weightometer described earlier (Ch. 3), some electronic sensing devices may be used to improve the accuracy. In one blending system, the material A is weighed continuously and sampled, as it passes from its feed belt to the blending belt. The weight is signalled to the sampling device and to a memory unit, which sends a delayed signal to the feeder delivering material B on to the blending belt. The feed rate is adjusted in accordance with the sampling information.

19.3.5 AUTOMATIC CONTROL IN THICKENING AND FILTRATION

Where continuous thickening is coupled with periodic filtration and a storage capacity inside the thickener is required, automation has been successfully employed to raise or lower the rakes in accordance with changes in torque signalled to a motor which lifts or lowers the raking mechanism.

19.3.6 AUTOMATIC CONTROL IN GRAVITY SEPARATION

In heavy media separation, the first step of automatic control is the continuous recording of the density of the bath media. The control of density is used in various systems to control changes in the use of diluting water and the return rate of cleaned medium (ferro-silicon or magnetite) from the densifier. Control of water provides an intermediate response, whereas trimming of densifier rakes (to increase or decrease the circulating load of solids) makes the adjustment of density more slowly. In one system, four zones of separating bath are monitored by density gauges. Another system provides pneumatic controls worked by signals originating from the return dense media. This monitors density as well as pressure and thus it affords a control level for height of both. In one system developed for the separation of diamonds, bath density is controlled by the differential pressures between two bubble pipes, and the viscosity is continuously monitored.

Efficient desliming of a finely ground pulp in a hydro-separator depends on the hydraulic current, i.e. volume of water admitted. The volume of water being admitted may be controlled by means of a magnetic coil, which senses the interface between a heavier magnetic-rich pulp and the supernatant silicious middlings overflowing to waste. Signals from the coils to a pulsed timer adjust the main water valve and maintain the interface at the optimum height in the hydro-separator. A sudden change in the ore causes the valve to remain in its open or closed position for much longer time, and the speed of the pump removing the underflow from the separator is varied automatically, until the condition returns to normal which is signalled by the return of the control water valve.

19.3.7 AUTOMATIC CONTROL IN ELECTROMAGNETISM

It may be used to control the permissible magnetite in an ilmenite concentrate produced by flotation. The success of control depends on close watch on the magnetite in the tailings from each primary magnetic separator. In the control system, the sample flows through a primary a.c. field and the magnitude of the induced secondary current in the pulp, is measured. If it rises above the permissible limit (e.g. 0.1 per cent), corrective action is taken. There may be a lag of up to 2 hr between this point and the flotation from which the escaped magnetite would be concentrated considerably to render the ilmenite product of poor grade.

19.3.8 AUTOMATIC CONTROL IN FLOTATION CIRCUITS

Automatic control in flotation circuits is mainly to automate the supply of flotation reagents to achieve a required metallurgical result. The various parameters that can be controlled, include pH, modifiers, collectors, Na_2S, $CuSO_4$, $NaCN$, etc., additions of which depend on the ore being floated. Monitoring of E_h, oxygen concentration and other parameters may also be helpful. The level of parameter may vary from circuit to circuit, and the values of parameters should be based on practical experience. The automation of reagents supply can be achieved in the following two directions:

a) Automation of the reagents supply according to the tonnage of the ore being processed or according to the proportion of solids in pulp.

b) Automation of the reagents supply according to the ionic concentration of reagents in the pulp.

For the first direction, special induction attachments are built into the conveyor weights, pulp flowmeters and pulp-solid flowmeters. The pulp discharge volume can be measured by electromagnetic flowmeters from the rate of flow of pulp in full pipe. The rate of pulp will be indicated by the emf arising between two electrodes, as the pulp moves between them in a magnetic field. The emf is measured on a compensative circuit by means of an electronic unit to which a secondary measuring instrument is connected.

By the combination of electromagnetic flowmeter, pulp density meter, and adjusting device, it becomes possible to control the supply of reagents according to the proportion of solid in the pulp. The amount of solid in the pulp is also measured with a piezometric flowmeter. A control circuit using piezometric flowmeter for the amount of solids in pulp is shown in Fig. 19.6.

The pulp passes under gravity through a special flowmeter cone and flows over its ring baffle and goes to process. Piezometric tube is mounted in the flowmeter cone. Compressed air is passed into piezometric tube through a filter, reductor, and rotameter at a pressure of about 400 mm Hg. A pressure (P) in the tube will develop depending on the pulp density (ρ) and the depth of overflow (h), i.e. $P = \rho h$. This pressure is measured by means of a differential suction gauge relative to the pressure in piezometric tube and transmitted kinetically to differential transformer coil. The electric signal of transformer coil is measured by the secondary meter which has a pneumatic adjusting device. This controls the performance of reagent feeder by means

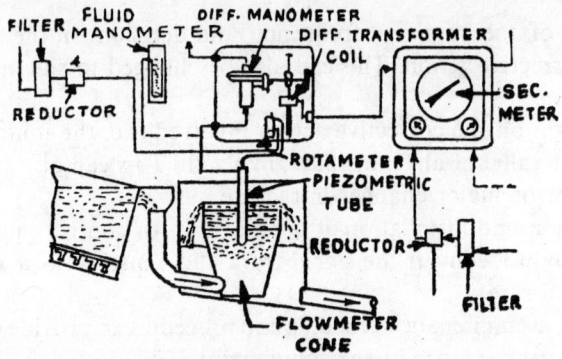

Fig. 19.6. A control circuit using piezometric flowmeter for
the control of pulp density.

of compressed air sent through filter and reductor, with a pneumatic executive mechanism which modifies the reagents supply to the process in accordance with the change in quantity of solid passing through flowmeter cone.

Some electrical devices are also developed, which regulate the reagents supply by altering the number of revolutions of the driving shaft by means of an electromagnetic clutch.

For the second direction, automatic pH meters are employed. The potentiometric method can also be used to measure the residual sulphide-ion concentration in the pulp with silver sulphide electrode. The supply of reagents, creating alkalinity (soda, lime, etc.) can be automated in accordance to pH.

An effective automatic control in a flotation circuit requires the integration of following aspects:

a) Development of suitable grinding, conditioning and flotation circuit capable of producing a suitable feed size, allowing adequate time for reactions to occur, and yielding an acceptable metallurgical result. The plant circuit cannot readily be changed at a short interval and only limited changes can be made in its operation.

b) Decision on chemical parameters to be controlled automatically, chemical parameters to be left to find their own level with fixed reagent addition, and chemical parameters to be ignored. Decision should also be taken as to whether feed forward or feedback control is to be used. These aspects can be changed at short notice with various degrees of ease and expense.

c) Decision on the sensing device or technique to be used for direct measurement or indirect measurement of each parameter. This can also be changed at short notice.

d) Decision on the tolerance and level of control for each parameter. This can be changed to suit a particular ore.

e) Decision on the location points of sensing device and for taking samples. This can also be changed to suit a particular ore.

f) Fixing of measurement frequency and decision on the real time frequency of corrective action. This can also be changed to suit minor or major changes in ore.

g) Decision on the corrective action involved (i.e. the control algorithm) and the exact value of any constants involved. This can also be changed to suit the minor or major changes in the ore.

h) Selection and installation of specific units of control equipment which are to be used to convert the signal from the sensor into a change in reagent addition rate.

i) Laying a maintenance schedule and procedure to provide the consistent and effective performance of the equipment.

j) Ensuring the acceptance and understanding of the system to the operator.

The measurement frequency may vary from continuous output to one every 15 min. Similarly, the need for corrective action to reagent addition rates may vary from once per minute to once per 30 min or so. Many types of control algorithms have been proposed. Trials are necessary to select the best suitable algorithm for a given circuit.

19.3.9 AUTOMATIC CONTROL IN CHEMICAL EXTRACTION

Automation of leaching circuits takes many forms. Control of pH in relation with maintenance of protective alkali in cyanidation of gold has been successfully adopted by use of a conductivity probe to control the addition of lime. A large-scale automatic control system includes indicators, and remote-control devices. The control can be used to check corrosion, thickening, blockage, and sequence of manipulations in circuits with an assembly of closely connected flow lines.

19.4 Computers in Mineral Processing[13,16,19]

Computers are increasingly used in ore-processing plants to receive, coordinate, and integrate the monitoring signals which are initiated at key points in the processing line. The computers then act on the combined information and ensure that the correction is made in respect of overall operation and not simply in response to a single variation. Computer-integrated corrections can be signalled to manual control points or applied directly to make the adjustment automatically. The present trend is towards increased operating control by computer, which is designed to maintain optimum conditions of the process. Computer can be employed for the following three types of work:

a) Simple recurrent problems.

b) Complex development problems.

c) Process control.

19.4.1 TYPES OF COMPUTERS

The computers employed are of two types, i.e. (a) digital computer, and (b) analog computer.

a) *Digital Computer*

This is a high-speed adding machine using numbers rather than physical quantities in processing of data. The control unit of digital computer determines the operating sequence of calculations to be made by calculating unit. The interal storage of the computer is called the memory unit, which holds the information required for job in hand, in the form of machine language (symbols, signs, processing rules, etc.). The digital control is only concerned with a step in decrease or increase in signal and thus would not work continuously to rebalance the flow line.

b) *Analog Computer*

This works by setting up a mathematical analogy of the problem in hand. Analog control is proportional to the dimension or function monitored. It uses signalled information (language of automation) to represent the numerical variables in a computation. Thus analog control can compare the pH of the pulp at selected points with the operating norms required, and initiate the mutiple adjustment needed to restore deviation smoothly.

The operational-digital technique of computer control combines the two methods, i.e. operational analog control and programmed digital control. The process controller provides feedback control which is based on input signals, coordination and adjusting signals.

19.4.2 APPLICATIONS OF COMPUTERS IN ORE PROCESSING

Computer control has been superimposed successfully on many automatically controlled processes. The simplest application of automation corrects only one variant. When several conditions react on quality of work and their automated sensing devices are to be coordinated, use of computer is required. There are two approaches in application of computer, i.e. (a) functional or black box approach, and (b) operational approach. In the former case, the system is not concerned with the mechanism of operation but with what is being done. The latter approach is concerned with the way in which things are being done.

In ore-processing plants, the language of automation is in the form of electrical signals, shaft rotation, variation of pneumatic or hydraulic pressure, change of voltage, current, magnetic flux, temperature change, change in noise level, density, pH, flow rate, metering of light beams, etc.

19.4.3 TYPES OF CONTROLS USED

There are two known types of control used in plants, i.e. (a) open-loop or open-end method, and (b) feedback or closed-loop system. In the former one, there is no feedback and constant check on performance is not made, whereas the latter system is used in mill control. The monitored detail responds to a difference between the operating norm represented by the correct signal and a working change which has varied the signal. Then it is followed by initiation of appropriate correction (automatic reset).

CHAPTER 20

Location, Layout and Selection of Equipments for Mineral Processing Plants

The successful application of ore treating techniques and processes depends to a great extent on the proper integration of location, layout, design, construction, and operation of plants on a sound economic basis. Since this subject is very vast, only general considerations associated with the subject are being discussed.

20.1 Site Location of the Plant

One of the most important parts of planning is the site location. If the plant is not located in the most economically favourable position, the competitive advantage of the processes will be wiped out. If the plant location is not thought carefully on the basis of various factors which must be considered for optimum plant location, the plant may even be inoperable. For example, if the plant involving flotation and/or leaching, is put up in the area where sufficient water is not available, the plant will be a failure. An ore-processing plant should essentially be located on firm ground near the mine site to prevent unnecessary expenditure on bulk handling of gangue material. In the case of non-ferrous ores like copper, lead and zinc, and precious metals like gold, and platinum where the tenor is very low, locating the processing plant near the mine site becomes absolutely essential. Measures like disposal of wastes and getting of water supply have to be worked out. If there is water shortage, it has to be brought to the site from nearby reservoir or river or by underground water reserves.

Preferably, the plant should be located slightly below the mine portal, so that advantage of gravity may be taken in transportation of ore to the mill. The delivery of the concentrates/products should also be considered in the location of ore processing plant. If the concentration ratio is low or the tonnage processed is large, the efficient handling of such products is quite important.

Power requirements in mineral dressing plants are substantial, since crushing, grinding and other units are power intensive Power should be available in sufficient supplies without frequent breakdowns, since most of the operations in ore processing work on continuous basis. If the ore needs transportation, it should be either by rail-road or rope-way trolleys as the road transportation of low priced bulky ore will not be economical. Other important factors affecting the selection of plant site for ore dressing plant are discussed as follows:

a) DISPOSAL OF TAILINGS

In ore processing, a major proportion of the ore is discarded in the form of tailings (waste), and thus disposal of tailings is of great importance in selection of site for processing plant. A million tonnes of tailing (a normal figure for ore plant, processing copper/zinc ores) will cover a sufficient area of ground. Further, the expansion of the plant poses this problem much more. Therefore, the site chosen should have sufficient space around for the ultimate disposal of such a huge quantity of material, otherwise in latter course, on disposal of tailings at even a distance of 20 km, the company will have to incur such an amount that all the economics of the processes will be offset.

b) POLLUTION POSSIBILITIES

In ore processing, the various chemicals used in leaching and froth flotation are discarded either in the form of dissolved compounds in water or with tailings. Many of the chemicals, such as cyanides, xanthates, etc., are highly toxic and poisonous. If pollution of ground water takes place, it is too serious, since ground water moves so slowly that pollution once introduced into a reservoir, may last indefinitely. Many streams become so polluted, that they become unusable. The discharge of tailings from leaching and froth flotation contain a huge amount of chemicals which are harmful even for plants. Therefore, due consideration should be given about the possibilities of pollution due to disposal of waste water, tailings, and in some cases even dust, which create the problems of air and water pollution.

c) FUTURE EXPANSION

Availability of sufficient space for the future expansion is equally important while selecting a site for locating a mineral processing plant. In the beginning, a company may start even with a small pilot plant of about 100 tonnes/day capacity. After working out the details, a plant capable of treating several thousand tonnes of ore per day may be installed. This plant may be enlarged further depending upon the ore reserves and demand of the product. Since most of the mines have such possibilities, a plant should be located with the view of future growth.

d) SAFETY OF THE LOCATION FROM FLOOD, LANDSLIDE, ETC.

This is also an important consideration. In the tropics, and places subject to sudden and torrential rains, there is always a danger of washouts, which may seriously damage the mill structure. If the mill location is unavoidable in such areas, the foundations of the plant should be protected by retaining walls and culverts. In the regions of heavy snowfalls and natural steep slopes, question of snowslides is of great importance. The mill should be located out of the reach of these slides. In some cases, the plants may be located even underground, due to either snowslide problem and lack of suitable surface site in steep mountains, or because of extreme climate conditions as in the arctic regions.

e) AVAILABILITY OF LABOUR

Before taking a decision on location of a plant in any particular place, a careful study should be made regarding the supply of available labour. The various factors to be studied include supply, kind, diversity, intelligence, wage scales, regulations, efficiency, and costs. The ore-processing plant requires particularly a large number of unskilled and semi-skilled labour. This factor should be given due consideration.

f) SITE CHARACTERISTICS

In the selection of a definite plant site in a specific area, the consideration should be given to the characteristics of the site. The cost of land for the mineral processing plant may also be substantial. The most important is the nature of the subsoil, since the need of piling or other expensive foundations can materially affect the construction costs. In selection of the site for ore processing plant, following informations should be available:

   i) Location with reference to adjacent areas, and restrictions imposed thereon.
  ii) Building and safety codes.
 iii) Accessibility to transportation of material and persons.
  iv) Labour regulations and conditions.
   v) Type of soil (rock, gravel, sand, silt, clay, etc.).
  vi) Elevation above sea level.
 vii) Ground water level.
viii) Drainage conditions.
  ix) Atmospheric temperatures, wind velocities and rainfall.
   x) Special conditions such as earthquakes, cyclones, excessively cold, heat, or humidity.

The subject of site location for an ore-processing plant is of secondary importance, since the major factor controlling the site is the distance from mine due to the bulk of ore to be treated and producing small quantity of concentrate. However, exhaustive tests should be carried out with the proposed methods of treatment and the certainty of ore supply sufficient to keep the mill running for a considerable length of time after its erection.

20.2 Plant Layout

Plant layout may be defined as the analysis and arrangement of equipment work centres, floor area and facilities specified from process flowsheet to achieve an efficient functioning of the plant. This is an essential requirement for accurate estimation of pre-construction cost and for future detailed design involving flow of materials, and structural and electrical facilities. This subject is of great importance in mineral processing plant, since it involves the handling of enormous quantity of materials, and water in relatively less-clean atmosphere.

20.2.1 Objective of Plant Layout

The objective of planning a layout is to combine labour with the physical aspects to the plants (process equipment, handling equipments, services, etc.) in such a way that maximum output is obtained with high quality of products at the lowest cost of production and distribution.

20.2.2 Factors in Planning Layout

A rational design of a plant should include arrangement of processing areas, storage areas, and handling areas in an efficient manner. Following are the factors to be considered in planning layouts:

a) New site development or addition to the existing plant.
b) Provision for future expansion.
c) Efficient distribution of water, process steam, power and gas.
d) Weather conditions amenable to outdoor construction.
e) Safety considerations, such as possible hazards of fire, explosion, fumes, etc.
f) Building code requirements.
g) Waste disposal problems, e.g. disposal of tailings and effluent water is of great importance in ore-processing plant.
h) Most efficient utilisation of floor and elevation space.

20.2.3 Principles of Plant Layout

There are some guiding principles which can be used in making the decisions for the plant layout. These are discussed as follows:

a) *Storage Layout*

Storage facilities for ore, chemicals, intermediate products, by-products, and concentrate may be located in isolated areas or adjoining areas. The toxic and poisonous chemicals and radioactive ores should be stored in isolated areas. Storage adjoining the process equipments reduces the material handling, but it will be an obstacle to future expansion of the plant. The storage of materials should be arranged so as to facilitate and simplify the handling. In ore-processing plants, the ore may be stored at some elevated points from where it can be subsequently handled by gravity into intermediate storage or in processing. Liquid reagents can be stored in small containers, barrels, horizontal or vertical tanks, etc., provided indoors or outdoors.

b) *Equipment Layout*

In making a layout, sufficient space should be assigned for individual equipment. In order to have accessibility to the equipment for its maintenance, the equipment layout should not fit too closely into a building. It will not result in a good economy. Though, a slightly larger building than necessary will cost slightly more, the extra cost will in fact be small compared to the penalties due to inconvenience and the need for future expansion (required sometimes to adjust the operation). In ore processing, the various processes are essentially a series of operations carried out simultaneously. The various processes in ore processing include crushing, grinding, classification, flotation, leaching, thickening, filtration, etc. These operations are generally repeated several times and thus the equipments should be arranged into groups of the same kinds. The relative level of the several parts of the equipment and their accessories decide their placement. Although, gravity flow is always preferable, but it is not necessary, since liquids can be transported by pumps and solids can be moved by mechanical means (conveyor belts, etc.). However, gravity flow system requires a multistory layout, whereas the mechanical flow favours single-story layout which will compensate the cost of mechanical transportation.

Another essential part of planning a layout is an access for initial construction and maintenance. For example, overhead equipment should have space for lowering it. Space should also be provided for repair and replacement of equipment.

c) *Safety Consideration in Layout*

An ore-processing plant involves handling and processing a large quantity of material with dirty and slippery conditions and thus special attention has to be given for safety considerations to avoid accidents arising due to physical, environmental or personnel factors. Therefore, an acceptable layout should incorporate safety measures in working areas, provision of sufficient transport facilities, elimination of repeated material handling, safeguarding of process equipments and power transmitting equipment, and safety and protective measures in plant and buildings by proper illumination, ventilation, and mitigation of hazards due to dust and fumes.

d) *Plant Expansion*

While designing a plant layout, expansion programme should always be kept in mind. The question of multiplying the number of units or increasing the size of the prevailing unit or units should be studied thoroughly. An engineering judgement should be exercised keeping in view the penalty for bad judgement, scrapping of present serviceable equipments, and remodelling which may involve much greater losses compared to the losses due to rejected equipment. However, sometimes due to economics of larger plants, the replacement is imperative.

e) *Proper Utilisation of Floor Space*

The value of land is a considerable factor in making use of floor space. However, a rule of practising economy of floor space should be followed, which should be consistent with good housekeeping in the plant with proper consideration to line flow of materials, access to equipment, space to permit working on parts of equipment requiring frequent servicing, and safety and comfort of the operators.

f) *Utilities Servicing*

The distribution of various utilities such as water, air, steam, and power generally permits the designing to meet almost any condition. However, considerations are given for the proper placement of each of these services and practising a good design for an ease of operation and reduction in cost of maintenance. The pipes should not be laid on floor or between the floor and 2.2 m, where the operator passes or works. The unorderly arrangement of piping invites problems in operation of the plant.

g) *Building*

After a thorough study of quantitative factors, the selection of buildings should be considered. Usually, standard factory buildings can be used, but if none is satisfactory to handle the space and process requirements, then a competent architect may be consulted to design a building around the process, and not a beautiful structure in which a process must fit. In mineral processing industry, much thought must be given to the disposal of tailings, waste liquors, dusts, etc. The drainage of waste liquors may require the installation of extra equipment. Some part of the building may require air conditioning, e.g. instrumental analysis laboratory, which may require an elaborate set-up.

h) *Materials Handling*

In ore-processing industries, the consideration of materials handling equipments is quite important, since a huge quantity of raw ore as well as products is involved in the process. Whenever possible, the advantage of the topography of the site location should be taken. The flow of material should be smooth and without causing the hindrance to the movement of the persons and running of process equipments. There should be minimum materials handling and transport with adequate supply of materials including services. There should be unrestricted disposal of products, (concentrates), tailings, and waste liquors. The congestion due to inadequate transport facilities may wash out the advantages of improved techniques. Therefore, the planning of sufficient transport system is of great importance for an effective layout.

i) *Rail-roads and Roads*

Existing or possible future rail-roads and roads adjacent to the plant should be known in order to plan rail sidings and access roads within the plant. The rail roads and roadways systems of the correct capacity and at the right location should be provided for traffic. The factors in rail-track planning are:

  i) Existing and future off-site main rail facilities.
 ii) Permissible radius of curvature rail-road spurs.
iii) Provision for traffic handling.
 iv) Loading and unloading facilities for initial plant construction and subsequent operations.
  v) Rack stations for liquid handling.
 vi) Storage space for full and empty cars.
vii) Space for cleaning and car repairs.

  Major provisions in road planning are:

  i) A means of interplant movement for road traffic, both pedestrian and vehicular.
 ii) Heavier and wider roads for heavy traffic.
iii) Routing of heavy traffic outside the operational areas.
 iv) Roadways for access to initial construction, maintenance, and repair points.
  v) Roadways to isolated points, such as storage of cyanides, safety equipment, fuel tanks, etc.

20.2.4 EFFECTS OF LAYOUT

Plant layout includes the various items of diversified nature such as materials handling starting from the point of receiving, siding, transport, blending and reclamation, etc., within the plant, bridges and weigh bridges, arrangement of machines, buildings for various purposes (process, administration, research, etc.), distribution of services, methods of plant control and automation, maintenance and stores, drainage and fencing, welfare housing, transport of personnel, etc. Therefore, there is a great importance of a proper planning of layout to prevent the mistakes leading to losses. Sometimes, the mistakes in layout planning become unrectifiable bottlenecks and incurable problems and affect the efficiency and economics of operation. The various direct effects of layout on the working and economics are discussed briefly as follows:

a) *Effect on Cost and Economics*

The layout affects both, cost and economic growth. The plant is expanded in future for the increased demand of the product and the disposition of buildings is affected by means of transport (railways, roadways, and conveyors). A considerable saving in both space and money may result by proper selection of the transport facilities. Since the cost of services is about 20 per cent of the capital cost, a considerable saving may result by proper planning of layout.

The economics of a plant depend largely on the capacity, productivity, and the availability of various items. In turn, productivity is affected by the nature of the process, availability of raw materials at the right place and at the right time, rapid disposal of products and wastes (affected by the transport system), and operations at the maximum efficiency (depends on supervision and other factors). Further, high production demands proper operation and rapid repairs. Therefore, a layout affects the productivity and in turn the cost of production.

b) *Effect on Management*

From the management point of view, the layout should incorporate economy and ease in handling, lower costs of useful areas, minimising production delays, preventing bottlenecks, provision for better supervision, and control of process, avoidance of unnecessary and costly changes after the layout is formed, and safety provisions. From the view-point of a worker, layout should be such that a right quality and quantity of materials is presented to him at the right time and right place with good working conditions (safe, clean, proper lighting, ventilation, etc.).

c) *Effect on Quality of Products*

In addition to the quality of raw materials, the quality of products depends on the various factors, such as efficiency of its design, efficiency of the process, and quality and accuracy of workmanship. All these factors are related to the layout planning.

20.2.5 CRITERIA OF A GOOD LAYOUT

In planning a layout, following points should be considered:

a) Proper arrangement of the main production units, the services, and the auxiliary equipments within the limitations of selected site.

b) Low initial installation cost with ample scope of expansion.

c) Comfortable and safe working conditions.

d) Location of various sections with least wastage of efforts.

e) Transportation of materials with minimum handling.

f) Stores, maintenance and welfare sections to cater the various needs for an efficient and competitive work.

g) Provision for uninterrupted intake and stocking of bulk raw materials, their rapid movement and finally smooth disposal of the products.

h) Incorporation of technological improvements in the initial design with the foresight of promising techniques in future (also automation).

i) High standard of layout of buildings and equipments.

20.2.6 LAYOUT PATTERNS OF ORE-PROCESSING PLANTS

An ore-processing plant may consist of following sections:

a) Run-of-mine ore (or raw coal) reception and handling section.

b) Crushing and screening section.

c) Grinding and classification section.

    d) Concentration/leaching section along with dewatering (or solid-liquid separation) section.

    e) Section for disposal of products, by-products, and waste products (solids and liquids).

    f) Services and maintenance section.

    g) Control and research laboratory.

    h) Administrative and other buildings.

The section (a) may consist of only few ore bins (where the ore is transported from the nearby mines), or a complicated network of railway system, having unloading sidings in the storage yards and a reclaiming and blending system for preparing the ore (particularly in the large coal washeries). Therefore, this section will depend on the tonnage handled and the sources of ore or raw feed.

The section (b) should be housed separately. This section should have a coarse ore bin, primary crusher, primary screen, secondary and/or tertiary crushers with necessary feeders, screens (vibrating or stationary) in closed or open circuit (depending upon the flowsheet of plant), and a set of conveyor belts or bucket elevators. The crushing and screening section has excessive problem of dust and vibration and thus should be sufficiently strong in construction and with proper ventilations and dust removal facilities.

The ore from the section (b) is transferred to section (c) which consists of fine ore bins, usually built near and above the grinding mills. In this section, a number of grinding mills with classifiers or cyclones may be installed in several parallel circuits. The undersize may be collected in a sump and pumped to the concentration section (flotation, gravity, magnetic, etc.), where more than one unit may be installed to run in several ways to form roughing, cleaning, recleaning, scavenging, recirculation, and retreatment circuits.

At different points of the flowsheet, a number of automatic samplers are installed for sampling solids, pulps, and solutions. The dewatering and drying section for the concentrate or tailings may consist of, from simple drainage and working screens for a heavy media separation plant to a series of thickeners, filters, and driers for flotation plant.

The section (f) of services includes water, fuel, power, etc. This may include a thermal station for a self-operated power plant or a sub-station for purchased power, overhead water tanks and water retreatment plant, stores, maintenance shops and other provisions for similar services. Therefore, the size of services section will depend on the requirements of power and water and their sources. There should also be facilities for storing gases and liquids.

The buildings for laboratory and administration should be located at some distance from dusty and smoky atmosphere. Some portion of the buildings may also need air-conditioning.

20.2.7 METHODS OF LAYOUT PLANNING

For a detailed planning study, space requirements should be known for various raw materials, products, by-products, and process equipments. A trial plot plan can be made with the help of directional schematic flow pattern. A number of studies and references will be required before selecting the final plant layout. The different methods of making the layout planning are (a) unit area concept, (b) two-dimensional layouts, and (c) scale models. These are briefly discussed as follows:

a) *Unit Area Concept*

Unit area concept refers to the use of basic blocks to build an arrangement for layout plans. This method is particularly suitable for large plant layouts, such as plants for treatment of zinc or copper ores. Unit areas are generally delineated by means of distinct process phases and operational procedures, taking into consideration materials flow and safety requirements. Thus, the first task of planning a layout is the delineation of the shape and extent of a unit area and the interrelationship of each area in a *master plot plan*. Figure 20.1 is an example of this type of planning.

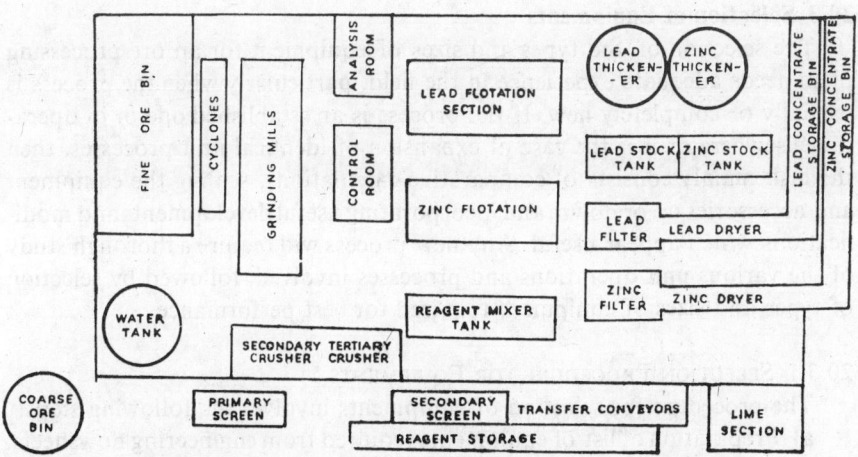

Fig. 20.1. Master plot plan for layout of lead/zinc concentrator.

b) *Two-dimensional Layouts*

In order to visualise the layout problem, two-dimensional scaled templates or small cutouts of unit areas and equipment are shifted within each area marked for them on the paper. This exercise is to be carried out by a group of engineers. This exercise will provide the basic plot plan, from which two dimensional diagrams can be prepared. The preparation of the·scale model can also be started simultaneously in the layout planning procedures.

c) *Scale Models (Three-dimensional Models)*

The scale models have the advantage over detailed two-dimensional method. A low-cost model can be made from blocks of wood and cardboard on the scaled paper. This low-cost model is mainly used to develop plot and elevation plans and cannot be used to represent piping and utilities layout.

A costly complete model can be used for the detailed layout of process piping, utilities, and control facilities. These models are made with a dimensional accuracy of $\pm$ 0.8 mm and give most scale-ups accurate to within 25–30 mm. These isometric layout diagrams to scale provide much more accurate cost estimation than two-dimensional models. These expensive scale models are also used during the construction and operator training period to great advantage. These models are usually placed permanently in control room, where an operator can trace lines quickly instead of walking and climbing over an extensive part of the plant. The various advantages of three-dimensional models are summarised:

   i) Selection of optimum design.

   ii) Effective planning in construction.

   iii) Savings in engineering design, construction, operating and maintenance cost.

   iv) Safer and rapid training of personnel.

20.3 Selection of Equipments

The selection of the types and sizes of equipment for an ore-processing plant needs adequate experience in the field, particularly when the process is partially or completely new. If the process is an established one or in operation elsewhere or it is the case of expansion of identical unit processes, then the task mainly consists of comparative calculations, scaling the equipment and accessories up or down, and incorporating useful developments and modifications which appear useful. Any new process will require a thorough study of the various unit operations and processes involved, followed by selection of types and sizes of equipment required for best performance.

20.3.1 SELECTION PROCEDURE FOR EQUIPMENTS

The procedure for selection of equipments involves the following steps:

   a) Preparation of list of equipments required from engineering flowsheets.

   b) Making necessary design calculations.

   c) Preparation of specification form for each major part of equipment using standard equipment if possible.

   d) Preparation of detailed specification sheets for the suppliers.

The standard forms are usually available from individual suppliers or from associations of manufacturers. If the design is to be used only for pre-construction cost estimation and plant layout work, a standard specification sheet as shown in Table 21.1 is sufficient. Use of published costs can be made in estimating the pre-construction costs.

20.3.2 STANDARD VERSUS SPECIAL EQUIPMENTS

Use of standard equipments such as crushers, grinders, pumps, thickeners, etc., is well recognised in the field of ore processing. Performance and service are demanded from all standard equipments. The experience of others is quite useful and should be used as fully as possible. Much valuable infor-

Table 20.1. Equipment specification sheet for pre-construction cost estimation[3-7]

1. Code No...................on Flow Sheet No......................
2. Date..
3. Name of equipment......................
4. Type..
5. Number required...........................
6. Materials to be processed or handled (type, composition)...............................
   ....
7. Operating conditions: Temperature...................Pressure......................
   Designed capacity (mass or volume/unit time).............................
8. Volumetric capacity (m³).......................................
9. Dimensions: Height..............Width/diameter.....................
   Length...............Floor area..............................
10. Principal design dimensions (filtering area, screening area, conveyor length, set and gape of crusher, etc.).
    ....
11. Recommended materials of construction.....................
12. Fitting required with specifications......................
13. Instrumentation requirements with estimated cost................................
    ....
14. Utility requirements:
    a) Electric motors: Type.................h.p..................kVA.
    b) Other electrical equipments: type.........................kVA.
    c) Steam pressure.................kg/cm², Quantity........kg/hr.
    d) Gas............m³/hr............, Compressed air.........m³/hr.
    e) Water...........m³/h Max. temp..................°C
15. Construction details...
16. Possible suppliers...
17. Estimated operating labour required................................
18. Remarks...

mation can be obtained from manufacturers having possibilities of placing orders.

Although, it is preferable to select the standard equipment, whenever possible, sometimes it may require special design and probably the use of special materials (particularly in chemical treatment of ores). In such cases, the requisite equipment should be designed on the basis of experience and training. The task in this case involves conversion of the specifications into a line picture or workshop drawing, from which the manufacturer can construct a three-dimensional piece of equipment. Most of the material handling and process equipments are standardised, and whenever such equipments serve the purpose, it should be selected in preference to special designs. In this instance, not only the first cost is substantially lower, but the duplication of equipment and repairs on old equipment will be much easier.

Before taking a decision on the design of special equipment, the complete survey to trade literature should be carried out. Standard equipment has been tried out and has passed the rigorous test service and gone through long periods of working, producing satisfactory results. The standard equipment is usually the result of many modifications of its original design. Standardisation does not mean only a minimum cost of manufacturing, but also that a machine has been constructed according to standard methods and in standard sizes with best thoughts given in its designing. Under such circumstances, it is possible to get the equipment under a guarantee of satisfactory performance. A new design is as much an experiment for the user as for the designer and it should withstand the test of satisfactory performance to acquire recognition. But when the design of new equipment is unavoidable or probability of yielding good results is quite high, there should be no hesitation in taking up the new designs.

20.3.3 SPECIFICATIONS

Before making a search of handbooks on the subject, trade literature files, and correspondence with manufacturers of equipment, a carefully written specification should be formulated with ranges of performance and other requirements. The specifications should contain all information deemed essential, including composition, physical and chemical characteristics of materials to be handled, type and quality of service available, service requirements on the equipment, packing, delivery requirements, and quotations. Manufacturers generally supply a form containing various questions, the reply of which helps the manufacturer to satisfy the demands. However, the time lost in correspondence may be saved by sending a well written specification to the manufacturer.

20.3.4 SPECIFICATION AND ENQUIRIES FOR COMPETITIVE QUOTATIONS

For all the large-scale equipment purchases, a competitive bidding is required for economic reasons. The competitive bidding (quotations) helps in review of the major items covering service requirements, principal components for the proposed flow cycle, accessories furnished, along with construction materials and cost ratio on an installed basis. In the final selection of the supplier, due importance should be given for delivery time, experience of supplier, reliability of the supplier and total costs.

20.3.5 EXAMPLES FOR SELECTION OF SOME ORE-PROCESSING EQUIPMENTS

From the above discussion, it is clear that a design engineer should be thoroughly familiar with various operations and processes to select the required equipment with possible alternatives. In many instances, the cost of equipment is an important factor in final selection. For example, it would be foolish to select an automatic basket centrifuge filter costing Rs. 1,50,000 in preference to a wooden filter press costing Rs. 10,000 for a small-scale operation, where ample labour for cleaning the press is available. The selection of some equipments employed in ore processing is discussed as follows:

a) *Selection of Size Reduction Equipments*

Size reduction is a general term used for a multitude of specific operations, such as crushing, grinding, pulverising, defiberising, etc. Usually, different mills perform different jobs and one particular type is best for a particular problem. Selection should be in terms of performance and cost, particularly operating cost. The purchase price has little effect on the overall cost of machine during its operation life. Low maintenance costs and reliability are the indications of a suitable equipment.

The particle size of the feed (particularly the maximum value), the hourly tonnage required, and the particle size and other characteristics of the product should be known before the selection of suitable equipment. The size-reduction machine should be so selected that it produces a maximum amount of the product with minimum of power and minimum of wear on the working parts. Size-reduction machine generally works in conjunction with accessory machines to separate the feed into component parts. The accessory machines, include screens and air classifiers for separating dry particles, and rake classifier, cyclone classifiers, etc., for wet grinding. Magnetic separator is used to keep the tramp iron away from entering the mill, and sometimes, in removing magnetic particles from the crushed product. The following factors are determinants in the selection of size-reduction equipments:

i) Physical properties of materials/ores, such as hardness, structure (brittle, fibrous, tough, or soft), moisture content, and specific gravity.

ii) Size of feed and product.

iii) Tonnage to be crushed/ground.

iv) Speed of the size-reduction mill.

v) Physical properties of size-reduction equipment, i.e., shape and character of lining, and shape and character of grinding medium.

An extensive compilation of information and data on the construction, design, capacity, and power requirements of various size-reduction machines is published in Mineral Dressing Handbooks and Chemical Engineering Handbooks. These handbooks also incorporate the industrial applications of various size-reduction mills with their operating characteristics. Figure 20.2 illustrates the applications of size-reduction equipments showing several applications for each type. Figure 20.3 shows the comparative installed costs of various size-reduction equipments.

b) *Screens*

Screens are employed to separate solids from solids, usually in coarse range. The various screens employed in ore processing are discussed in Ch. 5. In selection of screens the first consideration is the size of opening. However, the corrosive and abrasive characteristics of the material are also of great importance. In determining a suitable screen surface, the following characteristics of the process and screen should be considered:

Product		*Screen*
Size	. . .	Size of opening
Weight and abrasive nature	. . .	Wire diameter or thickness
Capacity	. . .	Size of screen deck

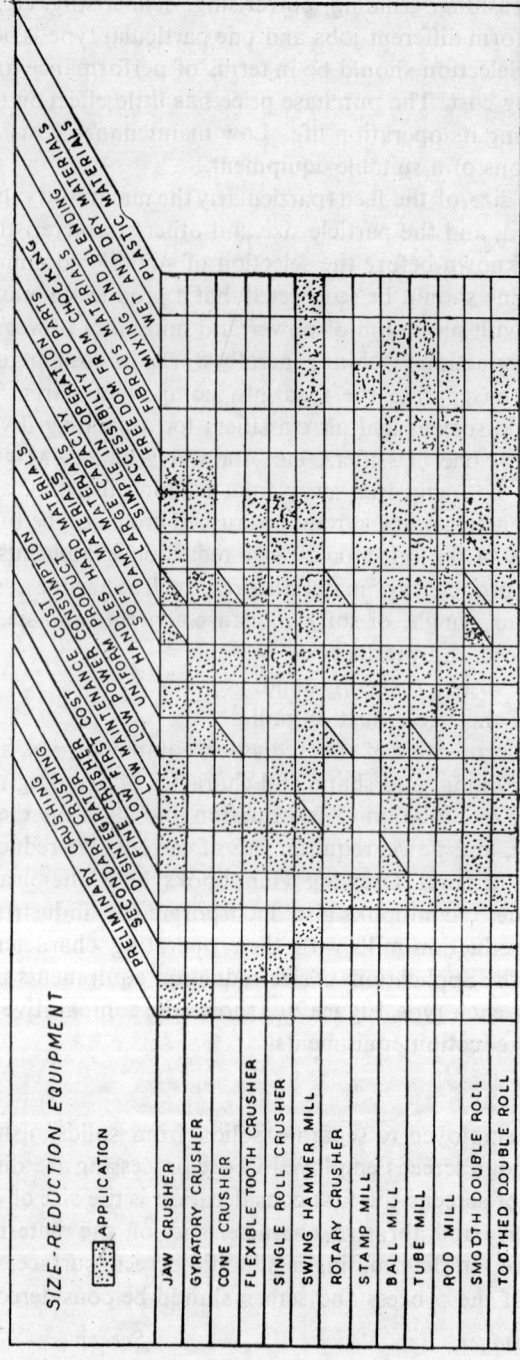

Fig. 20.2. Applications and other factors for different size-reduction equipments.

Material dry or wet   ...   Plain or twilled (weave)
Chemical characteristics   ...   Kind of metal or alloy

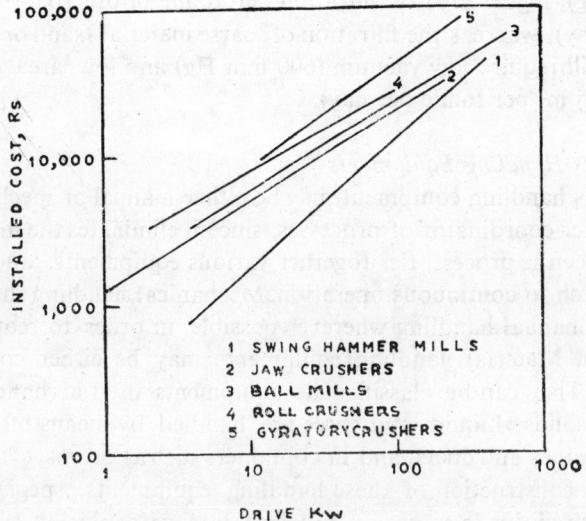

1 SWING HAMMER MILLS
2 JAW CRUSHERS
3 BALL MILLS
4 ROLL CRUSHERS
5 GYRATORY CRUSHERS

Fig. 20.3. Comparative installed costs of various size reduction equipments.

Further, it should be noted that a square opening has a larger area than a round opening when the diameter of the latter is equal to the side of a square.

c) *Flotation Cells*

The selection of flotation cells largely depends on the characteristics of minerals to be flotated and the capacity required. Easily flotable minerals such as graphite, galena, molybdenite, etc., can be treated easily by shallow flotation shells of less volume, whereas minerals not floatable easily, will require deep cells, with more time of conditioning and flotation. A number of manufacturers make a variety of designs to meet the requirements for the treatment of specific ore. However, the results may be first ascertained by testing the ore on a laboratory flotation cell. Various aspects of flotation cells have already been discussed in Ch. 13.

d) *Filters*

The various types of filters and their application have been discussed in Ch. 15. Filters of continuous type require little labour and have the advantage over batch-type pressure filters by giving a continuous discharge. Further, in certain filters, the cake can be dried to a greater extent after washing by permitting hot air to be sucked through the cake before discharge. Before selection of filters, filter areas required should be estimated, for which the procedures are described in Chemical Handbooks. The main factor considered in selection of filters is type of material such as cyanide slime, flotation concentrate, gravity concentrate, and sand. In the first two cases, the material is fine and consists of slime and thus it will require higher vacuum (250 mm Hg), higher area of filtration per tonne of ore (0.03–0.15 m² per tonne per day), whereas the filtration of coarse material (sand or gravity concentrate) will require low vacuum (600 mm Hg) and low area of filtration (0.001–0.005 m² per tonne per day).

e) *Materials Handling Equipments*

Materials handling equipment may be either manual or mechanical. The latter is better coordinator of processes, since it eliminates the manual work, serves to pace the process, ties together various equipments, and frequently converts batch to continuous operation. Mechanical handling should be substituted for manual handling wherever possible, in order to reduce the cost of operation. Material handling equipments may be either continuous or batch type. They can be classified as equipments used in handling gases, liquids and solids. Liquids and gases are handled by means of pumps and blowers; in pipes and ducts; and in containers such as drums, cylinders, tank cars, etc. In construction of these handling equipments, special materials may be required due to corrosion, fire, heat damage, explosion, pollution and poison problems together with special requirements. These problems influence the design and selection of these equipments.

Corrosion is generally the most difficult of these hazards and its solution is generally based on (i) the cheapest type of equipment available, (ii) the use of a high-first-cost, corrosion resistant material in the best type of handling equipment, or (iii) the use of containers which protect the equipment. Fire and explosion hazards are reduced by remote handling or closed container conveyances.

The selection of materials handling equipment thus depends on (i) the cost, and (ii) the work to be done. Following are the factors to be considered in selection of materials-handling equipments:

    i) Chemical nature of the material to be handled.

    ii) Physical nature of the material to be handled.

    iii) Type of movement to be made, i.e. horizontal, vertical or the combination of the two.

    iv) Distance of movement.

    v) Quantity of movement required per unit time.

    vi) Nature of feed to handling equipment.

    vii) Nature of discharge from handling equipment.

    viii) Nature of flow of material, i.e. intermittent or continuous.

Belt conveyors: A belt conveyor consists of a continuous rubber, leather, or wire screen belt supported on idler pulleys, usually arranged to trough the belt and driven by application of power to a head pulley. Most belts are in widths ranging from 30 to 150 cm. The large-size belts may have a speed as high as 200 m/min depending upon the type of material and loading. Belt conveyors give a very high capacity for materials handling. These possess the advantages, such as low maintenance, low power consumption, low cost, and continuous discharge. Belt conveyors can be used on horizontal or inclined runs for handling all types of solids ranging from fine powders to large lumps.

Chain conveyors: These ara dragged through shallow trenches or troughs and serve to carry such materials as hot ashes and hot cement clinker. When the chain conveyors are attached to platforms, slats, aprons, and pans, these serve to move both packaged and bulk materials at speeds 10–30 m/min. Apron conveyors are frequently used as feeders for handling coarse material to and from crushers. Such equipment at low speeds reduces breakage to a minimum. For the transportation of bulk materials over a varying path (anywhere from horizontal to vertical), buckets supported on chains and rollers are generally used. Due to flexibility of bucket conveyor, both in path and in bucket material, this conveyor is particularly suitable for handling of abrasive and difficult materials. *Roller conveyors* are used when gravity can supply the motive power. In the selection of *screw conveyors*, the manufacturers should be consulted. The *cut flight* type is preferred for materials which tend to pack or when placed in a trough having perforated lining for removing foreign materials from grains. There are many other types of conveyors which are employed for specific purposes.

Drag-line scrapers: These employ bucket-like scoops or disks which are moved back and forth by steel cables to drag loose materials from a large storage area, usually outdoors, towards a central elevator or conveyor hopper for subsequent delivery to the plant. This type of conveyor is used for storing and reclaiming materials such as coal or stone.

Pneumatic conveyors: The air can be used for the sweeping of comparatively light or powdered materials through ducts. Pneumatic conveyors are used for handling phosphate rock and other free-flowing materials.

Solid-pump: This can be used for handling of pulverised materials such as feldspar and portland cement. It makes use of screw feeder pressure in addition to aeration. The solid pump is more costly than the pneumatic conveyor, but employs smaller pipes, and does not require cyclone for receiving discharged material. The solid-pump offers more flexibility in many cases.

Interfloor elevators: These are used for handling of barrels, trays, and other heavy items. These are available in various forms. These employ in general, continuous chains operating vertically carrying platforms or arms to support the items. Some of them are devised for automatic pick-up and discharge.

Solid feeders: A solid feeder may be considered as a device that will maintain a reasonably uniform flow of bulk material, implying a metering function. It deals with solids in bulk, solid–liquid, and solid–gas mixtures which may be free-flowing, lumpy, sticky, corrosive, erosive, plastic or pasty. Almost any type of materials-handling equipment that can move bulk-loads can be employed as solid feeders. Many of them have belt, aprons, screws, flights and vibrating conveyors. These belong to the class of volumetric feeders, which meter their loads in volumes per unit time. But the weight per unit volume of bulk material can vary widely. For more precision feeding, gravimetric feeders are favoured.

Pumps: The severe pumping required in ore processing plants demands continuous heavy duty for long periods with freedom from forced shutdowns. It should provide (a) flexible operating characteristics, (b) ease of control, (c) availability in wide choice of materials of construction, (d) interchangeability of pumps and parts, (e) handling of solids and abrasives in suspension, and (f) a design that can tolerate some erosion and corrosion.

In ore-processing plants, the pumps are used for many purposes, in transferring liquids, solids suspended in water, pulp slurries, etc., from one point to another. Pump transportation includes both long and short distances, horizontal and vertical, and under varying pressure heads. The various types of pumps include (a) reciprocating pumps (delivery of liquid by the displacement of piston or plunger), (b) diaphragm pumps (delivery of liquids by suction), (c) centrifugal pumps (flow is produced by centrifugal force and is free from pulsations), and (d) rotary gear pumps (liquid is trapped by the gear teeth and carried from intake to discharge). While selecting pumps, following information should be known:

a) Capacity and head (maximum, desirable range, and future requirements.

b) Desirable operating characteristics (constant head and capacity or variable capacity with constant head, or constant capacity with variable heads, etc.).

c) Nature and size of solid in suspension.

d) Corrosion data for suitable materials of construction.

e) Viscosity range.

f) Type of power (steam or electricity).

g) Load factor.

h) Efficiencies.

i) Costs.

As in the case of process equipments, materials-handling equipment makers provide informations to assist in making the decision.

Bibliography

1. Jones, M.J., *Tenth International Mineral Processing Congress, 1973*, The Institution of Mining and Metallurgy, 1974.
2. Jones, M.J., Mineral Processing and Extractive Metallurgy, *Proceedings of Ninth Commonwealth Mining and Metallurgy Congress*, 1969, Vol. 3, The Institution of Mining and Metallurgy, 1970.
3. Donald, P. Eckman, *Industrial Instrumentation*, Wiley Eastern Ltd., 1966.
4. Roberts, A. Mineral Processing, *Proceedings of the Sixth International Congress*, May 26-June 2, 1963, Pergamon Press, 1965.
5. Donald, O.R. and Burt, C.M., *AIME World Symposium on Mining and Metallurgy of Lead and Zinc*, Vol. I, The American Institute of Mining, Metallurgical, and Petroleum Engineers, Inc., New York, 1970.
6. Gaudin, A.M., *Principles of Mineral Dressing*, McGraw-Hill Book Company Inc., 1971.
7. Richards, R.H. and Locke, C.E., *Text Book of Ore Dressing*, McGraw-Hill Book Company Inc., New York, 1940.
8. Pryer, E.J., *Mineral Processing*, Applied Science Publishers Ltd., London, 1974.
9. Venkatachalam, S. and Degalleesan, S.N. *Laboratory Experiments in Mineral Engineering*, Oxford and IBH Publishing Co., New Delhi, 1982.
10. Gilchrist, J.D., *Extraction Metallurgy*, Pergamon Press, New York, 1980.
11. Brown, G.G., et al., *Unit Operations*, John Wiley and Sons Inc., New York, 1950.
12. Brown, J.H., *Unit Operations in Mineral Engineering*, International Academic Services Ltd., Canada, 1979.
13. *Eighth International Mineral Dressing Congress*, Leningrad, Institute, Mekhanobr, 1968.
14. Roberts, A., *Mineral Processing*, Pergamon Press, Oxford, 1965.
15. *Proceedings of the Ninth Commonwealth Mining and Metallurgy Congress*, Vol. 3, Mineral Processing and Extractive Metallurgy, 1969.
16. *Proceedings of Eleventh International Mineral Processing Congress*, Cagliari, Italy, 1975.

17. *Proceedings of Twelth International Mineral Processing Congress*, Sao Paulo, Brazil, 1977.

18. *Proceedings of Thirteenth International Mineral Processing Congress*, Warsaw, Poland, 1979.

19. Arbiter, N., *Seventh International Mineral Processing Congress*, Gordon and Breach, New York, 1963.

20. Taggort, A.F., *Elements of Ore Dressing*, Wiley, New York, 1951.

21. Taggort, A.F., *Hand Book of Mineral Dressing*, Wiley, New York, 1945.

22. Rabone, P., *Flotation Plant Practice*, Mining Publications, London, 1957.

23. Fuerstenau, M.G., *Flotation*, A.M. Gaudin Memorial Volume, Volume 1 and 2, AIME, New York, 1976.

24. Gaudin, A.M. *Flotation*, McGraw-Hill, New York, 1957.

25. Glembotski, V.A., Klassen, V.I., and Plaksin, I.N., *Flotation*, Primary Sources, New York, 1963.

26. Ramney, M.W., *Flotation Agents and Processes Technology and Applications*, Noyes Data Corporation, U.S.A., 1980.

27. Sutherland, K.L. and Wark, T.W., *Principles of Flotation*, Australasiant Institute of Mining and Metallurgy, Melbourne, 1955.

28. Dean, R.S. and Davis, C.W., *Magnetic Separation of Ores*, U.S.B.M. Bulletin 425, 1941.

29. Frass, F., *Electrostatic Separation of Granular Materials*, U.S.B.M. Bulletin, 603, 1962.

30. Gokhale, K.V.G.K. and Rao, T.C., *Ore Deposits of India—Their Distribution and Processing*, Thomson Press (India) Ltd., Delhi, 1973.

31. Israelson, A.F., 'Magnetic Separation of Minerals', *Mining Magazine*, Sept. 1978, pp. 211.

32. *Mining Equipment International*, Oct. 1972.

33. Kearu, J.M., 'Sedimentation, Theory, Equipments and Methods', *World Mining*, Nov. 1979.

34. *Recent Developments in Mineral Dressing*, The Institution of Mining and Metallurgy, London, 1953.

35. *Nordberg Dynapactor Impact Crushers*, Rexnord Inc., Wisconsin, U.K., Process Machinery Division, Bulletin 414.

36. *Jaw Crushers*, Rexnord Process Machinery Div., Wisconsin, U.K., Catalog No. 452.

37. *Dorr Oliver Thickening Equipment and Systems*, Dorr Oliver Inc., Bulletin Thic-2.

38. 'Nordberg Grinding Mills', Rexnord Inc., Process Machinery Division, Wisconsin, U.K., Bulletin No. 437.

39. *SALA Flotation*, Sala International, Sweden, Leaflet.

40. *Secondary Impact Crushers*, Appareies Dragon, France, Bulletin GB 1631.

41. *Thickeners*, Sala International, Sweden, Leaflet 318 E.

42. 'Nordberg Grinding Mill Types', Rexnord Inc., Wisconsin, U.K., A leaflet.

43. *Personal Communications from Various Ore-Processing Plants in India.*

44. *Silver Jubilee Symposium*, Indian Institute of Metals, Jan., 1972, Delhi.

45. Wetzlar, H.F., *Applied Ore Microscopy*, The MacMillan Company, New York, 1966.

46. V.N.S. Mathur, 'Mineral Beneficiation—What Why and How, *Metals and Minerals Review*.

47. Tupkary, R.H., *Introduction to Modern Iron Making*, 1984, Khanna Publishers, Delhi.

48. Khoda, R. and V.N.S. Mathur, *J. Mines, Metals and Fuels*, August, 1972, pp. 235.

49. Boldt, J. and P. Queneau, *The Winning of Nickel*, Methewn and Co. Ltd., London.

50. Pinkney, E.T., *Mineral Processing and Extractive Met.*, Institute of Min. Metal, London, 1970, pp. 397.

51. Ray, M.R.W., *Trans. Institution of Min. and Met.*, Vol. 76, 1967, pp. 101.

52. Ray, M., *Proceedings of the International Congress of Min. and Met.*, Paris, Met. Section, Vol 11, Nov. 3, 1965, pp. 55.

53. Caron, M.H., *J. Metals*, Vol. 2, 1950, pp. 67.

54. Baheti, O.P., S.K. Jain and Dharmendra Kumar, *J. Institution of Engineers (India)*, Vol. 60, pt. MM-2 Nov., 1979, pp. 17.

55. Jena, P.K., et. al., *IIM Trans.* Vol. 28, No. 6, Dec. 1975, pp. 483.

56. Rao, P. Kanta, et al., *Symposia on Mineral Based Industries in the Eastern Region*, Bhubaneshwar, Dec., 1974, pp. 94.

57. Vilbrandt, F.C. and C.E. Dryden, *Chemical Engg. Plant Design*, McGraw-Hill Book Co. Inc., New York, 1959.

58. Mathur, V.N.S., *The Layout of Mineral Dressing Plant*.

59. Mathur, V.N.S., *Layout of Mineral Dressing Plant*, Part II, Minerals and Industries, Vol. IV, No. 2, pp. 3.

60. Mathur, V.N.S., Some Aspects of Design of Pilot Plants for Mineral Beneficiation Testing, *Minerals and Metals Review*, Vol. VIII, May, 1969, No. 6, pp 3.

61. Mathur, V.N.S. and A.V. Ulabhaje, Mineralogy and Metal Extraction, *Metals and Minerals Review*, Sept. 1969, Vol. VIII, No. 10, pp. 3.

62. Ulabhaje, A.V. and V.N.S. Mathur, Ore Microscopy and Mineral Beneficiation, *Metals and Minerals Review*, Feb. 1971, Vol. X, No. 3, pp. 39.

63. Friedrich Otmar, The History of Ore Microscopy, *Applied Ore Microscopy*, MacMillan Co., N.Y., 1966, pp 3.

64. Ramdohr, Paul and Gerhard Rehward, The selection of ore specimens and the preparation of polished section, *Applied Ore Microscopy*, MacMillan Co., N.Y., 1966, pp. 319.

65. Rehwald, Gerhard, The Application of Ore Microscopy in Beneficiation of Ores of the Precious Metals and Non-Ferrous Metals, *Applied Ore*

Microscopy, MacMillan Co., N.Y., 1966, pp. 441.

66. Ehrenberg, H., Investigation of Mill Products with the Ore Dressing Microscope, A guide to the Control of Ore Dressing Operations, *Applied Ore Microscopy*, MacMillan Co., N.Y., 1966, pp. 557.

67. Mathur, V.N.S., Mineral Beneficiation and Its Educational Perspective, *Metals and Minerals Review*, May, 1968. Vol. VII, No. 6, pp. 19.

68. Mathur, V.N.S., Some Considerations in Pilot Plant Investigations in Mineral Beneficiations, *Metals and Minerals Review*, March, 1970, Vol. IX, No. 3, pp. 11.

69. Mathur, V.N.S., *Metals and Minerals Review*, July, 1968, Vol. VII, No. 8, pp. 41.

70. Mathur, V.N.S., *The Banaras Metallurgist*, 1969, pp. 11.

71. Mathur, V.N.S., *Metals and Minerals Review*, March, 1969, Vol. VIII, No. 4, pp 8.

72. Malenbaum, Wilfred, *Proceedings of the Council of Economics*, 107th Annual Meeting, American Institute of Min. Met. and Petroleum Engineers, Inc., Feb. 26 to March 2, 1978, pp. 185.

73. Coburn, James L. and Dennis K. Mortenson, *Mining Engg.*, Vol. 29, No. 2, Feb. 1977, pp. 88.

74. Flavel, Malcolm D., *Mining Engg.*, Vol. 29, 1977, pp. 65.

75. Cohen, Henry E., *Economics of Mineral Engg.*, An International Seminar organised by the United States in Cooperation with the Govt. of Turkey, Ankara, April, 1976, pp. 18.

76. Mathur, V.N.S. and A.V. Ulabhaje, *The New Sketch, Coal and Mining Organisation*, 26 January, 1973, special number.

77. *Mineral Beneficiation Pilot Plant*, N.M.L., C.S.I.R. (India).

78. *Mining Engg.*, Vol. 29, Nov. 1977, pp. 26.

79. Schapper, Mark A., *Australian Mining*, April, 1977, pp. 44.

80. Burt, R.O., Fine sizing of minerals, *Mining Mag.* 128 (6) June, 1973, pp. 463–465.

81. Hepworth, W.E., Design and Performance of a Special Multi-deck Screen, *Quarry Management and Products*, Feb. 1974, pp. 39.

82. *British patent* 124904, 1970.

83. Shoemaker, R.S., Minerals Processing in 1973., *Min. Congr. J.*, 60 (2), Feb. 1974, pp 24–29.

84. Falstrom, P.H., *Lead and Zinc Flotation Practice at Boliden in the Mining and Concentrating of Lead and Zinc*, N.Y., AIME, 1970, pp. 668–694.

85. McDermott, W.F. et al., The dollars and sense of autogenous grinding, *Min. Engg.*, 24 (11) Nov. 1972, pp 46–50.

86. Anon Erwo, Milling limits overgrind–uses less power, *Wld. Mining*, 27 (11), Oct. 1974, pp. 6–66.

87. Chaston, I.R.M., Heavy Media Cyclone Plant Design and Practice for Diamond Recovery in Africa. In *Proc. 10th Int. Min and Met. Congress*, London Inst. of Min. and Met., 1975, pp. 257.

88. Burt, R.O. and Ottley, D.J., Fine Gravity Concentration Using the Bartles–Mozley Concentrator, *Int. J. Miner. Processing*, 1, 1974, pp. 347.

89. Muller, L.D. and Sayles, O.P., Processing Dry Granular Materials. *Min. Engg.*, 23 (3) Mar. 1971, pp. 54.

90. Douglas, E. and Walsh, T., New Type of Dry Heavy Medium Separator. *Trans. Inst. Min and Met.* 75, Sept. 1966, C 226–232.

91. Jones, G.H., Wet Magnetic Separator for Feebly Magnetic Minerals, *Proceeding 5th Int. Min. Proc. Cong.*, London IIM, 1960, pp. 317.

92. Ealy, G.K., Concentration of Copper Oxides by Flotation at Nacimiento, *Min. Cong. J.*, 59 (3) Mar. 1973, pp. 63.

93. Glembotsky, T. et al., Selective Separation of Fine Mineral Slimes Using the Method of Electroflotation, *Proc. XI Int. Min. Proc. Congr.*, Cagliari, 1974.

94. *Workshop on Hydrocyclones*, Hindustan Zinc Ltd., Udaipur, 25-27th Dec., 1980.

95. G.S. Dobby, J.A. Finch: "Flotation Column Scale-Up and Simulation," *Proc. 17th Annual Meeting of Canadian Mineral Processors*, Jan. 22-24, Can. Inst. Min. Metal., Ottawa 1985, pp 614-638.

96. A.L. Mular, M.A. Anderson: *Design and Installation of Concentration and Dewatering Circuits,* Society of Mining Engineers (SME), Littleton, Colo., 1986, p. 588.

97. R.J. Hunter: *Zeta Potential in Colloid Science,* Academic Press, London 1981, p. 11.

98. M.C. Fuerstenau, J.D. Miller, M.C. Kuhn: *Chemistry of Flotation,* Society of Mining Engineers (SME), New York 1985.

99. G.W. Poling: "Selection and Sizing of Flotation Machines," in A.L. Mular, R.B. Bhappu (eds.): *Mineral Processing Plant Design,* 2nd ed. Society of Mining Engineers (SME)-AIME, New York 1980, pp. 887-906.

100. I.S. Blagov et al.: "State and Development of Coal Flotation in the USSR," *Proc. 9th Int. Coal Preparation Congress,* New Delhi, India, 1982, pp. C1-C5.

101. G. Barbery: "Engineering Aspects of Flotation in the Minerals Industry: Flotation Machines, Circuits and their Simulation," *Scientific Basis of Flotation, NATO-Advanced Study Institute on Scientific Basis of Flotation, Preprints,* Cambridge, UK, 1982.

Index